Industrial and Applied Mathematics

The Industrial and Applied Mathematics series publishes high-quality research-level monographs, lecture notes and contributed volumes focusing on areas where mathematics is used in a fundamental way, such as industrial mathematics, bio-mathematics, financial mathematics, applied statistics, operations research and computer science.

More information about this series at http://www.springer.com/series/13577

Abul Hasan Siddiqi

Functional Analysis
and Applications

Springer

Abul Hasan Siddiqi
School of Basic Sciences and Research
Sharda University
Greater Noida, Uttar Pradesh
India

ISSN 2364-6837 ISSN 2364-6845 (electronic)
Industrial and Applied Mathematics
ISBN 978-981-13-3830-4 ISBN 978-981-10-3725-2 (eBook)
https://doi.org/10.1007/978-981-10-3725-2

Printed on acid-free paper

This Springer imprint is published by the registered company Springer Nature Singapore Pte Ltd. part of Springer Nature
The registered company address is: 152 Beach Road, #21-01/04 Gateway East, Singapore 189721, Singapore

To
My wife Azra

Preface

Functional analysis was invented and developed in the twentieth century. Besides being an area of independent mathematical interest, it provides many fundamental notions essential for modeling, analysis, numerical approximation, and computer simulation processes of real-world problems. As science and technology are increasingly refined and interconnected, the demand for advanced mathematics beyond the basic vector algebra and differential and integral calculus has greatly increased. There is no dispute on the relevance of functional analysis; however, there have been differences of opinion among experts about the level and methodology of teaching functional analysis. In the recent past, its applied nature has been gaining ground.

The main objective of this book is to present all those results of functional analysis, which have been frequently applied in emerging areas of science and technology.

Functional analysis provides basic tools and foundation for areas of vital importance such as optimization, boundary value problems, modeling real-world phenomena, finite and boundary element methods, variational equations and inequalities, inverse problems, and wavelet and Gabor analysis. Wavelets, formally invented in the mid-eighties, have found significant applications in image processing and partial differential equations. Gabor analysis was introduced in 1946, gaining popularity since the last decade among the signal processing community and mathematicians.

The book comprises 15 chapters, an appendix, and a comprehensive updated bibliography. Chapter 1 is devoted to basic results of metric spaces, especially an important fixed-point theorem called the Banach contraction mapping theorem, and its applications to matrix, integral, and differential equations. Chapter 2 deals with basic definitions and examples related to Banach spaces and operators defined on such spaces. A sufficient number of examples are presented to make the ideas clear. Algebras of operators and properties of convex functionals are discussed. Hilbert space, an infinite-dimensional analogue of Euclidean space of finite dimension, is introduced and discussed in detail in Chap. 3. In addition, important results such as projection theorem, Riesz representation theorem, properties of self-adjoint, positive, normal, and unitary operators, relationship between bounded linear operator and bounded bilinear form, and Lax–Milgram lemma dealing with the existence of solutions of abstract variational problems are presented. Applications and generalizations of the Lax–Milgram lemma are discussed in Chaps. 7 and 8. Chapter 4 is devoted to the Hahn–Banach theorem, Banach–Alaoglu theorem, uniform boundedness principle, open mapping, and closed graph theorems along with the concept of weak convergence and weak topologies. Chapter 5 provides an extension of finite-dimensional classical calculus to infinite-dimensional spaces, which is essential to understand and interpret various current developments of science and technology. More precisely, derivatives in the sense of Gâteau, Fréchet, Clarke (subgradient), and Schwartz (distributional derivative) along with Sobolev spaces are the main themes of this chapter. Fundamental results concerning existence and uniqueness of solutions and algorithm for finding solutions of optimization problems are described in Chap. 6. Variational formulation and existence of solutions of boundary value problems representing physical phenomena are described in Chap. 7. Galerkin and Ritz approximation methods are also included. Finite element and boundary element methods are introduced and several theorems concerning error estimation and convergence are proved in Chap. 8. Chapter 9 is devoted to variational inequalities. A comprehensive account of this elegant mathematical model in terms of operators is given. Apart from existence and uniqueness of solutions, error estimation and finite element methods for approximate solutions and parallel algorithms are discussed. The chapter is mainly based on the work of one of its inventors, J. L. Lions, and his co-workers and research students. Activities at the Stampacchia School of Mathematics, Erice, Italy, are providing impetus to researchers in this field. Chapter 10 is devoted to rudiments of spectral theory with applications to inverse problems. We present frame and basis theory in Hilbert spaces in Chap. 11. Chapter 12 deals with wavelets. Broadly, wavelet analysis is a refinement of Fourier analysis and has attracted the attention of researchers in mathematics, physics, and engineering alike. Replacement of the classical Fourier methods, wherever they have been applied, by emerging wavelet methods has resulted in drastic improvements. In this chapter, a detailed account of this exciting theory is presented. Chapter 13 presents an introduction to applications of wavelet methods to partial differential equations and image processing. These are emerging areas of current interest. There is still a wide scope for further research. Models and algorithms for removal of an unwanted component (noise) of a signal are discussed in detail. Error estimation of a given image with its wavelet representation in the Besov norm is given. Wavelet frames are comparatively a new addition to wavelet theory. We discuss their basic properties in Chap. 14. Dennis Gabor, Nobel Laureate of Physics (1971), introduced windowed Fourier analysis, now called Gabor analysis, in 1946. Fundamental concepts of this analysis with certain applications are presented in Chap. 15.

In appendix, we present a resume of the results of topology, real analysis, calculus, and Fourier analysis which we often use in this book. Chapters 9, 12, 13, and 15 contain recent results opening up avenues for further work.

The book is self-contained and provides examples, updated references, and applications in diverse fields. Several problems are thought-provoking, and many lead to new results and applications. The book is intended to be a textbook for graduate or senior undergraduate students in mathematics. It could also be used for an advance course in system engineering, electrical engineering, computer engineering, and management sciences. The proofs of theorems and other items marked with an asterisk may be omitted for a senior undergraduate course or a course in other disciplines. Those who are mainly interested in applications of wavelets and Gabor system may study Chaps. 2, 3, and 11 to 15. Readers interested in variational inequalities and its applications may pursue Chaps. 3, 8, and 9. In brief, this book is a handy manual of contemporary analytic and numerical methods in infinite-dimensional spaces, particularly Hilbert spaces.

I have used a major part of the material presented in the book while teaching at various universities of the world. I have also incorporated in this book the ideas that emerged after discussion with some senior mathematicians including Prof. M. Z. Nashed, Central Florida University; Prof. P. L. Butzer, Aachen Technical University; Prof. Jochim Zowe and Prof. Michael Kovara, Erlangen University; and Prof. Martin Brokate, Technical University, Munich.

I take this opportunity to thank Prof. P. Manchanda, Chairperson, Department of Mathematics, Guru Nanak Dev University, Amritsar, India; Prof. Rashmi Bhardwaj, Chairperson, Non-linear Dynamics Research Lab, Guru Gobind Singh Indraprastha University, Delhi, India; and Prof. Q. H. Ansari, AMU/KFUPM, for their valuable suggestions in editing the manuscript. I also express my sincere thanks to Prof. M. Al-Gebeily, Prof. S. Messaoudi, Prof. K. M. Furati, and Prof. A. R. Khan for reading carefully different parts of the book.

Greater Noida, India   Abul Hasan Siddiqi

Contents

About the Author

Abul Hasan Siddiqi is a distinguished scientist and professor emeritus at the School of Basic Sciences and Research, Sharda University, Greater Noida, India. He has held several important administrative positions such as Chairman, Department of Mathematics; Dean Faculty of Science; Pro-Vice-Chancellor of Aligarh Muslim University. He has been actively associated with International Centre for Theoretical Physics, Trieste, Italy (UNESCO's organization), in different capacities for more than 20 years; was Professor of Mathematics at King Fahd University of Petroleum and Minerals, Saudi Arabia, for 10 years; and was Consultant to Sultan Qaboos University, Oman, for five terms, Istanbul Aydin University, Turkey, for 3 years, and the Institute of Micro-electronics, Malaysia, for 5 months. Having been awarded three German Academic Exchange Fellowships to carry out mathematical research in Germany, he has also jointly published more than 100 research papers with his research collaborators and five books and edited proceedings of nine international conferences. He is the Founder Secretary of the Indian Society of Industrial and Applied Mathematics (ISIAM), which celebrated its silver jubilee in January 2016. He is editor-in-chief of the *Indian Journal of Industrial and Applied Mathematics*, published by ISIAM, and of the Springer's book series Industrial and Applied Mathematics. Recently, he has been elected President of ISIAM which represents India at the apex forum of industrial and applied mathematics—ICIAM.

Chapter 1
Banach Contraction Fixed Point Theorem

Abstract The main goal of this chapter is to introduce notion of distance between two points in an abstract set. This concept was studied by M. Fréchet and it is known as metric space. Existence of a fixed point of a mapping on a complete metric space into itself was proved by S. Banach around 1920. Application of this theorem for existence of matrix, differential and integral equations is presented in this chapter.

Keywords Metric space · Complete metric space · Fixed point · Contraction mapping · Hausdorff metric

1.1 Objective

The prime goal of this chapter is to discuss the existence and uniqueness of a fixed point of a special type of mapping defined on a metric space into itself, called contraction mapping along with applications.

1.2 Contraction Fixed Point Theorem by Stefan Banach

Definition 1.1 Let

$$d(\cdot, \cdot) : X \times X \to R$$

be a real-valued function on $X \times X$, where X is a nonempty set. $d(\cdot, \cdot)$ is called a *metric* and (X, d) is called a *metric space* if $d(\cdot, \cdot)$ satisfies the following conditions:

1. $d(x, y) \geq 0 \ \forall \, x, y \in X$, $d(x, y) = 0$ if and only if $x = y$.
2. $d(x, y) = d(y, x)$ for all $x, y \in X$,
3. $d(x, y) \leq d(x, z) + d(z, y)$ for all $x, y, z \in X$.

Remark 1.1 $d(x, y)$ is also known as the *distance* between x and y belonging to X. It is a generalization of the distance between two points on real line.

© Springer Nature Singapore Pte Ltd. 2018
A. H. Siddiqi, *Functional Analysis and Applications*, Industrial and Applied Mathematics,
https://doi.org/10.1007/978-981-10-3725-2_1

It may be noted that positivity condition:

$$d(x, y) \geq 0 \,\forall\, x, y \in X$$

follows from second part of condition (i). $d(x, y) \leq d(x, z) + d(z, y)$ by (iii). Choosing $x = y$, we get:

$d(x, x) \leq d(x, z) + d(z, x)$ or $0 \leq 2d(x, z)$ for $x, z \in X$

because

$d(x, x) = 0$ and $d(x, z) = d(z, x)$.

Hence for all $x, z \in X, d(x, z) \geq 0$, namely positivity.

Remark 1.2 A subset Y of a metric space (X, d) is itself a metric space. (Y, d) is a metric space if $Y \subseteq X$ and

$$d_1(x, y) = d(x, y) \,\forall\, x, y \in Y$$

Examples of Metric Spaces

Example 1.1 Let

$$d(\cdot, \cdot) : R \times R \to R \tag{1.1}$$

be defined by

$$d(x, y) = |x - y| \,\forall\, x, y \in R.$$

Then $d(\cdot, \cdot)$ is a metric on R (distance between two points of R) and (R, d) is a metric space.

Example 1.2 Let R^2 denote the Euclidean space of dimension 2. Define a function $d(\cdot, \cdot)$ on R^2 as follows: $d(x, y) = ((u_1 - u_2)^2 + (v_1 - v_2)^2)^{1/2}$, where $x = (u_1, u_2), y = (v_1, v_2)$.

$d(\cdot, \cdot)$ is a metric on R^2 and (R^2, d) is a metric space,

Example 1.3 Let R^n denote the vector space of dimension n. For $u = (u_1, u_2, \ldots, u_n) \in R^n$ and $v = (v_1, v_2, \ldots, v_n) \in R^n$. Define $d(\cdot, \cdot)$ as follows:

(a) $d(u, v) = (\sum_{k=1}^{n} |u_k - v_k|^2)^{1/2}$.

(R^n, d) is a metric space.

Example 1.4 For a number p satisfying $1 \leq p < \infty$, let ℓ_p denote the space of infinite sequences $u = (u_1, u_2, \ldots, u_n, ..)$ such that the series $\sum_{k=1}^{\infty} |u_k|^p$ is convergent.

$(\ell_p, d(\cdot, \cdot))$ is a metric space, where $d(\cdot, \cdot)$ is defined by

$$d(u, v) = \left(\sum_{k=1}^{\infty} |u_k - v_k|^p \right)^{1/p},$$

$u = (u_1, u_2, \ldots, u_k, ..), v = (v_1, v_2, \ldots, v_k, ..) \in \ell_p.$

$d(\cdot, \cdot)$ is distance between elements of ℓ_p.

Example 1.5 Suppose $C[a, b]$ represents the set of all real continuous functions defined on closed interval $[a, b]$. Let $d(\cdot, \cdot)$ be a function defined on $C[a, b] \times C[a, b]$ by:

(a) $d(f, g) = \sup_{a \leq x \leq b} |f(x) - g(x)|, \forall f, g \in C[a, b]$

(b) $d(f, g) = \left(\int_a^b |f(x) - g(x)|^2 dx \right)^{1/2}, \forall f, g \in C[a, b].$

$(C[a, b], d(\cdot, \cdot))$ is a metric space with respect to metrices given in (a) and (b).

Example 1.6 Suppose $L_2[a, b]$ denote the set of all integrable functions f defined on $[a, b]$ such that $\lim |f|^2 \, dx$ is finite. Then, $(L_2[a, b], d(\cdot, \cdot))$ is a metric space if

$$d(f, g) = (\int_a^b |f(x) - g(x)|^2 dx)^{1/2}, \ f, g \in L_2[a, b]$$

$d(\cdot, \cdot)$ is a metric on $L_2[a, b]$.

Definition 1.2 Let $\{u_n\}$ be a sequence of points in a metric space (X, d) which is called a *Cauchy sequence* if for every $\varepsilon > 0$, there is an integer N such that

$$d(u_m, u_n) < \varepsilon \ \forall \, n, m > N.$$

It may be recalled that a sequence in a metric space is a function having domain as the set of natural numbers and the range as a subset of the metric space. The definition of Cauchy sequence means that the distance between two points u_n and u_m is very small when n and m are very large.

Definition 1.3 Let $\{u_n\}$ be a sequence in a metric space (X, d). It is called *convergent* with limit u in X if, for $\varepsilon > 0$, there exists a natural number N having property

$$d(u_n, u) < \varepsilon \ \forall \, n > N$$

If $\{u_n\}$ converges to u, that is, $\{u_n\} \to u$ as $n \to \infty$, then we write, $\lim_{n \to \infty} u_n = u$.

Definition 1.4 Let every Cauchy sequence in a metric space (X, d) is convergent. Then (X, d) is called a *complete metric space*.

Complete Metric Spaces

Example 1.7 (a) Spaces R, R^2, R^n, ℓ_p, $C[a, b]$ with metric (a) of Example 1.5 and $L_2[a, b]$ are examples of complete metric spaces.
(b) $(0, 1]$ is not a complete metric space.
(c) $C[a, b]$ with integral metric is not a complete metric space.
(d) The set of rational numbers is not a complete metric space.
(e) $C[a, b]$ with (b) of Example 1.5 is not complete metric space.

Definition 1.5 (a) A subset M of a metric space (X, d) is said to be *bounded* if there exists a positive constant k such that $d(u, v) \leq k$ for all u, v belonging to M.
(b) A subset M of a metric space (X, d) is *closed* if every convergent sequence $\{u_n\}$ in M is convergent in M.
(c) If every bounded sequence $M \subset (X, d)$ has a convergent subsequence then the subset M is called *compact*.
(d) Let $T : (X, d) \rightarrow (Y, d)$. T is called *continuous* if $u_n \rightarrow u$ implies that $T(u_n) \rightarrow Tu$; that is, $d(u_n, u) \rightarrow 0$ as $n \rightarrow \infty$ implies that $d(T(u_n), Tu) \rightarrow 0$.

Remark 1.3 1. It may be noted that every bounded and closed subset of (R^n, d) is a compact subset.
 2. It may be observed that each closed subset of a complete metric space is complete.
   As we see above, a metric is a distance between two points. We introduce now the concept of distance between the subsets of a set, for example, distance between a line and a circle in R^2. This is called Hausdorff metric.

Distance Between Two Subsets (Hausdorff Metric)

Let X be a set and $H(X)$ be a set of all subsets of X. Suppose $d(\cdot, \cdot)$ be a metric on X. Then distance between a point u of X and a subset M of X is defined as

$$d(u, M) = \inf\{d(u, v)/v \in M\}$$
$$or = \inf_{v \in M}\{d(u, v)\}$$

Let M and N be two elements of $H(X)$. Distance or metric between M and N denoted by (M,N) is defined as

$$d(M, N) = \sup_{u \in M} \inf_{v \in N} d(u, v)$$
$$= \sup_{u \in M} d(u, N).$$

It can be verified that

$$d(M, N) \neq d(N, M)$$

where

$$d(N, M) = \sup_{v \in N} \inf_{u \in M} d(v, u)$$
$$= \sup_{u \in M} \inf_{u \in N} d(u, v).$$

Definition 1.6 The Hausdorff metric or the distance between two elements M and N of a metric (X, d), denoted by $h(M, N)$, is defined as

$$h(M, N) = \max\{d(M, N), d(N, M)\}$$

Remark 1.4 If $H(X)$ denotes the set of all closed and bounded subsets of a metric space (X, d) then $h(M, N)$ is a metric. If $X = R^2$ then $H(R^2)$ the set of all compact subsets of R^2 is a metric space with respect to $h(M, N)$.

Contraction Mapping

Definition 1.7 (*Contraction Mapping*) A mapping $T : (X, d) \rightarrow (X, d)$ is called a *Lipschitz continuous mapping* if there exists a number α such that

$$d(Tu, Tv) \leq \alpha d(u, v) \,\forall\, u, v \in X.$$

If α lies in $[0, 1)$, that is, $0 \leq \alpha < 1$, then T is called a *contraction mapping*. α is called the contractivity factor of T.

Example 1.8 Let $T : R \rightarrow R$ be defined as $Tu = (1+u)^{1/3}$. Then finding a solution to the equation $Tu = u$ is equivalent to solving the equation $u^3 - u - 1 = 0$. T is a contraction mapping on $I = [1, 2]$, where the contractivity factor is $\alpha = (3)^{1/3} - 1$.

Example 1.9 (a) Let $Tu = 1/3u, 0 \leq u \leq 1$. Then T is a contraction mapping on [0, 1] with contractivity factor $1/3$.
(b) Let $S(u) = u + b$, $u \in R$ and b be any fixed element of R. Then S is not a contraction mapping.

Example 1.10 Let $I = [a, b]$ and $f : [a, b] \rightarrow [a, b]$ and suppose that $f'(u)$ exist and $|f'(x)| < 1$. Then f is a contraction mapping on I into itself.

Definition 1.8 (*Fixed Point*) Let T be a mapping on a metric space (X, d) into itself. $u \in X$ is called a fixed point if

$$Tu = u$$

Theorem 1.1 (Existence of Fixed Point-Contraction Mapping Theorem by Stefan Banach) *Let (X, d) be a complete metric space and let T be a contraction mapping on (X, d) into itself with contractivity factor α. Then there exists only one point u in X*

such that $Tu = u$, that T has a unique fixed point. Furthermore, for any $u \in (X, d)$, the sequence $x, T(x), T^2(x), \ldots, T^k(x)$ converges to the point u; that is

$$\lim_{k \to \infty} T^k = u$$

Proof We know that $T^2(x) = T(T(x)), \ldots, T^k(x) = T(T^{(k-1)}(x))$, and

$$
\begin{aligned}
d(T^m(x), T^n(x)) &\le \alpha d(T^{m-1}(x), T^{(n-1)}(x)) \\
&\le \alpha^m d(x, T^{n-m}(x)) \\
&\le \alpha^m \sum_{k=1}^{n-m} d(T^{k-1}(x), T^k(x)) \\
&\le \alpha^m \sum_{k=1}^{n-m} \alpha^{k-1} d(x, T(x))
\end{aligned}
$$

This we obtain by applying contractivity $(k - 1)$ times. It is clear that

$$d(T^m(x), T^n(x)) \to 0 \text{ as } m, n \to \infty$$

and so $T^m(x)$ is a Cauchy sequence in a complete metric space (X, d). This sequence must be convergent, that is

$$\lim_{m \to \infty} T^m x = u$$

We show that u is a fixed point of T, that is, $T(u) = u$. In fact, we will show that u is unique. $T(u) = u$ is equivalent to showing that $d(T(u), u) = 0$.

$$
\begin{aligned}
d(T(u), u) &= d(u, T(u)) \\
&\le d(u, T^k(x)) + d(T^k(x), T(u)) \\
&\le d(u, T^k(x)) + \alpha d(u, T^{k-1}(x)) \to 0 \text{ as } k \to \infty
\end{aligned}
$$

It is clear that

$$\lim_{k \to \infty} d(u, T^k(x)) = 0$$

as $u = \lim_{k \to \infty} T^k(x)$ and $\lim_{k \to \infty} d(u, T^{k-1}(x)) = 0$ $(u = \lim_{k \to \infty} T^k(x))$.
Let v be another element in X such that $T(v) = v$. Then

$$d(u, v) = d(T(u), T(v)) \le \alpha d(u, v)$$

This implies $d(u, v) = 0$ or $u = v$ (Axiom (i) of the metric space).
Thus, T has a unique fixed point.

1.3 Application of Banach Contraction Mapping Theorem

1.3.1 Application to Matrix Equation

Suppose we want to find the solution of a system of n linear algebraic equations with n unknowns

$$\left.\begin{array}{l} a_{11}x_1 + a_{12}x_2 + \cdots + a_{1n}x_n = b_1 \\ a_{21}x_1 + a_{22}x_2 + \cdots + a_{2n}x_n = b_2 \\ \cdots\cdots\cdots\cdots\cdots\cdots\cdots\cdots\cdots\cdots\cdots \\ a_{n1}x_1 + a_{n2}x_2 + \cdots + a_{nn}x_n = b_n \end{array}\right\} \tag{1.2}$$

Equivalent matrix formulation $Ax = b$, where

$$A = \begin{pmatrix} a_{11} & a_{12} & \cdots & a_{1n} \\ a_{21} & a_{22} & \cdots & a_{2n} \\ \cdots & \cdots & \cdots & \cdots \\ a_{n1} & a_{n2} & \cdots & a_{nn} \end{pmatrix}$$

$x = (x_1, x_2, ..., x_n)^T$, $y = (y_1, y_2, ..., y_n)^T$
The system can be written as

$$\left.\begin{array}{l} x_1 = (1 - a_{11})x_1 - a_{12}x_2 \cdots - a_{1n}x_n + b_1 \\ x_2 = -a_{21}x_1 - (1 - a_{22})x_2 \cdots - a_{2n}x_n + b_2 \\ \cdots\cdots\cdots\cdots\cdots\cdots\cdots\cdots\cdots\cdots\cdots\cdots\cdots \\ x_n = -a_{n1}x_1 - a_{n2}x_2 \cdots + (1 - a_{nn})x_n + b_n \end{array}\right\} \tag{1.3}$$

By letting $\alpha_{ij} = -a_{ij} + \delta_{ij}$ where

$$\delta_{ij} = \begin{cases} 1, & for\ i = j \\ 0, & for\ i \neq j \end{cases}$$

Equation (1.2) can be written in the following equivalent form

$$x_i = \sum_{j=1}^{n} \alpha_{ij}x_j + b_i, \ i = 1, 2, \ldots n \tag{1.4}$$

If $x = (x_1, x_2, \ldots, x_n) \in R^n$, then Eq. (1.1) can be written in the equivalent form

$$x - Ax + b = x \tag{1.5}$$

Let $Tx = x - Ax + b$. Then the problem of finding the solution of system $Ax = b$ is equivalent to finding fixed points of the map T.

Now, $Tx - Tx' = (I - A)(x - x')$ and we show that T is a contraction under a reasonable condition on the matrix.

In order to find a unique fixed point of T, i.e., a unique solution of system of equations (1.1), we apply Theorem 1.1. In fact, we prove the following result.

Equation (1.1) has a unique solution if

$$\sum_{j=i}^{n} |\alpha_{ij}| = \sum_{j=1}^{n} |-a_{ij} + \delta_{ij}| \le k < 1, \ i = 1, 2, \ldots n$$

For $x = (x_1, x_2, \ldots, x_n)$ and $x' = (x'_1, x'_2, \ldots, x'_n)$, we have

$$d(Tx, Tx') = d(y, y')$$

where

$$y = (y_1, y_2, \ldots, y_n) \in R^n$$
$$y' = (y'_1, y'_2, \ldots, y'_n) \in R^n$$
$$y_i = \sum_{j=1}^{n} \alpha_{ij} x_j + b_i$$
$$y'_i = \sum_{j=1}^{n} \alpha_{ij} x'_j + b_i \ i = 1, 2, \ldots n$$

We have

$$
\begin{aligned}
d(y, y') &= \sup_{1 \le i \le n} |y_i - y'_i| \\
&= \sup_{1 \le i \le n} |\sum_{j=1}^{n} \alpha_{ij} + b_i - \sum_{j=1}^{n} \alpha_{ij} x'_j - b_i| \\
&= \sup_{1 \le i \le n} \left| \sum_{j=1}^{n} \alpha_{ij}(x_j - x'_j) \right| \\
&= \sup_{1 \le i \le n} \sum_{j=1}^{n} |\alpha_{ij}| |x_j - x'_j| \\
&\le \sup_{1 \le i \le n} |x_j - x'_j| \sup_{1 \le i \le n} \sum_{j=1}^{n} |\alpha_{ij}| \\
&\le k \sup_{1 \le i \le n} |x_j - x'_j|
\end{aligned}
$$

Since $\sum_{j=1}^{n} |\alpha_{ij}| \le k < 1$ for $i = 1, 2, \ldots, n$ and $d(x, x') = \sup 1 \le j \le n |x_j - x'_j|$, we have $d(Tx, Tx') \le kd(x, x'), 0 \le k < 1$; i.e, T is a contraction mapping on R^n into itself. Hence, by Theorem 1.1, there exists a unique fixed point $x^\star$ of T in R^n; i.e., $x^\star$ is a unique solution of system (1.1).

1.3.2 Application to Integral Equation

Here, we prove the following existence theorem for integral equations.

Theorem 1.2 *Let the function $H(x, y)$ be defined and measurable in the square $A = \{(x, y)/a \le x \le b, a \le y \le b\}$. Further, let*

$$\int_a^b \int_a^b |H(x, y)|^2 < \infty$$

and $g(x) \in L_2(a, b)$. Then the integral equation

$$f(x) = g(x) + \mu \int_a^b H(x, y) f(y) \, dy \tag{1.6}$$

possesses a unique solution $f(x) \in L_2(a, b)$ for every sufficiently small value of the parameter μ.

Proof For applying Theorem 1.1, let $X = L_2$, and consider the mapping T

$$T : L_2(a, b) \to L_2(a, b)$$
$$Tf = h$$

where $h(x) = g(x) + \mu \int_a^b H(x, y) f(y) dy \in L_2(a, b)$.

This definition is valid for each $f \in L_2(a, b)$, $h \in L_2(a, b)$, and this can be seen as follows. Since $g \in L_2(a, b)$ and μ is scalar, it is sufficient to show that

$$\psi(x) = \int_a^b K(x, y) f(y) dy \in L_2(a, b)$$

By the Cauchy–Schwarz inequality

$$\left| \int_a^b H(x, y) f(y) dy \right| \leq \int_a^b |H(x, y) f(y)| dy$$

$$\leq \left(\int_a^b |H(x, y)|^2 dy \right)^{1/2} \left(\int_a^b |f(y)|^2 \right)^{1/2}$$

Therefore

$$|\psi(x)|^2 = \left(\left| \int_a^b H(x, y) f(y) dy \right| \right)^2$$

$$\leq \left(\int_a^b |H(x, y)|^2 dy \right) \left(\int_a^b |f(y)|^2 dy \right)$$

or

$$\int_a^b |\psi(x)|^2 dx \leq \int_a^b (\int_a^b |H(x, y)|^2 dy)(\int_a^b |f(y)|^2 dy) dx$$

By the hypothesis,

$$\int_a^b \int_a^b |H(x, y)|^2 dx dy < \infty$$

and

$$\int_a^b |f(y)|^2 dy < \infty$$

Thus

$$\psi(x) = \int_a^b H(x, y) f(y) dy \in L_2(a, b)$$

We know that $L_2(a, b)$ is a complete metric space with metric

$$d(f, g) = \left(\int_a^b |f(x) - g(x)|^2 dx \right)^{1/2}$$

Now we show that T is a contraction mapping. We have $d(Tf, Tf_1) = d(h, h_1)$, where

$$h_1(x) = g(x) + \mu \int_a^b H(x, y) f_1(y) dy$$

$$d(h, h_1) = |\mu| (\int_a^b \left[\left[\int_a^b K(x, y)[f(y) - f_1(y)] dy \right] \right]^2 dx)^{1/2}$$

$$\leq |\mu| \left(\int_a^b \int_a^b |H(x, y)|^2 dx\, dy \right)^{1/2} \left(\int_a^b |f(y) - f_1(y)|^2 dy \right)^{1/2}$$

by using Cauchy–Schwarz–Bunyakowski inequality.

Hence, $d(Tf, Tf_1) \leq |\mu| \left(\int_a^b \int_a^b |K(x, y)|^2 dx dy \right)^{1/2} d(f, f_1)$. By definition of the metric in L_2, we have

$$d(f, f_1) = \left(\int_a^b |f(y) - f_1(y)|^2 dy \right)^{1/2}$$

If

$$|\mu| < 1 / \left(\int_a^b \int_a^b |H(x, y)|^2 dx dy \right)^{1/2}$$

then

$$d(Tf, Tf_1) \leq k d(f, f_1)$$

where

$$0 \leq k = |\mu| \left(\int_a^b \int_a^b |H(x, y)|^2 dx\, dy \right)^{1/2} < 1$$

Thus, T is a contraction and, so T has a unique fixed point, say, there exists a unique $f^\star \in L_2[a, b]$ such that $Tf^\star = f^\star$. Therefore, $f^\star$ is a solution of equation (1.6).

1.3.3   Existence of Solution of Differential Equation

We prove Picard theorem applying contraction mapping theorem of Banach.

Theorem 1.3 *Picard's Theorem Let $g(x, y)$ be a continuous function defined on a rectangle $M = \{(x, y)/a \leq x \leq b, c \leq y \leq d\}$ and satisfy the Lipschitz condition of order 1 in variable y. Moreover, let (u_0, v_0) be an interior point of M. Then the differential equation*

$$\frac{dy}{dx} = g(x, y) \tag{1.7}$$

has a unique solution, say $y = f(x)$ which passes through (u_0, v_0).

Proof We examine in the first place that finding the solution of equation (1.6) is equivalent to the problem of finding the solution of an integral equation. If $y = f(x)$ satisfies (1.6) and satisfies the condition that $f(u_0) = v_0$, then integrating (1.6) from u_0 to x, we have.

$$f(x) - f(u_0) = \int_{u_0}^{x} g(t, f(t))dt$$

$$f(x) = v_0 + \int_{u_0}^{x} g(t, f(t))dt \tag{1.8}$$

Thus, solution of (1.6) is equivalent to a unique solution of (1.7).

Solution of (1.7): $|g(x, y_1) - g(x, y_2)| \leq q|y_1 - y_2|, q > 0$ as $g(x, y)$ satisfies the Lipschitz condition of order 1 in the second variable y. $g(x, y)$ is bounded on M; that is, there exists a positive constant k such that $|g(x, y)| \leq m \forall (x, y) \in M$. This is true as $f(x, y)$ is continuous on a compact subset M of R^2.

Find a positive constant p such that $pq < 1$ and the rectangle $N = \{(x, y)/ - p + u_0 \leq x \leq p + u_0, -pm + v_0 \leq y \leq pm + v_0\}$ is contained in M.

Suppose X is the set of all real-valued continuous functions $y = f(x)$ defined on $[-p + u_0, p + u_0]$ such that $d(f(x), u_0) \leq mp$. It is clear that X is a closed subset of $C[u_0 - p, u_0 + p]$ with sup metric (Example 1.5(a)). It is a complete metric space by Remark 1.3.

Remark 1.5 Define a mapping $T : X \rightarrow X$ by $Tf = h$, where $h(x) = v_0 + \int_{u_0}^{x} g(t, f(t)dt)$. T is well defined as

$$d(h(x), v_0) = \sup \left| \int_{u_0}^{x} g(t, f(t))dt \right| \leq m(x - u_0) \leq mp$$

$h(x) \in X$.

For $f, f_1 \in X$

$$d(Tf, Tf_1) = d(h, h_1) = \sup \left| \int_{x_0}^{x} [g(t, f(t)) - g(t, f_1(t))]dt \right|$$

$$\leq q \int_{x_0}^{x} |f(t) - f_1(t)|dt$$

$$\leq qpd(f, f_1)$$

or

$$d(Tf, Tf_1) \leq \alpha d(g, g_1)$$

where $0 \leq \alpha = qp < 1$

Therefore, T is a contraction mapping or complete metric space and by virtue of Theorem 1.1, T has a unique fixed point. This fixed point say f^* is the unique solution of equation (1.6).

For more details, see [Bo 85, Is 85, Ko Ak 64, Sm 74, Ta 58, Li So 74].

1.4 Problems

Problem 1.1 Verify that (R^2, d), where $d(x, y) = |x_1 - y_1| + |x_2 - y_2|$ for all $x = (x_1, x_2)$, $y = (y_1, y_2) \in R^2$, is a metric space.

Problem 1.2 Verify that (R^2, d), where $d(x, y) = \max\{|x_1 - y_1|, |x_2 - y_2|\}$ for all $x = (x_1, x_2), y = (y_1, y_2) \in R^2$, is a metric space.

Problem 1.3 Verify that (R^2, d), where $d(x, y) = (\sum_{i=1}^{n} |x_i - y_i|^2)^{1/2}$ for all $x, y \in R$, is a metric space.

Problem 1.4 Verify that $(C[a, b], d)$, where $d(f, g) = \sup_{a \leq x \leq b} |f(x) - g(x)|$ for all $f, g \in C[a, b]$, is a complete metric space.

Problem 1.5 Verify that $(L_2[a, b], d)$, where $d(f, g) = (\int_a^b |f(x) - g(x)|^2)^{1/2}$, is a complete metric space.

Problem 1.6 Prove that $\ell_p, 1 \leq p \leq \infty$ is a complete metric space.

Problem 1.7 Let $m = \ell_\infty$ denote the set of all bounded real sequences. Then check that m is a metric space. Is it complete?

Problem 1.8 Show that $C[a, b]$ with integral metric defined on it is not a complete metric space.

Problem 1.9 Verify that $h(\cdot, \cdot)$, defined in Definition 1.6, is a metric for all closed and bounded sets A and B.

Problem 1.10 Let $T: R \to R$ be defined by $Tu = u^2$. Find fixed points of T.

Problem 1.11 Find fixed points of the identity mapping of a metric space (X, d).

Problem 1.12 Verify that the Banach contraction theorem does not hold for incomplete metric spaces.

Problem 1.13 Let $X = \{x \in R / x \geq 1\} \subset R$ and let $T: X \to X$ be defined by $Tx = (1/2)x + x^1$. Check that T is a contraction mapping on (X, d), where $d(x, y) = |x - y|$, into itself.

Problem 1.14 Let $T \to R^+ \to R^+$ and $Tx = x + e^x$, where R^+ denotes the set of positive real numbers. Check that T is not a contraction mapping.

Problem 1.15 Let $T: R^2 \to R^2$ be defined by $T(x_1, x_2) = (x_2^{1/3}, x_1^{1/3})$. What are the fixed points of T? Check whether T is continuous in a quadrant?

Problem 1.16 Let (X, d) be a complete metric space and T a continuous mapping on X into itself such that for some integer n, $T^n = T \circ T \circ T \cdots \circ T$ is a contraction mapping. Then show that T has a unique fixed point in X.

Problem 1.17 Let (X, d) be a complete metric space and T be a contraction mapping on X into itself with contractivity factor α, $0 < \alpha < 1$. Suppose that u is the unique fixed point of T and $x_1 = Tx$, $x_2 = Tx_1$, $x_3 = Tx_2, \ldots$, $x_n = T(T^{n-1}x) = T^n x, \ldots$ for any $x \in X$ is a sequence. Then prove that

1. $d(x_m, u) \leq (\frac{\alpha^m}{1-\alpha}) \; \forall \, m$
2. $d(x_m, u) \leq \frac{\alpha}{1-\alpha} d(x_{m-1}, x_m) \; \forall \, m$

Problem 1.18 Prove that every contraction mapping defined on a metric space X is continuous, but the converse may not be true.

Chapter 2
Banach Spaces

Abstract The chapter is devoted to a generalization of Euclidean space of dimension n, namely R^n (vector space of dimension n), known as Banach space. This was introduced by a young engineering student of Poland, Stefan Banach. Spaces of sequences and spaces of different classes of functions such as spaces of continuous differential integrable functions are examples of structures studied by Banach. The properties of set of all operators or mappings (linear/bounded) have been studied. Geometrical and topological properties of Banach space and its general case normed space are presented.

Keywords Normed space · Banach space · Toplogical properties · Properties of operators · Spaces of operators · Convex sets · Convex functionals · Dual space · Reflexive space · Algebra of operators

2.1 Introduction

A young student of an undergraduate engineering course, Stefan Banach of Poland, introduced the notion of magnitude or length of a vector around 1918. This led to the study of structures called *normed space* and special class, named *Banach space*. In subsequent years, the study of these spaces provided foundation of a branch of mathematics called functional analysis or infinite-dimensional calculus. It will be seen that every Banach space is a normed linear space or simply normed space and every normed space is a metric space. It is well known that every metric space is a topological space. Properties of linear operators (mappings) defined on a Banach space into itself or any other Banach space are discussed in this chapter. Concrete examples are given. Results presented in this chapter may prove useful for proper understanding of various branches of mathematics, science, and technology.

© Springer Nature Singapore Pte Ltd. 2018

A. H. Siddiqi, *Functional Analysis and Applications*, Industrial and Applied Mathematics, https://doi.org/10.1007/978-981-10-3725-2_2

2.2 Basic Results of Banach Spaces

Definition 2.1 Let X be a vector space over R. A real-valued function $|| \cdot ||$ defined on X and satisfying the following conditions is called a norm:

(i) $||x|| \geq 0$; $||x|| = 0$ if and only if $x = 0$.
(ii) $||\alpha x|| = |\alpha| \, ||x||$ for all $x \in X$ and $\alpha \in R$.
(iii) $||x + y|| \leq ||x|| + ||y|| \; \forall \, x, y \in X$.

$(X, || \cdot ||)$, vector space X equipped with $|| \cdot ||$ is called a normed space.

Remark 2.1 (a) Norm of a vector is nothing but length or magnitude of the vector. Axiom (i) implies that norm of a vector is nonnegative and its value is zero if the vector is itself is zero.

(b) Axiom (ii) implies that if norm of $x \in X$ is multiplied by $|\alpha|$, then it is equal to the norm of αx, that is $|\alpha| \, ||x|| = ||\alpha x||$ for all x in X and $\alpha \in R$.

(c) Axiom (iii) is known as the triangle inequality.

(d) It may be observed that the norm of a vector is the generalization of absolute value of real numbers.

(e) It can be checked (Problem 2.1) that normed space (X, d) is a metric space with metric:

$$d(x, y) = ||x - y||, \forall x \text{ and } y \in X.$$

Since $d(x, 0) = ||x - 0|| = ||x||$ so that the norm of any vector can be treated as the distance between the vector and the origin or the zero vector of X. The concept of Cauchy sequence, convergent sequence, completeness introduced in a metric space can be extended to a associate normed space. A metric space is not necessarily a normed space (see Problem 2.1).

(f) Different norms can be defined on a vector space; see Example 2.4.

(g) A norm is called *seminorm* if the statement $||x|| = 0$ if and only if $x = 0$ is dropped.

Definition 2.2 A normed space X is called a *Banach space*, if its every Cauchy sequence is convergent, that is $||x_n - x_m|| \to 0$ as $n, m \to \infty \; \forall x_n, x_m \in X$ implies that $\exists \, x \in X$ such that $||x_n - x|| \to 0$ as $n \to \infty$).

Remark 2.2 (i) Let $(X, || \cdot ||)$ be a normed space and Y be a subspace of vector X. Then, $(Y, || \cdot ||)$ is a normed space.

(ii) Let Y be a closed subspace of a Banach space $(X, || \cdot ||)$. Then, $(Y, || \cdot ||)$ is also a Banach space.

2.2.1 Examples of Normed and Banach Spaces

Example 2.1 Let R denote the vector space of real numbers. Then, $(R, ||.||)$ is a normed space, where $||x|| = |x|, x \in R$ ($|x|$ denotes the absolute value of real number x).

Example 2.2 The vector space R^2 (the plane where points have coordinates with respect to two orthogonal axes) is a normed space with respect to the following norms:

1. $||a||_1 = |x| + |y|$, where $a = (x, y) \in R^2$.
2. $||a||_2 = \max\{|x|, |y|\}$.
3. $||a||_3 = (x^2 + y^2)^{1/2}$.

Example 2.3 The vector space C of all complex numbers is a normed space with respect to the norm $||z|| = |z|, z \in C$ ($|\cdot|$ denotes the absolute value of the complex number).

Example 2.4 The vector space R^n of all n-tuples $x = (u_1, u_2, \ldots, u_n)$ of real numbers is a normed space with respect to the following norms:

1. $||x||_1 = \sum_{k=1}^{n} |u_k|$.
2. $||x||_2 = \left(\sum_{k=1}^{n} |u_k|^2 \right)^{1/2}$.
3. $||x||_3 = \left(\sum_{k=1}^{n} |u_k|^p \right)^{1/p}$ where $1 \leq p < \infty$.
4. $||x||_4 = \max\{|u_1|, |u_2| \ldots |u_n|\}$.

Notes

1. R^n equipped with the norm defined by (3) is usually denoted by ℓ_p^n.
2. R^n equipped with the norm defined by (4) is usually denoted by ℓ_∞^n.

Example 2.5 The vector space m of all bounded sequences of real numbers is a normed space with the norm $||x|| = \sup_n |x_n|$. (Sometimes ℓ^∞ or ℓ_∞ is used in place of m).

Example 2.6 The vector space c of all convergent sequences of real numbers $u = \{u_k\}$ is a normed space with the norm $||u|| = \sup_n |u_n|$.

Example 2.7 The vector space of all convergent sequences of real numbers with limit zero is denoted by c_0 which is a normed space with respect to the norm of c.

Example 2.8 The vector space $\ell_p, 1 \leq p < \infty$, of sequences for which $\sum_{k=1}^{\infty} |u_k|^p < \infty$ is a normed space with norm $||x||_p = \left[\sum_{k=1}^{\infty} |u_k|^p \right]^{1/p}$

Example 2.9 The vector space $C[a, b]$ of all real-valued continuous functions defined on $[a, b]$ is a normed space with respect to the following norms:

(a) $||f||_1 = \int_a^b |f(x)| dx.$

(b) $||f||_2 = \left(\left(\int_a^b |f(x)| \right)^2 dx \right)^{1/2}.$

(c) $||f||_3 = \sup_{a \le x \le b} |f(x)|.$

(where $|| \cdot ||_3$ is called *uniform convergence norm*).

Example 2.10 The vector space $P[0, 1]$ of all polynomials on $[0, 1]$ with the norm

$$||x|| = \sup_{0 \le t \le 1} |x(t)|$$

is a normed space.

Example 2.11 Let M be any nonempty set, and let $B(M)$ the class of all bounded real-valued functions defined on M. Then, $B(M)$ is a normed space with respect to the norm

$$||f|| = \sup_{t \in M} |f(t)|.$$

Note If M is the set of positive integers, then m is a special case of $B(M)$.

Example 2.12 Let M be a topological space, and let $BC(M)$ denote the set of all bounded and continuous real-valued functions defined on M. Then, $B(M) \supseteq BC(M)$, and $BC(M)$ is a normed space with the norm $||f|| = \sup_{t \in A} |f(t)| \forall f \in BC(M)$.

Note If M is compact, then every real-valued continuous function defined on A is bounded. Thus, if M is a compact topological space, then the set of all real-valued continuous functions defined on M, denoted by $C(M) = BC(M)$, is a normed space with the same norm as in Example 2.12. If $M = [a, b]$, we get the normed space of Example 2.9 with $|| \cdot ||_3$.

Example 2.13 Suppose $p \ge 1$ (p is not necessarily an integer). $\mathscr{L}_p$ denotes the class of all real-valued functions $f(t)$ such that $f(t)$ is defined for all t, with the possible exception of a set of measure zero (almost everywhere or a.e.), is measurable and $|f(t)|^p$ is Lebesgue integrable over (∞, ∞). Define an equivalence relation in $\mathscr{L}_p$ by stating that $f(t) \sim g(t)$ if $f(t) = g(t) a.e.$ The set of all equivalence classes into which $\mathscr{L}_p$ is thus divided is denoted by L_p or L^p. L_p is a vector space and a normed space with respect to the following norm:

$$||f^1|| = \left(\int\limits_{-\infty}^{\infty} |f^1(t)|^p dt \right)^{1/p}$$

Notes

1. In place of $(-\infty, \infty)$, one can consider $(0, \infty)$ or any finite interval (a, b) or any measurable set E.
2. $f^{(1)}$ represents the equivalence class $[f]$.
3. $\mathscr{L}_p$ is not a normed space if the equality is considered in the usual sense. However, $\mathscr{L}_p$ is a seminormed space, which means $||f|| = 0$, while $f \neq 0$.
4. The zero element of $\mathscr{L}_p$ is the equivalence class consisting of all $f \in L_p$ such that $f(t) = 0$ a.e.

Example 2.14 Let $[a, b]$ be a finite or an infinite interval of the real line. Then, a measurable function $f(t)$ defined on $[a, b]$ is called *essentially bounded* if there exists $k \geq 0$ such that the set $\{t/f(t) > k\}$ has measure zero; i.e., $f(t)$ is bounded a.e. on $[a, b]$. Let $L_\infty[a, b]$ denote the class of all measurable and essentially bounded functions $f(t)$ defined on $[a, b]$. L_∞ is in relation to L_∞ just as we define L_p in relation to ℓ_p. L_∞ or L_∞ is a normed space with the norm $||f^{(1)}|| = \overset{0}{\underset{t}{\sup}} |f(t)|$ [essential least upper bound of $|f(t)|$ or essential supremum of $|f(t)|$ over the domain of definition of f].

Example 2.15 The vector space $BV[a, b]$ of all functions of bounded variation on $[a, b]$ is a normed space with respect to the norm $||f|| = |f(a)| + V_a^b(x)$, where $V_a^b(x)$ denotes the total variation of $f(t)$ on $[a, b]$.

Example 2.16 The vector space $C^\infty[a, b]$ of all infinitely differentiable functions on $[a, b]$ is a normed space with respect to the following norm:

$$||f||_{n,p} = \left(\int\limits_a^b \sum_{i=0}^n |D^i f(t)|^p dt \right)^{1/p}, \ 1 \leq p \leq \infty$$

where D_i denotes the ith derivative.

Note The vector space $C^\infty[a, b]$ can be normed in infinitely many ways.

Example 2.17 Let $C^k(\Omega)$ denote the space of all real functions of n variables defined on Ω (an open subset of R^n) which are continuously differentiable up to order k. Let $\alpha = (\alpha_1, \alpha_2, \ldots, \alpha_n)$ where α's are nonnegative integers, and $|\alpha| = \sum_{i=1}^n \alpha$. Then for $f \in C^k(\Omega)$, the following derivatives exist and are continuous:

$$D^\alpha f = \frac{\partial^{|\alpha|} f}{\partial t_1^{\alpha_1}, \ldots, \partial t_n^{\alpha_n}} |\alpha| \leq k$$

$C^k(\Omega)$ is a normed space under the norm

$$\|f\|_{k,\alpha} = \max_{0 \le |\alpha| \le k} \sup |D^\alpha f|$$

Example 2.18 Let $C_0^\infty(\Omega)$ denote the vector space of all infinitely differentiable functions with compact support on Ω (an open subset of R^n). $C_0^\infty(\Omega)$ is a normed space with respect to the following norm:

$$\|f\|_{k,p} = \left(\int_\Omega \sum_{|\alpha| \le k} |D^\alpha f(t)|^p dt \right)^{1/p}$$

where Ω and $D^\alpha f$ are as in Example 2.17.

For compact support, see Definition A.14(7) of Appendix A.3.

Example 2.19 The set of all absolutely continuous functions on $[a, b]$, which is denoted by $AC[a, b]$, is a subspace of $BV[a, b]$. $AC[a, b]$ is normed space with the norm of Example 2.15.

Example 2.20 The class $Lip_\alpha[a, b]$, the set of all functions f satisfying the condition

$$\|f\|_\alpha = \sup_{t>0,x} \left(\frac{|f(x+t) - f(x)|}{t^\alpha} \right) < \infty$$

is a normed space for a suitable choice of α.

2.3   Closed, Denseness, and Separability

2.3.1   *Introduction to Closed, Dense, and Separable Sets*

Definition 2.3 (a) Let X be a normed linear space. A subset Y of X is called a closed set if it contains all of its limit points. Let Y' denote the set of all limit points of Y, then $Y \cup Y'$ is called the closure of Y and is denoted by $\bar{Y}$, that is, $\bar{Y} = Y \cup Y'$.

(b) Let $S_r(a) = \{x \in X / \|x - a\| < r, \ r > 0\}$. $S_r(a)$ is called the open sphere with radius r and center a of the normed space X. If $a = 0$ and $r = 1$, then it is called the unit sphere.

(c) Let $\bar{S}_r(a) = \{x \in X / \|x - a\| \le r, \ r > 0\}$, then $\bar{S}_r(a)$ is called the closed sphere with radius r and center a. $\bar{S}_1(0)$ is called the unit closed sphere.

Remark 2.3 (i) It is clear that $Y \subset \bar{Y}$. It follows immediately that Y is closed if and only if $Y = \bar{Y}$.

(ii) It can be verified that $C[-1, 1]$ is not a closed subspace in $L_2[-1, 1]$.

(iii) The concept of closed subset is useful while studying the solution of equation.

Definition 2.4 (*Dense Subsets*) Suppose A and B are two subsets of X such that $A \subset B$. A is called *dense* in B if for each $v \in B$ and every $\varepsilon > 0$, there exists an element $u \in A$ such that $||v - u||_X < \varepsilon$. If A is dense in B, then we write $\bar{A} = B$.

Example 2.21 (i) The set of rational numbers Q is dense in the set of real numbers R, that is $\bar{Q} = R$.

(ii) The space of all real continuous functions defined on $\Omega \subset R$ denoted by $\bar{C}(\Omega)$ is dense in $L_2(\Omega)$, that is, $\bar{C}(\Omega) = L_2(\Omega)$.

(iii) The set of all polynomials defined on Ω is dense in $L_2(\Omega)$.

Definition 2.5 (*Separable Sets*) Let X possess a countable subset which is dense in it, then X is called a *separable* normed space.

Example 2.22 (i) Q is a countable dense subset of R. There R is a separable normed linear space.

(ii) R^n is separable.

(iii) The set of all polynomials with rational coefficients is dense and countable in $L_2(\Omega)$. Therefore, $L_2(\Omega)$ is separable.

It may be observed that a normed space may contain more than one subset which is dense and countable.

Definition 2.6 Normed linear spaces $(X, || \cdot ||_1)$ and $(Y, || \cdot ||_2)$ are called *isometric* and *isomorphic* if following conditions are satisfied: There exists a one-to-one mapping T of X and Y having properties:

(i)

$$d_2(Tx, Ty) = ||Tu_1 - Tu_2||_2 = ||u_1 - u_2||_1 = d(u_1, u_2) \ or \ ||Tx||_2 = ||x||_1$$

T is linear, namely

(ii)

$$T(x + y) = Tx + Ty, \forall x, y \in X$$

(iii)

$$T(\alpha x) = \alpha Tx, \forall x \in X, \ \alpha \in R \ or \ C$$

If X and Y are isometric and isomorphic, then we write $X = Y$. It means that two isometric and isomorphic spaces can be viewed as the same in two different guises. Elements of two such spaces may be different, but topological and algebraic properties will be same. In other words distance, continuity, convergence, closedness, denseness, separability, etc., will be equivalent in such spaces. We encounter such situations in Sect. 2.5.

Definition 2.7 (a) Normed spaces $(X, ||\cdot||_1)$ and $(X, ||\cdot||_2)$ are called *topologically equivalent*, or *equivalently* two norms $|| \cdot ||_1$ and $|| \cdot ||_2$ are *equivalent* if there exist constants $k_1 > 0$ and $k_2 > 0$ such that

$$k_1||x||_1 \leq ||x||_2 \leq k_2||x||_1.$$

(b) A normed space is called *finite-dimensional* if the underlying vector space has a finite basis; that is, it is finite-dimensional. If underlying vector space does not have finite basis, the given normal space is called *infinite-dimensional*.

Theorem 2.1 *All norms defined on a finite-dimensional vector space are equivalent.*

Theorem 2.2 *Every normed space X is homeomorphic to its open unit ball* $S_1(0) = \{x \in X /||x|| < 1\}$.

2.3.2   Riesz Theorem and Construction of a New Banach Space

If M is a proper closed subspace of a normed space X, then a theorem by Riesz tells us that there are points at a nonzero distance from Y. More precisely, we have

Theorem 2.3 (Riesz Theorem) *Let M be a proper closed subspace of a normed space X, and let* $\varepsilon > 0$. *Then, there exists an* $x \in X$ *with* $||x|| = 1$ *such that* $d(x, M) \geq 1 - \varepsilon$.

Riesz theorem can be used to prove that a normed space is finite-dimensional if and only if its bounded closed subsets are compact.

The following result provides us the most useful method of forming a new Banach space from a given Banach space:

Theorem 2.4 *Let M be a closed subspace of a Banach space X. Then, the factor or quotient vector space X/M is a Banach space with the norm*

$$||x + M|| = \inf_{x \in M}\{||u + x||\} \text{ for each } u \in X$$

2.3.3   Dimension of Normed Spaces

R, R^n, ℓ_p^n, C are examples of finite-dimensional normed spaces. In fact, all real finite-dimensional normed spaces of dimension n are isomorphic to R^n.

$C[a, b]$, ℓ_p, L_p, $P[0, 1]$, $BV[a, b]$, $C^k(\Omega)$, etc., are examples of infinite-dimensional normed spaces.

2.3.4 Open and Closed Spheres

1. Consider a normed space R of real numbers, an open sphere with radius $r > 0$ and center a which we denote by $S_r(a)$. This is nothing but an open interval $(a - r, a + r)$. The closed sphere $\bar{S_r}(a)$ is the closed interval $[a - r, a + r]$. The open unit sphere is $(-1, 1)$, and the closed unit sphere is $[-1, 1]$.

2. (a) Consider the normed space R^2 (plane) with the norm (1) (see Example 2.2). Let

$$x = (x_1, x_2) \in R^2, a = (a_1, a_2) \in R^2$$

Then,

$$S_r(a) = \{x \in R_2/||x - a|| < r\} = \{x \in R_2/\{|x_1 - a_1| + |x_2 - a_2|\}\}$$

and

$$\overline{S_r(a)} = \{x \in R^2/\{|x_1 - a_1| + |x_2 - a_2|\} \leq r\}$$
$$S_1 = \{x \in R^2 | \{|x_1| + |x_2|\} < 1$$
$$\bar{S_1} = \{\{x \in R^2 ||x_1| + |x_2|\}\} \leq 1$$

Figure 2.1 illustrates the geometrical meaning of the open and closed unit spheres.
S_1 = Parallelogram with vertices $(-1, 0)$, $(0, 1)$, $(1, 0)$, $(0, -1)$.
S_1 = Parallelogram without sides AB, BC, CD, and DA. Surface or boundary of the closed unit sphere is lines AB, BC, CD, and DA.

(b) If we consider R^2 with respect to the norm (2) in Example 2.2, Fig. 2.2 represents the open and closed unit spheres. The rectangle ABCD = S_1, where S_1 is the rectangle without the four sides.

Fig. 2.1 The geometrical meaning of open and closed unit spheres in Example 2.2 (1)

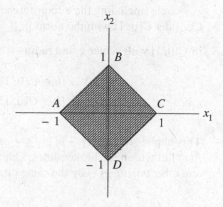

Fig. 2.2 The geometrical meaning of open and closed unit spheres in Example 2.2 (2)

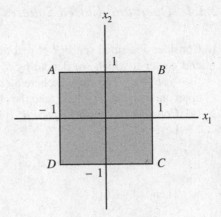

$$S_1 = \{x = (x_1, x_2) \in R^2 / \max(|x_1|, |x_2|) \leq 1\}$$
$$S_1 = \{x = (x_1, x_2) \in R^2 / \max(|x_1|, |x_2|) < 1\}$$

Surface or boundary S is given by

$$S = \{x = (x_1, x_2) \in R_2 / \max(|x_1|, |x_2|) = 1\}$$
$$= Sides\ AB,\ BC,\ CD\ and\ DA$$

(c) Figure 2.3 illustrates the geometrical meaning of the open and closed unit spheres in R^2 with respect to norm (3) of Example 2.2.

$$\bar{S}_1 = \{x = (x_1, x_2) \in R^2 / (x_1^2 + x_2^2)^{1/2} \leq 1\}$$
$$S_1 = \{x = (x_1, x_2) \in R^2 / (x_1^2 + x_2^2)^{1/2} < 1\}$$

S is the circumference of the circle with center at the origin and radius 1. S_1 is the interior of the circle with radius 1 and center at the origin. $\bar{S}_1$ is the circle (including the circumference) with center at the origin and radius 1.

3. Consider $C[0, 1]$ with the norm $\|f\| = \sup_{0 \leq t \leq 1} |f(t)|$. Let $\bar{S}_r(g)$ be a closed sphere in $C[0, 1]$ with center g and radius r. Then,

$$\bar{S}_r(g) = \{h \in C[0, 1] / \|g - h\| \leq r\}$$
$$= \{h \in C[0, 1] / \sup_{0 \leq x \leq 1} |g(x) - h(x)| \leq r\}$$

This implies that $|g(x) - h(x)| \leq r$ or $h(x) \in [g(x) - r, g(x) + r]$; see Fig. 2.4. $h(x)$ lies within the broken lines. One of these is obtained by lowering $g(x)$ by r and other raising $g(x)$ by the same distance r. It is clear that $h(x) \in S_r^\star(g)$. If $h(x)$

Example 2.24 Let four mappings be defined on R^2 as follows:

1. $T_1(x, y) = (\alpha x, \alpha y)$, where $x, y \in R$ and α is a fixed positive real number.
2. $T_2(x, y) = (x \cos(\theta) - y \sin(\theta), x \sin(\theta) + y \cos(\theta))$; that is, T is a mapping which rotates each point of the plane about the origin and through an angle (θ).
3. $T_3(x, y) = x$ for all $(x, y) \in R^2$.
4. $T_4(x, y) = y$ for all $(x, y) \in R^2$.

$$OP = (x^2 + y^2)^{1/2}$$
$$OQ = \alpha(x^2 + y^2)^{1/2}$$

where O is the origin in R^2, and P and Q are arbitrary points in R^2.

If (x, y) is any point P in the plane, then its image $(\alpha x, \alpha y)$ under T is point Q which is colinear with P and the origin and α times as far from the origin as P.
If we take the norm (3) of Example 2.2, then

$$d((x, y), (0, 0)) = ||(x, y) - (0, 0)|| = ||(x, y)|| = (x^2 + y^2)^{1/2}$$

T_1, T_2, T_3, and T_4 are linear operators.

Example 2.25 Let $A = (a_{ij})$ be an $m \times n$ matrix. Then, the equations

$$\eta_i = \sum_{j=1}^{n} a_{ij}\xi_j, \, i = 1, 2, 3, \ldots, m$$

define an operator T on R^n into R^m.

$$T: R^n \to R^m$$

$Tx = y$ where $x = (\xi_1, \xi_2, \ldots, \xi_n) \in R^n$, $y = (\eta_1, \eta_2, \ldots, \eta_m) \in R^m$

1. T is linear if

   (a) $T(x + x') = Tx + Tx'$
   (b) $T(\alpha x) = \alpha T(x)$

   LHS of (a)

$$= T(x + x') = T((\xi_1 + \xi_1'), (\xi_2 + \xi_2'), \ldots, (\xi_n + \xi'n))$$

where $x = (\xi_1, \xi_2, \ldots, \xi_n)$, $x = (\xi_1', \xi_2', \ldots, \xi_n')$. Thus,

$$T(x + x') = \sum_{j=1}^{n} a_{ij}(\xi_j + \xi_j') = \sum_{j=1}^{n} a_{ij}(\xi_j) + \sum_{j=1}^{n} a_{ij}(\xi_j')$$
$$= Tx + Tx' = RHS \, of \, (a)$$

LHS of (*b*)

$$T(\alpha x) = T(\alpha(\xi_1, \xi_2, \ldots, \xi_n))$$
$$= T(\alpha\xi_1, \alpha\xi_1, \ldots, \alpha\xi_1)$$
$$= \sum_{j=1}^{n} a_{ij}(\alpha\xi_j) = \alpha \sum_{j=1}^{n} a_{ij}(\xi_j) = \alpha Tx = RHS \ of \ (b)$$

2. T is *uniformly continuous* and hence *continuous* if $A \neq 0$.

Let $x' = (\xi_1, \xi_2', \ldots, \xi_n') \in R^n$ be fixed and let $y' = Tx'$. Then, for an arbitrary $x = (\xi_1, \xi_2, \ldots, \xi_n) \in R^n$, if $y = Tx$, we have

$$\|y - y'\|^2 = \|Tx - Tx'\|^2 = \left\| \sum_{j=1}^{n} a_{ij}(\xi_j) - \sum_{j=1}^{n} a_{ij}(\xi_j') \right\|^2$$

$$= \| \sum_{j=1}^{n} a_{ij}(\xi_j - \xi_j') \|^2$$

$$= \sum_{i=1}^{m} \left| \sum_{j=1}^{n} a_{ij}(\xi_j - \xi_j') \right|^2$$

Also,

$$\|x - x'\| = \left(\sum_{j=1}^{n} |(\xi_j - \xi_j')|^2 \right)^{1/2}$$

By applying the Cauchy–Schwartz–Bunyakowski's inequality, we get

$$\|y - y'\|^2 \leq \sum_{i=1}^{m} \left(\sum_{j=1}^{n} |a_{ij}|^2 \right) \left(\sum_{j=1}^{n} |\xi_j - \xi_j'|^2 \right)$$

$$= \sum_{i=1}^{m} \left(\sum_{j=1}^{n} |a_{ij}|^2 \right) \left(\sum_{j=1}^{n} |\xi_j - \xi_j'|^2 \right)$$

$$= \sum_{i=1}^{m} \sum_{j=1}^{n} |a_{ij}|^2 \|x - x'\|^2$$

$$\|y - y\| \leq c \|x - x'\|^2$$

where $c^2 = \sum_{i=1}^{m} \sum_{j=1}^{n} |a_{ij}|^2$

Thus, for a given $\varepsilon > 0$, we may choose $\delta = \varepsilon/c$, provided $c \neq 0$ such that $||x - x'|| < \varepsilon/c$ implies $||Tx - Tx'|| = ||y - y'|| < c\varepsilon/c = \varepsilon$. This proves that T is uniformly continuous provided A is a nontrivial matrix.

Remark 2.5 We shall see in Problem 2.6 that all linear operators defined on finite-dimensional spaces are continuous.

Example 2.26 Let $T : P_n[0, 1] \rightarrow P_n[0, 1]$ be defined as follows:

$$Tp(t) = \frac{d}{dt}p(t)(\text{first derivative of the polynomial})$$

$P_n[0, 1]$ denotes the space of all polynomials on $[0, 1]$ of degree less than or equal to n which is a finite-dimensional normed space.

T is a linear operator on $P_n[0, 1]$ into itself as the following relations hold good by the well-known properties of the derivative: $T(p_1 + p_2) = Tp_1 + Tp_2$ and $T(\alpha p_1) = \alpha Tp_1$ for p_1 and $p_2 \in P_n[0, 1]$ and for α real number. We have

$$Tx^n = \frac{d}{dx}x^n = nx^{n-1}$$

Thus, $||x_n|| = 1$, but $||Tx_n|| = n$. In view of this, T is not bounded and by Theorem 2.5; it is not continuous.

Example 2.27 Let X be a normed space under a norm $|| \cdot ||$. Then, a mapping T, defined on X in the following way:

$$Tx = ||x||$$

is a nonlinear operator. If we consider $X = C[0, 1]$, with sup norm, then

$$Tf = \sup_{0 \leq t \leq 1} |f(t)|$$

$$T(f_1 + f_2) = \sup_{0 \leq t \leq 1} (|(f_1 + f_2)(t)|) = \sup_{0 \leq t \leq 1} (|(f_1(t) + f_2(t)|)$$

which may not be equal to $[\sup_{0 \leq t \leq 1} |f_1(t)| + \sup_{0 \leq t \leq 1} |f_2(t)|]$; that is, $T(f_1 + f_2) \neq Tf_1 + Tf_2$.

Example 2.28 Let T be defined on c as

$$Tx = \sigma$$

where $\sigma = \{\sigma_n\}$ or σ is the nth $(C, 1)$ mean or Cesáro mean of order 1 of the sequence $x = \{x_n\} \in c$. Then, T is a linear transformation on c into itself.

Example 2.29 Let $x = \{x_n\} \in m$, then $Tx = y$, where

$$y = \left\{\frac{x_n}{n}\right\}$$

is a linear transformation of m into c_0.

Example 2.30 Let T be a mapping on $B(M)$ defined as follows:

$$Tf = f(t_0)$$

where t_0 is a fixed point of M, then T is a linear mapping on $B(M)$.

Example 2.31 Let T be a mapping on $C[0, 1]$ which maps every function of $C[0, 1]$ to its Riemann integral, i.e., $Tf = \int_0^t f(t)dt$. Then, T is linear by the well-known properties of the Riemann integral. We define an operator D on $C[0, 1]$ by taking the derivative of functions belonging to $C[0, 1]$; i.e.,

$$Df = \frac{d}{dt}f(t)$$

It is clear that D is not defined for all $f(t) \in C[0, 1]$, and even if Df exists, it may not belong to $C[0, 1]$. If, however, D is defined over a subspace of $C[0, 1]$ which consists of all functions having first continuous derivatives, the range of D is also contained in $C[0, 1]$. D is linear but not continuous.

Example 2.32 Consider $\ell_2(\ell_p, p = 2)$, the mapping S defined below maps ℓ_2 into itself. For $\{x_n\} \in \ell_2$, $Sx_n = \alpha x_{n+1} + \beta x_{n-1}$, where α and β are real constants. S is a linear operator on ℓ_2.

Example 2.33 Let A be the subset of $C[a, b]$ consisting of elements $f(t)$ which have first and second derivatives continuous on $[a, b]$ such that $f(a) = f'(a) = 0$. Let p and q belong to A. Let us define S on $C[a, b]$ by $Sf = g$, where $g(t) = f''(t) + p(t)f'(t) + q(t)f(t)$. S is a linear operator on $C[a, b]$ into A.

Example 2.34 Let $H(s, t)$ be a continuous function on the square $a \leq s \leq b, a \leq t \leq b$ and $f \in C[a, b]$. Then, S, defined by $Sf = g$, where $g(s) = \int_a^b H(s, t)f(t)dt$, is a linear operator on $C[a, b]$ into itself. S_1, defined by $S_1 f = g$, where $g(s) = f(s) - \int_a^b K(s, t)f(t)dt$, is also a linear operator on $C[a, b]$ into itself.

Remark 2.6 Some well-known initial and boundary value problems can be expressed in terms of S and S_1.

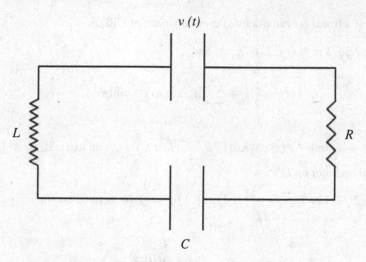

Fig. 2.5 LCR circuit

Example 2.35 Consider the LCR circuit as shown in Fig. 2.5. The differential equation for the charge f on the condenser is

$$L\frac{d^2f}{dt^2} + R\frac{df}{dt} + \frac{f}{c} = v \tag{2.1}$$

where $v(t)$ is the applied voltage at time t. If the initial conditions are assumed to be $f(0) = 0$, $\frac{df(0)}{dt} = 0$ and $R^2 > 4L/C$, then the solution of this differential equation is given by

$$f(s) = \int_0^s H(s-t)v(t)dt$$

where

$$H(s) = \frac{e^{\lambda_1 s} - e^{\lambda_2 s}}{L(\lambda_1 - \lambda_2)}$$

and λ_1, λ_2 are the distinct real roots of $L\lambda^2 + R\lambda + 1/C = 0$. If $v \in C[0, \infty)$, then so is f and, in this case, the previous equation may be written as $f = Sv$ where S is a linear operator on $C[0, \infty)$.

Thus, Eq. (2.1) can be written in the form

$$v = Tf$$

where T is a linear operator defined on a subspace of $C[0, \infty)$.

Example 2.36 Let $f(x) \in L_1(-\pi, \pi)$ and

$$f(x) \sim \frac{1}{2}a_0 + \sum_1^\infty (a_n \cos(nx) + \sin(nx))$$

where $a_n = (1/\pi) \int\limits_{-\pi}^{\pi} f(x) \cos nx dx, \ b_n = (1/\pi) \int\limits_{-\pi}^{\pi} f(x) \sin nx dx$. If $Tf = \{a_n\}$, T is a linear operator on L_1.

Example 2.37 Let $T : \ell_p \to \ell_p$, $1/p + 1/q = 1$ be defined by $Tx = y$, where $x = (x_1, x_2, \ldots, x_n, \ldots) \in \ell_p$

$$y = (y_1, y_2, \ldots y_n, \ldots) \in \ell_q \text{ with } y_i = \sum_{j=1}^\infty a_{ij} x_j$$

and $\{a_{ij}\}$ is an infinite matrix satisfying the condition $\sum\limits_{i=1}^\infty \sum\limits_{j=1}^\infty |a_{ij}|^q < \infty$, $q > 1$. Then, T is a linear operator on ℓ_p into ℓ_q.

Example 2.38 Let $X = \left\{ x = \{x_i\} / \sum_1^\infty |x_i| < \infty, \ x_i's \ are \ real \right\}$. X is a normed space with respect to the norm $||x|| = \sup\limits_i |x_i|$. Let $Tx = \sum\limits_{i=1}^\infty x_i$. Then, T is a linear functional on X but it is not continuous:

$$T(x + y) = \sum_{i=1}^\infty (x_i + y_i) = \sum_{i=1}^\infty x_i + \sum_{i=1}^\infty y_i = Tx + Ty$$

and

$$T(\alpha x) = \sum_{i=1}^\infty \alpha x_i = \alpha \sum_{i=1}^\infty x_i = \alpha T(x).$$

T is linear.

Let $x = (1, 1, \ldots, 1, 0, 0, \ldots)$ be an element of X having the first n entries 1 and the rest 0. Then, $||x|| = \sup |x_i| = 1$ and $Tx = \sum\limits_{i=1}^\infty x_i = n$. Therefore, T is not bounded. In view of Theorem 2.5, it is not continuous.

2.4.2 Properties of Linear Operators

The following results provide the interesting properties of linear operators:

Proposition 2.1 *1. If S is a linear operator, then $S0 = 0$.*
2. If $x_1, x_2, \ldots, x_n$ belong to a normed space U and $\alpha_1, \alpha_2, \ldots, \alpha_n$ are real numbers, then for any linear operator S defined on U

$$S\left(\sum_{i=1}^{n} \alpha_i x_i\right) = \sum_{i=1}^{n} \alpha(Sx_i)$$

3. Any operator S on a normed space is continuous if and only if $x_n \to x_0$ implies $Sx_n \to Sx_0$ [$x_n \in X$ and $x_0 \in X$].
4. The null space of a nonzero continuous linear operator is a closed subspace.

Theorem 2.5 *Let U and V be normed spaces and S a linear operator on U into V. Then, the following statements are equivalent:*

1. S is continuous.
2. S is continuous at the origin.
3. S is bounded.
4. $S\overline{S_1}$ is a bounded subset of Y.

Theorem 2.6 *For a bounded linear operator S, the following conditions are equivalent:*

1. $||S|| = \sup\{\frac{||Tx||}{||x||}/x \neq 0\}$ or $\sup\limits_{x \neq 0}\{\frac{||Sx||}{||x||}\}$.
2. $||S|| = \inf\{k/||Sx|| \leq k||x||\}$
3. $||S|| = \sup\{||Sx|| \, ||x|| \leq 1\}$
4. $||S|| = \sup\{||Sx||/||x|| = 1\}$ or $\sup\limits_{||x||=1} ||Sx||$

Theorem 2.7 *The set $\mathscr{B}[U, V]$ of all bounded linear operators on a normed space U into a normed space V is a normed space. If V is a Banach space, then $\mathscr{B}[U, V]$ is also a Banach space.*

Remark 2.7 1. It can be verified that the right-hand side of (1) in Theorem 2.6 exists and that $||Sx|| \leq ||S|| \, ||x||$ holds good for bounded S.
2. In view of Theorem 2.5, the concepts of continuity and boundedness for linear operators are equivalent.

Remark 2.8 In Theorem 2.7, if we take V as the Banach space of real numbers, then we get the following result. "The set of all bounded linear functionals defined on a normed space U is a Banach space." This is usually denoted by U^* or U and is called the *dual space*, the *conjugate space*, the *adjoint space*, or *first dual space* of the normed space U. U^{**} denotes the space of bounded linear functionals of U^* and is called the *second dual of U*. For the representation of the elements of U^* for some well-known spaces, see Sect. 2.4.

Proof (*Proof of Proposition* 2.1)

1. $S(0+0) = S_0 + S_0$, by the additivity property of S. Also, $LHS = S(0)$. Thus, $S0 = S0 + S0$. This implies that $S0 = 0$.
2. Since S is linear, the result is true for $n = 1, 2$. Suppose it is true for $n = k$, i.e.,

$$S \left| \sum_{i=1}^{k} \alpha_i x_i \right| = \sum_{i=1}^{k} \alpha_i S(x_i)$$

Then clearly,

$$S((\alpha_1 x_1 + \alpha_2 x_2 + \cdots + \alpha_k x_k) + \alpha_{k+1} x_{k+1})$$
$$= \alpha_1 S x_1 + \alpha_2 S x_2 + \cdots + \alpha_k S x_k + \alpha_{k+1} S x_{k+1}$$

that is, the result is true for $n = k + 1$. By the principle of finite induction, the result is true for all finite n.
3. This is a special case of Theorem A.6(2) of Appendix A. In fact, the linearity is not required.
4. Let N be the null space of $S \neq 0$ ($S \equiv 0$) such that

$$x, y \in N \Rightarrow Sx = 0 \text{ and } Sy = 0 \Rightarrow S(x+y) = 0 \Rightarrow x + y \in N.$$

$$x \in N \Rightarrow Sx = 0 \Rightarrow \alpha(Sx) = 0 \Rightarrow S(\alpha x) = 0 \Rightarrow \alpha x \in N.$$

Thus, N is a subspace. Let $x_n \in N$ and $x_n \to x$. In order to show that N is closed, we need to show that $x \in N$. Since $x_n \in N$, $Sx_n = 0 \; \forall \; n$. As S is continuous $Sx_n \to Sx$, so $\lim_{n \to \infty} Sx_n = Sx = 0 \Rightarrow x \in X$

Proof (*Proof of Theorem* 2.5) We shall prove that (1) $\Leftrightarrow$ (2), (2) $\Leftrightarrow$ (3), (3) $\Leftrightarrow$ (4) which means that all are equivalent.

1. (1) $\Leftrightarrow$ (2): Suppose S is continuous, which means that S is continuous at every point of X. Thus, (1) $\Rightarrow$ (2) is obvious. Now suppose that S is continuous at the origin, that is, for $x_n \in X$ with $x_n \to 0$, $Sx_n \to 0$. This implies that if $x_n \to x$, then $S(x_n - x) = Sx_n - Sx \to 0$. That is, for $x_n \to x$ in X, $Sx_n \to Sx$ in Y. By Proposition 2.1, S is continuous and (2) $\Rightarrow$ (1).
2. (2) $\Leftrightarrow$ (3): Suppose that S is continuous at the origin. Then $x_n \to 0 \Rightarrow Sx_n \to 0$ by Proposition 2.1. Suppose further that S is not bounded. This implies that for each natural number n, we can find a vector x_n such that

$$||Sx_n|| > n||x_n|| \text{ or } \frac{1}{n||x_n||}||Sx_n|| > 1$$

or

$$\left\| S\left(\frac{1}{n||x_n||}x_n\right)\right\| > 1$$

By choosing $y_n = \frac{x_n}{n||x_n||}$, we see that $y_n \to 0$ but $Sy_n \not\to 0$. This contradicts the first assumption. Hence, S is bounded whenever S is continuous at the origin, i.e., $(2) \Rightarrow (3)$.

To prove the converse, assume that S is bounded, i.e., there exists $k > 0$ such that $||Sx|| \le k||x||$. It follows immediately that if $x_n \to 0$, then $Sx_n \to 0$. Thus, $(3) \Rightarrow (2)$.

3. $(3) \Leftrightarrow (4)$: Suppose S is bounded, that is, there exists $k > 0$ such that $||Sx|| \le k||x||$. From this, it is clear that if $||x|| \le 1$, then $||Sx|| \le k$. This means that the image of closed unit sphere $\bar{S}_1$ under S is a bounded subset of V. Thus, $(3) \Rightarrow (4)$. For the converse, assume that $S\bar{S}_1$ is a bounded subset of V. This means that $S\bar{S}_1$ is contained in a closed sphere of radius k centered at the '0' of V. If $x = 0$, then $Sx = 0$ and $||Sx|| \le k$, and the result is proved. In case $x \ne 0, \frac{x}{||x||} \in \bar{S}_1$ and so $||S(\frac{x}{||x||})|| \le k$ or $||Sx|| \le k||x||$. This shows that $(4) \Rightarrow (3)$.

Proof (*Proof of Theorem* 2.6)

1. $(1) \Leftrightarrow (2)$: First, we prove that if $||S||$ is given by (1), then it is equal to $\inf\{k||Sx|| \le k||x||\}$

$$||S|| = \sup\left\{\frac{||Sx||}{||x||}/x \ne 0\right\}$$

or

$$||S|| \ge \frac{||Sx||}{||x||} \, for \, x \ne 0$$

or

$$||Sx|| \le ||S|| \, ||x|| \, as \, S0 = 0$$

Thus, $||S||$ is one of the k's satisfying the relation $||Sx|| \le k||x||$. Hence,

$$||S|| \ge \inf\{k/||Sx|| \le k||x||\} \quad (2.2)$$

On the other hand, for $x \ne 0$ and for a k satisfying the relation $||Sx|| \le k||x||$, we have $\frac{||Sx||}{||x||} \le k$. This implies that

$$\sup\left\{\frac{||Sx||}{||x||}/x \ne 0\right\} \le k$$

or $||S|| \leq k$. Since this relation is satisfied by all k and $||S||$ is independent of x and k, we get

$$||S|| \leq \inf\{k/||Sx|| \leq k||x||\} \tag{2.3}$$

By Eqs. (2.2) and (2.3), we have

$$||S|| = \inf\{k/||Sx|| \leq k||x||\}$$

This proves that $(1) \Leftrightarrow (2)$.
2. $(1) \Leftrightarrow (3)$: Let

$$||S|| = \sup\left\{\frac{||Sx||}{||x||} / x \neq 0\right\}$$

Then,

$$||Sx|| \leq ||S|| \, ||x|| \leq ||S|| \; if \; ||x|| \leq 1$$

Thus,

$$\sup\{||Sx||/||x|| \leq 1\} \leq ||S|| \tag{2.4}$$

Since $||S|| = \sup\{\frac{||Sx||}{||x||}/x \neq 0\}$, for any $\varepsilon > 0$, there exists an $x_1 \neq 0$, such that

$$\frac{||Sx_1||}{||x_1||} > ||S|| - \varepsilon$$

By taking $y = x_1/||x_1||$, $(||y|| = 1)$, we find that

$$\sup\{||Sx||/||x|| \leq 1\} \geq ||Sy|| = \left\|S\left(\frac{x}{||x_1||}\right)\right\|$$

$$= \frac{1}{||x_1||}||Sx_1|| > ||S|| - \varepsilon$$

Therefore,

$$\sup\{||Sx|| \, ||x|| \leq 1\} \geq ||S|| \tag{2.5}$$

By Eqs. (2.4) and (2.5), we find that $(1) \Leftrightarrow (3)$.
3. $(1) \Leftrightarrow (4)$: Suppose

$$||S|| = \sup\left\{\frac{||Sx||}{||x||} / x \neq 0\right\}$$

Then, as in the previous case

$$||Sx|| \leq ||S|| \, ||x|| \leq ||S|| \; if \; ||x|| = 1$$

or

$$\sup\left\{\frac{||Sx||}{||x||}||x|| = 1\right\} \leq ||S|| \qquad (2.6)$$

Further,

$$\frac{||Sx_1||}{||x_1||} > ||S|| - \varepsilon$$

and

$$\sup\left\{\frac{||Sx||}{||x||}||x|| = 1\right\} \geq ||Sy||$$

where $y = \frac{x_1}{||x_1||}$ or

$$\sup\left\{\frac{||Sx||}{||x||}||x|| = 1\right\} > ||Sy|| - \varepsilon$$

Thus,

$$\sup\left\{\frac{||Sx||}{||x||}||x|| = 1\right\} \geq ||S|| \qquad (2.7)$$

By Eqs. (2.6) and (2.7), we have (1) $\Leftrightarrow$ (4).

Proof (Proof of Theorem 2.7)

1. We introduce in $\mathscr{B}[U, V]$ the operations of addition (+) and scalar multiplication (·) in the following manner: For $T, S \in \mathscr{B}[U, V]$

   a. $(T + S)(x) = Tx + Sx$.
   b. $(\alpha T)x = \alpha Tx$, where α is a real number. It can be easily verified that the following relations are satisfied by the above two operations:
      i. If $T, S \in \mathscr{B}[U, V]$, then $T + S \in \mathscr{B}[U, V]$.
      ii. $(T + S) + W = T + (S + W)$, for all $T, S, U \in \mathscr{B}[U, V]$.
      iii. There is an element $0 \in \mathscr{B}[U, V]$ such that $0 + T = T$ for all $T \in \mathscr{B}[U, V]$.
      iv. For every $T \in \mathscr{B}[U, V]$, there is a $T_1 \in \mathscr{B}[U, V]$ such that $T + T_1 = 0$.
      v. $T + S = S + T$.
      vi. $\alpha T \in \mathscr{B}[U, V]$ for all real α and $T \in \mathscr{B}[U, V]$.
      vii. $\alpha(T + S) = \alpha T + \alpha S$.

    viii. $(\alpha + \beta)T = \alpha T + \beta T$.

    ix. $(\alpha\beta)T = \alpha(\beta T)$.

    x. $1T = T$.

These relations mean that $B[U, V]$ is a vector space.

2. Now we prove that

$$||T|| = \sup \left| \frac{||Tx||}{||x||} / x \neq 0 \right|$$

is a norm on $\mathscr{B}[U, V]$ and so $\mathscr{B}[U, V]$ is a normed space with respect to this norm or equivalent norms given by Theorem 2.6. It is clear that $||T||$ exists. Since $\{\frac{||Tx||}{||x||} / x \neq 0\}$ is a bounded subset of real numbers, its least upper bound or sup must exist.

  (a) $||T|| \geq 0$ as $||T||$ is the sup of nonnegative numbers. Let $||T|| = 0$. Then, $||Tx|| = 0$ or $Tx = 0$ for all $x \in X$ or $T = 0$, which is the zero element of $\mathscr{B}[U, V]$. Conversely, assume that $T = 0$. Then, $||Tx|| = ||0x|| = ||0|| = 0$. This means that $||T|| = \sup \left| \frac{||Tx|||}{||x||} / x \neq 0 \right| = 0$. Thus, $||T|| = 0$ if and only if $T = 0$.

  (b)

$$||\alpha T|| = \sup \left\{ \frac{||(\alpha T)x||}{||x||} / x \neq 0 \right\}$$

$$= \sup \left\{ \frac{||\alpha(Tx)||}{||x||} / x \neq 0 \right\}$$

$$= \sup \left\{ \frac{|\alpha| \, ||Tx||}{||x||} / x \neq 0 \right\}$$

(By applying property (2) of the norm on V)

$$= |\alpha| \sup \left\{ \frac{||Tx||}{||x||} / x \neq 0 \right\} = |\alpha| \, ||T||$$

$$||T + S|| = \sup \left\{ \frac{||(T + S)(x)||}{||x||} / x \neq 0 \right\}$$

$$= \sup \left\{ \frac{||Tx + Sx||}{||x||} / x \neq 0 \right\}$$

By the property (3) of the norm on V, we have

$$||Tx + Sx|| \leq ||Tx|| + ||Sx||$$

This implies that

$$||T + S|| \leq \sup \left| \frac{||Tx + Sx||}{||x||} / x \neq 0 \right|$$

$$= \sup \left| \frac{||Tx||}{||x||} + \frac{||Sx||}{||x||} / x \neq 0 \right|$$

$$= \sup \left| \frac{||Tx||}{||x||} \right| + \sup \left| \frac{||Sx||}{||x||} \right|$$

(by using Appendix A, Remark A.1.D)

$$= ||T|| + ||S||$$

or

$$||T + S|| \leq ||T|| + ||S||$$

This proves that $\mathscr{B}[U, V]$ is a normed space.

(c) Now, we prove that $\mathscr{B}[U, V]$ is a Banach space provided V is a Banach space. Let T_n be a Cauchy sequence in $\mathscr{B}[U, V]$. This means that for $\varepsilon > 0$, there exists a natural number N such that $||T_n - T_m|| < \varepsilon$ for all $n, m > N$. This implies that, for any fixed vector $x \in$, we have

$$||T_n x - T_m x|| \leq ||(T_n - T_m)|| \, ||x|| < \varepsilon ||x||, \; for \; n, \; m > N$$

that is, $T_n x$ is a Cauchy sequence in V. Since V is a Banach space, $\lim_{n \to \infty} T_n x = Tx$, say. Now, we verify that (a) T is a linear operator on U into V, (b) T is bounded on X into V, and (c) $||T_m - T|| \leq \varepsilon \; for m > N$.

(i) Since T is defined for arbitrary $x \in U$, it is an operator on U into V.

$$T(x + y) = \lim_{n \to \infty} T_n(x + y)$$

$$= \lim_{n \to \infty} [T_n x + T_n y] \; as \; all \; T_n's \; are \; linear$$

$$= \lim_{n \to \infty} T_n x + \lim_{n \to \infty} T_n y$$

$$= Tx + Ty$$

$$T(\alpha x) = \lim_{n \to \infty} T_n(\alpha x) = \lim_{n} \alpha T_n x \; since \; T_n's \; are \; linear$$

$$= \alpha \lim_{n \to \infty} T_n x = \alpha Tx$$

(ii) Since T_n's are bounded, there exists $M > 0$ such that $||T_n x|| \leq M$ for all n. This implies that for all n and any $x \in X$, $||T_n x|| \leq ||T_n|| \, ||x|| \leq M ||x||$. Taking the limit, we have

$$\lim_{n \to \infty} ||T_n x|| \leq M \, ||x||$$

or

$$||Tx|| \leq M \, ||x||$$

[$\lim_{n \to \infty} T_n x = Tx$. Since norm is a continuous function $lim_{n \to \infty} ||T_n x|| = ||Tx||$] This proves that T is bounded.

(iii) Since T_n is a Cauchy sequence, for each $\varepsilon > 0$ there exists a positive integer N such that $||T_n - T_m|| < \varepsilon$ for all $n, m > N$. Thus, we have

$$||T_n x - T_m x|| \leq ||T_n - T_m|| \, ||x|| < \varepsilon ||x|| \, |for \, n, m > N$$

or

$$\lim_{n \to \infty} ||T - T_m|| = \sup \left| \frac{||(T - T_m)x||}{||x||} /x \neq 0 \right| \leq \varepsilon \, \forall \, m > N,$$

that is, $T_m \to T$ as $m \to \infty$.

2.4.3   Unbounded Operators

Example 2.39 Suppose that T is an operator defined as

$$Tf = \frac{df}{dx} \; on \; C^1[0, 1]$$

Let $f_n = \sin nx$. Then,

$$||f_n|| = \sup_{0 \leq x \leq 1} |f_n x| = 1 \forall n$$

and

$$Tf_n = n \cos nx, \; ||Tf_n|| = n$$

Since $||f_n|| = 1$ and Tf_n increases indefinitely for $n \to \infty$, there is no finite constant k such that $||Tf|| \in k ||f|| \, \forall f \in C^1[0, 1]$. Thus, T is an unbounded operator.

Definition 2.9 An operator $T : X \to Y$ is called *bounded below* if and only if there exists a constant $m > 0$ such that

$$||Tx|| \geq m ||x||_X, \; \forall \, x \in X$$

The differential operator considered in Example 2.31 is unbounded on $C_0[0, 1] = \{f \in C[0, 1]/f(0) = f(1) = 0\}$ as well. However, it is bounded below on $C_0[0, 1]$ as follows: For $f \in C[0, 1]$

$$f(x) = \int_0^x \frac{df}{ds} ds \leq \sup_{0 \leq x \leq 1} |\frac{df}{dx}|$$

or $||Tf|| \geq \sup_{0 \leq x \leq 1} |f(x)| = ||f||$, that is, T is bounded below.

Example 2.40 Suppose $T: U \to V$ is continuous, where X is a Banach space. Let λ be a scalar. Then, the operator $\lambda I - T$, where I is the identity operator, is bounded below for sufficiently large λ.

Verification

$$||(\lambda I - T)x|| = ||\lambda I x - Tx||$$
$$\geq ||(\lambda x)|| - ||Tx||$$

Since T is bounded, $||Tx|| \leq k||x||$. Hence, $||(\lambda I x - Tx)|| \geq (|\lambda| - k)||x||$. This implies that $(\lambda I - T)$ is bounded below for sufficiently large λ.

An important feature of linear bounded below operators is that even though they may not be continuous, they always have a continuous inverse defined on their range. More precisely, we have following theorem.

Theorem 2.8 *Let S be a linear bounded below operator from a normed space U into a normed space V. Then, S has a continuous inverse S^{-1} from its range $R(S)$ into U. Conversely, if there is a continuous inverse $S^{-1}: R(S) \to U$, then there is a positive constant m such that*

$$||Sx||_Y \geq m||x||_X \text{ for every } x \in U.$$

2.5 Representation of Bounded and Linear Functionals

Representation of elements of the dual space X of a normed linear space which are bounded and linear functionals has been studied extensively. In the first place, we introduce the concept of a reflexive Banach space X that can be identified in a natural way with the space of all bounded and linear functionals defined on its dual space X^*. A Banach space X is called the *reflexive Banach space* if the second dual of X is equal to X; that is, $(X^*)^* = X^{**} = X$ (equality in the sense of isometric and isomorphic explained earlier). A mapping $J: X \to X^{**}$ defined by $J(x) = Fx$, where $Fx(f) = f(x) \ \forall f \in X^*$, is called the *natural embedding*. It can be verified that J is isometry, linear, and one-one. X is reflexive if J is onto. The Hungarian mathematician Riesz found a general representation of a bounded and linear functional on a Hilbert space which is discussed in Chap. 3. Here, we give a representation of the elements of the dual spaces of R^n, ℓ_p, $L_2[0, 1]$ and L_p. We also indicate a close relationship between the basis of a normed space X and the basis of its dual X^*.

Example 2.41 Any element F of the dual space of Rn can be written in the form

$$F(x) = \sum_{i=1}^{n} \alpha_i x_i \tag{2.8}$$

where $x = (x_1, x_2, \ldots, x_n) \in R_n$, $\alpha = (\alpha_1, \alpha_2, \ldots, \alpha_n) \in R^n$. In fact, $F(x) = \langle x, y \rangle = \sum_{i=1}^{n} y_i x_i$; that is, for every $x \in R_n$, there exists a unique $y \in R_n$ such that the value of a bounded and linear functional F can be expressed as the "inner product". ($\langle x, y \rangle$ is called the inner product of x and y and will be studied in Chap. 3).

Example 2.42 Let $X = L_2[0, 1]$ and define functional F on X by

$$F(f) = \int_0^1 f(x)g(x)\mathrm{d}x \; for \; all \; f \in L_2[0, 1] \tag{2.9}$$

and g is an arbitrary function. F is linear and bounded and hence an element of $(L_2[0, 1])^*$.

Example 2.43 $F \in (L_p)^*$ can be defined by

$$F(f) = \int_a^b f(x)g(x)\mathrm{d}x, f \in L_p, g \in L_q, \frac{1}{p} + \frac{1}{q} = 1 \tag{2.10}$$

Example 2.44 Let φ_i be a basis of X. Define a bounded linear functional F_i by

$$F_i(x) = \alpha_i \in R, \; where \; x = \sum_{i=1}^{n} \alpha_i \varphi_i \tag{2.11}$$

We can prove that F_i is linear and bounded. The relationship between F_i and φ_i is characterized by

$$F_i(\varphi_j) = \delta_{ij} \; \forall \; i, j \tag{2.12}$$

Since φ_j are linearly independent, F_i are also linearly independent. Thus, the set F_i forms a basis for the dual space U^*. F_i is called the *dual basis* or *conjugate basis*.

Example 2.45 Let M be the n-dimensional subspace of $L_2[0, 1]$, spanned by the set $\varphi_{i_{i=1}}^{n}$ and F be the bounded and linear functional defined by Eq. (2.9). Then for $f \in M$, we have

$$F(f) = \int_0^1 f(x)g(x)\mathrm{d}x$$

$$= \int_0^1 \left(\sum_{i=1}^n \alpha_i \varphi_i(x) \right) g(x)\mathrm{d}x$$

$$= \sum_{i=1}^n \alpha_i \int_0^1 g(x)\varphi_i(x)\mathrm{d}x$$

or

$$F(f) = \sum_{i=1}^n \alpha_i \beta_i, \text{ where } \beta_i = \int_0^1 g(x)\varphi_i(x)\mathrm{d}x \tag{2.13}$$

Example 2.46 a. $\ell_p, L_p, 1 < p < \infty$, are examples of reflexive Banach space.
b. Every finite-dimensional normed space say M is a reflexive Banach space.
c. $\ell_1, c, L_1(a, b)$ and $C[a, b]$ are examples of nonreflexive Banach space.

2.6 Space of Operators

Definition 2.10 a. A vector space A, in which multiplication is defined having the following properties, is called an *algebra*

i. $x(yz) = (xy)z \ \forall \ x, \ y, \ z \in A$.
ii. $x(y + z) = xy + xz \ \forall \ x, \ y, \ z \in A$.
iii. $(x + y)z = xz + yz \ \forall \ x, \ y, \ z \in A$.
iv. $\alpha(xy) = (\alpha x)y = x(\alpha y)$, for every scalar α and $x, \ y \in A$.

An algebra, A, is called *commutative* if

$$xy = yx \ \forall \ x, \ y \in A$$

An element $e \in X$ is called the identity of A if

$$ex = xe = x \ \forall \ x \in A$$

b. A Banach algebra A is a Banach space which is also an algebra such that

$$||xy|| \le ||x|| \, ||y|| \ \forall \ x, \ y \in A.$$

Remark 2.9 If A has an identity e, then

$$||e|| \geq 1, [||x|| = ||xe|| \leq ||x|| \, ||e|| \Rightarrow ||e|| \geq 1]$$

In such cases, we often suppose that $||e|| = 1$.

Remark 2.10 In a Banach algebra, the multiplication is jointly continuous; that is, if $x_n \to x$ and $y_n \to y$, then $x_n y_n \to xy$.
Verification We have

$$||x_n y_n - xy|| = ||x_n(y_n - y) + (x_n - x)y||$$
$$\leq ||x_n|| ||y_n - y|| + ||x_n - x|| ||y|| \to 0 \text{ as } n \to \infty$$

which gives the desired result.

Remark 2.11 Every element x of a Banach algebra, A satisfies $||x^n|| \leq ||x||^n$.
Verification The result is clearly true for $n = 1$. Let it be true for $n = m$. Then,

$$||x^{m+1}|| = ||x^m x|| \leq ||x^m|| ||x||$$
$$\leq ||x||^m \, ||x|| = ||x||^{m+1}$$

or $||x^{m+1}|| \leq ||x||^{m+1}$; i.e., the result is true for $m + 1$, and thus by the principle of induction, the desired result is true for every n.

Theorem 2.9 *If x is an element of a Banach algebra A, then the series $\sum_1^\infty x^n$ is convergent whenever $||x|| < 1$.*

The following lemma is required in the proof.

Lemma 2.1 *Let $x_1, x_2, \ldots, x_n, \ldots$ be a sequence of elements in a Banach space A. Then, the series $\sum_{n=1}^\infty x^n$ is convergent in A; that is, $s_n = \sum_{k=1}^n x_k$ is convergent in A whenever the series of real numbers $\sum_{n=1}^\infty ||x_k||$ is convergent.*

Proof (*Proof of Lemma 2.1*) We have that $\left\| \sum_{k=p}^q x_k \right\| \leq \sum_{k=p}^q x^n \, ||x_k||$. If $\sum_{n=1}^\infty ||x_k||$ is convergent. Then, $\sum_{k=p}^q ||x_k|| \to 0$ as $p, q \to \infty$ and in turn, the remainder of $\sum_{n=1}^\infty x_n$ tends to zero. Therefore, $\sum_{n=1}^\infty x_n$ is convergent.

Proof (*Proof of Theorem 2.9*) Since $||x^n|| \leq ||x||^n$ for all n, therefore, $\sum_{n=1}^\infty ||x_n||$ is less than equal to the sum of a convergent geometric series. By Lemma 2.1, $\sum_1^\infty x^n$ is convergent.

Definition 2.11 Let $B(A)$ denote the set of all bounded linear operators on a normed space X into itself and $S, T \in B(A)$.

The multiplication in $B(A)$ is defined as follows:

$$ST(x) = S(T(x)), \ \forall x \in A$$

$T \in B(A)$ is called *invertible* if there exists $T^1 \in B(A)$ called *inverse* of T such that

$$TT^1 = T^1 T = I$$

In view of Theorem 2.7, $B(X)$ is a normed space and it is a Banach space whenever A is a Banach space.

Theorem 2.10 *Let A be a Banach space. Then, $B(X)$ is a Banach algebra with respect to the multiplication defined in Definition 2.11 with identity I such that $||I|| = 1$.*

Proof By Theorem 2.7, $B(A)$ is a Banach space with respect to the norm

$$||T|| = \sup_{||x||=1} \{||T(x)||\}$$

$$= \sup_{x \neq 0} \left\{ \frac{||T(x)||}{||x||} \right\}$$

$B(X)$ is an algebra as the following results can be easily checked:

$$U(S + T) = US + TS$$
$$(S + T)U = SU + TU$$
$$U(ST) = (US)T \ \forall \ U, S, T \in B(X)$$

$\alpha(ST) = (\alpha S)T = S(\alpha T)$ for every scalar and all $S, T \in B(A)$. We have

$$||ST|| = \sup_{x \neq 0} \left\{ \frac{||ST(x)||}{||x||} \right\}$$

$$= \sup_{x \neq 0} \left\{ \frac{||S(Tx)||}{||T(x)||} \frac{||T(x)||}{||x||} \right\}$$

By definition,

$$||S|| = \sup_{x'=0} \left\{ \frac{||S(x')||}{||x'||} \right\}$$

If $x' = T(x)$, then clearly $T(x) \neq 0$; otherwise $||ST|| = 0$, and the desired result is trivial. In case $T(x) \neq 0$, we get

$$\frac{S[T(x)]}{||T(x)||} \leq ||S||$$

and

$$||ST|| \leq ||S|| \sup_{x \neq 0} \left\{ \frac{||T(x)||}{||x||} \right\} = ||S|| \, ||T||$$

Therefore, B(A) is a Banach algebra.

$$I(x) = x, \ \forall \, x \in X$$
$$||I|| = \sup_{||x||=1} ||I(x)|| = \sup_{||x||=1} \{||x||\} = 1.$$

Theorem 2.11 *If X is a Banach space and $T \in B(A)$ such that $||T|| < 1$, then the operator $I - T$ is invertible.*

Proof We have

$$(I - T)(I + T + T_2 + \cdots + T_n)$$
$$= (I + T + T^2 + \cdots + T^n) - (T + T^2 + \cdots + T^{n+1})$$
$$= I - T^{n+1}$$

and $||T^{n+1}|| \leq ||T||^{n+1} \to 0 \ as \ n \to \infty$, by the hypothesis $||T|| < 1$. Hence,

$$\lim_{n \to \infty} (I - T)(P_n) = I$$

where

$$P_n = \sum_{k=0}^{n} T^k$$

Since $||T|| < 1$, and B(A) is a Banach algebra, by Theorem 2.9, $\lim_{n \to \infty} P_n$ exists (say $P = \lim_{n \to \infty} P_n$); therefore,

$$(I - T)P = I$$

In fact,

$$(I - T)P = (I - T)P_n + (I - T)(P - P_n)$$
$$= I + (I - T)(P - P_n)$$

But

$$||(I - T)(P - P_n)|| \leq ||I - T|| \, ||P - P_n|| \to 0 \ as \ n \to \infty$$

therefore,

$$(I - T)P = I$$

Thus, $(I - T)$ is invertible and $(I - T)^1 = P$, where

$$P = I + T + T_2 + \cdots T_n + \cdots$$

Example 2.47 The set of all $n \times n(n$ finite) matrices is a Banach algebra. For

$$M = \begin{bmatrix} a_{11} & \cdots & a_{1n} \\ \cdot & \cdot & \cdot \\ \cdot & \cdot & \cdot \\ \cdot & \cdot & \cdot \\ a_{1n} & \cdots & a_{nn} \end{bmatrix}$$

we can consider the following equivalent norms (I is the identity matrix):

a. $||M|| = \sup |a_{ij}|$ where $||I|| = 1$
b. $||M|| = \sum_{i,j} |a_{ij}|$ where $||I|| = n$
c. $||M|| = \left(\sum_{i,j} |a_{ij}|^2 \right)^{1/2}$ where $||I|| = \sqrt{n}$

Example 2.48 Let $T \in B(X)$ be defined as follows:

$$e^T = 1 + T + \frac{T^2}{2!} + \cdots \frac{T^n}{n!} + \cdots$$

Then,

$$||e^T||_{B(X)} \leq ||I||_{B(X)} + ||T||_{B(X)} + \cdots \frac{||T^n||_{B(X)}}{n!} + \cdots$$

Since B(A) is a Banach algebra, $||T^n|| \leq ||T||^n$ for all n

$$||e^T|| \leq 1 + k + \frac{k^2}{2!} + \cdots \frac{k^n}{n!} + \cdots$$

where

$$k = ||T||_{B(A)}$$

or

$$||e^T||_{B(X)} \leq e^k$$

Therefore,

$$e^T \in B(A)$$

Example 2.49 The operation equation

$$Tx = y$$

has a unique solution $x = T^{-1}y$ if T^{-1} exists and it is bounded.

It may be observed that finding the inverse T^{-1} could be quite difficult, but the inverse of a neighboring operator T_0 (T_0 is close to T in the sense that $||T - T_0|| < \frac{1}{||T_0^{-1}||}$) can be found as follows: Given this condition it can be shown that

$$||T^{-1} - T_0^{-1}|| < \frac{||T_0^{-1}||^2 ||T - T_0||}{1 - ||T_0^{-1}|| \, ||T - T_0||}$$

hence if $x_0 = T_0^{-1}y$, we have

$$||x - x_0|| = ||T^{-1}y - T_0^{-1}y|| \leq ||T^{-1} - T_0^1|| \, ||y||$$

which provides an estimate for the error $||x - x_0||$. Let T be a square matrix and the elements of the matrix are subject to error which are at most of magnitude ε. If T_0 is the matrix actually used for computation, we can assess the possible error in the solution.

For more detailed account of themes discussed in Sects. 2.1–2.5, see [8, 12, 63, 64, 67, 117, 144, 158, 182].

2.7 Convex Functionals

We discuss here basic properties of convex sets, affine operators, and convex functionals.

2.7.1 Convex Sets

Definition 2.12 Let U be a normed space, and let K its subset. Then,

a. K is called *convex* if $\alpha x + \beta y \in K$ whenever $x, y \in M$, where $\alpha \geq 0, \beta \geq 0$ and $\alpha + \beta = 1$.

b. K is called *affine* if $\alpha x + \beta y \in K$ whenever $x, y \in M$, where $\alpha + \beta = 1$.

Remark 2.12   a. A subset K of a normed space is convex if and only if for all $x_i \in$ M, $i = 1, 2, \ldots, n$, real $\alpha_i \geq 0$, $i = 1, 2, \ldots, n$; $\sum_{i=1}^{n} \alpha_i = 1$, $\sum_{i=1}^{n} \alpha_i x_i \in M$.

b. Every affine set is convex, but the converse may not be true.

c. The normed space U, null set ϕ, and subsets consisting of a single element of X are convex sets. In R^2, line segments, interiors of triangles, and ellipses are its convex subsets. A unit ball in any normed space is its convex subset.

d. The closure of a convex subset of a normed space U is convex.

e. If T is a linear operator on the normed linear space U, then the image of a convex subset of X under T is also convex.

Properties of Convex Sets

a. Let A_i be a sequence of convex subsets of a normed space X. Then,

   i. $\cap_{i=1}^{n} A_i$ is a convex subset of X.

   ii. If $A_i \subseteq A_{i+1}$, $i = 1, 2, \ldots$, then $\cup_{i=1}^{\infty} A_i$ is convex.

   iii. $\liminf_{i \to \infty} A_i = \cup_{j=1}^{\infty} \cap_{i=1}^{\infty} A_i$ is convex.

b. For any subsets A and B of a normed space U, suppose that

$$A + B = \{x + y / x \in A, y \in B, \text{ and } \alpha A = \alpha x / x \in A \text{ and } \alpha \text{ real}\}$$

If A and B are convex, then (a) $A \pm B$ and αA, (b) $A = \alpha A + (1 - \alpha)A$ for $0 \leq \alpha \leq 1$, and (c) $(\alpha_1 +)A = \alpha_1 A + \alpha_2 A$ for $\alpha_1 \geq 0$, $\alpha_2 \geq 0$ are convex.

Definition 2.13 Let K be a subset of a normed space U. Then, the intersection of all convex subsets of X containing M is the smallest convex subset of X containing K. This is called the *convex hull* or the *convex envelope* of K and is usually denoted by $C_0 K$. [$C_0 M = \cap \{A_i / A_i$'s are convex and $K \subseteq A_i\}$]. The closure of the convex hull of M, that is, $\overline{C_0 K}$, is called the *closed convex hull* of M.

Properties of Convex Hull

a. The convex hull of the subset K of a normed space U consists of all vectors of the form $\sum_{i=1}^{n} \alpha_i x_i$, where the $x_i \in K$, $\alpha_i \geq 0$ are real for $i = 1, 2, \ldots, n$ and $\sum_{i=1}^{n} \alpha_i = 1$.

b. Let $A = \cap \{M / K \supset N, M convex and closed\}$. In other words, A is the intersection of all closed and convex sets containing N. Let N_c be equal to the convex hull of N. Then, $A = \bar{N}_c$ [Mazur Lemma (for the proof see Dunford and Schwartz DuSc58]).

c. Let A be a compact subset of a Banach space, then $C_0 A$ is compact.

Definition 2.14 Let U be a normed space and F a nontrivial fixed functional on U. Then,

a. $H = \{x \in U / F(x) = c, c \text{ is a real constant}\}$ is called a *hyperplane* in U.

b. $H_1 = \{x \in U / F(x) \leq c, c \text{ is a real constant}\}$ and $H_2 = \{x \in U / F(x) \geq c, c \text{ is a real constant}\}$ are called *half-spaces* determined by H, where H is defined in (i).

Example 2.50 Lines are hyperplanes in R^2, and planes are hyperplanes in R^3.

Definition 2.15 a. A hyperplane H is called the *support* of U at $x_0 \in V$ if $x_0 \in H$ and V is a subset of one of the half-spaces determined by H.

b. A point x_0 of a convex set K is called an extreme point if there do not exist points $x_1, x_2 \in K$ and $\alpha \in (0, 1)$ such that $x_0 = \alpha x_1 + (1 - \alpha)x_2$.

Remark 2.13 The extreme points of a closed ball are its boundary points. A half space has no extreme point even if it is closed.

2.7.2 Affine Operator

Definition 2.16 An operator T on a normed space U into a normed space V is called *affine* if for every $x \in U$, $Tx = Sx + b$, where S is a linear operator and b is a constant in V.

Remark 2.14 If $Y = R$, then $Sx = \alpha x$, for $x \in X$ and real α. In this case, affine operators (affine functionals) are described by $Tx = \alpha x + b$. In this case, T is called an *affine functional*.

Theorem 2.12 *An operator T is affine if and only if*

$$T\left(\sum_{i=1}^{n} \alpha_i x_i\right) = \sum_{i=1}^{n} \alpha_i (Tx_i) \tag{2.14}$$

for every choice of $x_i \in U$ and real α_i's such that $\sum_{i=1}^{n} \alpha_i = 1$.

The proof is a straightforward verification of axioms.

Convex Functionals

Definition 2.17 Let K be a convex subset of a normed space U. Then, a functional F defined on K is called a *convex functional* if

$$F[\alpha x + (1 - \alpha)y] \le \alpha Fx + (1 - \alpha)Fy \, \forall \, \alpha \in [0, 1] \, and \, \forall \, x, y \in K$$

F is called *strictly convex* if

$$F[\alpha x + (1 - \alpha)y] < \alpha Fx + (1 - \alpha)Fy \, \forall x, y \in K \, and \, \alpha \in [0, 1]$$

F is called *concave* if $-F$ is *convex*. If $X = R^n$, F is called a *convex* function.

Remark 2.15 If F is a convex functional on K, $x_1, x_2, \ldots, x_n \in K$, and $\alpha_1, \alpha_2, \ldots, \alpha_n$ are positive real numbers such that $\sum_{i=1}^{n} \alpha_i = 1$, then

$$F\left[\sum_{i=1}^{n} \alpha_i x_i\right] \leq \sum_{i=1}^{n} \alpha_i F(x_i) \tag{2.15}$$

Equation (2.15) is called Jensens inequality. Very often this relation is taken as the definition of a convex functional. By virtue of Theorem 2.12, every affine functional is convex. The converse of this may not be true. It may be noted that each linear functional is a convex functional.

Example 2.51   a. $f(x) = x^2, x \in (-1, 1)$ and $f(-1) = f(1) = 2$. The function defined in this manner is convex on $[-1, 1]$.

b. $f(x) = x^2$ is a convex function on $(-\infty, \infty)$.

c. $f(x) = sinx$ is a convex function on $[-pi, 0]$.

d. $f(x) = |x|$ is a convex function on $(-\infty, \infty)$.

e. $f(x) = e^x$ is a convex function on $(-\infty, \infty)$.

f. $f(x) = -\log x$ is a convex function on $(0,)$.

g. $f(x) = x^p$ for $p > 1$ and $f(x) = -x^p$, for $0 < p < 1$, are convex functionals on $[0,)$.

h. $f(x) = x \log x$ is a convex function on $(0, \infty)$.

i. $f(x, y) = x^2 + 3xy + 5y^2$ is a convex function on a subset of R^2.

j. $f(x_1, x_2, \ldots, x_n) = x_j$ is a convex function on R^n.

k. $f(x_1, x_2, \ldots, x_n) = \sum_{i=1}^{r} |x_i + b_i|, r \leq n$ is a convex function on R^n.

l. $f(x_1, x_2, \ldots, x_n) = \sum_{i=1}^{r} |x_i + b_i|^p, \alpha_i \geq 0, p \geq 1, r \leq n$ is a convex function on R_n^n.

m. $g(x) = \sup\{\sum_{i=1}^{n} x_i y_i / y = (y_1, y_2, \ldots, y_n)$ *belongs to a convex set of* $R^n\}$ is a convex function.

n. $g(x) = \inf\{\lambda \geq 0 / x \in \lambda M$, *where M is a convex subset of* $R^n\}$ is a convex function.

o. $g(x) = \inf\{\||x - y\|/y = (y_1, y_2, \ldots y_n)$ *belongs to a convex subset M of* $R^n\}$ is a convex function.

Definition 2.18 Let F be a functional on a normed space U into $\bar{R}$. In other words, let F be a real-valued function defined on U which may take the values $+\infty$ and $-\infty$. The set

$$epi \, F = \{(x, \alpha) \in U \times R / F(x) \leq \alpha\}$$

is called the *epigraph* of F.

Let K be a convex subset of U, then $F: K \to \bar{R}$ is called *convex* if for all x and y of K, we have

$$F(\lambda x + (1 - \lambda)y) \leq \lambda F(x) + (1 - \lambda)F(y) \, for \, all \, \lambda \in [0, 1]$$

while the expression on the right-hand side must be defined. $domF = \{x \in U/F(x) < \infty\}$ is called the *effective domain* of F.

It may be observed that the projection of *epi* F on U is the effective domain of F.

Theorem 2.13 *Let U be a normed space (we can take only a vector space). Then, $F: X \to \bar{R}$ is convex if and only if epi F is convex.*

Proof Let F be convex and (x, α) and (y, β) be in *epi* F. By the definition of *epi* F, $F(x) \leq \alpha < \infty$ and $F(y) \leq < \infty$. Then, we have

$$F(\lambda x + (1 - \lambda)y) \leq \lambda F(x) + (1 - \lambda)F(y) \leq \lambda \alpha + (1 - \lambda)\beta$$

This implies that

$$(\lambda x + (1 - \lambda)y, \lambda \alpha + (1 - \lambda)\beta) = (x, \alpha) + (1 - \lambda)(y, \beta) \in epiF$$

and so *epi* F is the convex subset of $U \times R$.

For the converse, suppose *epi* F is convex. Then, dom F is also convex. For x and $y \in dom\, F$, $F(x) \leq \alpha$ and $F(y) \leq \alpha$. By hypothesis, $\lambda(x, \alpha) + (1 - \lambda)(y, \beta) \in epiF$ for $\lambda \in [0, 1]$, which implies that

$$F(\lambda x + (1 - \lambda)y) \leq \lambda \alpha + (1 - \lambda)\beta$$

If $F(x)$ and $F(y)$ are finite, it is sufficient to take $\alpha = F(x)$ and $\beta = F(y)$. In case $F(x)$ or $F(y)$ are $-\infty$, it is sufficient to take α or β tending to $-\infty$ and we obtain that

$$F(\lambda x + (1 - \lambda)y) \leq \lambda F(x) + (1 - \lambda)F(y)$$

Thus, F is convex.

Properties of Convex Functionals

1. If F and G are convex functionals, then (a) $F + G$ and (b) $F \vee G = \max(F, G)$ are also convex functionals.
2. If F is a convex functional, then αF is also a convex functional, where α is a nonnegative real number.
3. If F_n is a sequence of convex functionals defined on the sequence of convex subsets K_n of a normed space U and $K = \cap K_n \neq \phi$, then the subset of M on which $Fx = \sup_n F_n x < \infty$ is convex and F is convex on it.
4. The limit of a sequence of convex functionals is convex.
5. Let $F: X \to R$ and $G: Y \to R$ be convex functionals defined on a normed space X and a subspace Y of R such that the range of $F \subseteq Y \subseteq R$ and G is increasing. Then, $G \circ F$ is a convex functional on X.
6. If F is a convex functional, then $G(x) = F(x) + H(x) + a$ (where H is a linear functional and a is a constant) is also a convex functional.

7. If $F(x) = G(x) + a$, where G is a linear functional and a is constant, then $|F|_p$ for $p \geq 1$ is also a convex functional.

Continuity of Convex Functionals

We may observe that convex functional is not necessarily continuous.

2.7.3 Lower Semicontinuous and Upper Semicontinuous Functionals

Definition 2.19 Let U be a metrizable space. A real-valued function f defined on X is called *lower semicontinuous* (lsc) at a point $x_0 \in U$ if $f(x_0) \geq \underline{\lim}_{n \to \infty} f(y_n)$ whenever $y_n \to x_0$ as $n \to \infty$. f is called lsc on U if f is lsc at every point of X. f is called *upper semicontinuous* (usc) at $x_0 \in U$ if $-f$ is lsc at x_0. f is called usc on U if it is usc at each point of U.

Remark 2.16 a. Every continuous function is lower and upper semicontinuous, but the converse may not be true.
  b. Every convex lsc function, defined on an open subset of a Banach space, is continuous.

Theorem 2.14 *Let U be a metrizable space and $F: U \to \bar{R}$. Then, F is lsc if and only if the epigraph of F is closed in $U \times R$.*

Proof Suppose epi F is closed and let $x_n \to x$. We want to show that F is lsc, i.e.,

$$F(x) \leq \underline{\lim}_{n \to \infty} F(x_n)$$

Let

$$\alpha = \underline{\lim}_{n \to \infty} F(x_n)$$

1. If $\alpha = \infty$, then $F(x) \leq \alpha \ \forall \, x \in X$ and so the desired result is true.
2. Let $-\infty < \alpha < \infty$. Choose a subsequence x_m such that $F(x_m) \to \infty$. Then, $(x_m, F(x_m)) \in epi \ F$. This implies that $F(x) \leq \alpha = \lim_{n \to \infty} F(x_n)$, and we get the desired result.
3. Let $\alpha = -\infty$. Choose a subsequence x_m such that $F(x_m)$ converges decreasingly to $-\infty$ as $m \to \infty$. Then $(x_m, F(x_m)) \in epiF$ form $\geq m_0$, since $F(x_m) \geq F(x_{m0})$ for $m \geq m_0$ and $(x_{m0}, F(x_{m0})) \to (x, F(x_{m0}))$ for $m \to \infty$. Hence, $(x, F(x_{m0}) \in epiF$ or $F(x) \geq F(x_{m0})$ for every m_0. But then, $F(x) \geq \alpha = -\infty$, and this is a contradiction. Hence, $F(x) = \alpha = -\infty$.

For the converse, suppose F is lsc, and $(x_n, \alpha_n) \to (x, \alpha)$ as $n \to \infty$ with

$$F(x) \leq \underline{\lim}_{n \to \infty} F(x_n) < \underline{\lim}_{n \to \infty} \alpha_n = \alpha.$$

This implies that $(x, \alpha) \in epi\ F$, and $epi\ F$ is closed in $X \times R$. Thus, we have proved the desired result.

Advanced discussion on Convex analysis can be found in [140, 154, 161–164].

2.8 Problems

2.8.1 Solved Problems

Problem 2.1   1. Show that every normed space is a metric space but that the converse may not be true.

2. Show that

  (a) $||x|| - ||y|| \leq ||x - y||$.
  (b) $|||x|| - ||y||| \leq ||x + y||$ for all $x,\ y$ belonging to a normed space X.

3. Prove that the mappings

$$\phi(x, y) = x + y \text{ on } X \times X \text{ into } X$$

and

$$\psi(\alpha, x) = \alpha x \text{ on } R \times X \text{ into } X$$

where X is a normed space and R is the field of real numbers, are continuous.

4. Prove that a norm is a real-valued continuous function.

Solution 2.1   1. Let $(X, ||\cdot||)$ be a normed space. Let $d(x, y) = ||x - y||, x, y \in X$. $d(\cdot, \cdot)$ is well defined as $x - y \in X$ and so $||x - y||$ is defined.

 (i) $d(x, y) \geq 0$ as $||x - y|| \geq 0$, by the first property of the norm. $d(x, y) = 0 \Leftrightarrow ||x - y|| = 0 \Leftrightarrow x - y = 0$ (by the first condition on the norm). Therefore, $d(x, y) = 0$ if and only if $x = y$.

 (ii) $d(x, y) = ||x - y|| = || - (y - x)|| = |-1|\ ||y - x|| = ||y - x|| = d(y, x)$.

 (iii) $d(x, y) = ||x - z + z - y|| \leq ||x - z|| + ||z - y||$ (by the third condition on the norm).

Thus, $d(x, y) \leq d(x, z) + d(z, y)$. Therefore, all the conditions of the metric are satisfied. So, $d(\cdot, \cdot)$ is a metric.

In order to show that the converse is not true, consider the following example: Let X be a space of all complex sequences $\{x_i\}$ and $d(\cdot, \cdot)$ a metric defined on X as follows:

$$d(x, y) = \sum_{i=1}^{\infty} \frac{1}{2^i} \frac{|x_i - y_i|}{1 + |x_i - y_i|}$$

where $x = (x_1, x_2, \ldots, x_n, \ldots)$ and $y = (y_1, y_2, \ldots, y_n, \ldots)$. In fact, $d(\cdot, \cdot)$ satisfies all the three conditions of the metric.

(i) $d(x, y) \geq 0, d(x, y) = 0$ if and only if $x = y$.
Verification

$$\frac{|x_i - y_i|}{1 + |x_i - y_i|} \geq 0 \; \forall \, i,$$

and so

$$\frac{1}{2^i} \frac{|x_i - y_i|}{1 + |x_i - y_i|} \geq 0 \; \forall \, i.$$

Consequently,

$$d(x, y) = \sum_{i=1}^{\infty} \frac{1}{2^i} \frac{|x_i - y_i|}{1 + |x_i - y_i|} \geq 0$$

If $d(x, y) = 0$, then

$$\sum_{i=1}^{\infty} \frac{1}{2^i} \frac{|x_i - y_i|}{1 + |x_i - y_i|} = 0$$

$$\Rightarrow \frac{1}{2^i} \frac{|x_i - y_i|}{1 + |x_i - y_i|} = 0 \; \forall \, i$$
$$\Rightarrow |x_i - y_i| = 0 \; \forall \, i$$
$$\Rightarrow x_i = y_i \; \forall \, i$$

$x = (x_1, x_2, \ldots, x_n, \ldots), \; y = (y_1, y_2, \ldots, y_n, \ldots).$

Conversely, if $x = y$, then $|x_i - y_i| = 0 \; \forall \, i$ and $d(x, y) = 0$.
2. $d(x, y) = d(y, x)$
Verification

$$d(x, y) = \sum_{i=1}^{\infty} \frac{1}{2^i} \frac{|x_i - y_i|}{1 + |x_i - y_i|}$$

$$= \sum_{i=1}^{\infty} \frac{1}{2^i} \frac{|(-1)||x_i - y_i|}{1 + |(-1)||x_i - y_i|}$$

$$= \sum_{i=1}^{\infty} \frac{1}{2^i} \frac{|y_i - x_i|}{1 + |y_i - x_i|}$$

3. $d(x, y) \leq d(x, z) + d(z, y)$ where $x = (x_1, x_2, \ldots, x_n \ldots)$, $y = (y_1, y_2, \ldots,$
$y_n, \ldots)$, $z = (z_1, z_2, \ldots, z_n \ldots)$.

Verification

1.

$$d(x, y) = \sum_{i=1}^{\infty} \frac{1}{2^i} \frac{|x_i - y_i|}{1 + |x_i - y_i|}$$

$$= \sum_{i=1}^{\infty} \frac{1}{2^i} \frac{|x_i - z_i + z_i - y_i|}{1 + |x_i - z_i + z_i - y_i|}$$

$$\leq \sum_{i=1}^{\infty} \frac{1}{2^i} \frac{|x_i - z_i|}{1 + |x_i - z_i|} + \sum_{i=1}^{\infty} \frac{1}{2^i} \frac{|z_i - y_i|}{1 + |z_i - y_i|}$$

by applying the inequality given in Theorem A.12 of Appendix A.4. Thus,

$$d(x, y) \leq d(x, z) + d(z, y)$$

Therefore, (X, d) is a metric space. This metric space cannot be a normed space because if there is a norm $|| \cdot ||$ such that $d(x, y) = ||x - y||$, then

$$d(\alpha x, \alpha y) = ||\alpha x - \alpha y|| = |\alpha| \, ||x - y|| = |\alpha| d(x, y)$$

must be satisfied. But for the metric under consideration, this relation is not valid as

$$d(\alpha x, \alpha y) = \sum_{i=1}^{\infty} \frac{1}{2^i} \frac{|\alpha| \, |x_i - y_i|}{1 + |\alpha| \, |x_i - y_i|}$$

and

$$|\alpha| d(x, y) = |\alpha| \sum_{i=1}^{\infty} \frac{1}{2^i} \frac{|x_i - y_i|}{1 + |x_i - y_i|}$$

2. (a) $||x|| = ||x - y + y|| \leq ||x - y|| + ||y||$, or

$$||x|| - ||y|| \leq ||x - y||$$

Also,

$$||y|| - ||x|| \leq ||x - y||$$

or

$$-(||x|| - ||y||) \leq ||x - y||$$

Thus, we get

$$|||x|| - ||y||| \leq ||x - y|| \qquad (2.16)$$

(b) In (a), replace y by −y. Then, we get

$$||x|| - ||-y|| \leq ||x + y|| \qquad (2.17)$$

Since

$$||-y|| = ||y||$$
$$||x|| - ||y|| \leq ||x + y|| \qquad (2.18)$$

which is the desired result.

3. In order to prove that ϕ is continuous, we need to show for $\varepsilon > 0$, $\exists \delta > 0$, such that $||\phi(x, y) - \phi(x', y')|| < \varepsilon$, whenever $||(x, y) - (x', y')|| < \delta$, or equivalently

$$||(x - x', y - y')|| = ||x - x'|| + ||y - y'|| < \delta$$

and we have

$$||\phi(x, y) - \phi(x', y')|| = ||(x + y) - (x' + y')||$$
$$= ||(x - x') + (y - y)||$$
$$\leq ||x - x'|| + ||y - y'|| < \delta = \varepsilon$$

This shows that ϕ is continuous. Similarly, ψ is continuous as

$$||\psi(\alpha, x) - \psi(\beta, x)|| = ||\alpha x - \beta x'|| = ||\alpha x - \alpha x' + \alpha x' - \beta x||$$
$$\leq |\alpha| \, ||x - x'|| + |\alpha - \beta| \, ||x|| < \varepsilon$$

whenever

$$||(\alpha, x) - \beta(\beta, x')|| = ||(\alpha - \beta, x - x')|| = |\alpha - \beta| + ||x - x'|| < \delta$$

4. By Solved Problem 2.1(2), for all x, $||x|| - ||y|| \leq ||x - y||$. This implies that if $||x - y|| < \delta$, then $||x|| - ||y|| \leq \varepsilon$, where $\varepsilon = \delta$. This shows that $||x||$ is a real-valued continuous function.

Alternative Solution If $x_n \to x$ as $n \to \infty$; i.e., $||x_n - x|| \to 0$ as $n \to \infty$, then $||x_n|| \to ||x||$ in view of the above relation. Thus, by Theorem A.6(2) of Appendix A.3, $||x||$ is a real-valued continuous function.

Problem 2.2 Prove that the set X^c of all convergent sequences in a normed space X is a normed space.

Solution 2.2 We can easily verify that X^c is a vector space. For $x = (x_1, x_2, \ldots, x_n, \ldots) \in X^c$, let $||x|| = \sup_n ||x_n||$. The right-hand side is well defined as every convergent sequence is bounded.

a. $||x|| \geq 0$ as $||x_n|| \geq 0$. $||x|| = 0$ if and only if $x = 0$. $||x|| = 0 \Leftrightarrow \sup_n ||x_n|| = 0 \Leftrightarrow ||x_n|| = 0 \; \forall \, n \Leftrightarrow x = 0$.

b. $||\alpha x|| = |\alpha| \, ||x||$; $||\alpha x|| = \sup_n ||\alpha x_n|| = |\alpha| \sup_n ||x_n|| = |\alpha| \, ||x||$.

c. $||x + y|| = \sup_n ||x_n + y_n|| \leq \sup_n ||x_n|| + \sup_n ||y_n|| = ||x|| + ||y||$.

Problem 2.3   a. Show that $C[a, b]$ is a normed space with respect to the norm
$$||f|| = (\int_a^b |f(t)|^2 dt)^{1/2}$$ but not a Banach space.

b. Show that $R_n, \ell_p, c, C[a, b]$ (with sup norm) and $BV[a, b]$ are Banach spaces.

Solution 2.3 1. (a)

$$||f|| \geq 0 \; as \; \int_a^b |f(t)|^2 \geq 0.$$

$$||f|| = 0 \Leftrightarrow |f(t)|^2 = 0 \Leftrightarrow f(t) = 0.$$

[It may be observed that the continuity of f implies the continuity of $|f|$ and so

$$\int_a^b |f(t)|^2 dt = 0$$

implies $f(t) = 0$ on $[a, b]$.]

(b) $||\alpha f|| = \left(\int_a^b |\alpha f(t)|^2 dt \right)^{1/2} = |\alpha| \left(\int_a^b |f(t)|^2 dt \right)^{1/2} = |\alpha| \, ||f||$

(c)

$$||f + g|| = \left(\int_a^b |f + g|^2 dt \right)^{1/2} \leq \left(\int_a^b |f(t)|^2 dt \right)^{1/2} + \left(\int_a^b |g(t)|^2 \right)^{1/2}$$

$$= ||f|| + ||g||$$

by Minkowski's inequality (Theorem A.14(b) of Appendix A.4).

Thus, $C[a, b]$ is a normed space as it is well known that $C[a, b]$ is a vector space. [In fact, for operations of addition and scalar multiplication, defined by

$$(f + g)(t) = f(t) + g(t)$$
$$(\alpha f)(t) = \alpha f(t)$$

the axioms of vector space follow from the well-known properties of continuous functions].

In order to show that $C[a, b]$ is not a Banach space with the integral norm, we consider the following example: Take $a = -1, b = 1$

$$f_n(t) = \begin{cases} 0, & -1 \le t \le 0 \\ nt, & 0 < t \le 1/n \\ 1, & 1/n < t \le 1 \end{cases}$$

We can show that $\{f_n(t)\}$ is a Cauchy sequence in $C[-1, 1]$, but it cannot converge to an element of $C[-1, 1]$. (See Solved Problem 3.4).

2.   i. R^n is a Banach space. R^n is a vector space with the operations

$$\forall x = (x_1, x_2, \ldots, x_n), y = (y_1, y_2, \ldots, y_n) \in R^n$$
$$x + y = (x_1 + y_1, x_2 + y_2, \ldots, x_n + y_n)$$
$$\alpha x = (\alpha x_1, \alpha x_2, \ldots, \alpha x_n)$$

We verify axioms of the norm of $\|\cdot\|_4$ of Example 2.4. R^n is a Banach space with respect to all norms defined on it.

$$\|x\| = \max(|x_1|, |x_2|, \ldots, |x_n|), \ x = (x_1, x_2, \ldots, x_n) \in R^n$$
$$\|x\| \ge 0 \ is \ obvious. \ \|x\| = 0 \Leftrightarrow |x_i| = 0 \ \forall \ i \Leftrightarrow x = (x_1, x_2, \ldots, x_n) = 0$$
$$\|\alpha x\| = \max(|\alpha x_1|, |\alpha x_2|, \ldots, |\alpha x_n|)$$
$$= |\alpha| \max(|x_1|, \ldots, |x_n|)$$
$$= |\alpha| \ \|x\|$$
$$\|x + y\| = \max(|x_1 + y_1|, |x_2 + y_2|, \ldots, |x_n + y_n|)$$
$$\le \max(|x_1| + |y_1|, |x_2| + |y_2|, \ldots, |x_n| + |y_n|)$$
$$\le \max(|x_1|, \ldots, |x_n|) + max(|y_1|, \ldots, |y_n|)$$
$$= \|x\| + \|y\|$$

ii. $\ell_p = \{x = \{x_i\} / \sum_{i=1}^{\infty} |x_i|^p < \infty$ where $1 \le p < \infty$.

ℓ_p is a vector space with respect to operations of addition and scalar multiplication defined as follows:

$$x = (x_1, x_2, \ldots, x_n, \ldots), \ y = (y_1, y_2, \ldots, y_n, \ldots) \in \ell_p$$
$$x + y = (x_1 + y_1, x_2 + y_2, \ldots, x_n + y_n \ldots)$$
$$\alpha x = (\alpha x_1, \alpha x_2, \ldots, \alpha x_n, \ldots)$$

By Minkowski's inequality and the fact that

$$\left(\sum_{i=1}^{\infty} |x_i|^p \right)^{1/p} < \infty, \left(\sum_{i=1}^{\infty} |y_i|^p \right)^{1/p} < \infty$$

we have

$$\left(\sum_{i=1}^{\infty} |x_i + y_i|^p \right)^{1/p} \leq \left(\left(\sum_{i=1}^{\infty} |x_i|^p \right)^{1/p} \right) + \left(\sum_{i=1}^{\infty} |y_i|^p \right)^{1/p}$$

This shows that $x + y \in \ell_p$. Since $\sum_{i=1}^{\infty} |\alpha x_i|^p < \infty$, $\forall$ scalar α, we see that $\alpha x \in \ell_p$.

All the axioms of vector space can be verified easily. Now, we verify the norm axioms for the norm defined by $||x|| = \left(\sum_{i=1}^{\infty} |x_i|^p \right)^{1/p}$ on ℓ_p.

It is clear that $||x|| \geq 0$.

$$||x|| = 0 \Leftrightarrow |x_i|^p = 0 \ \forall \ i \Leftrightarrow x_i = 0 \ \forall \ i \Leftrightarrow x = 0$$
$$||\alpha x|| = \left(\sum_{i=1}^{\infty} |\alpha x_i|^p \right)^{1/p}$$
$$= \left(|\alpha|^p \sum_{i=1}^{\infty} |x_i|^p \right)^{1/p}$$
$$= |\alpha| \ ||x||$$
$$||x + y|| = \left(\sum_{i=1}^{\infty} |x_i + y_i|^p \right)^{1/p}$$

exists and

$$\left(\left(\sum_{i=1}^{\infty} |x_i + y_i|^p\right)^{1/p}\right) \le \left(\sum_{i=1}^{\infty} |x_i|^p\right)^{1/p} + \left(\sum_{i=1}^{\infty} |y_i|^p\right)^{1/p}$$

in view of the Minkowski's inequality and the fact that $x = \{x_i\} \in \ell_p$, and $y = \{y_i\} \in \ell_p$. This shows that $||x + y|| \le ||x|| + ||y||$.

In order to show that ℓ_p is a Banach space, we need to show that every Cauchy sequence in ℓ_p is convergent in it.

Let $x_n = \{a_i^n\} \in \ell_p$ be a Cauchy sequence. This implies that

$$||x_n - x_m|| = \left(\sum_{i=1}^{\infty} |a_i^n - a_i^m|^p\right)^{1/p} < \varepsilon \text{ for } n, m \ge N \qquad (2.19)$$

Equation (2.19) implies that

$$|a_i^n - a_i^m| < \varepsilon \text{ for } n, \ m \ge N \text{ and } \forall \ i \qquad (2.20)$$

For fixed i, Eq. (2.20) implies that a_i^n is a Cauchy sequence of real numbers and, by the Cauchy criterion of real sequences, it must converge, say, to a_i; i.e.,

$$\lim_{n \to \infty} a_i^n = a_i \qquad (2.21)$$

From Eq. (2.19), we have

$$\sum_{i=1}^{k} |a_i^n - a_i^m|^p < \varepsilon^p, \text{ for every } k$$

Making $m \to \infty$ in this relation, we have

$$\sum_{i=1}^{k} |a_i^n - a_i^m|^p < \varepsilon^p, \text{ for } n \ge N$$

Making $k \to \infty$, we get

$$\sum_{i=1}^{\infty} |a_i^n - a_i^m|^p < \varepsilon^p, \text{ for every } k$$

This implies that

$$-(x_n - x) = x - x_n \in \ell_p$$

where $x = (a_1, a_2, \ldots, a_n, \ldots)$; hence $x = (x - x_n) + x_n \in \ell_p$

Furthermore,

$$||x_n - x|| = \left(\sum_{i=1}^{\infty} |a_i^n - a_i|^p \right)^{1/p} \leq \varepsilon \, for \, n \geq N$$

i.e.

$$||x_n - x|| \to 0 \, as \, n \to \infty$$

Therefore, x_n is convergent to $x \in \ell_p$, which shows that ℓ_p is a Banach space.

iii. $c = \{x_n / \lim_{n \to \infty} x_n = x\}$ is a vector space with operations of addition and scalar multiplication defined by

$$x + y = (x_1 + y_1, x_2 + y_2, \ldots, x_n + y_n, \ldots)$$
$$\alpha x = (\alpha x_1, \alpha x_2, \ldots, \alpha x_n, \ldots)$$

where

$$x = (x_1, x_2, \ldots, x_n, \ldots)$$
$$y = (y_1, y_2, \ldots, y_n, \ldots)$$

belong to c and α is a scalar. For $x \in c$, $||x|| = \sup_n |x_n|$. This is a norm. We shall now show that c is a Banach space. Let $x_n = \{a_i^n\}$ be a Cauchy sequence in c. This implies that for $\varepsilon > 0 \, \exists N$ such that

$$||x_n - x_m|| = \sup_i |a_i^n - a_i^m| < \varepsilon \, for \, n, m \geq N$$

This implies that

$$|a_i^n - a_i^m| < \varepsilon \, for \, n, m \geq N. \tag{2.22}$$

This means that $\{a_i^n\}$ is a Cauchy sequence of real members and, by the Cauchy criterion, for fixed i

$$\lim_{m \to \infty} a_i^m = a_i. \tag{2.23}$$

By Eqs. (2.22) and (2.23), we obtain

$$|a_i - a_i^n| < \varepsilon \, for \, n \geq N \tag{2.24}$$

Since $x_n \in c$, we have $\lim_i a_i^n = a_n$, which implies that

$$|a_n - a_i^n| < \varepsilon \, for \, i \geq N(\varepsilon, n)$$

Therefore,

$$
\begin{aligned}
|a_m - a_k| &= |a_m - a_m^n + a_m^n - a_n + a_n - a_k^n + a_k^n - a_k| \\
&\leq |a_m - a_m^n| + |a_m^n - a_n| + |a^n - a_k^n| + |a_k^n - a_k| \\
&< \varepsilon + \varepsilon + \varepsilon + \varepsilon = 4\varepsilon
\end{aligned}
$$

for $n \geq N$ and $m, k \geq N(\varepsilon, n)$. Thus, the sequence $\{a_i\}$ satisfies the Cauchy criterion. There exists $\lim a_i = a$, and hence, $x = \{a_i\} \in c$. We find from Eq. (2.19) that

$$||x_n - x|| = \sup_i |a_i - a_i^n| \to 0 \, as \, n \to \infty$$

i.e., $x_n \to x$ as $n \to \infty$ in c. Therefore, c is a Banach space.

iv. In order to prove that $C[a, b]$ with $||f|| = \sup_{a \leq t \leq b} |f(t)|$ is a Banach space, we need to show that every Cauchy sequence in $C[a, b]$ is convergent. Let $\{f_n(t)\}$ be a Cauchy sequence, that is, for $\varepsilon > 0, \exists N$ such that

$$||f_n - f_m|| = \sup_t |f_n(t) - f_m(t)| \leq \varepsilon \, for \, n, m \geq N,$$

This implies that $\{f_n(t)\}$ converges uniformly to a continuous function, say, $f(t)$ in $C[a, b]$. Therefore, $||f_n - f|| = \sup_t |f_n(t) - f(t)| \to 0$ as $n \to \infty$; i.e., $f_n \to f$ in $C[a, b]$ and so $C[a, b]$ is a Banach space.

v. $BV[a, b] = \{f, \, a \, real - valued \, function \, defined \, on \, [a, b]|$ $\sup_P \sum_{k=1}^n |f(t_k) - f(t_{k-1})| < \infty$, where P is an arbitrary partition of $[a, b]$; i.e., the set of points t_k such that $a = t_0 < t_1 < t_2 \ldots t_n = b\}$ is equal to the class of functions of bounded variation defined on [a, b].
Let, for $f \in BV[a, b]$

$$||f|| = |f(a)| + V_a^b(f) \tag{2.25}$$

where $V_a^b(f) = \sup_P \sum_{k=1}^n |f(t_k) - f(t_{k-1})|$ is called the *total variation of the function $f(t)$*.

$||f|| \geq 0 \, as \, |f(a)| \geq 0 \, and \, V_a^b(f) \geq 0$
$||f|| = 0 \Leftrightarrow |f(a)| = 0, \, and \, V_a^b(f) = 0 \Leftrightarrow f(a) = 0 \, and \, |f(t_k) - f(t_{k1})| = 0 \, for$

all $t_k \Leftrightarrow f(t_k) = f(t_{k1})$, *and* $f(a) = 0 \Leftrightarrow f(t_k) = 0 \forall \Leftrightarrow f = 0$

$$\|\alpha f\| = |\alpha f(a)| + \sup_P \sum_{k=1}^{n} |\alpha[f(t_k) - f(t_{k-1})]|$$

$$|\alpha| [|f(a)| + \sup_P \sum_{k=1}^{n} |f(t_k) - f(t_{k-1})|]$$

$$= |\alpha| \, \|f\|$$

Let $f, g \in BV[a, b]$ and $h = f + g$.
For any partition of the interval $[a, b]$: $a = t_0 < t_1 < t_2 < \cdots < t_n = b$, $|h(a)| = |f(a) + g(a)| \leq |f(a)| + |g(a)|$ and $|h(t_k) - h(t_{k-1})| \leq |f(t_k) - f(t)| + |g(t) - g(t_{k-1})|$ where $k = 1, 2, 3, \ldots, n$. Therefore

$$\|h\| = |h(a)| + \sup_P \sum_{k=1}^{n} |h(t_k) - h(t_{k-1})|$$

$$\leq |f(a)| + |g(a)| + \sup_P \sum_{k=1}^{n} |f(t_k) - f(t_{k-1})| + \sup_P \sum_{k=1}^{n} |g(t_k) - g(t_{k-1})|$$

$$= \|f\| + \|g\|$$

Thus, $\|f\|$ defined by Eq. (2.25) is a norm on $BV[a, b]$.
Now, we shall show that $BV[a, b]$ is a Banach space. Let $\{f_n\}$ be a Cauchy sequence in $BV[a, b]$. $\{f_n(t)\}$ is a Cauchy sequence of real numbers as

$$|f_m(t) - f_n(t)| \leq |[f_m(t) - f_n(t)] - [f_m(a) - f_n(a)]| + |f_m(a) - f_n(a)|$$

$$\leq |f_m(a) - f_n(a)| + V_a^b(f_m - f_n)$$

$$= \|f_m - f_n\|$$

Hence, $\{f_n(t)\}$ must be convergent, say to $f(t) \forall \, t \in [a, b]$. $f(t) \in BV[a, b]$ as

$$\sup_P \sum_{k=1}^{n} |f(t_k) - f(t_{k-1})| = \lim_{n \to \infty} \sup_P \sum_{k=1}^{n} |f_n(t_k) - f_n(t_{k-1})| < \infty$$

$$\|f_n - f\| = |f_n(a) - f(a)| + \sup_P \sum_{k=1}^{n} |f_n(t_k) - f(t_k)|$$

$$+ \sup_P \sum_{k=1}^{n} |f_n(t_{k-1}) - f(t_{k-1})|$$

$$\to 0 \text{ as } n \to \infty.$$

Hence, $f_n \to f$ in $BV[a, b]$, and consequently $BV[a, b]$ is a Banach space.

See also [146],

2.8.2 Unsolved Problems

Problem 2.4   a. Let L be the space of real-valued Lipschitz functions of order 1 defined on $[0, 1]$; that is, the class of functions f such that

$$\sup_{(x,y)\in[0,1]\times[0,1], x\neq y} \frac{|f(x)-f(y)|}{x-y} = K(f) < \infty$$

Show that L is a subspace of $C[0, 1]$.

  b. Let $\|f\|_1 = \sup\limits_{0\leq t\leq 1} |f(t)| + K(f) = \|f\| + K(f)$. Show that $\| \cdot \|_1$ is a norm on L.

  c. Show that $(L, \| \cdot \|_1)$ is a Banach space.

Problem 2.5 Let $C_1[0, 1]$ denote the vector space of all real-valued functions defined on $[0, 1]$ having first-order continuous derivatives. Show that the expressions given below are equivalent norms.

$$\|f\|_1 = \sup_{0\leq t\leq 1} |f(t)| + |f'(t)|$$

$$\|f\|_1 = \sup_{0\leq t\leq 1} |f(t)| + \sup_{0\leq t\leq 1} |f'(t)|$$

Problem 2.6 Prove that all linear operators, defined on a finite-dimensional normed space into an arbitrary normed space, are bounded.

Problem 2.7 Prove that the natural embedding J is an isometric isomorphism of X into X^{**}.

Problem 2.8 Let $p(x) = \|f\| \, \|x\|$, where f is a bounded linear functional on a normed space X. Show that

(a) $p(x+y) \leq p(x) + p(y) \ \forall \ x \text{ and } y$.
(b) $p(\alpha x) = \alpha p(x) \ \forall \ \alpha \geq 0 \text{ and } x$.

Problem 2.9 Let $C_1[0, 1]$ denote the vector space of all real-valued functions defined on $[0, 1]$ which have a first-order continuous derivative.

  a. Show that $p(f)$ defined by the expression

$$p^2(f) = f^2(0) + \int_0^1 (f'(t))^2 dt$$

is a norm, and that convergence in this norm implies uniform convergence.

  b. Show that if $C^1[0, 1]$ is equipped with the norm

$$\|f\| = \sup_{0\leq t\leq 1} |f(t)|$$

then, the functional F defined on $C^1[0, 1]$ by $F(f) = f(0)$ is linear but not continuous.

Problem 2.10 (a) Let X be the vector space of all the sequences $\{x_n\}$ of complex numbers such that the series $\sum_{n \geq 1} n!|x_n|$ is convergent. Assume that

$$||(x_n)|| = \sum_{n \leq 1} n!|x_n|$$

Show that $(X, || \cdot ||)$ is a normed space.
(b) Let $T : \ell_p \to X$ be defined by

$$\cdot T(\{x_n\}) = \frac{x_n}{n!}.$$

Show that T is a bounded linear operator and find its norm. Deduce that X is a Banach space.

Problem 2.11 Every finite-dimensional normed space is a Banach space.

Problem 2.12 Prove that a normed space is homeomorphic to its open unit ball.

Problem 2.13 Prove that ℓ_p, $1 \leq p \leq \infty$, is a Banach space.

Problem 2.14 Let X and Y be two real Banach spaces such that

(a) $A : D(A) \subseteq X \to Y$ is a bounded linear operator.
(b) $D(A)$ is dense in X; that is, $\overline{D(A)} = X$. Show that A can be uniquely extended to a bounded linear operator defined on the whole space X.

Problem 2.15   a. Let x_n and y_n be two sequences in a normed space X, and $\{\lambda_n\}$ be a sequence of real numbers. Further, suppose that $x_n \to x$ and $y_n \to y$, as $n \to \infty$, in X and $\lambda_n \to \lambda$ in R as $n \to \infty$. Show that $x_n + y_n \to x + y$ and $\lambda_n x_n \to \lambda x$ as $n \to \infty$.
   b. If $\{x_n\}$ is a Cauchy sequence in a normed space, show that $\{\lambda x_n\}$ is a Cauchy sequence.
   c. Show that a metric $d(\cdot, \cdot)$ induced by a norm on a normed space X is translation invariant, i.e., $d(x + b, y + b) = d(x, y)$, where b is a fixed element of X, and $d(\alpha x, \alpha y) = |\alpha| d(x, y)$, where α is a scalar.

Problem 2.16 Give an example of a seminorm that is not a norm.

Problem 2.17 Show that $P[0, 1]$ with $||f|| = \sup_{t \in [0,1]} |f(t)|$ is a normed space but not a Banach space.

Problem 2.18 Let $X = \{x_i / x_i = 0 \text{ for } i > n\}$. Show that X is a normed space with $||x|| = \sup_i |x_i|$, but it is not a Banach space.

Problem 2.19 If X and Y are Banach spaces, show that $X \times Y$ is also a Banach space.

Problem 2.20 Let X and Y be two real normed spaces and T an operator on X into Y such that

a. $T(x + y) = T(x) + T(y) \; \forall \, x, y \in X$.
b. T is bounded on the unit ball of X. Prove that T is a continuous linear operator.

Problem 2.21 Let X and Y be normed spaces and T an operator of X into Y. Show that T is a continuous linear operator if and only if

$$T\left(\sum_{i=1}^{\infty} \alpha_i x_i\right) = \sum_{i=1}^{\infty} \alpha_i T(x_i)$$

for every convergent series $\sum_{i=1}^{\infty} \alpha_i x_i$.

Problem 2.22 Show that $R^n, C, C[0, 1], \ell_p, L_p$ are separable normed spaces.

Problem 2.23 Let $T: C[0, 1] \to C[0, 1], \; T(f) = \int_0^1 K(s, t) f(t) dt$, where K(s, t) is continuous on the unit square $0 \le s \le 1, \; 0 \le t \le 1$. *Compute* $\|T\|$.

Problem 2.24 Show that if T is a linear operator on a normed space X into a normed space Y, then T^{-1} is linear provided it exists.

Problem 2.25 Show that the dual of R^n is R^n.

Problem 2.26 Show that the dual of $\ell_p, 1 < p < \infty$, is ℓ_q, where $1/p + 1/q = 1$. Also, show that ℓ_p is reflexive for $1 < p < \infty$, but ℓ_1 is nonreflexive.

Problem 2.27 Prove that the set $M_{m,n}$ of all $m \times n$ matrices is a normed space with respect to the following equivalent norms:

a. $\|A\|_\infty = \max_{i,j} |a_{ij}| \; and \; A = (a_{ij}) \in M_{m,n}$.
b. $\|A\|_1 = \sum_{i=1}^{m} \sum_{j=1}^{n} |a_{ij}| \; and \; A = (a_{ij}) \in M_{m,n}$.
c. $\|A\|_2 = \{\sum_{i=1}^{m} \sum_{j=1}^{n} |a_{ij}|^2\}^{1/2} \; and \; A = (a_{ij}) \in M_{m,n}$.

Problem 2.28 Prove that the space of all real sequences converging to 0 is a normed space.

Problem 2.29 Let T be a functional defined on $X \times Y$, the product of normed spaces X and Y, such that

a. $T(x + x', y + y') = T(x, y) + T(x, y') + T(x', y) + T(x', y')\forall\ (x, y) \in X \times Y, (x', y') \in X \times Y$.

b. $T(\alpha x, \beta y) = \alpha\beta T(x, y)\forall\ (x, y) \in X \times Y$ with α and β being scalars, i.e., T is bilinear. Show that T is continuous if and only if there exists $k > 0$ such that $||T(x, y)|| \le k||x||\ ||y||$

Let

$$||T|| = \sup_{||x||\le 1, ||y||\le 1} |T(x, y)|.$$

Show that $||T(x, y)|| \le ||T||\ ||x||\ ||y||$.

Problem 2.30 A linear operator T defined on a normed space U into a normed space V is called *compact* or *(completely continuous)* if, for every bounded subset M of U, its image $T(M)$ is relatively compact; i.e., the closure $\overline{T(M)}$ is compact.

a. Show that the operator in Problem 2.23 is compact.
b. Show that if S and T are compact operators, then $S + T$ and αT, where α is scalar, are compact.
c. Show that every compact operator is bounded and hence continuous.

Problem 2.31 Show that the identity operator defined on an infinite-dimensional normed space X is not compact.

Problem 2.32 Let $T: U \to U$ (U is a normed space) be defined by the relation $T(x) = F(x)z$, where z is a fixed element of U and $F \in U^*$. Show that T is compact.

Problem 2.33 Show that the space of all compact operators on a Banach space X into a Banach space Y is a Banach space which is a closed subspace of $\mathscr{B}[X, Y]$.

Problem 2.34 Let $T: \ell_p \to \ell_p$, $p \ge 1$, be defined by $T(x) = y$, where $y = (x_1/1,\ x_2/2,\ x_3/3,\ \ldots,\ x_n/n, \ldots)$ for $x = (x_1, x_2, x_3, \ldots, x_n, \ldots) \in \ell_p$. Show that T is compact.

Problem 2.35 Let T be a linear operator with its domain $D(T)$ and range $R(T)$ contained in a normed space U. A scalar λ such that there exists an $x \in D(T)$, $x \ne 0$, satisfying the equation $T(x) = \lambda x$, is called an *eigenvalue* or characteristic value or proper value of T. The corresponding x is called an *eigenvector* of T. If λ is an eigenvalue of T, the null space of the operator $\lambda I - T, N(\lambda I - T)$, is called the eigenmanifold (eigenspace) corresponding to the eigenvalue λ. The dimension of the eigenmanifold is called the multiplicity of the eigenvalue λ. The set of λ such that $(\lambda I - T)$ has a continuous inverse defined on its range is said to be the resolvent set of T and is denoted by $\rho(T)$. The set of all complex numbers that are not in the resolvent set is said to be the *spectrum* of T and denoted by $\sigma(T)$. $R_\lambda(T) = (\lambda I - T)^{-1}$ is called the *resolvent* of T. Let $T: U \to U$, U being Banach space, and

$$r_\sigma(T) = \sup |\lambda/\lambda \in \sigma(T)|$$

is called the *spectral radius* of T.

(a) Let $T: X \to X$, X being a normed space, be a linear operator on X, and let $\lambda_1, \lambda_2, \lambda_3, \ldots$ be a set of distinct eigenvalues of T. Further, let $\{x_n\}$ be an eigenvector associated with $_n$, $n = 1, 2, \ldots$. Show that the set $x_1, x_2, \ldots, x_n$ is linearly independent.

(b) Show that the set of eigenvalues of a compact linear operator $T: X \to X$ on a normed space X is countable and the only possible point of accumulation is $\lambda = 0$.

(c) Find $\sigma(T)$, where T is defined as in Problem 2.23 on $L_2[0, 1]$ into itself.

(d) Let $T: X \to X$, X being a Banach space, be linear and continuous. If $|\lambda| \geq ||T||$, then λ is in the resolvent set of T. Moreover, $(\lambda I - T)^{-1}$ is given by $(\lambda I - T)^{-1} = \sum_{n=0}^{\infty} \lambda^{-n-1} T^n$ and $||(\lambda I - T)^{-1}|| \leq (|\lambda| - ||T||)^{-1}$. Show that the spectrum of T is contained in the circle, $\{\lambda | \lambda| \leq \lim \sup(||T^n||)^{1/n}\}$, which is contained in $(\lambda/|\lambda| \leq ||T||)$.

(e) Show that for a bounded linear operator T, $\sigma(T)$ is closed and $\rho(T)$ is open.

(f) Let $T: C[a, b] \to C[a, b]$ be as in Problem 2.23, then $(\lambda I - T)^{-1} u = v$, $\lambda \neq 0$, has a unique solution $u = \lambda \sum_{n=0}^{\infty} \lambda^{-n} - T^n v$.

Problem 2.36 Inverse Operator Theorem: Let T be a bounded linear operator on a normed space U onto a normed space V. If there is a positive constant m such that

$$m||x|| \leq ||T(x)|| \ \forall \, x \in X$$

then show that $T^{-1}: V \to U$ exists and is bounded. Conversely, show that if there is a continuous inverse $T^{-1}: Y = R(T) \to U$, then there is a positive constant m such that the above inequality holds for all x in U.

Problem 2.37 Let $T: C[a, b] \to C[a, b]$ be defined by $f(t) \to g(t) = \int_0^t f(x) \mathrm{d}x$. Find $R(T)$ and $T^{-1}: R(T) \to C[a, b]$. Is T^{-1} linear and bounded.

Chapter 3
Hilbert Spaces

Abstract In this chapter, we study a special class of Banach space in which the underlying vector space is equipped with a structure, called an inner product or a scalar product providing the generalization of geometrical concepts like angle and orthogonally between two vectors. The inner product is nothing but a generalization of the dot product of vector calculus. Hilbert space method is a powerful tool to tackle problems of diverse fields of classical mathematics like linear equations, variational methods, approximation theory, differential equations.

Keywords Inner product space (pre-Hilbert space) ·
Cauchy–Schwarz–Bunyakowski inequality · Hilbert space · Parallelogram law ·
Polarization identity · Orthogonal complements · Projection theorem · Projection
on convex sets · Orthonormal systems Fourier expansion · Bessel's inequality ·
Projections · Orthogonal bases · Riesz representation theorem · Duality of Hilbert
space · Adjoint operator · Self-adjoint operator · Positive operator · Unitary
operator · Normal operator · Adjoint of an unbounded operator · Bilinear forms ·
Lax–Milgram lemma

3.1 Introduction

In Chap. 1, we have studied concept of distance between two elements of an abstract set X as well as the distance between two subsets of X. Chapter 2 is devoted to concepts of magnitude of vectors. An important notion of angle and related concept of orthogonality or perpendicularity between two vectors is missing in the previous chapter. David Hilbert, a Professor at Göttingen, Germany, introduced the space of sequences denoted by $l_2 = \{\{x\} / \sum_{n=1}^{\infty} |x_n|^2 < \infty\}$ which is the first example of a structure whose axiomatic definition was given around 1927 by Von Neumann and now known as *Hibert space*. A Hilbert space is a special class of Banach space in which the underlying vector space is equipped with a structure known as inner product or scalar product. This concept enables us to study geometric properties like Pythagorean theorem and parallelogram law of classical geometry, and vector space equipped with inner product that is completely known as Hilbert space is a powerful

© Springer Nature Singapore Pte Ltd. 2018

A. H. Siddiqi, *Functional Analysis and Applications*, Industrial and Applied Mathematics,
https://doi.org/10.1007/978-981-10-3725-2_3

apparatus to solve problems of different fields such as linear equations, minimization problems, variational methods, approximation theory, differential equations. The concept of orthogonality leads to the celebrated projection theorem and theory of Fourier series extending numerous classical results concerning trigonometric Fourier series. The structure of inner product has consequences of vital importance that every real Hilbert space is reflexive. Besides these results, the Hilbert space exhibits some interesting properties of linear operators which are very useful for the study of certain systems occurring in physics and engineering. The chapter is concluded by the famous Lax–Milgram lemma proved by Abel prize recipient P.D. Lax and A.N. Milgram in 1954 which is applied to prove existence of solution of different types of boundary value problems.

3.2   Fundamental Definitions and Properties

3.2.1   Definitions, Examples, and Properties of Inner Product Space

Definition 3.1 Suppose H is a real or complex vector space. Then, a mapping, denoted by $\langle \cdot, \cdot \rangle$, on $H \times H$ into the underlying field R or C, is said to be an *inner product* of any two elements $x, y \in H$ if the following conditions are satisfied:

1. $\langle x + x', y \rangle = \langle x, y \rangle + \langle x', y \rangle \ \forall x, x', y \in H$.
2. $\langle \alpha x, y \rangle = \alpha \langle x, y \rangle, \forall x, y \in H$ and α belongs to R or C.
3. $\overline{\langle x, y \rangle} = \langle y, x \rangle \ \forall x, y \in H$ and $-$ denotes complex conjugate.
4. $\langle x, x \rangle \geq 0, \ \forall \ x \in X$; and $\langle x, x \rangle = 0$ if and only if $x = 0$. $\langle x, x \rangle$ is denoted by $||x||^2$, that is, $||x|| = (\langle x, x \rangle)^{1/2}$.

$(X, \langle \cdot, \cdot \rangle)$ [X equipped with inner product $\langle \cdot, \cdot \rangle$] is called an *inner product space* or *pre-Hilbert space*.

Remark 3.1 (a) Sometimes symbol $(\cdot, \cdot)$ is used to denote the inner product $\langle \cdot, \cdot \rangle$.
(b) For real inner product $\langle \cdot, \cdot \rangle$, $\overline{\langle x, y \rangle} = \langle x, y \rangle = \langle y, x \rangle$.
(c) Since a complex number z is real if and only if $z = \overline{z}$, $\langle x, x \rangle = z = \overline{z} = \langle x, x \rangle$, that is $\langle x, x \rangle$ is real.
(d) $\langle \cdot, \cdot \rangle$ is linear in the first variable x by virtue of Definition 3.1(a) and (b). If underlying vector space is real, then $\langle \cdot, \cdot \rangle$ is also linear in the second variable, that is,

    (i) $\langle x, y + y' \rangle = \langle x, y \rangle + \langle x, y' \rangle$.
    (ii) $\langle x, \beta y \rangle = \overline{\beta} \langle x, y \rangle$.

(e) It may be checked that

$$\left\langle \sum_{k=1}^{n} \alpha_k x_k, \sum_{l=1}^{n} \beta_l y_l \right\rangle = \sum_{k=1}^{n} \sum_{l=1}^{n} \alpha_k \overline{\beta}_l \langle x_k, y_l \rangle$$

(f) $\langle x, 0 \rangle = 0, \ \forall \ x \in X$.

(g) If, for a given element $y \in X$, $\langle x, y \rangle = 0$, then y must be zero.

(h) The inner product, $\langle \cdot, \cdot \rangle$, is a continuous function with respect to the norm induced by it. See Remark 3.4 and Solved Problem 3.3.

Example 3.1 Let $x = (x_1, x_2)$ and $y = (y_1, y_2)$ belong to R^2. Then,

$$\langle x, y \rangle = x_1 y_1 + x_2 y_2$$

is an inner product on R^2 and $(R^2, \langle \cdot, \cdot \rangle)$ is an inner product space.

Example 3.2 For $x = (x_1, x_2, \ldots, x_n) \in R^n$, $y = (y_1, y_2, \ldots, y_n) \in R^n$ belonging to R^n, $n \geq 1$,

$$\langle x, y \rangle = x_1 y_1 + x_2 y_2 + \cdots + x_n y_n$$

is an inner product and $(R^n, \langle \cdot, \cdot \rangle)$ is an inner product space.

Example 3.3 Let $C[a, b]$ denote the vector space of continuous functions defined on $[a, b]$. $C[a, b]$ is an inner product space with respect to an inner product $\langle \cdot, \cdot \rangle$ defined as

(a) $\langle f, g \rangle = \int_a^b f(x)\overline{g(x)}dx, \ \forall \ f, g \in C[a, b]$. For $f, g \in C[a, b]$, another inner product can be defined as follows:

(b) $\langle f, g \rangle = \int_a^b f(x)\overline{g(x)}w(x)dx$, where $w(x) \in C[a, b]$ and $w(x) > 0$.

$w(x)$ is called a *weight function*. For $w(x) = 1$, (b) reduces to (a).

Example 3.4 Let

$$\ell_2 = \left\{ x = \{x_n\} / \sum_{n=1}^{\infty} |x_n|^2 < \infty \right\}$$

Then,

$$\langle x, y \rangle = \sum_{k=1}^{\infty} x_k \overline{y_k},$$

for $x = (x_1, x_2, \ldots, x_k, \ldots)$ and $y = (y_1, y_2, \ldots, y_k, \ldots)$ belonging to ℓ_2, an inner product, and $(\ell_2, \langle \cdot, \cdot \rangle)$, an inner product space. If sequences $\{x_k\}$ and $\{y_k\}$ are real, then $\langle x, y \rangle = \sum_{k=1}^{\infty} x_k y_k$ and ℓ_2 is called real inner product space.

This space was introduced by David Hilbert.

Example 3.5 Let $L_2(a, b) = \{ f : [a, b] \to R \text{ or } C / \int_a^b |f(x)|^2 dx < \infty \}$. Then,

$$\langle f, g \rangle = \int_a^b f(x)\overline{g}(x)dx$$

is an inner product on $L_2(a, b)$ and $L_2(a, b)$ is an inner product space equipped with this inner product. $L_2(a, b)$ is called the *space of finite energy*.

Remark 3.2 (a) An inner product space is called *finite-dimensional inner product space* if the underlying vector space is finite-dimensional. Sometime a finite-dimensional complex inner product space is called *Hermitian space* or *unitary space*. It is called *Euclidean space* if underlying field is real.
(b) All inner product spaces considered in this book will be defined over the field of complex numbers unless explicitly mentioned.

Theorem 3.1 (Cauchy–Schwarz–Bunyakowski Inequality)

$$|\langle x, y \rangle|^2 \leq \langle x, x \rangle \langle y, y \rangle \tag{3.1}$$

for all x, y belonging to an inner product space X.

Proof It is clear that $\langle x, y \rangle = 0$ if $y = 0$ and $\langle y, y \rangle = 0$ for $y = 0$. Therefore, (3.1) is satisfied for $y = 0$. Let us prove it for $y \neq 0$. For $y \neq 0$, $\langle y, y \rangle \neq 0$. Let

$$\lambda = \frac{\langle x, y \rangle}{\langle y, y \rangle}.$$

Then,

$$\frac{|\langle x, y \rangle|^2}{\langle y, y \rangle} = \frac{\langle x, y \rangle \overline{\langle x, y \rangle}}{\langle y, y \rangle}$$
$$= \lambda \overline{\langle x, y \rangle} = \lambda \langle y, x \rangle$$

by virtue of condition (3) of Definition 3.1. Also, $\lambda \langle y, x \rangle = \overline{\lambda} \langle x, y \rangle$ by the same reasoning. Thus,

$$\frac{|\langle x, y \rangle|^2}{\langle y, y \rangle} = \lambda \langle y, x \rangle = \lambda \langle x, y \rangle = |\lambda|^2 \langle y, y \rangle \tag{3.2}$$

By Definition 3.1(*a*), (*b*) and Remark 3.1, we get

$$0 \leq \langle x - \lambda y, x - \lambda y \rangle = \langle x, x \rangle + \langle x, -\lambda y \rangle + \langle -\lambda y, x \rangle + \langle -\lambda y, -\lambda y \rangle$$
$$= \langle x, x \rangle - \overline{\lambda} \langle x, y \rangle - \lambda \langle y, x \rangle + |\lambda|^2 \langle y, y \rangle \tag{3.3}$$

By (3.2) and (3.3), we get

$$0 \leq \langle x, x \rangle - \frac{|\langle x, y \rangle|^2}{\langle y, y \rangle} - \frac{|\langle x, y \rangle|^2}{\langle y, y \rangle} + \frac{|\langle x, y \rangle|^2}{\langle y, y \rangle}$$

or

$$0 \leq \langle x, x \rangle - \frac{|\langle x, y \rangle|^2}{\langle y, y \rangle} \tag{3.4}$$

or

$$|\langle x, y \rangle|^2 \leq \langle x, x \rangle \langle y, y \rangle.$$

Theorem 3.2 *Every inner product space defines a norm, and so every inner product space is a normed space. Let $(X, \langle \cdot, \cdot \rangle)$ be an inner product space, and then, $(X, ||\cdot||)$ is a normed space, where*

$$||x|| = \langle x, x \rangle^{1/2} \; \forall \, x \in X.$$

Proof (a) $||x|| = (\langle x, x \rangle)^{1/2}$. Let $||x|| = 0$, and then, $\langle x, x \rangle = 0$ implying $x = 0$ by Definition 3.1(d) and also $x = 0$ implies $\langle x, x \rangle = 0$ or $||x|| = 0$. By Definition 3.1(d) $\langle x, x \rangle \geq 0$. Thus, condition of the norm is satisfied.

(b)

$$||\alpha x|| = (\langle \alpha x, \alpha x \rangle)^{1/2} = (|\alpha|^2 \langle x, x \rangle)^{1/2}$$
$$= |\alpha| (\langle x, x \rangle)^{1/2} = |\alpha| \, ||x||$$

Here, we used $\alpha \times \overline{\alpha} = |\alpha|^2$ and $\langle \alpha x, \alpha x \rangle = \alpha \overline{\alpha} \langle x, x \rangle$ by Definition 3.1 and Remark 3.1(*d*).

(c) $||x + y|| \leq ||x|| + ||y|| \; \forall \, x, y \in X.$

$$lefthandside = ||x + y||^2 = \langle x + y, x + y \rangle$$
$$= \langle x, x \rangle + \langle x, y \rangle + \overline{\langle x, y \rangle} + \langle y, y \rangle$$

by Definition 3.1 and Remark 3.1.

or

$$||x + y||^2 \leq ||x||^2 + 2Re\langle x, y \rangle + ||y||^2$$

by Definition 3.1 and Appendix $A.4$. By Theorem 3.1.,

$$Re\langle x, y \rangle \leq |\langle x, y \rangle| \leq (\langle x, x \rangle)^{1/2}(\langle y, y \rangle)^{1/2}.$$

Therefore,

$$||x + y||^2 \leq ||x||^2 + 2||x|| \, ||y|| + ||y||^2 \tag{3.5}$$

or

$$||x + y|| \leq ||x|| + ||y||.$$

Thus, $||x||$ is a norm and $(X, || \cdot ||)$ is a normed space.

Remark 3.3 (a) In view of Theorem 3.2, Theorem 3.1 can be written as

$$|\langle x, y \rangle| \leq ||x|| \, ||y||.$$

(b) In Theorem 3.1, equality holds, that is,

$$|\langle x, y \rangle| = ||x|| \, ||y||,$$

if and only if x and y are linearly dependent.

Verification

Let $|\langle x, y \rangle| = ||x|| \, ||y||$, $x \neq 0$ and $y \neq 0$. If either $x = 0$ or $y = 0$, then x and y are linearly dependent; that is, the result follows. Since $x \neq 0$, $y \neq 0$, $\langle x, y \rangle \neq 0$ and $\langle y, y \rangle \neq 0$. If

$$\lambda = \frac{\langle x, y \rangle}{\langle y, y \rangle}, \ then$$
$$\lambda \neq 0$$

and

$$\langle x - \lambda y, x - \lambda y \rangle = \langle x, x \rangle - \frac{|\langle x, y \rangle|^2}{\langle y, y \rangle}$$
$$= ||x||^2 - \frac{||x||^2 ||y||^2}{||y||^2} = 0.$$

By Definition 3.1(d), it follows that $x - \lambda y = 0$. Hence, x and y are linearly dependent. Conversely, if x and y are linearly dependent, then we can write $x = \lambda y$. This implies that .

$$|\langle x, y \rangle| = |\langle \lambda y, y \rangle| = |\lambda| \, |\langle y, y \rangle| (By \; Definition \; 3.1)$$
$$= |\lambda| \, ||y|| \, ||y|| = (|\lambda| \, ||y||) \, ||y||$$
$$= ||x|| \, ||y||$$

Remark 3.4 (a) The norm $||x|| = |\langle x, x \rangle|^{1/2}$ is called the norm induced by an inner product.

(b) All notions discussed in Chap. 2 can be extended for inner product space as it is a normed space with respect to the norm given in part (a). For example, a sequence $\{x_n\}$ in an inner product space $(X, \langle \cdot, \cdot \rangle)$ is called a *Cauchy sequence* if, for $\varepsilon > 0$, there exists N such that

$$||x_n - x_m|| = [\langle x_n - x_m, x_n - x_m \rangle]^{1/2} < \varepsilon$$

for n and $m > N$, that is,

$$\lim_{n,m \to \infty} \langle x_n - x_m, x_n - x_m \rangle = 0.$$

Hilbert Spaces

Definition 3.2 If every Cauchy sequence in an inner product space $(H, \langle \cdot, \cdot \rangle)$ is convergent, that is, for $x_n, x_m \in H$ with $\langle x_n - x_m, x_n - x_m \rangle = 0$ as $n, m \to \infty$, we get there exists $u \in H$ such that $\langle x_n - u, x_n - u \rangle = 0$ as $n \to \infty$. Then, H is called a Hilbert space. Thus, an inner product space is called a *Hilbert space* if it is complete.

Example 3.6 (a) R^n, ℓ_2, and $L_2(a, b)$ are examples of Hilbert spaces.

(b) $C[a, b]$ and $P[a, b]$ are examples of inner product spaces but they are not Hilbert space.

An inner product $\langle \cdot, \cdot \rangle$ on $P[a, b]$ is defined as follows: For $f, g \in P[a, b]$,

$$\langle f, g \rangle = \int\limits_a^b f(t)g(t)dt.$$

Example 3.7 Let $Y = \{f : [a, b] \to R / f$ is absolutely continuous on $[a, b]$ with $f, \frac{df}{dt} \in L_2(a, b)$ and $f(a) = f(b) = 0\}$. Then, Y is a dense subspace of $L_2(a, b)$. Y is a Hilbert space with respect to

$$\langle f, g \rangle_Y = \langle f, g \rangle_{L_2(a,b)} + \left\langle \frac{df}{dx}, \frac{dg}{dx} \right\rangle_{L_2(a,b)}.$$

Example 3.8 Let Ω be an open subset of R^3 and $C_0^\infty(\Omega) = \{f : \Omega \to R / f$ infinitely differentiable with compact support in $\Omega\}$. For $f, g \in C_0^\infty(\Omega)$,

$$\langle f, g \rangle = \int\limits_{\Omega} \left(fg + \frac{\partial f}{x_1} \frac{\partial g}{x_1} + \frac{\partial f}{x_2} \frac{\partial g}{x_2} + \frac{\partial f}{x_3} \frac{\partial g}{x_3} \right) dx_1 dx_2 dx_3$$

is an inner product on $C_0^\infty(\Omega)$.

$(C_0^\infty(\Omega), \langle \cdot, \cdot \rangle)$ is inner product space but not a Hilbert space.

This inner product induces the norm:

$$\|f\| = \left(\int\limits_{\Omega} (|f|^2 + |\nabla f|^2) dx_1 dx_2 dx_3 \right)^{1/2}$$

However, $C_0^\infty(\Omega)$ is not a Hilbert space.

Remark 3.5 (a) David Hilbert discovered space ℓ_2 in 1910. In fact, abstract Hilbert space (Definitions 3.1 and 3.2) was introduced by J. von Neumann in 1927.

(b) In the beginning of the study of this interesting space, a separable complete inner product space was called *Hilbert space* and this continued up to 1930.

(c) It can be proved that every separable Hilbert space is isometric and isomorphic to ℓ_2.

3.2.2 Parallelogram Law

In classical geometry, the sum of the squares of the diagonals of a parallelogram is equal to the sum of the squares of its sides. We prove extension of this result in the following theorem.

Theorem 3.3 (Parallelogram Law)

$\|x + y\|^2 + \|x - y\|^2 = 2\|x\|^2 + \|y\|^2 \forall x, y$ *belonging to an inner product space X.*

Proof By Theorem 3.2, we have

$$\|x + y\|^2 = \langle x + y, x + y \rangle \ and$$
$$\|x - y\|^2 = \langle x - y, x - y \rangle$$

By Definition 3.1 and Remark 3.1(d), we get

$$\|x + y\|^2 = \langle x, x \rangle + \langle x, y \rangle + \langle y, x \rangle + \langle y, y \rangle$$

or

$$\|x + y\|^2 = \|x\|^2 + \langle x, y \rangle + \langle y, x \rangle + \|y\|^2 \tag{3.6}$$

Similarly, we can show that

$$||x - y||^2 = ||x||^2 - \langle x, y \rangle - \langle y, x \rangle + ||y||^2 \tag{3.7}$$

By adding Eqs. (3.5) and (3.6), we get

$$||x + y||^2 + ||x - y||^2 = 2||x||^2 + 2||y||^2.$$

This proves the theorem.

Remark 3.6 The parallelogram law is not valid for an arbitrary norm as demonstrated below. Let $X = C[0, 2\pi]$ be the Banach space where

$$||f|| = \sup_{0 \le t \le 2\pi} |f(t)| \ for \ f \in X.$$

This norm does not satisfy the parallelogram law. Let $f(t) = \max(\sin t, 0)$ and $g(t) = \max(-\sin t, 0)$. Then,

$$||f|| = 1, ||g|| = 1, ||f + g|| = 1, \ and \ ||f - g|| = 1$$

Thus,

$$2||f||^2 + 2||g||^2 = 4 \ and \ ||f + g||^2 + ||f - g||^2 = 2$$

Hence,

$$||f + g||^2 + ||f - g||^2 \ne 2||f||^2 + 2||g||^2$$

In fact, we see in Theorem 3.5 that any norm satisfying the parallelogram law is defined by an inner product.

Theorem 3.4 (Polarization Identity) *Let* $x, y \in X$ *(inner product space), and then*

$$\langle x, y \rangle = \frac{1}{4}\{||x + y||^2 - ||x - y||^2 + i||x + iy||^2 - i||x - iy||^2\}$$

Proof Using definition of inner product and Remark 3.1, we get

$$||x + y||^2 = \langle x, x \rangle + \langle x, y \rangle + \langle y, x \rangle + \langle y, y \rangle$$
$$-||x - y||^2 = -\langle x, x \rangle + \langle x, y \rangle + \langle y, x \rangle - \langle y, y \rangle$$
$$i||x + iy||^2 = i\langle x, x \rangle + \langle x, y \rangle + \langle y, x \rangle + i\langle y, y \rangle$$
$$-i||x + iy||^2 = -i\langle x, x \rangle + \langle x, y \rangle - \langle y, x \rangle - i\langle y, y \rangle$$

By adding these four relations, we obtain

$$\|x + y\|^2 - \|x - y\|^2 + i\|x + iy\|^2 - i\|x - iy\|^2 = 4\langle x, y \rangle$$

or

$$\langle x, y \rangle = \frac{1}{4}[\|x + y\|^2 - \|x - y\|^2 + i\|x + iy\|^2 - i\|x - iy\|^2].$$

Theorem 3.5 (Jordan-von-Neumann, 1935) *A normed space is an inner product space if and only if the norm of the normed space satisfies the parallelogram law.*

Theorem 3.6 *A Banach space is a Hilbert space if and only if its norm satisfies the parallelogram law.*
  For proof of above theorems, one may see, for example, Siddiqi [169].

3.3   Orthogonal Complements and Projection Theorem

3.3.1   *Orthogonal Complements and Projections*

Definition 3.3   1. Two vectors x and y in an inner product space are called *orthogonal*, denoted by $x \perp y$ if

$$\langle x, y \rangle = 0.$$

  2. A vector x of an inner product space X is called *orthogonal* to a nonempty subset A of X, denoted by $x \perp A$, if $\langle x, y \rangle = 0$ for each $y \in A$.
  3. Let A be a nonempty subset of an inner product space X. Then, the set of all vectors orthogonal to A, denoted by $A^\perp$, is called the *orthogonal complement* of A; that is,

$$A^\perp = \{x \in X / \langle x, y \rangle = 0 \ for \ each \ y \in A\}.$$

$A^{\perp\perp} = (A^\perp)^\perp$ will denote orthogonal complement of $A^\perp$.
  4. Two subsets A and B of an inner product space X are called orthogonal denoted by $A \perp B$ if $\langle x, y \rangle = 0 \forall \ x \in A$ and $\forall \ y \in B$.

Remark 3.7 1. Since $\langle x, y \rangle = \langle y, x \rangle$, $\langle x, y \rangle = 0$ implies that $\langle y, x \rangle = 0$ or $\langle y, x \rangle = 0$ and vice versa. Hence, $x \perp y$ if and only if $y \perp x$.
  2. In view of Remark 3.1 (6), $x \perp 0$ for every x belonging to an inner product space. By condition (4) of the definition of the inner product, 0 is the only vector orthogonal to itself.
  3. It is clear that $0^\perp = X$ and $X^\perp = 0$.
  4. It is clear that if $A \perp B$, then $A \cap B = 0$.
  5. Nonzero mutually orthogonal vectors, $x_1, x_2, x_3, \ldots, x_n$ of an inner product space are linearly independent.

Theorem 3.7 *Let X be an inner product space and A its arbitrary subset. Then, the following results hold good:*

1. $A^\perp$ *is a closed subspace of X.*
2. $A \cap A^\perp \subset \{0\}$. $A \cap A^\perp = 0$ *if and only if A is a subspace.*
3. $A \subset A^{\perp\perp}$
4. *If* $B \subset A$, *then* $B^\perp \supset A^\perp$.

Proof 1. Let $x, y \in A^\perp$. Then, $\langle x, z \rangle = 0 \,\forall\, z \in A$ and $\langle y, z \rangle = 0 \forall z \in A$. Since for arbitrary scalars α, β,

$$\langle \alpha x + \beta y, z \rangle = \alpha \langle x, z \rangle + \beta \langle y, z \rangle,$$

by Definition 3.1, we get $\langle \alpha x + \beta y, z \rangle = 0$, i.e.,

$$\alpha x + \beta y \in A^\perp.$$

So, $A^\perp$ is a subspace of X. For showing that $A^\perp$ is closed, let $\{x_n\} \in A^\perp$ and $x_n \to y$. We need to show that y must belong to $A^\perp$. By definition of $A^\perp$, for every $x \in X$, $\langle x, x_n \rangle = 0 \,\forall\, n$. This implies that

$$\lim_{n \to \infty} \langle x, x_n \rangle = \lim_{n \to \infty} \langle x_n, x \rangle = 0 \; (Remark \; 3.7).$$

Since $\langle \cdot, \cdot \rangle$ is a continuous function

$$\langle \lim_{n \to \infty} x_n, x \rangle = 0$$

or $\langle y, x \rangle = 0$. Hence, $y \in A^\perp$.
2. If $y \in A \cap A^\perp$ and $y \in A^\perp$, by Remark 3.7, $y \in \{0\}$. If A is a subspace, then $0 \in A$ and $0 \in A \cap A^\perp$. Hence, $A \cap A^\perp = \{0\}$.
3. Let $y \in A$, but $y \notin A^{\perp\perp}$. Then, there exists an element $z \in A^\perp$ such that $\langle y, z \rangle \neq 0$. Since $z \in A^\perp$, $\langle y, z \rangle = 0$ which is a contradiction. Hence, $y \in A^{\perp\perp}$.

Definition 3.4 The angle θ between two vectors x and y of an inner product space X is defined by the following relation:

$$\cos \theta = \frac{\langle x, y \rangle}{\|x\| \, \|y\|} \tag{3.8}$$

Remark 3.8 1. By the Cauchy–Schwarz–Bunyakowski inequality, the right-hand side of Eq. (3.8) is always less than or equal to 1, and so the angle θ is well defined, i.e., $0 \leq \theta \leq \pi$, for every x and y different from 0.
2. If $X = R^3$, $x = (x_1, x_2, x_3)$, $y = (y_1, y_2, y_3)$, then

$$\cos \theta = \frac{x_1 y_1 + x_2 y_2 + x_3 y_3}{(x_1^2 + x_2^2 + x_3^2)^{1/2}(y_1^2 + y_2^2 + y_3^2)^{1/2}}$$

This is a well-known relation in three-dimensional Euclidean space.

3. If $x \perp y$, then $\cos \theta = 0$; i.e., $\theta = \pi/2$. In view of this, orthogonal vectors are also called *perpendicular vectors*.

A well-known result of plane geometry is the sum of the squares of the base, and the perpendicular in a right-angled triangle is equal to the square of the hypotenuse. This is known as the *Pythagorean theorem*. Its infinite-dimensional analogue is as follows.

Theorem 3.8 *Let X be an inner product space and $x, y \in X$. Then for $x \perp y$, we have*

$$||x + y||^2 = ||x||^2 + ||y||^2.$$

Proof

$$||x + y||^2 = \langle x + y, x + y \rangle = \langle x, x \rangle + \langle y, x \rangle + \langle x, y \rangle + \langle y, y \rangle \ (by \ Definition \ 3.1).$$

Since $x \perp y$, $\langle x, y \rangle = 0$ and $\langle y, x \rangle = 0$ (by Definition 3.3 and Remark 3.7). Hence, $||x + y||^2 = ||x||^2 + ||y||^2$.

Example 3.9 Let $X = R^3$, the three-dimensional Euclidean space, and $M = \{x = (x_1, x_2, x_3)\}$ be its subspace spanned by a nonzero vector x. The orthogonal complement of M is the plane through the origin perpendicular to the vector x (Fig. 3.1).

Example 3.10 Let A be a subspace of R^3 generated by the set $\{(1, 0, 1), (0, 2, 3)\}$. A typical element of A can be expressed as

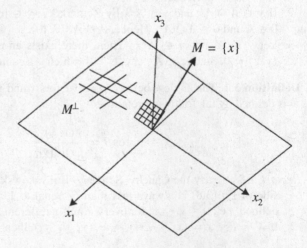

Fig. 3.1 Orthogonal complement in Example 3.9

$$x = (x_1, x_2, x_3) = \lambda(1, 0, 1) + \mu(0, 2, 3)$$
$$= \lambda i + 2\mu j + (\lambda + 3\mu)k$$
$$\Rightarrow x_1 = \lambda, \ x_2 = 2\mu, \ x_3 = \lambda + 3\mu$$

Thus, a typical element of A is of the form $\left(x_1, x_2, x_1 + \frac{3}{2}x_2\right)$. The orthogonal complement of A can be constructed as follows: Let $x = (x_1, x_2, x_3) \in A^{\perp}$. Then for $y = (y_1, y_2, y_3) \in A$, we have

$$\langle x, y \rangle = x_1 y_1 + x_2 y_2 + x_3 y_3 = x_1 y_1 + x_2 y_2 + x_3 \left(y_1 + \frac{3}{2}y_2\right)$$

$$(x_1 + x_3)y_1 + \left(x_2 + \frac{3}{2}x_3\right)y_2 = 0$$

Since y_1 and y_2 are arbitrary, we have

$$(x_1 + x_3) = 0 \ and \ \left(x_2 + \frac{3}{2}x_3\right) = 0$$

Therefore

$$A^{\perp} = \left\{x = (x_1, x_2, x_3)/x_1 = -x_3, x_2 = -\frac{3}{2}x_3\right\}$$
$$= \left\{x \in R^3/x = \left(-x_3, -\frac{3}{2}x_3, x_3\right)\right\}$$

3.4 Orthogonal Projections and Projection Theorem

As we know, an algebraic projection P on a vector space X is a linear operator P on X into itself such that $P^2 = P$. A projection P on an inner product space X is called an *orthogonal projection* if its range and null space are orthogonal, that is, $R(P) \perp N(P)(\langle u, v \rangle = 0 \ \forall \ u \in N(P), v \in N(P))$, where $R(P)$ and $N(P)$ denote, respectively, the range and null space. It can be easily verified that if P is an orthogonal projection, then so is $I - P$. Important properties of an orthogonal projection are contained in the following theorem:

Theorem 3.9 *Let P be an orthogonal projection of an inner product space X. Then,*

1. *Each element $z \in X$ can be written uniquely as*

$$z = x + y, \ where \ x \in R(P) \ and \ y \in N(P)$$

2. $\|z\|^2 = \|x\|^2 + \|y\|^2$
3. *Every orthogonal projection is continuous, and moreover, $\|P\| = 1$ for $P \neq 0$.*

4. $N(P)$ and $R(P)$ are closed linear subspaces of X.
5. $N(P) = R(P)^\perp$ and $R(P) = N(P)^\perp$.

Proof We know that for an algebraic projection P on X

$$X = R(P) + N(P)$$

where $R(P) \cap N(P) = \{0\}$. Thus, every element z in X can be expressed in the form $z = x + y$, $x \in R(P)$, $y \in N(P)$. To show the uniqueness, let

$$z = x_1 + y_1, \ x_1 \in R(P), \ y_1 \in N(P)$$
$$z = x_2 + y_2, \ x_2 \in R(P), \ y_2 \in N(P)$$

Then, $x_1 + y_1 = x_2 + y_2$ or $x_1 - x_2 = y_2 - y_1$.

1. Since $x_1 - x_2 \in R(P)$ as $R(P)$ is a subspace of X and $y_2 - y_1 \in N(P)$ as $N(P)$ is a subspace of X, we find that $x_1 - x_2 = y_2 - y_1 = 0$. Therefore, the representation is unique.
2. Since $z = x + y, x \in R(P)$, and $y \in N(P)$, we have

$$\begin{aligned} ||z||^2 = ||x + y||^2 &= \langle x + y, x + y \rangle \\ &= \langle x, x \rangle + \langle x, y \rangle + \langle y, x \rangle + \langle y, y \rangle \\ &= ||x||^2 + ||y||^2 \ as \ \langle x, y \rangle = 0 \ and \ \langle y, x \rangle = 0. \end{aligned} \tag{3.9}$$

3. By parts (1.) and (2.) for all $z \in X$, we have

$$||Pz||^2 = ||P(x + y)||^2 = ||Px||^2 + ||Py||^2, x \in R(P), y \in N(P).$$

$||Py||^2 = 0$ as $y \in N(P)$. Thus,

$$||Pz||^2 = ||Px||^2 = ||x||^2 \le ||z||^2.$$

This means that P is bounded and hence continuous. Let $P \ne 0$; then, for $x \in R(P)$, we have $Px = x$. Therefore, $||Px|| = ||x||$ which implies that $||P|| = 1$.
4. It is clear that $R(P)$ is the null space of $I - P$. By Proposition 2.1(4), the result follows.
5. Since $N(P) \perp R(P)$, $N(P) \subset R(P)^\perp$. To prove the assertion, it is sufficient to show that $N(P) \supset R(P)^\perp$. Let $z \in R(P)^\perp$. Then, there exists a unique $x_0 \in R(P)$ and $y_0 \in N(P)$ such that $z = x_0 + y_0$. Since $z \in R(P)^\perp$, we have $\langle z, u \rangle = 0$ for all $u \in R(P)$. Then, $0 = \langle z, u \rangle = \langle x_0, u \rangle$ for all $u \in R(P)$. This implies that $x_0 = 0$ and $z = y_0 \in N(P)$ which shows that $R(P)^\perp \subset N(P)$. Similarly, we can prove that $R(P) = N(P)^\perp$.

The existence of orthogonal projection is guaranteed by the following theorem:

Theorem 3.10 (Projection Theorem) *If M is a closed subspace of a Hilbert space X, then*

$$X = M \oplus M^{\perp} \tag{3.10}$$

Remark 3.9 1. Theorem 3.10 implies that a Hilbert space is always rich in projections. In fact, for every closed subspace M of a Hilbert space X, there exists an orthogonal projection on X whose range space is M and whose null space is $M^{\perp}$.
2. Equation (3.10) means that every $z \in X$ is expressible uniquely in the form $z = x + y$, where $x \in M$ and $y \in M^{\perp}$. Since $M \cap M^{\perp} = \{0\}$, in order to prove Theorem 3.10, it is sufficient to show that $X = M + M^{\perp}$.
Equation (3.10) is called the *orthogonal decomposition* of Hilbert space X.
3.   i. Let $X = R^2$. Then, Figure 3.2 provides the geometric meaning of the orthogonal decomposition of R^2 as

$$H = R^2, \ z \in R^2, \ z = x + y, x \in M, y \in M^{\perp}$$

   ii. Theorem 3.10 is not valid for inner product spaces (see Problem 3.20).
We require the following results in the proof of Theorem 3.10 (Fig. 3.2).

Lemma 3.1 *Let M be a closed convex subset of a Hilbert space X and $\rho = \inf\limits_{y \in M} ||y||$. Then, there exists a unique $x \in M$ such that $||x|| = \rho$.*

Lemma 3.2 *Let M be a closed subspace of a Hilbert space X, $x \notin M$, and let the distance between x and M be ρ, i.e., $\rho = \inf\limits_{u \in M} ||x - u||$. Then, there exists a unique vector $w \in M$ such that $||x - w|| = \rho$.*

Lemma 3.3 *If M is a proper closed subspace of a Hilbert space X, then there exists a nonzero vector u in X such that $u \perp M$.*

Fig. 3.2 Geometrical meaning of the orthogonal decomposition in R^2

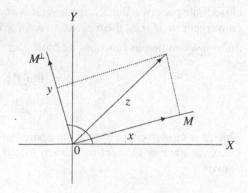

Lemma 3.4 *If M and N are closed subspaces of a Hilbert space X, such that M ⊥ N, then the subspace*

$$M + N = \{x + y \in X / x \in M \text{ and } y \in N\}$$

is also closed.

Remark 3.10 Lemma 3.2 can be rephrased as follows: Let M be a closed convex subset of a Hilbert space X, and for $x \in X$, let $\rho = inf_{u \in M}||x - u||$. Then, there exists a unique element $w \in M$ such that $\rho = ||x - w||$. w is called the *projection* of x on M, and we write $Px = w$ (Definition 3.5).

Proof (Proof of Lemma 3.1) Since M is a convex subset of X, $\alpha x + (1 - \alpha)y \in M$, for $\alpha = 1/2$ and every $x, y \in M$. By the definition of ρ, there exists a sequence of vectors $\{x_n\}$ in M such that $||x_n|| \to \rho$. For $x = x_n$ and $y = x_m$, we have

$$\frac{x_n + x_m}{2} \in M \text{ and } ||\frac{x_n + x_m}{2}|| \geq \rho.$$

Hence,

$$||x_n + x_m|| \geq 2\rho \tag{3.11}$$

By the parallelogram law for elements x_n and x_m

$$||x_n + x_m||^2 + ||x_n - x_m||^2 = 2||x_n||^2 + 2||x_m||^2$$

or

$$||x_n - x_m||^2 = 2||x_n||^2 + 2||x_m||^2 - ||x_n + x_m||^2$$
$$\geq 2||x_n||2 + 2||x_m||^2 - 4\rho^2 \text{ by Eq. } (3.11)$$

Since $2||x_n||^2 \to 2\rho^2$ and $2||x_m||^2 \to 2\rho^2$, we have $||x_n - x_m||^2 \to 2\rho^2 + 2\rho^2 - 4\rho^2 = 0$ as $n, m \to \infty$. Hence, $\{x_n\}$ is a Cauchy sequence in M. Since M is a closed subspace of a Banach space X, it is complete (see Remark 1.3). Thus, $\{x_n\}$ is convergent in M; i.e., there exists a vector x in M such that $\lim_{n \to \infty} x_n = x$. Since the norm is a continuous function, we have

$$\rho = \lim_{n \to \infty} ||x_n||$$
$$= \lim_{n \to \infty} x = ||x||.$$

Thus, x is an element of M with the desired property. Now, we show that x is unique. Let x' be another element of M such that $\rho = ||x'||$. Since $x, x' \in M$, and M is convex,

$$\frac{x + x'}{2} \in M.$$

By the parallelogram law for the elements $\frac{x}{2}$ and $\frac{x'}{2}$, we have

$$\left\|\frac{x}{2} + \frac{x'}{2}\right\|^2 + \left\|\frac{x'}{2} - \frac{x'}{2}\right\|^2 = \frac{2}{4}\|x\|^2 + \frac{2}{4}\|x'\|^2$$

or

$$\left\|\frac{x + x'}{2}\right\|^2 = \frac{\|x\|^2}{2} + \frac{\|x'\|^2}{2} - \frac{\|x - x'\|^2}{2}$$

$$< \frac{\|x\|^2}{2} + \frac{\|x'\|^2}{2} = \rho^2$$

or

$$\left\|\frac{x + x'}{2}\right\| < \rho$$

which contradicts the definition of ρ. Hence, x is unique.

Proof (*Proof of Lemma* 3.2) The set

$$N = x + M = \{x + v / v \in M\}$$

is a closed convex subset of X, and

$$\rho = \inf_{x+v \in N} \|0 - (x + v)\|$$

is the distance of 0 from N. Since $-v \in M$ for all

$$v \in M, \quad \rho = \inf_{v \in M} \|0 - (x + v)\|$$

By Lemma 3.1, there exists a unique vector $u \in N$ such that $\rho = \|u\|$. The vector $w = x - u = -(u - x)$ belongs to M [as we have $M = N - x = \{z - x / z \in N; x \notin M\}$ and M is a subspace, therefore, $-(z - x) \in M$]. Thus, $\|x - w\| = \|u\| = \rho$. w is unique. For if w is not unique, $w_1 \neq w$ is a vector in M such that $\|x - w_1\| = \rho$. This implies that $u_1 = (x - w_1)$ is a vector in N such that $\|u_1\| = \|x - w_1\| = \rho$. This contradicts that u is unique. Hence, w is unique.

Proof (*Proof of Lemma* 3.3) Let $x \notin M$ and $\rho = \inf_{v \in M} \|x - v\|$, the distance from x to M. By Lemma 3.2, there exists a unique element $w \in M$ such that $\|x - w\| = \rho$. Let $u = x - w$. $u \neq 0$ as $\rho > 0$. (If $u = 0$, then $x - w = 0$ and $\|x - w\| = 0$ implies that $\rho = 0$.)

Now, we show that $u \perp M$. For this, we show that for arbitrary $y \in M$, $\langle u, y \rangle = 0$. For any scalar α, we have $||u - \alpha y|| = ||x - w - \alpha y|| = ||x - (w + \alpha y)||$. Since M is a subspace, $w + \alpha y \in M$ whenever $w, y \in M$. Thus, $w + \alpha y \in M$ implies that

$$||u - \alpha y|| \geq \rho = ||u||,$$
$$or \ ||u - \alpha y||^2 - ||u||^2 \geq 0,$$
$$or \ \langle u - \alpha y, u - \alpha y \rangle - ||u||^2 \geq 0.$$

Since

$$\langle u - \alpha y, u - \alpha y \rangle = \langle u, u \rangle - \alpha \langle y, u \rangle$$
$$- \overline{\alpha} \langle u, y \rangle + \alpha \overline{\alpha} \langle y, y \rangle$$
$$= ||u|| - \alpha \langle u, y \rangle - \overline{\alpha} \langle y, u \rangle + |\alpha|^2 \langle y, y \rangle,$$

we have

$$- \overline{\alpha} \langle u, y \rangle - \alpha \overline{\langle u, y \rangle} + |\alpha|^2 ||y||^2 \geq 0 \qquad (3.12)$$

By putting $\alpha = \beta \langle u, y \rangle$ in Eq. (3.12), β being an arbitrary real number, we get

$$- 2\beta |\langle u, y \rangle|^2 + \beta^2 |\langle u, y \rangle|^2 ||y||^2 \geq 0 \qquad (3.13)$$

If we put $\alpha = |\langle u, y \rangle|^2$ and $b = ||y||^2$ in Eq. (3.12), we obtain

$$-2\beta a + \beta^2 ab \geq 0.$$

Or

$$\beta a (\beta b - 2) \geq 0 \ \forall \ real \ \beta \qquad (3.14)$$

If $a > 0$, Eq. (3.13) is false for all sufficiently small positive β. Hence, a must be zero, i.e., $a = |\langle u, y \rangle|^2 = 0$ or $\langle u, y \rangle = 0 \ \forall \ y \in M$.

Proof (*Proof of Lemma* 3.4) It is a well-known result of vector spaces that $M + N$ is a subspace of X. We show that it is closed; i.e., every limit point of $M + N$ belongs to it. Let z be an arbitrary limit point of $M + N$. Then, there exists a sequence $\{z_n\}$ of points of $M + N$ such that $z_n \to z$ (see Theorem A.2 of Appendix A.3). $M \perp N$ implies that $M \cap N = \{0\}$. So, every $z_n \in M + N$ can be written uniquely in the form $z_n = x_n + y_n$, where $x_n \in M$ and $y_n \in N$. By the Pythagorean theorem for elements $(x_m - x_n)$ and $(y_m - y_n)$, we have

$$||z_m - z_n||^2 = ||(x_m - x_n) + (y_m - y_n)||^2$$
$$= ||x_m - x_n||^2 + ||y_m - y_n||^2 \qquad (3.15)$$

(It is clear that $(x_m - x_n) \perp (y_m - y_n) \ \forall \ m, n$.) Since $\{z_n\}$ is convergent, it is a Cauchy sequence and so $||z_m - z_n||^2 \to 0$. Hence, from Eq. (3.15), we see that $||x_m - x_n|| \to 0$ and $||y_m - y_n|| \to 0$ as $m, n \to \infty$. Hence, $\{x_m\}$ and $\{y_n\}$ are Cauchy sequences in M and N, respectively. Being closed subspaces of a complete space, M and N are also complete. Thus, $\{x_m\}$ and $\{y_n\}$ are convergent in M and N, respectively, say $x_m \to x \in M$ and $y_n \to y \in N$. $x + y \in M + N$ as $x \in M$ and $y \in N$. Then,

$$z = \lim_{n \to \infty} z_n = \lim_{n \to \infty} (x_n + y_n) = \lim_{n \to \infty} x_n + \lim_{n \to \infty} y$$
$$= x + y \in M + N$$

This proves that an arbitrary limit point of $M + N$ belongs to it and so it is closed.

Proof (*Proof of Theorem* 3.10) By Theorem 3.9, $M \perp$ is also a closed subspace of X. By choosing $N = M^\perp$ in Lemma 3.4, we find that $M + M^\perp$ is a closed subspace of X. First, we want to show that $X = M + M^\perp$. Let $X \neq M + M^\perp$; i.e., $M + M^\perp$ is a proper closed subspace of X. Then by Lemma 3.3, there exists a nonzero vector u such that $u \perp M + M^\perp$. This implies that $\langle u, x + y \rangle = 0, \ \forall x \in M$, and $y \in M \perp$. If we choose $y = 0$, then $\langle u, x \rangle = 0, \forall \in M$; i.e., $u \in M \perp$. On the other hand, if we choose $x = 0$, then $\langle u, y \rangle = 0$ for all $y \in M \perp$; i.e., $u \in M^{\perp\perp}$. (Since M and $M^\perp$ are subspaces, this choice is possible.) Thus, $u \in M^\perp \cap M^{\perp\perp}$. By Theorem 3.7 for $A = M^\perp$, we obtain that $u = 0$. This is a contradiction as $u \neq 0$. Hence, our assumption is false and $X = M + M^\perp$. In view of Remark 3.9 (2), the theorem is proved.

Remark 3.11 Sometimes, the following statement is also added in the statement of the projection theorem. "Let M be a closed subspace of a Hilbert space X, and then $M = M^{\perp\perp}$ (Problem 3.8)."

Remark 3.12 1. Let $X = L_2(-1, 1)$. Then, $X = M \oplus M^\perp$, where $M = \{f \in L_2(-1, 1)/f(-t) = f(t) \ \forall \ t \in (-1, 1)$, i.e., the space of even functions$\}$; $M^\perp = \{f \in L_2(-1, 1)/f(-t) = -f(t) \forall t - (-1, 1)$, i.e., the space of odd functions$\}$.
2. Let $X = L_2[a, b]$. For $c \in [a, b]$, let $M = \{f \in L_2[a, b]/f(t) = 0$ almost everywhere in $(a, c)\}$ and $M^\perp = \{f \in L_2[a, b]/f(t) = 0$ almost everywhere in $(c, b)\}$. Then, $X = M \oplus M^\perp$.

Remark 3.13 The orthogonal decomposition of a Hilbert space, i.e., Theorem 3.10, has proved quite useful in potential theory (Weyl [195]). The applications of the results concerning orthogonal decomposition of Hilbert spaces can be found in spectral decomposition theorems which deal with the representation of operators on Hilbert spaces. For example, for a bounded self-adjoint operator, T, $\langle Tx, y \rangle$ is represented by an ordinary Riemann–Stieltjes integral. For details, see Kreyszig [117], and Naylor and Sell [144].

Example 3.11 Let $X = L_2[0, \pi]$ and $Y = \{f \in L_2[0, \pi] / f \text{ is constant}\}$. It can be checked that Y is a closed subspace of X and hence itself a Hilbert space. Every $h \in L_2[0, \pi]$ can be decomposed as

$$h = f + g, \ f \in Y \text{ and } g \in Y^{\perp} \text{ (by projection theorem)}$$

For example, $h = \sin x$ can be decomposed as $\sin x = a + (\sin x - a), a \in Y$. The constant a is determined from the orthogonality of $g = \sin x - a$ to every element c in Y.

$$\langle c, g \rangle = \int_0^\pi c(\sin x - a) dx = 0$$

$$= c(2 - a\pi)$$

Since c is arbitrary, we have $a = \frac{2}{\pi}$. Thus, the projection of $h = \sin x$ on Y is the function $f(x) = \frac{2}{\pi}$. Let us define an operator P on $L_2[0, \pi]$ by

$$Ph = P(f + g) = f, \ f \in Y$$
$$Pg = 0 \ g \in Y^{\perp}$$

$$\left\| \sin x - \frac{2}{\pi} \right\| = \left(\int_0^\pi \left(\sin x - \frac{2}{\pi} \right)^2 dx \right)^{1/2}$$

$$= \left[\left(\int_0^\pi \left(\sin^2 x + \frac{4}{\pi^2} - \frac{4}{\pi} \sin x \right) dx \right) \right]^{1/2}$$

$$= \left(\frac{\pi}{2} + \frac{4}{\pi} - \frac{8}{\pi} \right)^{1/2} \simeq 0.545$$

$$\| \sin x - c \| = \left[\int_0^\pi (\sin^2 x + c^2 - 2c \sin x) dx \right]^{1/2}$$

$$= \left[\frac{\pi}{2} + c^2 \pi - 4c \right]^{1/2}$$

3.5 Projection on Convex Sets

We discuss here the concepts of projection and projection operator on convex sets which are of vital importance in such diverse fields as optimization, optimal control, and variational inequalities.

Definition 3.5 (a) Let X be a Hilbert space and $K \subset X$ a nonempty closed convex set. For $x \in X$, by projection of x on K, we mean the element $z \in K$ denoted by $P_K(x)$ such that

$$\|x - P_K(x)\|_x \geq \|x - y\|_x \, \forall \, y \in K$$
$$or \quad \|x - z\| = \inf_{y \in K} \|x - y\| \tag{3.16}$$
$$or \quad \langle x - z, x - z \rangle \leq \langle x - y, x - y \rangle \, \forall \, y \in K$$

(b) An operator on X into K, denoted by P_K, is called the *projection operator* if $P_K(x) = z$, where z is the projection of x on K.

Theorem 3.11 (Existence of Projection on Convex Sets) *Let K be a closed convex subset of a Hilbert space X. Then for any x in X, there is a unique element in K closest to x; that is, there is a unique element z in K such that (3.15) is satisfied.*

Theorem 3.12 (Variational Characterization of Projection) *Let K be a closed convex set in a Hilbert space X. For any $x \in X$, $z \in K$ is the projection of x if and only if*

$$\langle x - z, y - z \rangle \leq 0 \, \forall \, y \in K \tag{3.17}$$

Proof (Proof of Theorem 3.11) The proof follows from Lemma 3.2 and Remark 3.10 if we observe that the set $x - K$ consisting of all elements of the form $x - y$ for all $y \in K$ is a closed convex set.

Proof (Proof of Theorem 3.12) Let z be the projection of $x \in X$. Then for any α, $0 \leq \alpha \leq 1$, since K is convex, $\alpha y + (1 - \alpha)z \in K$ for all $y \in K$. Now,

$$\|x - (\alpha y + (1 - \alpha)z)\|^2 = g(\alpha) \tag{3.18}$$

is a twice continuously differentiable function of α. Moreover

$$g'(\alpha) = 2\langle x - \alpha y - (1 - \alpha)z, z - y \rangle \tag{3.19}$$
$$g''(\alpha) = 2\langle z - y, z - y \rangle \tag{3.20}$$

Now, for z to be the projection of x, it is clear that $g'(0) \geq 0$, which is (3.17). In order to prove the converse, let (3.17) be satisfied for some element z in K. This implies that $g'(0)$ is nonnegative, and by (3.20) $g''(\alpha)$ is nonnegative. Hence, $g(0) \leq g(1)$ for all $y \in K$ such that (3.16) is satisfied.

Remark 3.14 A geometrical interpretation of the characterization (3.16) is given in Fig. 3.3. The left hand side is just the cosine of the angle θ between lines connecting the point $P_K(x) = z$ with the point x and with the arbitrary point $y \in K$, respectively, and we have that $\cos \theta \leq 0$ as $\theta \geq \frac{2}{\pi}$ necessarily holds. Conversely, for every other point $z' \in K$, $z' \neq P_K(x)$, there exists a point $y' \in K$ so that the angle between the lines connecting the point z' with points x and y' is less than $\frac{\pi}{2}$.

Fig. 3.3 Geometrical
interpretation of the
characterization of projection

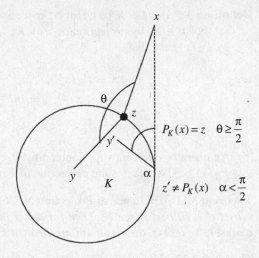

Theorem 3.13 *The projection operator P_K defined on a Hilbert space X into its nonempty closed convex subset K possesses the following properties:*

 (a) $\|P_K(u) - P_K(v)\| \le \|u - v\| \ \forall \, u, v \in X$ (3.21)

 (b) $\langle P_K(u) - P_K(v), u - v \rangle \ge 0 \forall \, u, v \in X$ (3.22)

 (c) P_K *does not satisfy* $\langle P_K(u) - P_K(v), u - v \rangle > 0 \forall \, u, v \in X.$ (3.23)

 (d) P_K *is nonlinear.*

 (e) P_K *is continuous.*

Proof (*Proof of Theorem* 3.13)

1. (a) In view of (3.17)

$$\langle P_K(u) - u, P_K(u) - v \rangle \le 0 \ \forall \, v \in K \tag{3.24}$$

Put $u = u_1$ in (3.24) to get

$$\langle P_K(u_1) - u_1, P_K(u_1) - v \rangle \le 0 \ \forall \, v \in K \tag{3.25}$$

Put $u = u_2$ in (3.24) to get

$$\langle P_K(u_2) - u_2, P_K(u_2) - v \rangle \le 0 \ \forall \, v \in K \tag{3.26}$$

Since $P_K(u_2)$ and $P_K(u_1) \in K$, choose $v = P_K(u_2)$ and $v = P_K(u_1)$, respectively, in (3.25) and (3.26); therefore, we get

$$\langle P_K(u_1) - u_1, P_K(u_1) - P_K(u_2) \rangle \leq 0 \qquad (3.27)$$

$$\langle P_K(u_2) - u_2, P_K(u_2) - P_K(u_1) \rangle \leq 0 \qquad (3.28)$$

By (3.27) and (3.28), we get

$$\langle P_K(u_1) - u_1 - P_K(u_2) + u_2, P_K(u_1) - P_K(u_2) \rangle \leq 0 \qquad (3.29)$$

or

$$\langle P_K(u_1) - P_K(u_2), P_K(u_1) - P_K(u_2) \rangle$$
$$\leq \langle u_1 - u_2, P_K(u_1) - P_K(u_2) \rangle$$

or

$$\|P_K(u_1) - P_K(u_2)\|^2 \leq \langle u_1 - u_2, P_K(u_1) - P_K(u_2) \rangle \qquad (3.30)$$

or

$$\|P_K(u_1) - P_K(u_2)\|^2 \leq \|u_1 - u_2\| \, \|P_K(u_1) - P_K(u_2)\| \qquad (3.31)$$

by the Cauchy–Schwarz–Bunyakowski inequality, or

$$\|P_K(u_1) - P_K(u_2)\| \leq \|u_1 - u_2\| \qquad (3.32)$$

2. It follows from (3.30) as $\|P_K(u_1) - P_K(u_2)\|^2 \geq 0 \; \forall \, u_1, u_2 \in X$.
3. If $K = X$, then $P_K = I$ (identity operator); then,

$$\langle u - v, u - v \rangle > 0 \; if \; u \neq v$$

However, if $K \neq X$, that is, $u_0 \in X$ and $u_0 \notin K$, then $P_K(u_0) \neq u_0$ but $P_K(P_K(u_0)) = P_K(u_0)$.
4. It is clear that P_K is nonlinear.
5. Continuity follows from (3.32) because $u_n \to u$ implies that $P_K(u_n) \to P_K(u)$ as $n \to \infty$.

3.6 Orthonormal Systems and Fourier Expansion

Fourier Expansion in Hilbert Space

Definition 3.6 A set of vectors $\{u_n\}$ in an inner product space is called orthonormal if

$$\langle u_i, u_j \rangle = \delta_{ij} \qquad (3.33)$$

where δ_{ij} is the Kronecker delta.

Theorem 3.14 *An orthonormal set of nonzero vectors is linearly independent.*

Proof (*Proof of Theorem* 3.14) Let $\{u_i\}$ be an orthonormal set. Consider the linear combination

$$\alpha_1 u_1 + \alpha_2 u_2 + \cdots + \alpha_n u_n = 0$$

For $u_1 \neq 0$,

$$0 = \langle 0, u_1 \rangle = \langle \alpha_1 u_1 + \alpha_2 u_2 + \cdots + \alpha_n u_n, u_1 \rangle$$
$$= \alpha_1 \langle u_1, u_1 \rangle + \alpha_2 \langle u_2, u_1 \rangle + \cdots + \alpha_n \langle u_n, u_1 \rangle.$$

Since $\langle u_i, u_j \rangle = \delta_{ij}$, we have $0 = \alpha_1 \langle u_1, u_1 \rangle$ or $\alpha_1 = 0$ as $u_1 \neq 0$. Similarly, we find that $\alpha_2 = \alpha_3 = \cdots = \alpha_n = 0$. This shows that $\{u_i\}$ is a set of linearly independent vectors.

Example 3.12 (i) In the first place, it may be noted that any nonzero set of orthogonal vectors of an inner product space can be converted into an orthonormal set by replacing each vector u_i with $u_i / \|u_i\|$.

(ii) Let us consider the set $\{\cos nx\}_{n=0}^{\infty}$ in $L_2[-\pi, \pi]$ which is orthogonal.

$$\frac{1}{\sqrt{2\pi}}, \ \frac{1}{\sqrt{\pi}} \cos x, \ \frac{1}{\sqrt{\pi}} \cos 2x$$

is a set of orthonormal vectors in $L_2[-\pi, \pi]$.

(iii) Let $X = L_2[-1, 1].u = 1$, $v = bx \in L_2[-1, 1]$, where b is a constant. u and v are orthogonal as

$$\langle u, v \rangle = \int_{-1}^{1} bx \, dx = \left[b \frac{x^2}{2} \right]_{-1}^{1} = 0$$
$$= 2 \ if \ u = v = 1$$

The functions $\tilde{u} = \frac{u}{\|u\|} = \frac{1}{\sqrt{2}}$ and $\tilde{v} = \frac{v}{\|v\|} = \sqrt{\frac{3}{2}} x$ are orthonormal.

Definition 3.7 a. An orthogonal set of vectors $\{\psi_i\}$ in an inner product space X is called an *orthogonal basis* if for any $u \in X$, there exist scalars α_i such that

$$u = \sum_{i=1}^{\infty} \alpha_1 \psi_i$$

If elements ψ_i are orthonormal, then it is called an *orthonormal basis*.
An orthonormal basis $\{\psi_i\}$ in a Hilbert space X is called *maximal* or *complete*

if there is no unit vector ψ_0 in X such that $\{\psi_0, \psi_1, \psi_2, \ldots\}$ is an orthonormal set. In other words, the sequence $\{\psi_i\}$ in X is complete if and only if the only vector orthogonal to each of ψ_i's is the null vector.

b. Let $\{\psi_i\}$ be an orthonormal basis in a Hilbert space X, and then, the numbers $\alpha_i = \langle u, \psi_i \rangle$ are called the Fourier coefficients of the element u with respect to the system $\{\psi_i\}$ and $\sum \alpha_i \psi_i$ is called the *Fourier series* of the element u.

Theorem 3.15 (Bessel's Inequality) *Let $\{\psi_i\}$ be an orthonormal basis in a Hilbert space X. For each $u \in X$,*

$$\sum_{i=1}^{\infty} |\langle u, \psi_i \rangle|^2 \leq ||u||^2 \tag{3.34}$$

Theorem 3.16 *Let $\{\psi_i\}$ be a countably infinite orthonormal set in a Hilbert space X. Then, the following statements hold:*

a. *The infinite series $\sum_{n=1}^{\infty} \alpha_n \psi_n$, where α_n's are scalars, converges if and only if the series $\sum_{n=1}^{\infty} |\alpha_n|^2$ converges.*

b. *If $\sum_{n=1}^{\infty} \alpha_n \psi_n$ converges and*

$$u = \sum_{n=1}^{\infty} \alpha_n \psi_n = \sum_{n=1}^{\infty} \beta_n \psi_n$$

then $\alpha_n = \beta_n \; \forall \, n$ and $||u||^2 = \sum_{n=1}^{\infty} |\alpha_n|^2$

Proof (Proof of Theorem 3.15) We have

$$||u - \sum_{i=1}^{N} \langle u, \psi_i \rangle \psi_i||^2 \geq 0$$

L.H.S.

$$= ||u||^2 - 2\sum_{i=1}^{N} \langle u, \psi_i \rangle \langle u, \psi_i \rangle + \sum_{i=1}^{N}\sum_{j=1}^{N} \langle u, \psi_i \rangle \langle u, \psi_i \rangle \langle \psi_i \psi_j \rangle$$

by using properties of inner product and we get

$$||u||^2 - \sum_{i=1}^{N} |\langle u, \psi_i \rangle|^2 \geq 0$$

by applying the fact that

$$\langle \psi_i, \psi_j \rangle = \delta_{ij} = \begin{cases} 0 \; if \; i \neq j \\ 1 \; if \; i = j \end{cases}$$

This gives us

$$\sum_{i=1}^{N} |\langle u, \psi_i \rangle|^2 \leq ||u||^2$$

Taking limit $N \to \infty$, we get

$$\sum_{i=1}^{\infty} |\langle u, \psi_i \rangle|^2 \leq ||u||^2$$

Proof (Proof of Theorem 3.16)

1. Let $\sum_{n=1}^{\infty} \alpha_n \psi_n$ be convergent, and assume that

$$u = \sum_{n=1}^{\infty} \alpha_n \psi_n$$

or equivalently, $\lim_{N \to \infty} \left\| u - \sum_{n=1}^{N} \alpha_n \psi_n \right\|^2 = 0$

$$\langle u, \psi_m \rangle = \left\langle \sum_{n=1}^{\infty} \alpha_n \psi_n, \psi_m \right\rangle$$

$$= \sum_{n=1}^{\infty} \alpha_n \langle \psi_n, \psi_m \rangle \; m = 1, 2, \ldots$$

$$= \alpha_m \; (by \; Definition \; 3.6)$$

By the Bessel inequality, we get

$$\sum_{m=1}^{\infty} |\langle u, \psi_m \rangle| = \sum_{m=1}^{\infty} |\alpha_m|^2 \leq ||u||^2$$

which shows that $\sum_{n=1}^{\infty} |\alpha_n|^2$ converges. Consider the finite sum $s_n = \sum_{i=1}^{n} \alpha_i \psi_i$. We have

$$||s_n - s_m||^2 = \left\langle \sum_{i=m+1}^{n} \alpha_i \psi_i, \sum_{i=m+1}^{n} \alpha_i \psi_i \right\rangle$$

$$= \sum_{i=m+1}^{n} |\alpha_i|^2 \to 0 \ as \ n, m \to \infty$$

This means that $\{s_n\}$ is a Cauchy sequence. Since X is complete, the sequence of partial sums $\{s_n\}$ is convergent in X and therefore the series $\sum_{i=1}^{\infty} \alpha_n \psi_n$ converges.

2. We first prove that $||u||^2 = \sum_{n=1}^{\infty} |\alpha_n|^2$. We have

$$||u||^2 - \sum_{n=1}^{N} |\alpha_n|^2 = \langle u, u \rangle - \sum_{n=1}^{N} \sum_{m=1}^{N} \langle \alpha_n \psi_n, \alpha_m \psi_m \rangle$$

$$= \left\langle u, u - \sum_{n=1}^{N} \alpha_n \psi_n \right\rangle + \left\langle \sum_{n=1}^{N} \alpha_n \psi_n, u - \sum_{n=1}^{N} \alpha_n \psi_n \right\rangle$$

$$\leq \left\| u - \sum_{n=1}^{N} \alpha_n \psi_n \right\| \left(||u|| + \left\| \sum_{n=1}^{N} \alpha_n \psi_n \right\| \right) = M$$

Since $\sum_{n=1}^{N} \alpha_n \psi_n$ converges to u, the M converges to zero, proving the result.

If $u = \sum_{n=1}^{\infty} \alpha_n \psi_n = \sum_{n=1}^{\infty} \beta_n \psi_n$, then

$$0 = \lim_{n \to \infty} \left[\sum_{n=1}^{N} (\alpha_n - \beta_n) \psi_n \right] \Rightarrow 0 = \sum_{n=1}^{\infty} |\alpha_n - \beta_n|^2$$

by part 1, implying that $\alpha_n = \beta_n$ for all n.

Theorem 3.17 (Fourier Series Representation) *Let Y be the closed subspace spanned by a countable orthonormal set $\{\psi_i\}$ in a Hilbert space X. Then, every element $u \in Y$ can be written uniquely as*

$$u = \sum_{i=1}^{\infty} \langle u, \psi_i \rangle \psi_i \tag{3.35}$$

Theorem 3.18 (Fourier Series Theorem) *For any orthonormal set $\{\psi_n\}$ in a separable Hilbert space X, the following statements are equivalent:*

a. *Every u in X can be represented by the Fourier series in X; that is*

$$u = \sum_{i=1}^{\infty} \langle u, \psi_i \rangle \psi_i \qquad (3.36)$$

b. *For any pair of vectors u, v ∈ X, we have*

$$\langle u, v \rangle = \sum_{i=1}^{\infty} \langle u, \psi_i \langle \psi_i, v \rangle \qquad (3.37)$$

$$= \sum_{i=1}^{\infty} \alpha_i \beta_i$$

where $\alpha_i = \langle u, \psi_i \rangle$ = *Fourier coefficients of u.* $\beta = \langle v, \psi_i \rangle$ = *Fourier coefficients of v. [Relation (3.38) is called the Parseval formula]*

c. *For any u ∈ X, one has*

$$\|u\|^2 = \sum_{i=1}^{\infty} |\langle u, \psi_i \rangle|^2 \qquad (3.38)$$

d. *Any subspace Y of X that contains* $\{\psi_i\}$ *is dense in X.*

Proof (*Proof of Theorem* 3.17) Uniqueness of (3.36) is a consequence of Theorem 3.16(2). For any $u \in Y$, we can write

$$u = \lim_{N \to \infty} \sum_{i=1}^{M} \alpha_i \psi_i, \ M \geq N$$

as Y is closed. From Theorems 3.10 and 3.16, it follows that

$$\left\| u - \sum_{i=1}^{M} \langle u, \psi_i \rangle \psi_i \right\| \leq \left\| u - \sum_{i=1}^{M} \alpha_i \psi_i \right\|$$

and as $N \to \infty$, we get the desired result.

Proof (*Proof of Theorem* 3.18) $(a) \Rightarrow (b)$. This follows from (3.36) and the fact that $\{\psi_i\}$ is orthonormal. $(b) \Rightarrow (c)$. Put $u = v$ in (3.37) to get (3.37). $(a) \Rightarrow (d)$. The statement (d) is equivalent to the statement that the orthogonal projection onto $\overline{S}$, the closure of S, is the identity. In view of Theorem 3.17, statement (d) is equivalent to statement (a).

Example 3.13 (Projections and Orthonormal Bases)

1. We now show that the Fourier analysis described above can be used to construct orthogonal projections onto closed subspaces of Hilbert spaces. Let $\{\psi_n\}$ be an orthonormal basis of a Hilbert space X. Then, each $u \in X$ can be written as

$$u = \sum_{i=1}^{\infty} \alpha_i \psi_i = \sum_{i=1}^{N} \alpha_i \psi_i + \sum_{i=N+1}^{\infty} \alpha_i \psi_i$$

where α_n's are Fourier coefficients.

Let the set $\{\psi_1, \psi_2, \ldots, \psi_N\}$ forms a basis for an N-dimensional subspace Y of X. The operator P defined by

$$Pu = \sum_{n=1}^{N} \langle u, \psi_n \rangle \psi_n$$

is an orthogonal projection of X onto Y. In fact, P is linear and $P^2 = P$; that is

$$P(u+v) = Pu + Pv : P(u+v) = \sum_{n=1}^{N} \langle u+v, \psi_n \rangle \psi_n$$

$$= \sum_{n=1}^{N} \langle u, \psi_n \rangle \psi_n + \sum_{n=1}^{N} \langle v, \psi_n \rangle \psi_n \ (by\ Definition\ 3.1(1))$$

$$= Pu + Pv$$

$$P(\lambda u) = \lambda P(u) : P(\lambda u) = \sum_{n=1}^{N} \langle \lambda u, \psi_n \rangle \psi_n$$

$$= \lambda \sum_{n=1}^{N} \langle u, \psi_n \rangle \psi_n \ (by\ Definition\ 3.1(2))$$

$$P^2 u = P(Pu) = \sum_{n=1}^{N} \langle Pu, \psi_n \rangle \psi_n$$

$$= \sum_{n=1}^{N} \langle \sum_{m=1}^{N} \langle u, \psi_m \psi_m \rangle \psi_n \rangle \psi_n$$

$$= \sum_{n=1}^{N} \langle u, \psi_n \rangle \sum_{m=1}^{N} \langle \psi_n, \psi_m \rangle \psi_m$$

$$= \sum_{n=1}^{N} \langle u, \psi_n \rangle \psi_n \ (by\ using\ orthonormality\ of\ \psi_n)$$

$$= Pu \ \forall u \in X$$

Thus, $P^2 = P$.

2. $R(P) = Y$. To show that $R(P) \perp N(P)$, let $v \in N(P)$, and $u \in R(P)$ such that

$$\langle v, u \rangle = \langle v, Pu \rangle = \left\langle v, \sum_{n=1}^{N} \langle u, \psi_n \rangle \psi_n \right\rangle$$

$$= \sum_{n=1}^{N} \langle v, \psi_n \rangle \langle u, \psi_n \rangle$$

$$= \left\langle \sum_{n=1}^{N} \langle v, \psi_n \rangle \psi_n, u \right\rangle$$

$$= \langle Pv, u \rangle$$

Since $v \in N(P)$, we have $Pv = 0$ and hence $N(P) \perp R(P)$.

Now, we show that the projection error, that is, $u - Pu$, is orthogonal to Y: For $v \in Y$,

$$\langle u - Pu, v \rangle = \left\langle \sum_{n=N+1}^{\infty} \alpha_n \psi_n, v \right\rangle$$

$$= \sum_{n=N+1}^{\infty} \alpha_n \langle \psi_n, v \rangle = 0 \; as \; N \to \infty.$$

Furthermore,

$$\|Pu\|^2 = \sum_{n=1}^{N} |\langle u, \psi_n \rangle|^2, \; and \; if \; v \in Y$$

$$\|u - v\|^2 = \langle u - v + Pu - Pu, u - v + Pu - Pu \rangle$$

$$= \|u - Pu\|^2 + \|v - Pu\|^2 v \in Y$$

Hence, to make $\|u - v\|^2$ as small as possible $u \notin Y$, we must take $v = Pu$; that is

$$\inf_{v \in Y} \|u - v\| = \|u - Pu\|$$

Remark 3.15 Results of Theorem 3.16 can be stated in alternative terms as follows: For an element $u \in X$, $\sum_{i=1}^{\infty} \langle u, \psi_i \rangle \psi_i$ converges to $u_0 = Pu$ in the closed subspace Y spanned by the ψ_i's. The vector $u - u_0 = u - Pu$ is orthogonal to Y.

3.7 Duality and Reflexivity

3.7.1 Riesz Representation Theorem

In this section, we prove a theorem which gives the representation of a bounded linear functional defined on a Hilbert space, and with the help of this theorem, the relationship between a Hilbert space and its dual is studied.

Theorem 3.19 (Riesz Representation Theorem) *If f is a bounded linear functional on a Hilbert space X, then there exists a unique vector $y \in X$ such that $f(x) = \langle x, y \rangle \forall x \in X$, and $||f|| = ||y||$.*

Proof 1. In the first place, we prove that there exists an element y such that $f(x) = \langle x, y \rangle \ \forall x \in X$.

(a) If $f = 0$, then $f(x) = 0 \ \forall \ x \in X$. Therefore, $y = 0$ is the vector for which $\langle x, 0 \rangle = 0 \ \forall \ x \in X$. We have seen that $\langle x, 0 \rangle = 0$. Thus, the existence of vector y is proved when $f = 0$.

(b) Let $f \neq 0$. By Proposition 2.1, the null space N of f is a proper closed subspace of X and by Lemma 3.3, and there exists a nonzero vector $u \in X$ such that $u \perp N$. We show that if α is a suitably chosen scalar, $y = \alpha u$ satisfies the condition of the theorem.

(c) If $x \in N \subset X$, then whatever be α, $f(x) = 0 = \overline{\alpha} \langle x, u \rangle$ as $u \perp N$. Thus, $f(x) = \langle x, \alpha u \rangle$. Hence, the existence of $y = \alpha u$ is proved for all $x \in N$.

(d) Since $u \perp N, u \notin N$; let $x = u \in X - N$. If $f(u) = \langle u, \alpha u \rangle = \overline{\alpha} ||u||^2$, then

$$\overline{\alpha} = \frac{f(u)}{||u||^2}.$$

Therefore, for

$$\alpha = \frac{\overline{f(u)}}{||u||^2},$$

the vector $y = \alpha u$ satisfies the condition of the theorem in this case, that is, $f(u) = \langle u, \alpha u \rangle$.

(e) Since $u \notin N$, $f(u) \neq 0$ and so $\frac{f(x)}{f(u)}$ is defined for any $x \in X$. Consider $x - \beta u$, where $\beta = \frac{f(x)}{f(u)}$. Then, $f(x - \beta u) = f(x) - \beta f(u)$. This implies that $x - \beta u \in N$. Every $x \in X$ can be written as $x = x - \beta u + \beta u$. Therefore, for each $x \in X$

$$f(x) = f(x - \beta u + \beta u) = f(x - \beta u) + f(\beta u)$$
$$= f(x - \beta u) + \beta f(u) \ (f \ is \ linear)$$
$$= \langle x - \beta u, \alpha u \rangle + \beta \langle u, \alpha u \rangle$$

by (iii) and (iv), where

$$\alpha = \frac{\overline{f(u)}}{||u||^2}.$$

Since $x - \beta u \in N$, by (iii), $f(x - \beta u) = \langle x - \beta u, \alpha u \rangle$ for every α and so for $\alpha = \frac{\overline{f(u)}}{||u||^2}$. This gives us $f(x) = \langle x - \beta u, \alpha u \rangle + \langle \beta u, \alpha u \rangle = \langle x, \alpha u \rangle$. Thus, for an arbitrary $x \in X$, there exists a vector $y = \alpha u$ such that $f(x) = \langle x, y \rangle$.

2. We now show that y is unique. Let y_1 and y_2 be two vectors such that $f(x) = \langle x, y_1 \rangle \ \forall x \in X$ and $f(x) = \langle x, y_2 \rangle \ \forall x \in X$. Then, $\langle x, y_1 \rangle = \langle x, y_2 \rangle \ \forall x \in X$ or $\langle x, y_1 - y_2 \rangle = 0 \ \forall x \in X$. By Remark 3.1(7), $y_1 - y_2 = 0$ or $y_1 = y_2$. This proves the uniqueness of the vector y.

3. For each $x \in X$, there exists a unique $y \in X$ such that

$$f(x) = \langle x, y \rangle \ or \ |f(x)| = |\langle x, y \rangle|$$

By the Cauchy–Schwarz–Bunyakowski inequality

$$|\langle x, y \rangle| \leq ||x|| \ ||y||$$

Hence, $|f(x)| \leq ||y|| \ ||x||$.
By the definition of the norm of a functional, we have

$$||f|| \leq ||y|| \tag{3.39}$$

If $||y|| = 0$ or $y = 0$, then $|f(x)| = |\langle x, 0 \rangle| = 0$, for all x and so

$$||f|| = \sup\{\frac{|f(x)|}{||x||} / x \neq 0\} = 0,$$

i.e., $||f|| = ||y||$. Suppose $y \neq 0$. Then by Theorem (2.6),

$$||f|| = \sup\{|f(x)| / ||x|| = 1\}$$
$$\geq |f\left(\frac{y}{||y||}\right)| \ as \ ||\frac{y}{||y||} = 1$$
$$= \langle \frac{y}{||y||}, y \rangle = \frac{1}{||y||} \langle y, y \rangle$$
$$= \frac{||y||^2}{||y||} = ||y|| \tag{3.40}$$

or $||f|| \geq ||y||$. By Eqs. (3.39) and (3.40), we have $||f|| = ||y||$.

Theorem 3.20 *Let y be a fixed element of a Hilbert space X. Then, the functional f_y defined below belongs to X^**

$$f_y(x) = \langle x, y \rangle \ \forall x \in X$$

The mapping $\psi : y \to f_y$ of X into X^* satisfies the following properties:

1. $||\psi(y)|| = ||y||$.
2. ψ *is onto.*
3. $\psi(y_1 + y_2) = \psi(y_1) + \psi(y_2)$.
4. $\psi(\alpha y) = \alpha \psi(y)$.
5. ψ *is one-one.*

Remark 3.16 If X is a real Hilbert space, then in property (4) of ψ, we have $\psi(\alpha y) = \alpha \psi(y)$. Then, Theorem (3.20) means that every real Hilbert space X can be identified with its dual, i.e., $X = X^*$. (See Definition 2.6 and Remark thereafter.)

Proof In order to show that $f_y \in X^*$, we need to verify that

(a) $f_y(x_1 + x_2) = f_y(x_1) + f_y(x_2)$
    Verification
    By the definition of the inner product

$$f_y(x_1 + x_2) = \langle x_1 + x_2, y \rangle = \langle x_1, y \rangle + \langle x_2, y \rangle$$

or

$$f_y(x_1 + x_2) = f_y(x_1) + f_y(x_2)$$

(b) $f_y(\alpha x) = \alpha f_y(x)$
    Verification $f_y(\alpha x) = \langle \alpha x, y \rangle = \alpha \langle x, y \rangle$ by the definition of the inner product, or $f_y(\alpha x) = \alpha f_y(x)$.
(c) f_y is bounded.
    Verification $|f_y(x)| = |\langle x, y \rangle| \leq ||x|| \, ||y||$, by the Cauchy–Schwarz–Bunyakowski inequality. Thus, there exists a $k = ||y|| > 0$ such that $|f_y(x)| \geq k||x||$. That is, f_y is bounded.
    Now, we verify the properties of ψ. $||\psi(y)|| = ||f_y||$. By the second part of Theorem 3.19, $||f_y|| = ||y||$. Hence, $||\psi(y)|| = ||y||$. Since Theorem (3.19) states that for every $f \in X^*$, there exists a unique y such that $f(x) = \langle x, y \rangle \forall x \in X$, and the mapping $\psi : y \to f_y$ is onto.

$$\psi(y_1 + y_2)(x) = f_{y_1+y_2}(x) = \langle x, y_1 + y_2 \rangle$$
$$= \langle x, y_1 \rangle + \langle x, y_2 \rangle \ (by \ Remark \ 3.1(4a))$$
$$= f_{y_1}(x) + f_{y_2}(x)$$
$$= \psi(y_1)(x) + \psi(y_2)(x)$$
$$= (\psi(y_1) + \psi(y_2))(x) \ \forall \ x \in X$$

Hence,

$$\psi(y_1 + y_2) = \psi(y_1) + \psi(y_2)$$
$$\psi(\alpha y)(x) = f_{\alpha y}(x) = \langle x, \alpha y \rangle$$
$$= \overline{\alpha}\langle x, y \rangle \ (by \ Remark \ 3.1(4b))$$

or

$$\psi(\alpha y)(x) = \overline{\alpha}\psi(y)(x) \ \forall \ x \in X$$

Therefore, $\psi(\alpha y) = \overline{\alpha}\psi(y)$.
Let $\psi(y_1) = \psi(y_2)$. Then,

$$f y_1(x) = f y_2(x) \ \forall \ x \in X$$

or

$$\langle x, y_1 \rangle = \langle x, y_2 \rangle$$

or

$$\langle x, y_1 - y_2 \rangle = 0.$$

By Remark 3.1(7), $y_1 - y_2 = 0$. Hence, $y_1 = y_2$. Conversely, if $y_1 = y_2$, then $\langle x, y_1 - y_2 \rangle = 0$ by Remark 3.1(4a) or $\langle x, y_1 \rangle = \langle x, y_2 \rangle$ or $\psi(y_1) = \psi(y_2)$. Hence, ψ is one-one.

Remark 3.17   1. Theorem 3.19 was proved by the Hungarian mathematician Riesz around 1910. It is one of the most important results of Hilbert space. It has several applications of vital importance. Applying this theorem, we prove in the next subsection that every Hilbert space is reflexive.
2. Since R^n, ℓ_2, $L_2(a, b)$ are Hilbert spaces, by Theorem 3.19, bounded linear functionals defined on R^n, ℓ_2, and $L_2(a, b)$ are, respectively, of the following forms:

(a) For $F \in (R^n)^*$, there exists a unique $a = (a_1, a_2, a_3, \ldots, a_n) \in R^n$ such that

$$F(x) = \sum_{i=1}^{n} x_i a_i, \ where \ x = (x_1, x_2, \ldots, x_n) \in R^n$$
$$= \langle x, a \rangle = \langle a, x \rangle$$

(b) For $F \in (\ell_2)^*$, there exists a unique $a = (a_1, a_2, a_3, \ldots, a_n \ldots) \in \ell_2$ such that

$$F(x) = \sum_{i=1}^{\infty} x_i \overline{a_i}, \ where \ x = (x_1, x_2, \ldots, x_n) \in \ell_2$$
$$= \langle x, a \rangle$$

(c) For $F \in (L_2)^*$, there exists a unique $g \in L_2$ such that

$$F(f) = \int_a^b f(t) \overline{g}(t) dt, \ f \in L_2$$
$$= \langle f, g \rangle$$

3.7.2 Reflexivity of Hilbert Spaces

Let us recall that a Banach space X is called a *reflexive* Banach space if it can be identified with its second dual $(X^*)^{star} = X^{**}$. This means that a Banach space X is reflexive if there exists a mapping J on X into X^{**} which is linear, norm preserving, 1-1, and onto. In view of Problem 2.7, to verify whether a Banach space is reflexive or not, it is sufficient to show that the natural mapping is an onto mapping.

Theorem 3.21 *Every Hilbert space X is reflexive.*

Proof To show that the natural embedding $J \colon x \to F_x$ of X into X^{**} is an onto mapping, we have to show that for an arbitrary element $F \in X^{**}$, there exists $z \in X$ such that $J_z = F$. Let g be a functional on X defined as follows:

$$g(x) = \overline{F(\psi(x))} \psi(x) = f_x \in X^*.$$

Properties of ψ are given by Theorem 3.20. Using the properties of ψ and the fact that $F \in X^{**}$, we see that

$$g(x_1 + x_2) = \overline{F(\psi(x_1 + x_2))}$$
$$= \overline{F(\psi(x_1))} + \overline{F(\psi(x_2))}$$
$$g(\alpha x) = \overline{F(\psi(x))} = \overline{\alpha} \overline{F(\psi(x))}$$
$$= \alpha F(\psi(x))$$
$$and \ |g(x)| = |\overline{F(\psi)}| = |F(\psi(x))| \le ||F|| \ ||\psi(x)||$$
$$= ||F|| \ ||x||.$$

Hence, g is a bounded linear functional on X. By Theorem 3.19, there exists a unique $z \in X$ such that

$$g(x) = \langle x, z \rangle \forall x \in X$$
$$or \ \overline{F(\psi(x))} = \langle x, z \rangle$$
$$F(\psi(x)) = \overline{\overline{F(\psi(x))}} = \overline{\langle x, z \rangle} = \langle z, x \rangle \tag{3.41}$$

From the definition of the natural imbedding J and Theorem 3.20,

$$J_z(\psi(x)) = \psi(x)(z) = f_x(z) = \langle z, x \rangle \tag{3.42}$$

From Eqs. (3.40) and (3.41), we have $J_z = F$ which is the desired result.

3.8   Operators in Hilbert Space

3.8.1   Adjoint of Bounded Linear Operators on a Hilbert Space

Definition 3.8 Let X be a Hilbert space and $T : X \to X$ be a bounded linear operator on X into itself. Then, the *adjoint operator* $T^\star$ is defined by

$$\langle Tx, y \rangle = \langle x, T^\star y \rangle \ \forall \ x, y \in X.$$

Remark 3.18 1. $T^\star$ always exists.
  Verification Let $\langle Tx, y \rangle = f_y(x)$.

$$f_y(x_1 + x_2) = \langle T(x_1 + x_2), y \rangle = \langle Tx_1, y \rangle + \langle Tx_2, y \rangle$$
$$= f_y(x_1) + f_y(x_2)$$
$$f_y(\alpha x) = \langle T(\alpha x), y \rangle = \alpha \langle Tx, y \rangle = \alpha f_y(x)$$

Now, $|f_y(x)| = |\langle Tx, y \rangle| \le ||Tx|| \ ||y||$ (by Cauchy–Schwarz–Bunyakowski inequality) $\le ||T|| \ ||x|| \ ||y|| \le k \ ||x|| \ for \ a \ fixed \ y \in X$. Thus, $f_y \in X^\star$ and by the Riesz theorem, there exists a $y^* \in X$ such that

$$\langle Tx, y \rangle = f_y(x) = \langle x, y^* \rangle \, x \in X$$

Thus, T induces a linear map $y \to y^*$ and we write $y^* = T^*y$, where T^* is defined on X into itself.

2. T^* is bounded, linear, and unique.
Verification

$$||T^*x||^2 = \langle T^*x,$$
$$T^*x \rangle = \langle T(T^*x), x \rangle \le ||T(T^*x)|| \, ||x||$$

by Cauchy–Schwarz–Bunyakowski inequality. Since T is bounded, there exists $k > 0$ such that

$$||T(T^*x)|| \le k||T^*x||$$

Hence,

$$||(T^*x)||^2 \le k||x|| \, ||T^*x||$$

or $||T^*x|| \le k||x||$. Hence, T^* is bounded. For all $z \in X$,

$$
\begin{aligned}
\langle z, T^*(x + y) \rangle &= \langle Tz, (x + y) \rangle \\
&= \langle Tz, x \rangle + \langle Tz, y \rangle \\
&= \langle z, T^*x \rangle + \langle z, T^*y \rangle \\
&= \langle z, T^*x + T^*y \rangle
\end{aligned}
$$

or

$$\langle z, [T^*(x + y)] - [T^*x + T^*y] \rangle = 0 \, \forall \, z \in X$$

This implies (see Remark 3.1(7)) that $T^*(x + y) = T^*x + T^*y \, \forall \, x, y \in X$.

$$
\begin{aligned}
\langle z, T^*(\alpha x) \rangle &= \langle Tz, \alpha x \rangle \\
&= \overline{\alpha} \langle Tz, x \rangle \\
&= \overline{\alpha} \langle z, T^*(x) \rangle \\
&= \langle z, \alpha T^*(x) \rangle
\end{aligned}
$$

or

$$\langle z, T^*(\alpha x) - \alpha T^*(x) \rangle = 0 \, \forall z \in X.$$

Hence,

$$T^\star(\alpha x) = \alpha T^\star(x) \forall x \in X \text{ and } \alpha \text{ scalar.}$$

Thus, $T^\star$ is linear.

Suppose that for a given T, there exist two operators $T_1^\star$ and $T_2^\star$ such that

$$\langle Tx, y \rangle = \langle x, T_1^\star y \rangle$$

and

$$\langle Tx, y \rangle = \langle x, T_2^\star y \rangle$$

Then, we have

$$\langle x, T_1^\star y \rangle - \langle x, T_2^\star y \rangle = 0$$

or

$$\langle x, T_1^\star y - T_2^\star y \rangle = 0 \,\forall\, x \in X$$

This implies that $T_1^\star y = T_2^\star y \,\forall\, y \in X$; i.e., $T_1^\star = T_2^\star$. Hence, the adjoint of T is unique.

3. The adjoint operator of $T^\star$ is denoted by $T^{\star\star}$.

Example 3.14 Let $X = R^n$ be a real Hilbert space of dimension n, and for all $x = (x_1, x_2, \ldots, x_n) \in R^n$, let T be defined as follows:

$Tx = y$ where $y = (y_1, y_2, \ldots, y_n)$; $y_i = \sum\limits_{j=1}^{n} a_{ij} x_j$ and (a_{ij}) is an $n \times n$ matrix;

and T is bounded linear operator on R^n into itself.

$$\langle Tx, y \rangle = \sum_{i=1}^{n} \left(\sum_{j=1}^{n} a_{ij} x_j \right) y_i$$

$$= \sum_{j=1}^{n} x_j \left(\sum_{i=1}^{n} a_{ij} y_i \right)$$

$$= \sum_{i=1}^{n} x_i \left(\sum_{j=1}^{n} a_{ij} y_j \right)$$

$$= \langle x, T^\star y \rangle$$

where $T^\star y = z$, $z = (z_1, z_2, \ldots, z_n)$

$$z_i = \sum_{j=1}^{n} a_{ji} y_j$$

Thus, if a bounded linear operator T on R^n is represented by an $n \times n$ matrix (a_{ij}), the adjoint T^* of T is represented by (a_{ji}), transpose matrix of (a_{ij}).

Example 3.15 Let $X = L_2(a, b)$ and let $T : X \to X$, defined by

$$Tf(t) = \int_a^b K(s, t) f(t) dt,$$

where $K(s, t)$, $a \leq s \leq b$, $a \leq t \leq b$, is a continuous function be a bounded linear operator. Then,

$$\langle Tf, g \rangle = \int_a^b (Tf(t)) g(\overline{s}) ds$$

$$= \int_a^b \left(\int_a^b K(s, t) f(t) dt \right) g(\overline{s}) ds$$

$$= \int_a^b f(s) \left(\int_a^b K(s, t) g(\overline{t}) dt \right) ds$$

$$= \int_a^b f(s) \left(\int_a^b \overline{K(s, t) g(t)} dt \right) ds$$

$$= \langle f, T^* g \rangle$$

where $T^* g = \int_a^b \overline{K(t, s)} g(t) dt$. Thus, the adjoint operator T^* of T is given by

$$T^* g = \int_a^b \overline{K(t, s)} g(t) dt.$$

Example 3.16 Let $X = \ell_2$ and $T : \ell_2 \to \ell_2$ be defined by $T(\alpha_1, \alpha_2, \alpha_3, \ldots, \alpha_n, \ldots) = (0, \alpha_1, \alpha_2, \ldots, \alpha_n, \ldots,)$. For

$$x = \{\alpha_k\}_{k=1}^{\infty} \in \ell_2 \text{ and } y = \{\beta\}_{k=1}^{\infty} \in \ell_2$$

$$\langle Tx, y \rangle = \sum_{k=1}^{n} \alpha_k \overline{\beta}_{k+1}$$

$$= \langle x, T^*y \rangle$$

where $T^*\{(\beta_1, \beta_2, \beta_3, \ldots)\} = (\beta_2, \beta_3, \ldots)$. Thus, the adjoint of T is T^*. T is known as the *right shift operator* and T^* the *left shift operator*.

Theorem 3.22 *Let T be a bounded linear operator on a Hilbert space X into itself. Then, its adjoint operator T^* has the following properties:*

1. $I^* = I$ *where I is the identity operator.*
2. $(T + S)^* = T^* + S^*$.
3. $(\alpha T)^* = \overline{\alpha} T^*$.
4. $(TS)^* = S^*T^*$.
5. $T^{**} = T$
6. $\|T^*\| = \|T\|$
7. $\|T^*T\| = \|T\|^2$
8. *If T is invertible, then so is T^* and $(T^*)^{-1} = (T^{-1})^*$.*

Proof 1. Since $Ix = x \; \forall \, x \in X$, we have $\langle Ix, y \rangle = \langle x, y \rangle$, and by the definition of I^*, $\langle Ix, y \rangle = \langle x, I^*y \rangle$. Hence, $\langle Ix, y \rangle = \langle x, I^*y \rangle$ or $\langle x, y - I^*y \rangle = 0 \; \forall x \in X$. Thus, $y - I^*y = 0$ or $I^*y = y$; i.e., $I = I^*$.

2. $(T + S)^* = T^* + S^* \Leftrightarrow (T + S)^*(x) = T^*(x) + S^*(x) \; \forall \, x \in X$. By the definition of $(T + S)^*$, $\forall \, x \in X$, we have

$$\langle (T + S)z, x \rangle = \langle z, (T + S)^*x \rangle.$$

Also,

$$\langle (T + S)z, x \rangle = \langle Tz + Sz, x \rangle$$
$$= \langle Tz, x \rangle + \langle Sz, x \rangle$$
$$= \langle z, T^*x \rangle + \langle z, S^*x \rangle$$

Thus, $\langle z, (T + S)^*x \rangle = \langle z, T^*x + S^*x \rangle \; \forall \, x$ and $z \in X$, which implies that $(T + S)^*x = T^*x + S^*x$.

3. $\langle (\alpha T)z, x \rangle = \langle z, (\alpha T)^*x \rangle$

$$\langle (\alpha T)z, x \rangle = \alpha \langle Tz, x \rangle = \alpha \langle z, T^*(x) \rangle$$
$$= \langle z, \overline{\alpha} T^*(x) \rangle$$

These two relations show that $\langle z, (\alpha T)^*x \overline{\alpha} T^*x \rangle = 0 \; \forall \, z$ and x. Hence, $(\alpha T)^* = \overline{\alpha} T^*$.

4. $(TS)(x) = T(S(x))$

$$\langle (TS)(x), y \rangle = \langle T(S(x)), y \rangle$$
$$= \langle Sx, T^\star y \rangle$$
$$= \langle x, S^\star(T^\star(y)) \rangle = \langle x, (S^\star T^\star)(y) \rangle$$

On the other hand,

$$\langle (TS)(x), y \rangle = \langle x, (TS)^\star y \rangle$$
$$\langle x, (S^\star T^\star)y \rangle = \langle x, (TS)^\star y \rangle$$

or $(TS)^\star = S^\star T^\star$.

5.

$$\langle Tx, y \rangle = \langle x, T^\star y \rangle$$
$$= \langle (T^\star)^\star x, y \rangle$$

or $\langle (T - T^{\star\star})x, y \rangle = 0 \ \forall \ y \in X, \ where \ T^{\star\star} = (T^\star)^\star$. Hence, $T = T^{\star\star}$.

6. $\|T\| = \|T^\star\| \Leftrightarrow \|T\| \le \|T^\star\|$ and $\|T^\star\| \le \|T\|$. We have

$$\|T^\star(x)\|^2 = \langle T^\star x, T^\star x \rangle = \langle T(T^\star(x)), x \rangle$$
$$\le \|T(T^\star(x))\| \, \|x\|$$
$$\le \|T^\star(x)\| \, \|T\| \, \|x\|$$

or

$$\|T^\star(x)\| \le \|T\| \, \|x\|$$

This implies that $\|T^\star\| \le \|T\|$ (see Theorem 2.6). Applying this relation to $T^\star$, we have $\|T^{\star\star}\| \le \|T^\star\|$. However, $T^{\star\star} = T$ and, therefore, $\|T\| \le \|T^\star\|$.

7. $\|T^\star T\| \le \|T\|^2$ as $\|T * T(x)\| \le \|T * \| \, \|T\| \, \|x\| = \|T\|^2 \, \|x\|$ (in view of Theorem 2.6 and the previous relation). On the other hand,

$$\|Tx\|^2 = \langle Tx, Tx \rangle = \langle T^\star Tx, x \rangle$$
$$\le \|T^\star Tx\| \, \|x\|$$
$$\le \|T^\star T\| \, \|x\|^2$$

or $\|Tx\| \le (\|T^\star T\|)^{1/2}\|x\|$ or $\|T\|^2 \le \|T^\star T\|$. Hence,

$$\|T^\star T\| = \|T\|^2.$$

8. Since $I^\star = I$ and $TT^{-1} = T^{-1}T = I$, $(TT^{-1})^\star = I^\star = I$; by (4), $(TT^{-1})^\star = (T^{-1})^\star T^\star$. Therefore, $(T^{-1})^\star T^\star = I$ and consequently

$$(T^\star)^{-1} = (T^{-1})^\star.$$

Remark 3.19   1. The definition of the adjoint operator may be extended to bounded linear operators defined on a Hilbert space X into another Hilbert space Y in the following manner:

$$\langle Tx, y \rangle_Y = \langle x, T^*y \rangle_X \ \forall \ x \in X, y \in Y$$
$$T: X \to Y$$
$$T^*: Y \to X$$

2. Let $T: X \to Y$ be a bounded linear operator on a Hilbert space X into another Hilbert space Y. Then, the null and range spaces of T and its adjoint T^* are related by the following relations:

    i. $(R(T))^\perp = N(T^\star)$.
   ii. $\overline{R(T)} = (N(T^\star))^\perp$.
  iii. $(R(T^\star))^\perp = N(T)$.
  iv. $\overline{R(T^\star)} = (N(T))^\perp$.

3.8.2   Self-adjoint, Positive, Normal, and Unitary Operators

Definition 3.9 Let T be a bounded linear operator on a Hilbert space X into itself. Then,

a. T is called *self-adjoint* or *Hermitian* if $T = T^*$.
b. A self-adjoint operator T is called a *positive operator* if $\langle Tx, x \rangle \geq 0 \ \forall \ x \in X$ is called *strictly positive* if $\langle Tx, x \rangle = 0$ only for $x = 0$. Let S and T be two self-adjoint operators on X. We say that $S \geq T$ if $\langle (S - T)x, x \rangle \geq 0 \ \forall x \in X$.
c. T is called *normal* if $TT^\star = T^\star T$.
d. T is called *unitary* if $TT^\star = T^\star T = I$, where I is the identity operator.

Example 3.17   1. T given in Example 3.14 is self-adjoint if the matrix (a_{ij}) is symmetric, i.e., $a_{ij} = a_{ji} \ \forall \ i$ and j.
2. T given in Example 3.15 is self-adjoint if

$$K(s, t) = \overline{K}(t, s) \ \forall \ s \text{ and } t.$$

3. The identity operator I is self-adjoint.
4. The null operator 0 is self-adjoint.

Example 3.18 Let $T: L_2(0, 1) \to L_2(0, 1)$ be defined as follows:

$$Tf(t) = tf(t)$$

T is a self-adjoint operator. T is bounded, linear

$$\langle Tf, g \rangle = \int\limits_0^1 Tf\overline{g}dt = \int\limits_0^1 tf(t)\overline{g(t)}dt$$

and

$$\langle f, Tg \rangle = \int\limits_0^1 f\,\overline{Tg}dt$$

$$= \int\limits_0^1 f(t)\overline{tg(t)}dt$$

$$= \int\limits_0^1 tf(t)\overline{g(t)}dt \ as \ t \in (0, 1)$$

Hence, $\langle Tf, g \rangle = \langle f, Tg \rangle$.

Example 3.19 Let $X = \ell_2$, and $T : \ell_2 \to \ell_2$ be defined as

$$T(\alpha_k) = \left\{ \frac{\alpha_k}{k} \right\}.$$

T is bounded, linear, and

$$\langle Tx, y \rangle = \sum_{k=1}^{\infty} \frac{\alpha_k}{k}\overline{\beta_k}$$

$$= \langle x, Ty \rangle$$

Thus, T is self-adjoint.

Example 3.20 T given in Example 3.14 is positive if the matrix (a_{ij}) is symmetric and positive.

Example 3.21 Let $T : X \to X$ be given by $T = 2iI$, where I is the identity operator. Then, T is normal.

$$TT^{\star} = (2iI)(2iI)^{\star} = -4I$$

and

$$T^{\star}T = (2iI)^{\star}(2iI) = -4I$$

Example 3.22 T given in Example 3.14 is unitary if the matrix a_{ij} coincides with the inverse matrix of a_{ij}.

Remark 3.20   1. Every self-adjoint operator is normal but the converse is not true, in general.
2. Every unitary operator is normal but the converse need not be true.
   Verification:

1. Let T be self-adjoint. Then, $T^{\star} = T \Rightarrow TT^{\star} = T^2$ and $T^{\star}T = T^2$. Hence, $T^{\star}T = TT^{\star}$ and T is normal. For the converse, consider T as given in Example 3.21. $T^{\star} = (2i)^{\star}I^{\star} = -2iI \neq T$. Hence, T is not self-adjoint.

2. Since T is unitary, $T^{\star}T = I = T^{\star}T$ and so T is normal. For the converse, we consider the above operator where $T^{\star}T = T^{\star}T = -4I \neq I$. Hence, T is normal but not unitary.

The theorems given below provide interesting properties of self-adjoint, positive, normal, and unitary operators.

Theorem 3.23 *Let X be a real Hilbert space. The set $A(X)$ of all self-adjoint operators of a Hilbert space X into itself is a closed subspace of the Banach space $\mathcal{B}(X)$ of all bounded linear operators of X into itself, and therefore, it is a real Banach space containing the identity operator.*

Theorem 3.24 *Let T_1 and T_2 be two self-adjoint operators on a Hilbert space X. Then, their product T_1T_2 is self-adjoint if and only if $T_1T_2 = T_2T_1$.*

Theorem 3.25 *An operator T on a Hilbert space X is self-adjoint if and only if $\langle Tx, x \rangle$ is real $\forall x \in X$.*

Theorem 3.26   *1. If T and S are two positive operators defined on a Hilbert space X such that $TS = ST$, then ST is positive.*
2. If T is a bounded linear operator, then $T^{\star}T$ and $TT^{\star}$ are positive.

Theorem 3.27 *The set of all normal operators on a Hilbert space X is a closed subset of $\mathcal{B}(X)$, the Banach space of bounded linear operators of X into itself, which contains the set $\mathscr{A}(X)$ of all self-adjoint operators and is closed under scalar multiplication.*

Theorem 3.28 *If T_1 and T_2 are normal operators on a Hilbert space X with the property that either commutes with the adjoint of the other, then, $T_1 + T_2$ and T_1T_2 are normal.*

Theorem 3.29 *A bounded linear operator T on a Hilbert space X is normal if and only if $||T^{\star}x|| = ||Tx||$ for every $x \in X$.*

Theorem 3.30 *If T is a normal operator, then $||T^2|| = ||T||^2$.*

Theorem 3.31 *Let T be a bounded linear operator on a Hilbert space X into itself and T^* be its adjoint. Further, suppose that*

$$A = \frac{T + T^*}{2} \text{ and } B = \frac{T - T^*}{2i} (T = A + iB)$$

and $T^* = A - iB$. A and B are called real and imaginary parts. T is normal if and only if $AB = BA$.

Theorem 3.32 *A bounded linear operator T on a Hilbert space X into itself is unitary if and only if it is an isometry of X onto itself.*

Proof (Proof of Theorem 3.23) Let A and B be self-adjoint operators

$$(\alpha A + \beta B)^* = (\alpha A)^* + (\beta B)^* \text{ (by Theorem 3.22(2))}$$
$$= -\alpha A^* + -\beta B^* \text{ (by Theorem 3.22(3))}$$

Thus, for real scalars α and β, $(\alpha A + \beta B)^* = \alpha A + \alpha B$; i.e., $\alpha A + \beta B$ is self-adjoint, and the set of all self-adjoint operators $A(X)$ is a subspace of $B(X)$.

To show that this subspace is closed, it is sufficient to show that if $\{T_n\}$ is a sequence of self-adjoint operators on X, $\lim_{n \to \infty} T_n = T$ is also self-adjoint. We have

$$\|T - T^*\| = \|T - T_n + T_n - T_n^* + T_n^* - T^*\|$$
$$\leq \|T - T_n\| + \|T_n - T_n^*\| + \|T_n^* - T^*\|$$
$$= \|T - T_n\| + \|T_n - T_n\| + \|T_n^* - T^*\|$$

as T_n's are self-adjoint. Thus, the limit of right-hand side tends to 0 as $n \to \infty$ and consequently $T = T^*$. $I \in A(X)$ as $I = I^*$ by Theorem 3.22(1).

Proof (Proof of Theorem 3.24) Let $T_1 T_2$ be self-adjoint. Then, $(T_1 T_2)^* = T_1 T_2$. By Theorem 3.22(4), $(T_1 T_2)^* = T_2^* T_1^*$ and as T_1 and T_2 are self-adjoint, $(T_1 T_2)^* = T_2 T_1$. Thus, $T_1 T_2 = T_2 T_1$. Conversely, let $T_1 T_2 = T_2 T_1$. Then,

$$\langle x, (T_1 T_2)^*(y) \rangle = \langle x, T_2^* T_1^*(y) \rangle$$
$$= \langle x, T_2 T_1(y) \rangle$$
$$= \langle x, T_1 T_2(y) \rangle$$

or $\langle x, ((T_1 T_2)^* - T_1 T_2)y \rangle = 0$. Hence, $T_1 T_2$ is self-adjoint. For proving Theorem 3.25, we require the following lemma:

Lemma 3.5 *If T is a bounded linear operator on a Hilbert space X, then $T = 0$ if $\langle Tx, x \rangle = 0 \, \forall \, x \in X$..*

Proof We have

$$\langle T(\alpha x + \beta y), \alpha x + \beta y \rangle - |\alpha|^2 \langle Tx, x \rangle - |\beta|^2 \langle Ty, y \rangle$$
$$= \alpha \overline{\beta} \langle Tx, y \rangle + \overline{\alpha} \beta \langle Ty, x \rangle$$
$$LHS = \alpha \overline{\alpha} \langle Tx, x \rangle + \beta \overline{\alpha} \langle Ty, x \rangle + \alpha \overline{\beta} \langle Tx, y \rangle$$
$$+ \beta \overline{\beta} \langle Ty, y \rangle - |\alpha|^2 \langle Tx, x \rangle - |\beta|^2 \langle Ty, y \rangle$$
$$= \alpha \overline{\beta} \langle Tx, y \rangle + \overline{\alpha} \beta \langle Ty, x \rangle = RHS.$$

By putting $\alpha = i$ and $\beta = 1$ in the above equation, we get

$$i \langle Tx, y \rangle - i \langle Ty, x \rangle = 0 \qquad (3.43)$$

and by putting $\alpha = 1$ and $\beta = 1$ in the above equation, we get

$$\langle Tx, y \rangle + \langle Ty, x \rangle = 0 \qquad (3.44)$$

From Eqs. (3.42) and (3.43), we find that $\langle Tx, y \rangle = 0 \; \forall \; x$ and y. Hence, $Tx = 0 \; \forall \; x$, i.e., $T = 0$.

Proof (*Proof of Theorem* 3.25) Let T be self-adjoint. Then, $\langle Tx, x \rangle = \langle x, Tx \rangle = \langle T^*x, x \rangle = \langle Tx, x \rangle$. Hence, $\langle Tx, x \rangle$ is real $\forall \; x \in X$. Let $\langle Tx, x \rangle$ be real $\forall x \in X$. Then,

$$\langle Tx, x \rangle = \overline{\langle Tx, x \rangle} = \langle x, Tx \rangle = \langle T^*x, x \rangle$$

or

$$\langle (T - T^*)(x), x \rangle = 0 \; \forall \; x \in X$$

By Lemma 3.5, $T - T^* = 0$ or $T = T^*$; i.e., T is self-adjoint.

Proof (**Proof of Theorem** 3.26)

1. Since T is self-adjoint, $\langle T_2 x, x \rangle = \langle Tx, Tx \rangle \geq 0$; i.e., $T_2 \geq 0$. Let $T \neq 0$. We define a sequence of operators $\{T_n\}$ in the following manner:

$$T_1 = \frac{T}{\|T\|}, \; T_2 = T_1 - T_1^2$$
$$T_3 = T_2 - T_2^2, \; T_4 = T_3 - T_3^2$$
$$T_{n+1} = T_n - T_n^2$$

It can be verified that T_n's are self-adjoint, where $0 \leq T_n \leq 1$ and $\sum_{i=1}^{n} T_i^2 x \to T_i x$. For verification, one may see [8, pp. 415–416]. Since S commutes with T, it must commute with every T_n. This implies that

$$\langle TSx, y \rangle = ||T|| \langle ST_1 x, x \rangle$$

$$= ||T|| \left\langle S \lim_n \sum_{i=1}^{n} T_i^2 x, x \right\rangle$$

$$= ||T|| \lim_n \sum_{i=1}^{n} \langle ST_i^2 x, x \rangle$$

$$= ||T|| \lim_n \sum_{i=1}^{n} \langle ST_i x, T_i x \rangle$$

Since $S \geq 0$ implies that $\langle ST_i x, T_i x \rangle \geq 0$ for every i, we have $\langle TSx, x \rangle \geq 0$; i.e., TS is positive.

2. $\langle TT^* x, x \rangle = \langle T^* x, T^* x \rangle = ||T^* x||^2 \geq 0 \Rightarrow TT^*$ is positive. Similarly, $\langle T * Tx, x \rangle = \langle Tx, Tx \rangle$ as $T^{**} = T$. Hence,

$$\langle T^* Tx, x \rangle = ||Tx||^2 \geq 0,$$

i.e., $T^* T$ is positive.

Proof (*Proof of Theorem* 3.27) To prove the closedness, we show that the limit of the sequence of normal operators is a normal operator. Let $\lim_{n \to \infty} T_n = T$. Then, $\lim_{n \to \infty} T_n^* = T^*$.

$$||TT^* - T^* T|| = ||TT^* - T_n T_n^* + T_n T_n^* - T_n^* T_n + T_n^* T_n - T^* T||$$

$$\leq ||TT^* - T_n T_n^*|| + ||T_n T_n^* - T_n^* T_n|| + ||T_n^* T_n - T^* T|| \to 0$$

$$as \ n \to \infty$$

which implies that $TT^* = T^* T$; i.e., T is normal.

By Remark 3.20(1), every self-adjoint operator is normal, and so the set of normal operators contains the set of self-adjoint operators. $(\alpha T)(\alpha T)^* = \alpha \overline{\alpha}(TT^*) = \alpha \overline{\alpha}(T^* T)$ as T is normal. Thus,

$$(\alpha T)(\alpha T)^* = (\alpha T)^*(\alpha T),$$

and hence αT is normal.

Proof (*Proof of Theorem* 3.28) We have $T_1 T_2^* = T_2^* T_1$ and $T_2 T_1^* = T_1^* T_2$. Using this fact and the fact that T_1 and T_2 are normal, we get

$$(T_1 + T_2)(T_1 + T_2)^* = (T_1 + T_2)(T_1^* + T_2^*)$$

$$= T_1 T_1^* + T_2 T_1^* + T_1 T_2^* + T_2 T_2^*$$

and

$$(T_1 + T_2)^\star (T_1 + T_2) = (T_1^\star + T_2^\star)(T_1 + T_2)$$
$$= T_1^\star T_1 + T_2^\star T_1 + T_1^\star T_2 + T_2^\star T_2$$
$$= T_1^\star T_1 + T_1 T_2^\star + T_2 T_1^\star + T_2^\star T_2$$

which shows that $T_1 + T_2$ is normal. Similarly, it can be seen that $T_1 T_2$ is normal under the given condition.

Proof (*Proof of Theorem* 3.29)

$$\|T^\star x\| = \|Tx\| \Leftrightarrow \|T^\star x\|^2$$
$$= \|Tx\|^2 \Leftrightarrow \langle T^\star x, T^\star x \rangle$$
$$= \langle Tx, Tx \rangle \Leftrightarrow \langle T T^\star x, x \rangle$$
$$= \langle T^\star T x, x \rangle \Leftrightarrow \langle (T T^\star - T^\star T)x, x \rangle$$
$$= 0 \ \forall \ x \in X.$$

In view of this relation, Lemma 3.5 gives us $T T^\star = T^\star T$. Thus, T is normal if and only if $\|T^\star x\| = \|Tx\|$.

Proof (*Proof of Theorem* 3.30) By Theorem 3.29, $\|T^2 x\| = \|TTx\| = \|T^\star T x\|$ for every x, which implies that $\|T\|^2 = \|T^\star T\|$. By Theorem 3.22(7), $\|T^\star T\| = \|T\|^2$ and so $\|T\| = \|T\|^2$

Proof (*Proof of Theorem* 3.31) Let $AB = BA$, and then, we have

$$T T^\star = (A + iB)(A - iB)$$
$$= A^2 + B^2 + i(BA - AB)$$

and

$$T^\star T = (A - iB)(A + iB)$$
$$= A^2 + B^2 + i(AB - BA)$$

Since $AB = BA$, $T T^\star = T^\star T$; i.e., T is normal. Conversely, let $T T^\star = T^\star T$; i.e.,

$$A^2 + B^2 + i(BA - AB) = A^2 + B^2 + i(AB - BA)$$

or

$$AB - BA = BA - AB$$

or

$$2AB = 2BA$$

Hence, $AB = BA$.

We need the following Lemma for the proof of Theorem 3.32.

Lemma 3.6 *If T is a bounded linear operator on a Hilbert space X, then the following conditions are equivalent:*

1. $T^\star T = I$.
2. $\langle Tx, Ty \rangle = \langle x, y \rangle \; \forall \; x$ and y.
3. $||Tx|| = ||x|| \; \forall \; x$.

Proof (1) $\Rightarrow$ (2): Let $T^\star T = I$. Then, $\langle T^\star Tx, y \rangle = \langle x, y \rangle$ or $\langle Tx, Ty \rangle = \langle x, y \rangle$.(2) $\Rightarrow$ (3): Let $\langle Tx, Ty \rangle = \langle x, y \rangle \; \forall \; x$ *and* y. Then, for $y = x$, we have $||Tx||^2 = ||x||^2$ or $||Tx|| = ||x||$. (3) $\Rightarrow$ (1): Let $||Tx|| = ||x|| \; \forall \; x$. Then,

$$||Tx||^2 = ||x||^2 \Rightarrow \langle Tx, Tx \rangle = \langle x, x \rangle$$
$$\Rightarrow \langle T^\star Tx, x \rangle = \langle x, x \rangle$$

or

$$\langle (T^\star T - I)x, x \rangle = 0 \; \forall \; x$$

Then, by Lemma 3.5, we get $T^\star T - I = 0$ or $T^\star T = I$.

Proof (*Proof of Theorem* 3.32) Let T be unitary. Then by Lemma 3.6, T is an isometric isomorphism of X onto itself. Conversely, if T is an isometric isomorphism of X onto itself, then T^{-1} exists, and by Lemma 3.6, we have

$$T^\star T = I$$
$$\Rightarrow (T^\star T)T^{-1} = IT^{-1} \; or \; T^\star (TT^{-1}) = IT^{-1}$$

or

$$T^\star = T^{-1}$$

or

$$TT^\star = TT^{-1} = I$$

which shows that T is unitary.

Definition 3.10 Let T be an operator on a Hilbert space X into itself, and then, a nonzero vector x such that $T(x) = \lambda x$, λ being a scalar is called an *eigenvector* or *characteristic vector* or *proper vector* of T. The corresponding λ is called an *eigenvalue* or characteristic value or proper value.

Theorem 3.33 *The proper values of a self-adjoint operator are real numbers. Two proper vectors corresponding to two different proper values of a self-adjoint operator are orthogonal.*

Theorem 3.34 *The proper values of a unitary operator are complex numbers such that $|\lambda| = 1$.*

Proof (*Proof of Theorem* 3.33) Let T be self-adjoint and $T(x) = \lambda x$. We want to show that λ is real, i.e., $\lambda = \overline{\lambda}$.

$$\langle Tx, x \rangle = \langle x, T^{\star}x \rangle = \langle x, Tx \rangle$$

or

$$\langle \lambda x, x \rangle = \langle x, \lambda x \rangle$$

or

$$\lambda \langle x, x \rangle = \overline{\lambda} \langle x, x \rangle$$

or

$$\lambda = \overline{\lambda}$$

Let $\lambda_1 \neq \lambda_2$, $Tx = \lambda_1 x$ and $Ty = \lambda_2 y$. Then, $\langle Tx, y \rangle = \langle \lambda_1 x, y \rangle = \lambda_1 \langle x, y \rangle$. As $T = T^{\star}$,

$$\langle Tx, y \rangle = \langle x, T * y \rangle = \langle x, Ty \rangle = \langle x, \lambda_2 y \rangle = \langle \lambda_2 x, y \rangle$$

Thus, $\lambda_1 \langle x, y \rangle = \lambda_2 \langle x, y \rangle$. Since $\lambda_1 \neq \lambda_2$, $\langle x, y \rangle$ must be zero. Hence, $x \perp y$.

Proof (*Proof of Theorem* 3.34) Let $Tx = \lambda x$, $x \neq 0$. Then,

$$\langle Tx, Tx \rangle = \langle \lambda x, \lambda x \rangle = \lambda \overline{\lambda} \langle x, x \rangle$$
$$= |\lambda|^2 ||x||^2.$$

On the other hand, $\langle Tx, Tx \rangle = \langle x, T^{\star}Tx \rangle = ||x||^2$ as T is unitary, i.e., $T^{\star}T = I$. Hence, $|\lambda|^2 ||x|| = ||x|| \Rightarrow |\lambda| = 1$.

3.8.3 Adjoint of an Unbounded Linear Operator

In Sect. 3.7.2, we have studied the concept of the adjoint of a bounded linear operator on a Hilbert space. This concept may be extended to unbounded linear operators on a Banach space, in general, and on a Hilbert space, in particular. From the point of view of applications, the concept of adjoint of unbounded linear operators on Hilbert spaces is more useful and, therefore, we discuss it here in brief.

Definition 3.11 Let T be an unbounded linear operator on a Hilbert space X, and assume that the domain of T, $\mathscr{D}(T)$, is dense in X. The adjoint operator of T, T^*,is defined by

$$\langle Tx, y \rangle = \langle x, T^*y \rangle \ \forall \ x \in \mathscr{D}(T), y \in D(T^*)$$

where $\mathscr{D}(T^*) = \{y \in X / \langle Tx, y \rangle = \langle x, z \rangle$ for some $z \in X$ and all $x \in \mathscr{D}(T)\}$. For each such $y \in \mathscr{D}(T^*)$, the adjoint operator T^* of T is defined in terms of that z by $z = T^*y$.

Remark 3.21 If T is also closed, then $\mathscr{D}(T^*)$ is dense in X and T^* is closed. Moreover, $T^{**} = T$; that is, $\mathscr{D}(T^{**}) = \mathscr{D}(T)$, and the operators agree on these domains. For closed operators, see Sect. 4.7.

Example 3.23 Let $X = L_2(a, b)$ and T be defined by

$$(Tf)(t) = \frac{df}{dt}$$

and $\mathscr{D}(T) = \{f \in X / f$ is absolutely continuous with $\frac{df}{dt} \in X$ and $f(a) = 0\}$. Then,

$$\langle Tf, g \rangle = \int_a^b \frac{df}{dt} g(t) dt$$

$$= [f(t)g(t)]_a^b - \int_a^b f(t) \frac{dg}{dt} dt$$

$$= [f(b)g(b)] - \int_a^b f(t) \frac{dg}{dt} dt = \langle f, T^*g \rangle$$

where $(T^*g) = -\frac{dg}{dt}$ and $\mathscr{D}(T^*) = \{g \in X / g$ is absolutely continuous with $\frac{dg}{dt} \in X, g(b) = 0\}$.

Definition 3.12 If the linear operator T on a Hilbert space X is 1-1 on $\mathscr{D}(T)$, then the linear operator T^{-1} on X defined on $\mathscr{D}(T^{-1}) = R(T)$ by

$$T^{-1}(T(x)) = x, \ \forall \, x \in \mathscr{D}(T)$$

is called the *inverse* of T.

Theorem 3.35 *Let T be a 1-1 linear operator on a Hilbert space X such that T^{-1} exists and $\overline{\mathscr{D}(T)} = \mathscr{D}(T^{-1}) = X$. Then, $T^\star$ is also 1-1, and*

$$(T^\star)^{-1} = (T^{-1})^\star$$

Proof The theorem will be proved if we show that $(T^{-1})^\star T^\star = I$ and $T^\star (T^{-1})^\star = I$. For verification of the first relation, take any $y \in D(T^\star)$. Then for every $x \in D(T^{-1})$, we have $T^{-1}x \in D(T)$ and so

$$\langle T^{-1}x, T^\star y \rangle = \langle TT^{-1}x, y \rangle = \langle Ix, y \rangle = \langle x, y \rangle.$$

This implies that $T^\star y \in \mathscr{D}((T^{-1})^\star)$ and $(T^{-1})^\star T^\star y = (TT^{-1})^\star y = I^\star y = y$. In order to verify the second relation, take an arbitrary $y \in \mathscr{D}((T^{-1})^\star)$. Then, for every $x \in \mathscr{D}(T)$, we have $Tx \in \mathscr{D}(T^{-1})$ and therefore

$$\langle Tx, (T^{-1})^\star y \rangle = \langle T^{-1}Tx, y \rangle = \langle x, y \rangle.$$

This shows that $(T^{-1})^\star y \in \mathscr{D}(T^\star)$ and $T^\star (T^{-1})^\star y = y$.

Definition 3.13 An unbounded linear operator T on a Hilbert space X is called *symmetric* if $T^\star$ extends T in the sense that $T^\star = T$ on $\mathscr{D}(T)$ and $D(T^\star) \supseteq \mathscr{D}(T)$. In this case, $\langle T^\star x, y \rangle = \langle x, Ty \rangle \ \forall \, x, y \in \mathscr{D}(T^\star)$. A symmetric operator is self-adjoint if $\mathscr{D}(T^\star) = \mathscr{D}(T)$.

It can be checked that every self-adjoint operator T is symmetric.

Example 3.24 Let $X = L_2(a, b)$, and T be defined by

$$(Tf)(t) = \frac{d^2 f}{dt^2}$$

and $\mathscr{D}(T) = \{f \in X / f \text{ and } \frac{df}{dt} \text{ are absolutely continuous with } \frac{d^2 f}{dt^2} \in X \text{ and } f(a) = 0 = f(b)\}$. Then,

$$\langle Tf, g \rangle = \int\limits_a^b \frac{d^2 f}{dt^2} g(t) dt$$

$$= \left[\frac{df}{dt} g(t) \right]_a^b - \left[f(t) \frac{g(t)}{dt} \right]_a^b + \int\limits_a^b f(t) \frac{d^2 g}{dt^2} dt$$

$$= \left(\frac{df}{dt}\right)_{t=b} g(b) - \left(\frac{df}{dt}\right)_{t=a} g(a) + \int_a^b f(t)\frac{d^2 g}{dt^2}dt$$

so that $(T^\star g)(t) = \dfrac{d^2 g}{dt^2}$ where $\mathscr{D}(T^\star) = \{g \in X / g$ and $\frac{dg}{dt}$ are absolutely continuous with $\frac{d^2 g}{dt^2} \in X$ and $g(a) = g(b) = 0\} = \mathscr{D}(T)$. Hence, T is self-adjoint.

Example 3.25 Let $X = \ell_2$ and T be defined by

$$T(\alpha k) = \left\{\frac{\alpha_k}{k}\right\}$$

T is self-adjoint and one-one. The subspace $T\ell_2 = \mathscr{D}(T^{-1})$ is everywhere dense. $\mathscr{D}(T^{-1})$ is the set of all sequences $\{\beta_k\} \in \ell_2$ such that

$$\sum_{k=1}^{\infty} k^2 |\beta^2| < \infty.$$

Then, the inverse T^{-1} is defined on $\mathscr{D}(T^{-1})$ by $T^{-1}(\beta_k) = \{k\beta_k\}$. T^{-1} is linear.

Let $\{e_k\}$ be the basis of ℓ_2, where e_k is the vector of all the components which are zero except the kth one which is 1. T^{-1} is unbounded as

$$\|T^{-1}(e_k)\| = k\|e_k\|$$

and

$$\lim_{k\to\infty} \|T^{-1}e_k\| = \infty$$

By Theorem 3.35, $(T^{-1})^\star = (T^\star)^{-1} = T^{-1}$. Therefore, T^{-1} is self-adjoint.

3.9 Bilinear Forms and Lax–Milgram Lemma

3.9.1 Basic Properties

Definition 3.14 Let X be a Hilbert space. A mapping a $(\cdot, \cdot)\colon X \times X \to C$ on $X \times X$ into C is called a *sesquilinear functional* if the following conditions are satisfied:

1. $a(x_1 + x_2, y) = a(x_1, y) + a(x_2, y)$.
2. $a(\alpha x, y) = \alpha a(x, y)$.
3. $a(x, y_1 + y_2) = a(x, y_1) + a(x, y_2)$.
4. $a(x, \beta y) = \bar{\beta} a(x, y)$

Remark 3.22   1. The sesquilinear functional is linear in the first variable but not so in the second variable. A sesquilinear functional which is also linear in the second variable is called a *bilinear form* or a *bilinear functional*. Thus, a bilinear form $a(\cdot, \cdot)$ is a mapping defined on $X \times X$ into C which satisfies conditions (a) through (c) of Definition 3.14 and (d). $a(x, \beta y) = \bar{\beta} a(x, y)$.

2. If X is a real Hilbert space, then the concepts of sesquilinear functional and bilinear forms coincide.

3. An inner product is an example of a sesquilinear functional. The real inner product is an example of a bilinear form.

4. If $a(\cdot, \cdot)$ is a sesquilinear function, then $g(x, y) = \overline{a(y, x)}$ is a sesquilinear functional.

Definition 3.15 Let $a(\cdot, \cdot)$ be a bilinear form. Then,

1. $a(\cdot, \cdot)$ is called *symmetric* if $a(x, y) = \overline{a(y, x)} \ \forall \ (x, y) \in X \times X$.

2. $a(\cdot, \cdot)$ is called *positive* if $a(x, x) \geq 0 \ \forall x \in X$.

3. $a(\cdot, \cdot)$ is called *positive definite* if $a(x, x) \geq 0 \ \forall \ x \in X$ and $a(x, x) = 0$ implies that $x = 0$.

4. $F(x) = a(x, x)$ is called *quadratic* form.

5. $a(\cdot, \cdot)$ is called *bounded* or *continuous* if there exists a constant $M > 0$ such that $|a(x, y)| \leq M \|x\| \ \|y\|$.

6. $a(\cdot, \cdot)$ is said to be *coercive* (X-*coercive*) or X-*elliptic* if there exists a constant $\alpha > 0$ such that $a(x, x) \geq \alpha \|x\|^2 \ \forall x \in X$.

7. A quadratic form F is called real if $F(x)$ is real for all $x \in X$.

Remark 3.23   1. If $a(\cdot, \cdot) \colon X \times X \to R$, then the bilinear form $a(\cdot, \cdot)$ is symmetric if $a(x, y) = a(y, x)$.

2.

$$\|a\| = \sup_{x \neq 0, y \neq 0} \frac{|a(x, y)|}{\|x\| \ \|y\|} = \sup_{x \neq 0, y \neq 0} |a(\frac{x}{\|x\|}, \frac{y}{\|y\|})|$$
$$= \sup_{\|x\|=\|y\|=1} |a(x, y)|$$

It is clear that $|a(x, y)| \leq \|a\| \ \|x\| \ \|y\|$.

3. $\|F\| = \sup_{\|x\|=1} |F(x)|$

4. If $a(\cdot, \cdot)$ is any fixed sesquilinear form and $F(x)$ is an associated quadratic form on a Hilbert space X, then

$$a(x, y) = \frac{1}{4}[F(x + y) - F(x - y) + i F(x + iy) - i F(x - iy)].$$

Verification: By using linearity of the bilinear form a, we have

$$F(x + y) = a(x + y, x + y) = a(x, x) + a(y, x) + a(x, y) + a(y, y)$$

and

$$F(x - y) = a(x - y, x - y) = a(x, x) - a(y, x) - a(x, y) + a(y, y).$$

By subtracting the second of the above equation from the first, we get

$$F(x + y) - F(x - y) = 2a(x, y) + 2a(y, x) \qquad (3.45)$$

Replacing y by iy in Eq. (3.45), we obtain

$$F(x + iy) - F(x - iy) = 2a(x, iy) + 2a(iy, x)$$

or

$$F(x + iy) - F(x - iy) = 2\bar{i}a(x, y) + 2ia(y, x) \qquad (3.46)$$

Multiplying Eq. (3.46) by i and adding it to Eq. (3.45), we get the result.

Theorem 3.36 *If a bilinear form $a(\cdot, \cdot)$ is bounded and symmetric, then $||a|| = ||F||$, where F is the associated quadratic functional.*

Theorem 3.37 *Let T be a bounded linear operator on a Hilbert space X. Then, the complex-valued function $a(\cdot, \cdot)$ on $X \times X$ defined by*

$$a(x, y) = \langle x, Ty \rangle \qquad (3.47)$$

is a bounded bilinear form on X, and $||a|| = ||T||$. Conversely, let $a(\cdot, \cdot)$ be a bounded bilinear form on a Hilbert space X. Then, there exists a unique bounded linear operator T on X such that

$$a(x, y) = \langle x, Ty \rangle \ \forall (x, y) \in X \times X \qquad (3.48)$$

Corollary 3.1 *Let T be a bounded linear operator on a Hilbert space X. Then, the complex-valued function $b(\cdot, \cdot)$ on $X \times X$ defined by $b(x, y) = \langle Tx, y \rangle$ is a bounded bilinear form on X and $||b|| = ||T||$. Conversely, let $b(\cdot, \cdot)$ be a bounded bilinear form on X. Then, there is a unique bounded linear operator T on X such that*

$$b(x, y) = \langle Tx, y \rangle \ \forall (x, y) \in X \times X.$$

Corollary 3.2 *If T is a bounded linear operator on X, then*

$$||T|| = \sup_{||x||=||y||=1} |\langle x, Ty \rangle| = \sup_{||x||=||y||=1} |\langle Tx, y \rangle|.$$

Theorem 3.38 *Let T be a bounded linear operator on a Hilbert space X. Then, the following statements are equivalent:*

1. T is self-adjoint.
2. The bilinear form $a(\cdot, \cdot)$ on X defined by $a(x, y) = \langle Tx, y \rangle$ is symmetric.
3. The quadratic form $F(x)$ on X defined by $F(x) = \langle Tx, x \rangle$ is real.

Corollary 3.3 *If T is a bounded self-adjoint operator on X, then* $\|T\| = \sup\limits_{\|x\|=1} |\langle Tx, x \rangle|$.

The following lemmas are needed for the proof of Theorem 3.36.

Lemma 3.7 *A bilinear form $a(x, y)$ is symmetric if and only if the associated quadratic functional $F(x)$ is real.*

Proof If $a(x, y)$ is symmetric, then we have

$$
\begin{aligned}
F(x) &= a(x, x) \\
&= \overline{a(x, x)} \\
&= \overline{F(x)}
\end{aligned}
$$

This implies that $F(x)$ is real. Conversely, let $F(x)$ be real, and then by Remark 3.23(4) and in view of the relation

$$F(x) = F(-x) = F(ix)(F(x) = a(x, x), F(-x) = a(x, x) = a(-x, -x)$$

and

$$F(ix) = a(ix, ix) = i\bar{i}a(x, x) = a(x, x))$$

we obtain

$$
\begin{aligned}
a(x, y) &= \frac{1}{4}[F(x + y) - F(y - x) + iF(y + ix) - iF(y - ix)] \\
&= \frac{1}{4}[F(x + y) - F(x - y) + iF(x - iy) - iF(x + iy)] \\
&= \frac{1}{4}[F(x + y) - F(x - y) + iF(x + iy) - iF(x - iy)] \\
&= a(x, y)
\end{aligned}
$$

Hence, $a(\cdot, \cdot)$ is symmetric.

Lemma 3.8 *$a(\cdot, \cdot)$ is bounded if and only if the associated quadratic form F is bounded. If $a(\cdot, \cdot)$ is bounded, then $\|F\| \geq \|a\| \geq 2\|F\|$.*

Proof Suppose $a(\cdot, \cdot)$ is bounded. Then, we have

$$\sup_{\|x\|=1} |F(x)| = \sup_{\|x\|=1} |a(x, x)| \leq \sup_{\|x\|=\|y\|=1} |a(x, y)| = \|a\|$$

and, therefore, F is bounded and $||F|| \leq ||a||$. On the other hand, suppose F is bounded. From Remark 3.23(4) and the parallelogram law, we get

$$|a(x, y)| \leq \frac{1}{4}||F||(||x + y||^2 + ||x - y||^2 + ||x + iy||^2 + ||x - iy||^2)$$

$$= \frac{1}{4}||F||2(||x||^2 + ||y||^2 + ||x||^2 + ||y||^2)$$

$$= ||F||(||x||^2 + ||y||^2)$$

or

$$\sup_{||x||=||y||=1} |a(x, y)| \leq 2||F||$$

Thus, $a(\cdot, \cdot)$ is bounded and $||a|| \leq 2||F||$.

Proof (*Proof of Theorem* 3.36) By Lemma 3.7, F is real. In view of Lemma 3.8, we need to show that $||a|| \leq ||F||$. Let $a(x, y) = \gamma e^{i\alpha}$, where $\gamma \in R$, $\gamma \geq 0$, and $\alpha \in R$. Then by using Remark 3.23(4) and bearing in mind that the purely imaginary terms are zero, we get

$$|a(x, y)| = \gamma = a(e^{-i\alpha}\gamma, y) = \frac{1}{4}[F(x' + y) + F(x' - y)]$$

where $x' = \gamma e^{i\alpha}$. This implies that

$$|a(x, y)| = \frac{1}{4}||F||(||x' + y||^2 + ||x' - y||^2)$$

$$= \frac{1}{2}||F||(||x'||^2 + ||y||^2) \ (by\ the\ parallelogram\ law)$$

or

$$\sup_{||x'||=||y||=1} |a(x', y)| \leq ||F||$$

or

$$||a|| \leq ||F||$$

Proof (*Proof of Theorem* 3.37)

1. Let T be a bounded linear operator on X. Then, $a(x, y) = \langle x, Ty \rangle$ satisfies the following condition:

   (a) $a(x + x', y) = \langle x + x', Ty \rangle = \langle x, Ty \rangle + \langle x', Ty \rangle = a(x, y) + a(x', y)$
   (b) $a(\alpha x, y) = \langle \alpha x, Ty \rangle = \alpha \langle x, Ty \rangle = \alpha a(x, y)$
   (c) $|a(x, y)| = |\langle x, Ty \rangle| \leq ||T|| \ ||x|| \ ||y||$, by the Cauchy–Schwarz–Bunyakowski inequality.

This implies that $\displaystyle\sup_{\|x\|=\|y\|=1} |a(x, y)| \leq \|T\|$ or $\|a\| \leq \|T\|$. In fact, $\|a\| = \|T\|$ (see Eq. (3.51)).

2. For the converse, let $a(\cdot, \cdot)$ be a bounded bilinear form on X. For any $y \in X$, we define f_y on X as follows:

$$f_y(x) = a(x, y) \tag{3.49}$$

We have

$$f_y(x_1 + x_2) = a(x_1 + x_2, y) = a(x_1, y) + a(x_2, y) \, f_y(\alpha x) = a(\alpha x, y) = \alpha a(x, y)$$

and

$$|f_y(x)| = |a(x, y)| \leq \|a\| \, \|x\| \, \|y\|$$

or

$$\|f_y\| \leq \|a\| \, \|y\|$$

Thus, f_y is a bounded linear functional on X. By the Riesz representation theorem, there exists a unique vector $Ty \in X$ such that

$$f_y(x) = \langle x, Ty \rangle \forall x \in X \tag{3.50}$$

and

$$\|Ty\| = \|f_y\| \leq \|a\| \, \|y\| \tag{3.51}$$

The operator $T: y \to Ty$ defined by $T(y) = \langle x, Ty \rangle$ is linear in view of the following relations:

$$\langle x, T(\alpha y) \rangle = f_{\alpha y}(x) = \langle x, \alpha y \rangle = \overline{\alpha} \langle x, y \rangle$$
$$= \overline{\alpha} f_y(x) = \overline{\alpha} \langle x, Ty \rangle$$
$$\langle x, T(\alpha y) \rangle = \langle x, \alpha Ty \rangle$$

or

$$\langle x, T(\alpha y) - \alpha(T(y)) \rangle = 0 \, \forall \, x \in X$$
$$\Rightarrow T(\alpha y) = \alpha T(y)$$
$$\langle x, T(y + y') \rangle = f_{y+y'}(x) = \langle x, y + y' \rangle$$

$$= \langle x, y \rangle + \langle x, y' \rangle = f_y(x) + f'_y(x)$$
$$= \langle x, Ty \rangle + \langle x, Ty' \rangle = \langle x, T(y) + T(y') \rangle \; \forall \, x \in X$$

or

$$\langle x, [T(y + y')] - [T(y) + T(y')] \rangle = 0 \; \forall \, x \in X$$

which gives

$$T(y + y') = T(y) + T(y')$$

Equation (3.51) implies that $||T|| \le ||a||$. By Eqs. (3.49) and (3.50), we have

$$a(x, y) = f_y(x) = \langle x, Ty \rangle \; \forall \, (x, y) \in X \times X.$$

Then, for every fixed $y \in X$, we get $\langle x, Ty \rangle = \langle x, Sy \rangle$ or $\langle x, (T - S)y \rangle = 0$. This implies that $(T - S)y = 0 \; \forall \, y \in X$, i.e., $T = S$. This proves that there exists a unique bounded linear operator T such that $a(x, y) = \langle x, Ty \rangle$.

Proof (*Proof of Corollary* 3.1)

1. Define the function $a(x, y)$ on $X \times X$ by

$$a(x, y) = \overline{b(y, x)} = \langle x, Ty \rangle$$

By Theorem 3.37, $a(x, y)$ is a bounded bilinear form on X and $||a|| = ||T||$. Since we have $b(x, y) = a(y, x)$, b is also bounded bilinear on X and

$$||b|| = \sup_{||x||=||y||=1} |b(x, y)| \quad \sup_{||x||=||y||=1} |a(y, x)| = ||a|| = ||T||$$

2. If b is given, we define a bounded bilinear form a on X by

$$a(x, y) = \overline{b(y, x)}$$

Again, by Theorem 3.37, there is a bounded linear operator T on X such that

$$a(x, y) = \langle x, Ty \rangle \; \forall \, (x, y) \in X \times X$$

Therefore, we have $b(x, y) = \overline{a(y, x)} = \overline{\langle y, Tx \rangle} = \langle Tx, y \rangle \forall (x, y) \in X \times X$.

Proof (*Proof of Corollary* 3.2) By Theorem 3.37, for every bounded linear operator on X, there is a bounded bilinear form a such that $a(x, y) = \langle x, Ty \rangle$ and $||a|| = ||T||$. Then,

$$||a|| = \sup_{||x||=||y||=1} |a(x, y)| = \sup_{||x||=||y||=1} \langle x, Ty \rangle$$

From this, we conclude that $||T|| = \sup\limits_{||x||=||y||=1} |\langle x, Ty\rangle|$

Proof (*Proof of Theorem* 3.38) (1) $\Rightarrow$ (2): $F(x) = \langle Tx, x\rangle = \langle x, Tx\rangle = \langle Tx, x\rangle = F(x)$. In view of Lemma 3.7, we obtain the result.
(3) $\Rightarrow$ (2): By Lemma 3.7, $F(x) = \langle Tx, x\rangle$ is real if and only if the bilinear form $a(x, y) = \langle Tx, y\rangle$ is symmetric.
(2) $\Rightarrow$ (1): $\langle Tx, y\rangle = a(x, y) = a(y, x) = \langle Ty, x\rangle = \langle x, Ty\rangle$. This shows that $T^* = T$ that T is self-adjoint.

Proof (*Proof of Corollary* 3.3) Since T is self-adjoint, we can define a bounded bilinear form $a(x, y) = \langle Tx, y\rangle = \langle x, Ty\rangle$. By Corollary 3.2 and Theorem 3.36,

$$||T|| = ||a|| = ||F||$$
$$= \sup\limits_{||x||=1} |F(x)|$$
$$= \sup\limits_{||x||=1} |\langle Tx, x\rangle|$$
$$||T|| = \sup\limits_{||x||=1} |\langle Tx, x\rangle|$$

The following theorem, known as the Lax–Milgram lemma proved by PD Lax and AN Milgram in 1954 [119], has important applications in different fields.

Theorem 3.39 (Lax–Milgram Lemma) *Let X be a Hilbert space and* $a(\cdot, \cdot): X \times X \to R$ *a bounded bilinear form which is coercive or X-elliptic in the sense that there exists* $\alpha > 0$ *such that*

$$a(x, x) \leq \alpha ||x||_X^2 \ \forall \ x \in X.$$

Also, let $f: X \to R$ *be a bounded linear functional. Then, there exists a unique element* $x \in X$ *such that*

$$a(x, y) = f(y) \forall y \in X \tag{3.52}$$

Proof (*Proof of Theorem* 3.39) Since $a(\cdot, \cdot)$ is bounded, there exists a constant $M > 0$ such that

$$|a(u, v)| \leq M||u|| \ ||v||. \tag{3.53}$$

By Corollary 3.1 and Theorem 3.19, there exists a bounded linear operator T on X and $f_y \in X^*$ such that equation $a(u, v) = f(v)$ can be rewritten as

$$\langle \lambda Tu, v\rangle = \langle \lambda f_y, v\rangle \tag{3.54}$$

or

$$\langle \lambda T u - \lambda f_y, v \rangle = 0 \ \forall \, v \in X.$$

This implies that

$$\lambda T u = \lambda f_y \tag{3.55}$$

We will show that (3.55) has a unique solution by showing that for appropriate values of parameter $\rho > 0$, the affine mapping for $v \in X$, $v \to v - \rho(\lambda T v - \lambda f_y) \in X$ is a contraction mapping. For this, we observe that

$$\begin{aligned}
||v - \rho \lambda T v||^2 &= \langle v - \rho \lambda T v, v - \rho \lambda T v \rangle \\
&= ||v||^2 - 2\rho \langle \lambda T v, v \rangle + \rho^2 ||\lambda T v||^2 \ (by \ applying \ inner \ product \ axioms) \\
&\leq ||v||^2 - 2\rho \alpha ||v||^2 + \rho^2 M^2 ||v||^2
\end{aligned}$$

as

$$a(v, v) = \langle \lambda T v, v \rangle \geq \alpha ||v||^2 \ (by \ the \ coercivity) \tag{3.56}$$

and

$$||\lambda T v|| \leq M ||v|| \ (by \ boundedness \ of \ T)$$

Therefore,

$$||v - \rho \lambda T v||^2 (1 - 2\rho\alpha + \rho^2 M^2) ||v||^2 \tag{3.57}$$

or

$$||v - \rho \lambda T v|| \leq (1 - 2\rho\alpha + \rho^2 M^2)^{1/2} ||v|| \tag{3.58}$$

Let $Sv = v - \rho(\lambda T v - \lambda f_y)$. Then,

$$\begin{aligned}
||Sv - Sw|| &= ||(v - \rho(\lambda T v - \lambda f y)) - (w - \rho(\lambda T w - \rho f y))|| \\
&= ||(v - w) - \rho(\lambda T(v - w))|| \\
&\leq (1 - 2\rho\alpha + \rho^2 M^2)^{1/2} ||v - w|| \ (by \ 3.58) \tag{3.59}
\end{aligned}$$

This implies that S is a contraction mapping if $0 < 1 - 2\rho\alpha + \rho^2 M^2 < 1$ which is equivalent to the condition that $\rho \in (0, 2\alpha/M^2)$. Hence, by the Banach contraction fixed point theorem (Theorem 1.1), S has a unique fixed point which is the unique solution.

The following problem is known as the abstract variational problem.

Problem 3.1 Find an element x such that

$$a(x, y) = f(y) \; \forall \, y \in X$$

where $a(x, y)$ and f are as in Theorem 3.39.

In view of the Lax–Milgram lemma, the abstract variational problem has a unique solution.

Solution 3.1 Note: A detailed and comprehensive description of the Hilbert space theory and its applications is given by Helmberg [97], Schechter [166], and Weidmann [194].

See also [10, 11, 19, 54, 86, 93, 97, 100, 101, 118, 119].

3.10   Problems

3.10.1   Solved Problems

Problem 3.2 If $X \neq \{0\}$ is a Hilbert space, show that

$$\|x\| = \sup_{\|y\|=1} |\langle x, y \rangle|.$$

Solution 3.2 If $x = 0$, the result is clearly true as both sides will be 0. Let $x \neq 0$. Then

$$
\begin{aligned}
\|x\| &= \frac{\|x\|^2}{\|x\|} = \frac{\langle x, x \rangle}{\|x\|} = \left\langle x, \frac{x}{\|x\|} \right\rangle \\
&\leq \sup_{\|y\|=1} \langle x, y \rangle \\
&\leq \sup_{\|y\|=1} \|x\| \, \|y\|, \; (by \; the \; Cauchy-Schwarz-Bunyakowski \; inequality) \\
&= \|x\|
\end{aligned}
$$

This implies that $\|x\| = \sup\limits_{\|y\|=1} |\langle x, y \rangle|$.

Problem 3.3 Let $x_n \to x$ and $y_n \to y$ in the Hilbert space X and $\alpha_n \to \alpha$, where α_n's and α are scalars. Then, show that

(a) $x_n + y_n \to x + y$.
(b) $\alpha_n x_n \to \alpha x$.
(c) $\lim\limits_{n \to \infty} \langle x_n, y_n \rangle = \langle x, y \rangle$.

Solution 3.3 The last relation shows that the inner product is a continuous function.

1.

$$||(x_n + y_n) - (x + y)||^2 = \langle (x_n - x) + (y_n - y), (x_n - x) + (y_n - y) \rangle$$
$$= \langle (x_n - x), (x_n - x) \rangle + \langle (y_n - y), (y_n - y) \rangle$$
$$+ \langle (x_n - x), (y_n - y) \rangle + \langle (y_n - y), (x_n - x) \rangle$$

Since $x_n \to x$ and $y_n \to y$, $||x_n - x||^2 = \langle x_n - x, x_n - x \rangle \to 0$ as $n \to \infty$ and $||y_n - y||^2 = \langle y_n - y, y_n - y \rangle \to 0$ as $n \to \infty$. By the Cauchy–Schwarz–Bunyakowski inequality, we have

$$|\langle (x_n - x), (y_n - y) \rangle| \le ||x_n - x||, ||y_n - y|| \to 0 \ as \ n \to \infty$$

and

$$|\langle (y_n - y), (x_n - x) \rangle| \le ||y_n - y|| \, ||x_n - y|| \to 0 \ as \ n \to \infty$$

In view of these relations, we have $||(x_n + y_n) - (x + y)|| \to 0$ as $n \to \infty$; i.e., $x_n + y_n \to x + y$.

2. In view of these relations, we have $||(x_n + \bar{y}_n) - (x + y)|| \to 0$ as $n \to \infty$; i.e., $x_n + y_n \to x + y$.

$$||(\alpha_n x_n - \alpha x)||^2 = \langle \alpha_n x_n - \alpha x, \alpha_n x_n - \alpha x \rangle$$
$$= \langle \alpha_n x_n - \alpha x_n + \alpha x_n - \alpha x, \alpha_n x_n - \alpha x_n + \alpha x_n - \alpha x \rangle$$
$$= \langle \alpha_n x_n - \alpha x_n, \alpha n x n \alpha x_n \rangle + \langle \alpha_n x_n - \alpha x_n, \alpha x_n - \alpha x \rangle$$
$$+ \langle \alpha x_n - \alpha x, \alpha x_n - \alpha x + \alpha x_n \rangle - \langle \alpha x, \alpha_n x_n \alpha x_n \rangle$$
$$\le |\alpha_n - \alpha|^2 ||x_n||^2 + |\alpha_n - \alpha| \, ||x_n|| \, ||x_n - x|| |\alpha|$$
$$+ |\alpha|^2 ||x_n - x||^2 + |\alpha| ||x_n - x|| \, ||x_n|| |\alpha_n - \alpha| \to 0$$
$$as \ n \to \infty \ under \ the \ given \ conditions$$

Hence, $\alpha_n x_n \to \alpha x$ as $n \to \infty$

3.

$$|\langle x_n, y_n \rangle - \langle x, y \rangle| = |\langle x_n, y_n \rangle - \langle x, y_n \rangle + \langle x, y_n \rangle - \langle x, y \rangle|$$
$$= |\langle x_n - x, y_n \rangle + \langle x, y_n - y \rangle|$$
$$\le |\langle x_n - x, y_n \rangle| + |\langle x, y_n - y \rangle|$$
$$\le ||x_n - x|| \, ||y_n|| + ||x|| \, ||y_n - y||$$

Since $x_n \to x$ and $y_n \to y$, there exists $M > 0$ such that $||x_n|| \le M$, $||x_n - x|| \to 0$, and $||y_n - y|| \to 0$ as $n \to \infty$. Therefore, $|\langle x_n, y_n \rangle - \langle x, y \rangle| \to 0$ as $n \to \infty$; or

$$\langle x_n, y_n \rangle \to \langle x, y \rangle \; as \; n \to \infty$$

In view of Theorem A.6(2) of Appendix A.3, the inner product $\langle \cdot, \cdot rangle$ is continuous.

Problem 3.4 Show that R^n, ℓ_2, and $L_2(a, b)$ are Hilbert spaces, while $C[a, b]$ is not a Hilbert space.

Solution 3.4 R^n is a Hilbert space. For $x, y \in R^n$, we define the inner product by $\langle x, y \rangle = \sum_{i=1}^{n} x_i y_i$

1. (a) $\langle x + x', y \rangle = \sum_{i=1}^{n}(x_i + x_i'), y_i = \sum_{i=1}^{n} x_i y_i + \sum_{i=1}^{n} x_i' y_i = \langle x, x \rangle + \langle x', y \rangle$.

   (b) $\langle \alpha x, y \rangle = \sum_{i=1}^{n}(\alpha x_i) y_i = \alpha \sum_{i=1}^{n} x_i y_i = \alpha \langle x, y \rangle$

   (c) $\overline{\langle x, y \rangle} = \overline{\sum_{i=1}^{n} x_i y_i} = \sum_{i=1}^{n} \overline{x_i y_i} = \sum_{i=1}^{n} x_i y_i = \langle x, y \rangle$

   (d) $\langle x, x \rangle = \sum_{i=1}^{n} x_i x_i = \sum_{i=1}^{n} x_i^2$

   $$\Rightarrow \langle x, x \rangle \geq 0 \; and \; \langle x, x \rangle = 0 \Leftrightarrow \sum_{i=1}^{n} x_i^2$$

   $$\Leftrightarrow x_i = 0$$
   $$\Leftrightarrow x = (0, 0, \ldots)$$
   $$\Leftrightarrow x = 0$$

   Thus, $\langle \cdot, \cdot \rangle$ is an inner product on R^n and it induces the norm given by

   $$||x|| = \sqrt{\langle x, x \rangle} = \left(\sum_{i=1}^{n} x_i^2 \right)^{1/2}$$

   R^n is complete with respect to any norm and so R^n is a Hilbert space.

2. ℓ_2 is a Hilbert space: For $x = (x_1, x_2, \ldots, x_n, \ldots)$, $y = (y_1, y_2, \ldots, y_n, \ldots) \in \ell_2$, we define the inner product $\langle \cdot, \cdot \rangle$ on ℓ_2 by $\langle x, y \rangle = \sum_{i=1}^{n} x_i \overline{y_i}$

   (a) $\langle x + x', y \rangle = \sum_{i=1}^{n}(x_i + x_i')(\overline{y_i})$ where

   $$x' = (x_1', x_2', \ldots, x_n', \ldots)$$
   $$\langle x + x', y \rangle = \sum_{i=1}^{\infty} x_i \overline{y_i} + \sum_{i=1}^{\infty} x_i' \overline{y_i}$$
   $$= \langle x, y \rangle + \langle x', y \rangle$$

   (b) $\langle \alpha x, y \rangle = \sum_{i=1}^{\infty}(\alpha x_i)\overline{y_i} = \alpha \sum_{i=1}^{\infty} x_i \overline{y_i} = \alpha \langle x, y \rangle$

(c) $\overline{\langle x, y \rangle} = \sum_{i=1}^{\infty} \overline{x_i \overline{y_i}} = \sum_{i=1}^{\infty} \overline{x_i} \, \overline{\overline{y_i}} = \sum_{i=1}^{\infty} y_i \overline{x_i} = \langle y, x \rangle$

(d) $\langle x, x \rangle = \sum_{i=1}^{\infty} x_i \overline{x_i} = \sum_{i=1}^{\infty} |x_i|^2$

$\langle x, x \rangle \geq 0$ as all the terms of the series $\sum_{i=1}^{\infty} |x_i|^2$ are positive.

$\langle x, x \rangle = 0 \Leftrightarrow \sum_{i=1}^{\infty} |x_i|^2 = 0 \Leftrightarrow |x_i|^2 = 0 \Leftrightarrow x_i = 0 \, \forall \, i$

$\Leftrightarrow x = (0, 0, \ldots, 0, \ldots) = 0 \in \ell_2$

In view of the above relations, ℓ_2 is an inner product space with the inner product $\langle x, y \rangle = \sum_{i=1}^{\infty} x_i \overline{y_i}$. Let $x_n = \{a_k^{(n)}\}_{n=1}^{\infty}$, $k = 1, 2, 3, \ldots$ and $\{x_n\}$ be a Cauchy sequence in ℓ_2. From the relation $|a_k^{(m)} - a_k^{(n)}|^2 \leq \sum_{k=1}^{\infty} = ||x_m - x_n||^2$, we obtain that for every fixed $k \geq 1$ the sequence $\{a_k^{(n)}\}_{n=1}^{\infty}$ is a Cauchy sequence of complex numbers and, therefore, it converges to a complex number, say a_k. (We know that C is complete.) Let

$$x = \{a_k^{(n)}\}_{n=1}^{\infty}$$

For every $\varepsilon > 0$, we have $||x_m - x_n||^2 = \sum_{i=1}^{\infty} |a_k^{(m)} - a_k^{(n)}|^2 < \varepsilon^2$ for all $m \geq n(\varepsilon)$ and $n \geq n(\varepsilon)$, and therefore, for every fixed index $p \geq 1$, $\sum_{i=1}^{p} |a_k^{(m)} - a_k^{(n)}|^2 < \varepsilon^2 \, \forall \, m \leq n(\varepsilon)$ and $n \geq n(\varepsilon)$. Letting m to ∞, we obtain $\sum_{i=1}^{p} |a_k - a_k^{(n)}|^2 < \varepsilon^2 \, \forall \, n \leq n(\varepsilon)$ and $p \geq 1$. Now, we let p to ∞ and we get

$$\sum_{i=1}^{p} |a_k - a_k^{(n)}|^2 < \varepsilon^2 \, \forall \, n \leq n(\varepsilon) \tag{3.60}$$

The sequence $x - x_n = \{a_k - a_k^{(n)}\}_{k=1}^{\infty}$, therefore, belongs to ℓ_2, and since ℓ_2 is a vector space, the sequence $x = (x - x_n) + x_n \in \ell_2$. From Eq. (3.59), we conclude that $||x - x_n|| \leq \varepsilon \, \forall n \geq n(\varepsilon)$; i.e., $\{x_n\}$ is convergent in ℓ_2. Thus, ℓ_2 is a Hilbert space.

3. $L_2(a, b)$ is a Hilbert space: $L_2(a, b)$ is a vector space (see Appendix A.5). We define the inner product by $\langle f, g \rangle = \int_a^b f(t)\overline{g(t)}dt \, \forall \, f, g \in L_2(a, b)$.

(a) $\langle f + h, g \rangle = \int_a^b f + h(t)\overline{g(t)}dt = \int_a^b f(t)\overline{g(t)}dt + \int_a^b h(t)g(t)dt = \langle f, g \rangle + \langle h, g \rangle$.

(b) $\langle \alpha f, g \rangle = \int\limits_a^b \alpha f(t)\overline{g(t)}dt = \alpha \int\limits_a^b f(t)\overline{g(t)}dt = \alpha\langle f, g \rangle.$

(c) $\overline{\langle f, g \rangle} = \int\limits_a^b \overline{f(t)\overline{g(t)}}dt = \int\limits_a^b g(t)\overline{f(t)}dt = \langle g, f \rangle.$

(d) $\langle f, f \rangle = \int\limits_a^b f(t)\overline{f(t)} = \int\limits_a^b |f(t)|^2 dt \Rightarrow \langle f, f \rangle \geq 0 \, as \, |f(t)|^2 \geq 0 \, implying$

$\int\limits_a^b |f(t)|^2 dt \geq 0. \langle f, f \rangle = 0 \Leftrightarrow |f(t)| = 0 \, a.e. \Leftrightarrow f = 0 \, a.e.$

Thus, $\langle \cdot, \cdot \rangle$ is an inner product on $L_2(a, b)$ which is complete with respect to the norm

$$\|f\| = \left(\int\limits_a^b |f(t)|^2 dt \right)^{1/2}$$

induced by the above inner product. For $f \in L_2(a, b)$, we have

$$\int\limits_a^b |f(t)| dt = \langle |f|, 1 \rangle \leq \|f\| \, \|1\| \, (by \, the \, CSB \, inequality)$$

$$= (\sqrt{b - a})\|f\| \tag{3.61}$$

Suppose $\{f_n\}$ is a Cauchy sequence in $L_2(a, b)$. By induction, we construct an increasing sequence of natural numbers $n_1 < n_2 < \cdots n_k < \cdots$ such that

$$\|f_m - f_n\| < \frac{1}{2^k} \, \forall \, m \geq n_k \, and \, n \geq n_k.$$

Consider the sequence $\{f_{n_{k+1}} - f_{n_k}\}_{k=1}^\infty \in L_2(a, b)$. By Theorem D.7(3), we have

$$\int\limits_a^b \sum_{k=1}^\infty |f_{n_{k+1}}(t) - f_{n_k}(t)| dt = \sum_{k=1}^\infty \int\limits_a^b |f_{n_{k+1}} - f_{n_k}| dt$$

$$\leq (\sqrt{b - a}) \sum_{k=1}^\infty \|f_{n_{k+1}}(t) - f_{n_k}(t)\|$$

$$\leq (\sqrt{b - a}) \sum_{k=1}^\infty \frac{1}{2^k} = (\sqrt{b - a}) < \infty$$

By Beppo Levi's theorem (see Appendix A.4), the series $\sum\limits_{k=1}^\infty |f_{n_{k+1}}(t) - f_{n_k}(t)|$ converges for almost every $t \in [a, b]$ and so does the series

$$f_{n_\ell}(t) + \sum_{k=1}^{\infty} [f_{n_{k+1}}(t) - f_{n_k}(t)] = \lim_{\to \infty} f_{n_k}(t)$$

As a consequence, the function f defined a.e. on $[a, b]$ by $f(t) = \lim_{k \to \infty} f_{n_k}(t)$ is finite a.e. and Lebesgue measurable on $[a, b]$. Furthermore, by Theorem A.16(3), we have

$$\int_a^b |f(t) - f_{n_p}(t)|^2 dt \leq \int_a^b \left[\sum_{k=1}^{\infty} f_{n_{k+1}}(t) - f_{n_k}(t) \right]^2 dt$$

$$= \lim_{m \to \infty} \int_a^b \left[\sum_{k=1}^{m} f_{n_{k+1}}(t) - f_{n_k}(t) \right]^2 dt$$

$$= \lim_{m \to \infty} \left\| \left[\sum_{k=1}^{m} f_{n_{k+1}} - f_{n_k} \right] \right\|^2 dt$$

$$\leq \lim_{m \to \infty} \left(\sum_{k=1}^{m} \| f_{n_{k+1}} - f_{n_k} \| \right)^2 dt$$

$$\leq \lim_{m \to \infty} \left(\sum_{k=1}^{\infty} \frac{1}{2^k} \right)^2 = \left(\frac{1}{2^{p-1}} \right)^2$$

This implies that $(f - f_{n_\ell}) \in L_2(a, b)$ and $\| f - f_{n_p} \| \leq \frac{1}{2^{p-1}}$. Therefore, we have $f = (f - f_{n_p}) + f_{n_p} \in L_2(a, b)$. For $n \geq n_p$, we obtain

$$\| f - f_n \| \leq \| f - f_{n_p} \| + \| f_n - f_{n_p} \| \leq \frac{1}{2^{p-1}} + \frac{1}{2^p} + \frac{1}{2^p} < \frac{1}{2^{p-2}}$$

This shows that $\{f_n\}$ converges to $f \in L_2(a, b)$ and hence $L_2(a, b)$ is a Hilbert space.

4. $C[a, b]$ is an inner product space but not a Hilbert space: For $f, g \in C[a, b]$, we define the inner product by $\langle f, g \rangle = \int_a^b f(t) \overline{g(t)} dt$.

It can be seen, as in the case of $L_2(a, b)$, that $\langle \cdot, \cdot \rangle$ satisfies the axioms of the inner product. We now show that it is not complete with respect to the norm induced by this inner product; i.e., there exists a Cauchy sequence in $C[a, b]$, which is not convergent in $C[a, b]$. Let $a = -1, b = 1$ and

$$f_n(t) = \begin{cases} 0 & for \ -1 \leq t \leq 0 \\ nt & for \ 0 \leq t \leq 1/n \\ 1 & for \ 1/n \leq t \leq 1 \end{cases}$$

be a sequence of real-valued continuous function on $[-1, 1]$.

We prove the desired result by showing that this is a Cauchy sequence with respect to the norm induced by the inner product

$$||f|| = \left(\int_{-1}^{1} |f(t)|^2 dt \right)^{1/2}$$

but it is not convergent to a function $f \in C[-1, 1]$.

We have (take $m > n$)

$$f_m(t) - f_n(t) = 0 \; for \; -1 \le t \le 0$$
$$0 \le f_m(t) - f_n(t) \le 1 - f_n(t) \; for \; 0 \le t \le 1$$

and so

$$||f_m - f_n||^2 = \left(\int_{-1}^{1} |f_m(t) - f_n(t)|^2 dt \right)$$

$$= \int_{-1}^{0} |f_m(t) - f_n(t)|^2 dt + \int_{0}^{1} |f_m(t) - f_n(t)|^2 dt$$

$$\le \int_{0}^{1} |1 - f_n(t)|^2 dt$$

$$= \int_{0}^{1/n} |1 - f_n(t)|^2 dt + \int_{1/n}^{1} |1 - f_n(t)|^2 dt$$

$$= \int_{0}^{1/n} |1 - f_n(t)|^2 dt$$

as $1 - f_n(t) = 0$ for $\frac{1}{n} < t \le 1$ or

$$||f_m - f_n||^2 \le \int_{0}^{1/n} |1 - nt|^2 dt \le \int_{0}^{1/n} dt = \frac{1}{n}$$

or

$$||f_m - f_n|| \le \frac{1}{\sqrt{n}} \; for \; m > n$$

This shows that $\forall\, \varepsilon > 0$, $m > n \geq N(\varepsilon)[N(\varepsilon) = \frac{1}{\varepsilon}^2 + 1]$, where $[\frac{1}{\varepsilon}^2]$ denotes the integral part of $\frac{1}{\varepsilon}^2$, $\|f_m - f_n\| \leq \varepsilon$. Hence, fn is a Cauchy sequence in $C[-1, 1]$. Suppose $f_n \to f \in C[-1, 1]$; i.e., f is a real-valued continuous function on $[-1, 1]$ and $\varepsilon > 0$; $\exists N(\varepsilon)$ such that $\|f_n - f\| \leq \varepsilon$ for $n \geq N(\varepsilon)$. Let $-1 < \alpha < 0$. Since $(f_n - f)^2$ is positive

$$\int_{-1}^{\alpha} (f_n(t) - f(t))^2 dt \leq \int_{-1}^{1} (f_n(t) - f(t))^2 dt$$

$$= \|f_n - f\|^2 \leq \varepsilon^2 \ for \ n \geq N(\varepsilon)$$

or

$$\int_{-1}^{\alpha} (f^2)(t) dt \leq \varepsilon^2 \ as \ f_n(t) = 0 \ for \ 1 \leq t \leq \alpha, \ \forall \varepsilon \geq 0$$

Hence,

$$\int_{-1}^{\alpha} f^2(t) dt = 0 \tag{3.62}$$

The mapping $t \to f^2(t)$ is positive and continuous on $[-1, \alpha](as f \in C[-1, 1])$. Then, by Eq. (3.61), $f^2(t) = 0$ on $[-1, \alpha]$. Thus, $f(t) = 0$, $-1 \leq t \leq \alpha$, where $-1 < \alpha < 0$. Let $0 < \beta < 1$ and $N_1 = [\frac{1}{\beta}] + 1$. For $n \geq \upsilon = \sup\{N(\varepsilon), N_1\}$, we have $f_n(t) = 1$ for $t \in [\beta, 1]$, $\|f_n - f\| \leq \varepsilon$. Then,

$$\int_{\beta}^{1} (1 - f(t))^2 dt = \int_{\beta}^{1} (f_n(t) - f(t))^2 dt$$

$$\leq \int_{-1}^{1} (f_n(t) - f(t))^2 dt \leq \varepsilon^2 \ \forall \, \varepsilon$$

and, consequently

$$\int_{\beta}^{1} (1 - f(t))^2 dt = 0 \tag{3.63}$$

The function $t \rightarrow (1 - f(t))^2$ is positive on $[\beta, 1]$ as f is continuous on $[-1, 1)]$. Then, Eq. (3.62) implies that $[1 - f(t)]^2 = 0 \, \forall \, t \in [\beta, 1]$ or $f(t) = 1$ for $t \in [\beta, 1]$, where $0 < \beta < 1$. In short, we have shown that

$$f(t) = \begin{cases} 0 & for \ -1 \le t < 0 \\ 1 & for \ 0 < t \le 1 \end{cases} \tag{3.64}$$

This is a contradiction to the fact that f is continuous f given by Eq. (3.63) is discontinuous. Hence, our assumption that $f_n \rightarrow f \in C[-1, 1]$ is false, and so $C[-1, 1]$ is not a Hilbert space.

Problem 3.5 Let V be a bounded linear operator on a Hilbert space H_1 into a Hilbert space H_2. Then, $||I - V^*V|| < 1$ if and only if $\forall \, x \in H_1$, $0 < \inf_{||x||=1} = ||Vx|| \le$ $\sup_{||x||=1} ||Vx|| = ||V|| < \sqrt{2}$, where I is the identity operator on H_1.

Solution 3.5 Since $I - V^*V$ is self-adjoint by Corollary 3.3

$$||I - V^*V|| = \sup_{||x||=1} |\langle x, (I - V^*V))x \rangle|$$
$$= \sup_{||x||=1} |1 - ||Vx||^2| < 1$$

Therefore, the condition that $||I - V^*V|| < 1$ is equivalent to the two conditions

$$\sup_{||x||=1} (1 - ||Vx||^2) = 1 - \inf_{||x||=1} ||Vx||^2 < 1$$

and

$$\sup_{||x||=1} (||Vx||^2 - 1) = \sup_{||x||=1} ||Vx||^2 - 1 < 1$$

which is equivalent to

$$0 < \inf_{||x||=1} ||Vx||^2 \le \sup_{||x||=1} ||Vx||^2 < 2$$

3.10.2 Unsolved Problems

Problem 3.6 a. Give an example of a nonseparable Hilbert space.
  b. If A is any orthonormal set of a separable Hilbert space X, show that A is countable.

Problem 3.7 Prove that every separable Hilbert space is isometrically isomorphic to ℓ_2.

Problem 3.8 If M is a closed subset of a Hilbert space X, prove that $M = M^{\perp\perp}$

Problem 3.9 If A is an arbitrary subset of a Hilbert space X, then prove that

1. $A^{\perp} = (A^{\perp\perp}) \perp$.
2. $A^{\perp\perp} = [\overline{A}]$.

Problem 3.10 Show that for a sequence $\{x_n\}$ in an inner product space, the conditions $\|x_n\| \to \|x\|$ and $\langle x_n, x \rangle \to \|x\|^2$ imply $\|x_n - x\| \to 0$ as $n \to \infty$.

Problem 3.11 Let $e_n = (0, 0, \ldots, 1, 0, \ldots)$, where 1 is in the nth place. Find $F\{e_n\}$ for $F \in (\ell_2)^*$.

Problem 3.12 Let $K = \{x = (x_1, x_2, \ldots, x_n, \ldots) \in \ell_2 / |x_n| \leq \frac{1}{n}$ for all n $\}$. Show that K is a compact subset of ℓ_2.

Problem 3.13 In an inner product space, show that $\|x + y\| = \|x\| + \|y\|$ if and only if $\|x - y\| = \|\|x\| - \|y\|\|$.

Problem 3.14 (a) Show that the vectors $x_1 = (1, 2, 2)$, $x_2 = (2, 1, -2)$, $x_3 = (2, -2, 1)$ of R^3 are orthogonal.

(b) Show that the vectors $(1/3, 2/3, 2/3)$, $(2/3, 1/3, -2/3)$, $(2/3, -2/3, 1/3)$ of R^3 are orthonormal.

(c) Show that ℓ_p, $p \neq 2$ is not an inner product space and hence not a Hilbert space.

Problem 3.15 Consider R^2 with norm $\|x\|_1 = |x_1| + |x_2|$.

(a) Show that every closed convex set in $(R^2, \|\cdot\|_1)$ has a point of minimum norm.
(b) Show, by an example, that this may not be unique.
(c) What happens with $\|x\|_\infty = \max |x1|, |x2|$?

Problem 3.16 Let $P : L_2(-a, a) \to L_2(-a, a)$, where $0 < a \leq \infty$, be defined by $y = Px$, where $y(t) = \frac{1}{2}[x(t) + x(-t)]$. Find the range and domain of P. Show that P is an orthogonal projection.

Problem 3.17 Let $T : L_2(-\infty, \infty) \to L_2(-\infty, \infty)$ be defined by

$$Tf(t) = f(t) \ for \ t \geq 0$$
$$= -f(t) \ for \ t < 0$$

Show that T is a unitary operator.

Problem 3.18 For any elements x, y, z of an inner product space, show that

$$\|z - x\|^2 + \|z - y\|^2 = \frac{1}{2}\|x - y\|^2 + 2\|z - \frac{1}{2}(x + y)^2\|^2$$

Problem 3.19 Show that in an inner product space, $x \perp y$ if and only if we have $\|x + \alpha y\| = \|x - \alpha y\|$ for all scalars α.

Problem 3.20 Let $E[-1, 1] = \{f \in C[-1, 1]/f(-t) = f(t),\ for\ all\ t \in [-1, 1]\}$ and $O[1, 1] = \{f \in C[-1, 1]/f(-t) = -f(t)\ for\ all\ t \in [-1, 1]\}$. Show that $C[-1, 1] = E[-1, 1] \oplus O[-1, 1]$.

(a) Let $X = R^2$. Find $M^\perp$ if $M = \{x\}$, $x = (x_1, x_2) \neq 0 \in X$.
(b) Show that Theorems 3.10 and 3.19 are not valid for inner product spaces.
(c) Let $H = \{f \in L_2[a, b]/f\ is\ absolutely\ continuous\ with\ f' \in L_2[a, b]\}$.

   Show that for $f \in L_2[a, b]$, there is a $g \in H$ such that $F(f) = \int_a^b gt dt$ is
   a solution of the differential equation $g(t) - g''(t) = f(t)$ with the boundary
   condition $g'(a) = g'(b) = 0$.

Problem 3.21 Let H be an inner product space and y a fixed vector of H. Prove that the functional $f(x) = \langle x, y \rangle\ \forall\ x \in H$ is a continuous linear functional.

Problem 3.22 Show that every bounded linear functional F on ℓ_2 can be represented in the form

$$F(x) = \sum_{i=1}^{\infty} x_i \overline{y_i}$$

where $x = (x_1, x_2, \ldots, x_n, \ldots) \in l_2$ and fixed $y = (y_1, y_2, \ldots, y_n, \ldots) \in \ell_2$.

Problem 3.23 Let $T : C^2 \to C^2$ be defined by

$$T(x) = (x_1 + ix_2, x_1 - ix_2),\ where\ x = (x_1, x_2)$$

Find T^*. Show that $T^*T = TT^* = 2I$. Find $\frac{1}{2}(T + T^*)$ and $\frac{1}{2i}(T - T^*)$

Problem 3.24 If $||x_n|| \leq 1$, $n = 1, 2, \ldots$, in a separable Hilbert space H, then discuss the properties of Fx_{n_k}, where $F \in H^*$ and x_{n_k} is a subsequence of $\{x_n\}$.

Problem 3.25 Show that T^4 is a self-adjoint operator provided T is a self-adjoint operator.

Problem 3.26 Let T be an arbitrary operator on a Hilbert space H into itself, and if α and β are scalars such that $|\alpha| = |\beta|$. Show that $\alpha T + \beta T^*$ is normal.

Problem 3.27 Show that the unitary operators on a Hilbert space H into itself form a group.

Problem 3.28 Let T be a compact normal operator on a Hilbert space X, and consider the eigenvalue problem $(\lambda I - T)(x) = y$ on X. If λ is not an eigenvalue of T, then show that this equation has a unique solution $x = (\lambda I - T)^{-1} y$.

Problem 3.29 Prove the results (a) through (d) of Remark 3.19(2).

Problem 3.30 If $S, T : X \to X$, where X is a complex inner product space such that $\langle Sx, x \rangle = \langle Tx, x \rangle\ \forall\ x \in X$, then show that $S = T$.

Problem 3.31 Give an example of a self-adjoint operator without eigenvalues.

Problem 3.32 If H is a finite-dimensional inner product space, show that every isometric isomorphism on H into itself is unitary.

Problem 3.33 Let X be a Hilbert space. If T is a self-adjoint operator (not necessarily bounded) mapping a subset of X into X such that $D_T = X$, prove that the operator $U = (T - i)(T - i)^{-1}$ is defined on all of X and is unitary.

Problem 3.34 Let T be a compact normal operator on a Hilbert space H. Show that there exists an orthonormal basis of eigenvectors $\{e_n\}$ and corresponding eigenvalues $\{\mu_n\}$ such that if $x = \sum_n \langle x, e_n \rangle e_n$ is the Fourier expansion for x, then $T(x) = \sum_n \mu_n \langle x, e_n \rangle e_n$.

Chapter 4
Fundamental Theorems

Abstract This chapter is devoted to five fundamental results of functional analysis, namely Hahn–Banach theorem, Banach–Alaoglu theorem, uniform boundedness principle, open mapping theorem, and closed graph theorem. These theorems have played significant role in approximation theory, Fourier analysis, numerical analysis, control theory, optimization, mechanics, mathematical economics, and differential and partial differential equations.

Keywords Hahn–Banach theorem · Topological Hahn–Banach theorem · Strong topology · Weak topology · Weak convergence · Weak convergence in Hilbert spaces · Weak star topology · Graph of a linear operator · Open mapping

4.1 Introduction

There are five basic results of functional analysis known as Hahn–Banach theorem, Banach–Alaoglu theorem, uniform boundedness principle, open mapping and closed graph theorem. These theorems have been very useful in approximation theory, Fourier analysis, numerical analysis, control theory, optimization, mechanics, mathematical economics, and differential equations. Topological Hahn–Banach theorem tells us that properties of functional defined on a subspace are preserved if the functional is extended over the whole space. It is well known that existence and uniqueness of solutions of problems in science and technology depend on topology and associated convergence of the underlying space. In this chapter, we present strong, weak, and weak* topologies and related convergence such as convergence of convex sets [138], Γ convergence [6, 57], and two-scale convergence [101]. The uniform boundedness principle indicates that the uniform boundedness of a sequence of bounded linear operators is implied by the boundedness of the set of images under the given bounded linear operators. The converse of this result is known as the Banach–Steinhaus theorem which is given in the solved example (Problem 4.2). It states that uniform boundedness of a sequence of bounded linear operators, along with pointwise convergence on a dense subset of the entire space. A one-to-one continuous operator from a Banach space into another has inverse that is continuous. In other

© Springer Nature Singapore Pte Ltd. 2018   145
A. H. Siddiqi, *Functional Analysis and Applications*, Industrial and Applied Mathematics,
https://doi.org/10.1007/978-981-10-3725-2_4

words, the images of open sets are open sets. This result is called open mapping theorem. The concept of graph of linear operator is defined, and it is proved that a linear operator with a closed graph is continuous. We call this result the closed graph theorem. This chapter deals with these five theorems along with their consequences. Interested readers may go through references [Hu 68], [130, 168], [Si 71], [174].

4.2 Hahn–Banach Theorem

Theorem 4.1 (Hahn–Banach theorem) *Let X be a real vector space, M a subspace of X, and p a real function defined on X satisfying the following conditions:*

1. $p(x + y) \leq p(x) + p(y)$
2. $p(\alpha x) = \alpha p(x)$

$\forall \, x, y \in X$ *and positive real α.*

Further, suppose that f is a linear functional on M such that $f(x) \leq p(x) \, \forall \, x \in M$. Then, there exists a linear functional F defined on X for which $F(x) = f(x) \, \forall \, x \in M$ and $F(x) \leq p(x) \, \forall \, x \in X$. In other words, there exists an extension F of f having the property of f.

Theorem 4.2 (Topological Hahn–Banach theorem) *Let X be a normed space, M a subspace of X, and f a bounded linear functional on M. Then, there exists a bounded linear functional F on X such that*

1. $F(x) = f(x) \, \forall \, x \in M$
2. $\|F\| = \|f\|$

In other words, there exists an extension F of f which is also bounded linear and preserves the norm.

The proof of Theorem 4.1 depends on the following lemma:

Lemma 4.1 *Let X be a vector space and M its proper subspace. For $x_0 \in X - M$, let $N = [M \cup \{x_0\}]$. Furthermore, suppose that f is a linear functional on M and p a functional on X satisfying the conditions in Theorem 4.1 such that $f(x) \leq p(x) \, \forall \, x \in M$. Then, there exists a linear functional F defined on N such that $F(x) = f(x) \, \forall \, x \in M$ and $F(x) \leq p(x) \, \forall \, x \in N$.*

In short, this lemma tells us that Theorem 4.1 is valid for the subspace generated or spanned by $M \cup \{x_0\}$.

Proof Since $f(x) \leq p(x)$, *for* $x \in M$, and f is linear, we have for arbitrary $y_1, y_2 \in M$

$$f(y_1 - y_2) = f(y_1) - f(y_2) \leq p(y_1 - y_2)$$

or

$$f(y_1) - f(y_2) \leq p(y_1 + x_0 - y_2 - x_0)$$
$$\leq p(y_1 + x_0) + p(-y_2 - x_0)$$

by condition (1) of Theorem 4.1.

Thus, by regrouping the terms of y_2 on one side and those of y_1 on the other side, we have

$$-p(-y_2 - x_0) - f(y_2) \leq p(y_1 + x_0) - f(y_1) \qquad (4.1)$$

Suppose y_1 is kept fixed and y_2 is allowed to vary over M, then (4.1) implies that the set of real numbers $\{-p(-y_2 - x_0) - f(y_2)/y_2 \in M\}$ has upper bounds and hence the least upper bound (See Remark $A.1(A)$).

Let $\alpha = \sup\{-p(-y_2 - x_0) - f(y_2)/y_2 \in M\}$. If we keep y_2 fixed and y_1 is allowed to vary over M, (4.1) implies that the set of real numbers $\{p(y_1 + x_0) - f(y_1)/y_1 \in M\}$ has lower bounds and hence the greatest lower bound (See Remark $A.1(A)$).

Let $\beta = \inf\{p(y_1 + x_0) - f(y_1)/y_1 \in M\}$. From (4.1), it is clear that $\alpha \leq \beta$. As it is well known that between any two real numbers there is a always a third real number, let γ be a real number such that

$$\alpha \leq \gamma \leq \beta \qquad (4.2)$$

It may be observed that if $\alpha = \beta$, then $\gamma = \alpha = \beta$. Therefore, for all $y \in M$, we have

$$-p(-y - x_0) - f(y) \leq \gamma \leq p(y + x_0) - f(y) \qquad (4.3)$$

From the definition of N, it is clear that every element x in N can be written as

$$x = y + \lambda x_0 \qquad (4.4)$$

where $x_0 \neq M$ or $x_0 \in X - M$, λ is a uniquely determined real number and y a uniquely determined vector in M.

We now define a real-valued function on N as follows:

$$F(x) = F(y + \lambda x_0) = f(y) + \lambda \gamma \qquad (4.5)$$

where γ is given by inequality (4.2) and x is as in Eq. (4.4).

We shall now verify that the well-defined function $F(x)$ satisfies the desired conditions, i.e.,

1. F is linear.
2. $F(x) = f(x) \ \forall \ x \in M$.

3. $F(x) \leq p(x) \ \forall \ x \in N$.

  a. F is linear: For

$$z_1, z_2 \in N(z_1 = y_1 + \lambda_1 x_0, z_2 = y_2 + \lambda_2 x_0)$$
$$F(z_1 + z_2) = F(y_1 + \lambda_1 x_0 + y_2 + \lambda_2 x_0)$$
$$= F((y_1 + y_2) + (\lambda_1 + \lambda_2)x_0)$$
$$= f(y_1 + y_2) + (\lambda_1 + \lambda_2)\gamma$$
$$= f(y_1) + f(y_2) + \lambda_1 \gamma + \lambda_2 \gamma$$

     as f is linear; or

$$F(z_1 + z_2) = [f(y_1) + \lambda_1 \gamma] + [f(y_2) + \lambda_2 \gamma]$$
$$= F(z_1) + F(z_2)$$

    Similarly, we can show that $F(\mu z) = \mu F(z) \ \forall \ z \in N$ and for real μ.

  b. If $x \in M$, then λ must be zero in (4.4) and then Eq. (4.5) gives $F(x) = f(x)$.

  c. Here, we consider three cases. (See (4.4)).

Case 1: $\lambda = 0$ We have seen that $F(x) = f(x)$ and as $f(x) \leq p(x)$, we get that $F(x) \leq p(x)$.

Case 2: $\lambda > 0$ From (4.3), we have

$$\gamma \leq p(y + x_0) - f(y) \tag{4.6}$$

Since N is a subspace, $\frac{y}{\lambda} \in N$. Replacing y by $\frac{y}{\lambda}$ in (4.6), we have

$$\gamma \leq p\left(\frac{y}{\lambda} + x_0\right) - f\left(\frac{y}{\lambda}\right)$$
$$or \quad \gamma \leq p\left(\frac{1}{\lambda}(y + \lambda x_0)\right) - f\left(\frac{y}{\lambda}\right).$$

By condition (2) of Theorem 4.1,

$$p\left(\frac{1}{\lambda}(y + \lambda x_0)\right) = \frac{1}{\lambda}p(y + \lambda x_0), \ for \ \lambda > 0,$$

and

$$f\left(\frac{y}{\lambda}\right) = \frac{1}{\lambda}f(y)$$

as f is linear. Therefore,

$$\lambda \gamma \leq p(y + \lambda x_0) - f(y)$$

or $f(y) + \lambda \gamma \le p(y + \lambda x_0)$. Thus, from (4.4) and (4.5), we have $F(x) \le p(x) \ \forall \ x \in N$.

Case 3: $\lambda < 0$ From (4.3), we have

$$- p(-y - x_0) - f(y) \le \gamma \qquad (4.7)$$

Replacing y by y/λ in (4.7), we have

$$-p\left(\frac{-y}{\lambda} - x_0\right) - f\left(\frac{y}{\lambda}\right) \le \gamma$$

or

$$-p\left(\frac{-y}{\lambda} - x_0\right) \le \gamma + f\left(\frac{y}{\lambda}\right)$$

$$= \gamma + \frac{1}{\lambda} f(y)$$

as f is linear, i.e.,

$$-p\left(\frac{-y}{\lambda} - x_0\right) \le \gamma + \frac{1}{\lambda} f(y) \qquad (4.8)$$

Multiplying (4.8) by λ, we have

$$-\lambda p\left(\frac{-y}{\lambda} - x_0\right) \ge \lambda\gamma + \frac{1}{\lambda} f(y)$$

(the inequality in (4.8) is reversed as λ is negative), or

$$(-\lambda)p\left(\left(\frac{-1}{\lambda}\right)(y + \lambda x_0)\right) \ge F(x)$$

Since $\frac{-1}{\lambda} > 0$, by condition (2) of Theorem 4.1, we have

$$p\left(\left(\frac{-1}{\lambda}\right)(y + \lambda x_0)\right) = \frac{-1}{\lambda} p(y + \lambda x_0)$$

and so

$$(-\lambda)\left(\frac{-1}{\lambda}\right) p(y + \lambda x_0) \ge F(x)$$

or

$$F(x) \le p(x) \ \forall \ x \in N$$

Proof (*Proof of Theorem* 4.1) Let S be the set of all linear functionals F such that $F(x) = f(x) \; \forall x \in M$ and $F(x) \leq p(x) \; \forall \, x \in X$. That is to say, S is the set of all functionals F extending f and $F(x) \leq p(x)$ over X. S is nonempty as not only does f belong to it but there are other functionals also which belong to it by virtue of Lemma 4.1. We introduce a relation in S as follows.

For $F_1, F_2 \in S$, we say that F_1 is in relation to F_2 and we write $F_1 < F_2$ if $DF_1 \subset DF_2$ and $F_2/DF_1 = F_1$ (Let DF_1 and DF_2 denote, respectively, the domain of F_1 and F_2. F_2/DF_1 denotes the restriction of F_2 on the domain of F_1). S is a partially ordered set. The relation $<$ is reflexive as $F_1 < F_1$. $<$ is transitive, because for $F_1 < F_2$, $F_2 < F_3$, we have $DF_1 \subset DF_2$, $DF_2 \subset DF_3$. $F_2/DF_1 = F_1$ and $F_3/DF_2 = F_2$, which implies that $DF_1 \subset DF_3$ and $F_3/DF_1 = F_1$. $<$ is antisymmetric. For $F_1 < F_2$

$$DF_1 \subset DF_2$$
$$F_2/DF_1 = F_1$$

For $F_2 < F_1$

$$DF_2 \subset DF_1$$
$$F_1/DF_2 = F_2$$

Therefore, we have

$$F_1 = F_2$$

We now show that every totally ordered subset of S has an upper bound in S. Let $T = F_\sigma$ be a totally ordered subset of S. Let us consider a functional, say F defined over $\bigcup_\sigma DF_\sigma$. If $x \in \bigcup_\sigma DF_\sigma$, there must be some σ such that $x \in DF_\sigma$, and we define $F(x) = F_\sigma(x)$. F is well defined, and its domain $\bigcup_\sigma DF_\sigma$ is a subspace of X.

$\bigcup_\sigma DF_\sigma$ is a subspace: Let $x, y \in \bigcup_\sigma DF_\sigma$. This implies that $x \in DF_{\sigma_1}$, and $y \in DF_{\sigma_2}$. Since T is totally ordered, either $DF_{\sigma_1} \subset DF_{\sigma_2}$ or $DF_{\sigma_2} \subset DF_{\sigma_1}$. Let $DF_{\sigma_1} \subset DF_{\sigma_2}$. Then, $x \in DF_{\sigma_2}$ and so $x + y \in DF_{\sigma_2}$, or $x + y \in \bigcup_\sigma DF_\sigma$. Let $x \in \bigcup_\sigma DF_\sigma$, then $x \in DF_{\sigma_1}$ which implies that $\mu x \in \bigcup_\sigma DF_\sigma \; \forall$ real μ. This shows that $\bigcup_\sigma DF_\sigma$ is a subspace. F is well defined: Suppose $x \in DF_\sigma$ and $x \in DF_\nu$. Then by the definition of F, we have $F(x) = F_\sigma(x)$ and $F(x) = F_\nu(x)$. By the total ordering of T, either F_σ extends F_ν or vice versa, and so $F_\sigma(x) = F_\nu(x)$ which shows that F is well defined.

It is clear from the definition that F is linear, $F(x) = f(x)$ for $x \in Df = M$, and $F(x) \leq p(x) \; \forall \, x \in D_F$. Thus, for each $F_\sigma \in T$, $F_\sigma < F$; i.e., F is an upper bound of T. By Zorn's lemma (Theorem A.2 of the Appendix A.1), there exists a maximal element $\hat{F}$ in S; i.e., $\hat{F}$ is a linear extension of f, $\hat{F}(x) \leq p(x)$, and $F < \hat{F}$ for every $F \in S$. The theorem will be proved if we show that $D_{\hat{F}} = X$. We know that $D_F \subset X$. Suppose there is an element $x \in X$ such that $x_0 \notin D_{\hat{F}}$. By Lemma 4.1, there exists $\hat{F}$

such that $\hat{F}$ is linear, $F(x) = \hat{F}(x) \ \forall \ x \in D_{\hat{F}}$, and $\hat{F}(x) \leq p(x)$ for $x \in [D_F \cup x_0]$ ($\hat{F}$ is also an extension of f). This implies that $\hat{F}$ is not a maximal element of S which is a contradiction. Hence, $D_F = X$.

Proof (Proof of Theorem 4.2) Since f is bounded and linear, we have $|f(x)| \leq ||f|| \ ||x||$, $\forall \ x$ (see Remark 2.7).

If we define $p(x) = ||f|| \ ||x||$, then $p(x)$ satisfies the conditions of Theorem 4.1 (see Problem 2.8). By Theorem 4.1, there exists F extending f which is linear and $F(x) \leq p(x) \ \forall \ x \in X$.

We have $-F(x) = F(-x)$ as F is linear, and so by the above relation,

$$-F(x) \leq p(-x) = ||f|| \ || - x|| = ||f|| \ ||x|| = p(x).$$

Thus, $|F(x)| \leq p(x) = ||f|| \ ||x||$ which implies that F is bounded and

$$||F|| = \sup_{||x||=1} |F(x)| \leq ||f|| \tag{4.9}$$

On the other hand, for $x \in M$, $|f(x)| = |F(x)| \leq ||F|| \ ||x||$, and so

$$||f|| = \sup_{||x||=1} |f(x)| \leq ||F|| \tag{4.10}$$

Hence, by Eqs. (4.9) and (4.10), $||f|| = ||F||$.

Remark 4.1 The Hahn–Banach theorem is also valid for normed spaces defined over the complex field.

The proofs of the following important results mainly depend on Theorem 4.2:

Theorem 4.3 *Let w be a nonzero vector in a normed space X. Then, there exists a continuous linear functional F, defined on the entire space X, such that $||F|| = 1$ and $F(w) = ||w||$.*

Theorem 4.4 *If X is a normed space such that $F(w) = 0 \ \forall \ F \in X^\star$, then $w = 0$.*

Theorem 4.5 *Let X be a normed space and M its closed subspace. Further, assume that $w \in X - M$ ($w \in X$ but $w \notin M$). Then, there exists $F \in X^\star$ such that $F(m) = 0$ for all $m \in M$, and $F(w) = 1$.*

Theorem 4.6 *Let X be a normed space, M its subspace, and $w \in X$ such that $d = \inf_{m \in M} ||w - m|| > 0$ (It may be observed that this condition is satisfied if M is closed and $w \in X - M$). Then, there exists $F \in X^\star$ with $||F|| = 1$, $F(w) \neq 0$, and $F(m) = 0$ for all $m \in M$.*

Theorem 4.7 *If $X^\star$ is separable, then X is itself separable.*

Proof (*Proof of Theorem* 4.3) Let $M = [\{w\}] = \{m/m = \lambda w, \lambda \in R\}$ and $f : M \to R$ such that $f(m) = \lambda ||w||$. f is linear $[f(m_1 + m_2) = (\lambda_1 + \lambda_2) \, ||w||$, where $m_1 = \lambda_1 w$ and $m_2 = \lambda_2 w$ or $f(m_1 + m_2) = (\lambda_1 + \lambda_2)||w|| = \lambda_1||w|| + \lambda_2||w|| = f(m_1) + f(m_2)]$. Similarly, $f(\mu m) = \mu f(m) \forall \mu \in R$. f is bounded ($|f(m)| = ||\lambda w|| = ||m||$ and so $|f(m)| \le k||m||$ where $0 \le k \le 1$) and $f(w) = ||w||$ (If $m = w$, then $\lambda = 1$). By Theorem 2.6

$$||f|| = \sup_{m \in M, \, ||m||=1} |f(m)| = \sup_{||m||=1} |\lambda| \, ||w|| = \sup_{||m||=1} ||m|| = 1$$

Since f, defined on M, is linear and bounded (and hence continuous) and satisfies the conditions $f(w) = ||w||$ and $||f|| = 1$, by Theorem 4.2, there exists a continuous linear functional F over X extending f such that $||F|| = 1$ and $F(w) = ||w||$.

Proof (*Proof of Theorem* 4.4) Suppose $w \ne 0$ but $F(w) = 0$ for all $F \in X^*$. Since $w \ne 0$, by Theorem 4.3, there exists a functional $F \in X^*$ such that $||F|| = 1$ and $F(w) = ||w||$. This shows that $F(w) \ne 0$ which is a contradiction. Hence, if $F(w) = 0 \, \forall \, F \in X^*$, then w must be zero.

Proof (*Proof of Theorem* 4.5) Let $w \in X - M$ and $d = \inf_{m \in M} ||w - m||$. Since M is a closed subspace and $w \notin M$, $d > 0$. Suppose N is the subspace spanned by w and M; i.e., $n \in N$ if and only if

$$n = \lambda w + m, \; \lambda \in R, \; m \in M \tag{4.11}$$

Define a functional on N as follows:

$$f(n) = \lambda$$

f is linear and bounded: $f(n_1 + n_2) = \lambda_1 + \lambda_2$, where $n_1 = \lambda_1 w + m$ and $n_2 = \lambda_2 w + m$. So $f(n_1 + n_2) = f(n_1) + f(n_2)$. Similarly, $f(\mu n) = \mu f(n)$ for real μ. Thus, f is linear. In order to show that f is bounded, we need to show that there exists $k > 0$ such that $|f(n)| \le k||n|| \, \forall \, n \in N$. We have

$$||n|| = ||m + \lambda w|| = \left\| -\lambda(\frac{-m}{\lambda} - w) \right\| = |\lambda| \left\| \frac{-m}{\lambda} - w \right\|.$$

Since $-m\lambda \in M$ and $d = \inf_{m \in M} ||w - m||$, we see that $\left\| \frac{-m}{\lambda} - w \right\| \ge d$. Hence, $||n|| \ge |\lambda| d$, or $|\lambda| \le ||n||/d$. By definition, $|f(n)| = |\lambda| \le ||n||/d$, or $|f(n)| \le k||n||$, where $k \ge 1/d > 0$. Thus, f is bounded. $n = w$ implies that $\lambda = 1$, and therefore, $f(w) = 1$. $n = m \in M$ implies that $\lambda = 0$ (see Eq. 4.11). Therefore, from the definition of f, $f(m) = 0$. Thus, f is bounded linear and satisfies the conditions $f(w) = 1$ and $f(m) = 0$. Hence, by Theorem 4.2 there exists F defined over X such that F is an extension of f, and F is bounded linear, i.e., $F \in X^*$, $F(w) = 1$, and $F(m) = 0$ for all $m \in M$.

Proof (*Proof of Theorem* 4.6) Let N be the subspace spanned by M and w (see (4.11)). Define f on N as $f(n) = \lambda d$. Proceeding exactly as in the proof of Theorem 4.5, we can show that f is linear and bounded on $N[|f(n)| = |\lambda|d \leq ||n||]$, $f(w) = d \neq 0$, and $f(m) = 0$ for all $m \in M$. Since $|f(n)| \leq ||n||$, we have

$$||f|| \leq 1 \tag{4.12}$$

For arbitrary $\varepsilon > 0$, by the definition of d, there must exist an $m \in M$ such that $||w - m|| < d + \varepsilon$. Let

$$z = \frac{w - m}{||w - m||}.$$

Then,

$$||z|| = \frac{w - m}{||w - m||} = 1$$

and

$$f(z) = \frac{1}{||w - m||} f(w - m) = d/||w - m||$$

[By definition, $f(n) = \lambda d$; if $n = w - m$, then $\lambda = 1$, and so

$$f(w - m) = d$$

or

$$f(z) > \frac{d}{d + \varepsilon} \tag{4.13}$$

By Theorem 2.6, $||f|| = \sup_{||m||=1} |f(m)|$. Since $||z|| = 1$, Eq. 4.13 implies that

$$||f|| > \frac{d}{d + \varepsilon}.$$

Since $\varepsilon > 0$ is arbitrary, we have

$$||f|| \geq 1 \tag{4.14}$$

From (4.12) and (4.14), we have $||f|| = 1$. Thus, f is bounded and linear, $f(m) = 0 \ \forall \ m \in M$, $f(w) \neq 0$, and $||f|| = 1$. By Theorem 4.2, there exists $F \in X^*$ such that $F(w) \neq 0$, $F(m) = 0$ for all $m \in M$, and $||F|| = 1$.

Proof (*Proof of Theorem* 4.7) Let $\{F_n\}$ be a sequence in the surface of the unit sphere S of $X^*[S = \{F \in X^*/||F|| = 1\}]$ such that $\{F_1, F_2, \ldots, F_n\}$ is a dense subset of S. By Theorem 2.6, $||F|| = \sup\limits_{||v||=1} |F(v)|$ and so for $\varepsilon > 0$, there exists $v \in X$ such that

$$||v|| = 1$$

and

$$(1 - \varepsilon)||F|| \leq |F(v)| \tag{4.15}$$

Putting $\varepsilon = \frac{1}{2}$ in (4.15), there exists $v \in X$ such that

$$||v|| = 1$$

and

$$\frac{1}{2}||F|| = |F(v)|.$$

Let $\{v_n\}$ be a sequence such that

$$||v_n|| = 1$$

and

$$\frac{1}{2}||F_n|| = |F_n(v_n)|.$$

Let M be a subspace spanned by $\{v_n\}$. Then, M is separable by its construction. In order to prove that X is separable, we show that $M = X$. Suppose $X \neq M$; then, there exists $w \in X$, $w \notin M$. By Theorem 4.6, there exists $F \in X^*$ such that

$$||F|| = 1$$
$$F(w) \neq 0 \tag{4.16}$$

and

$$F(m) = 0 \; \forall \, m \in M.$$

In particular, $F(v_n) = 0 \; \forall \, n$ where

$$\frac{1}{2}||F_n|| \leq |F_n(v_n)| = |F_n(v_n) - F(v_n) + F(v_n)|$$
$$\leq F_n(v_n) - F(v_n) + F(v_n)$$

Since $||v_n|| = 1$ and $F(v_n) = 0 \ \forall \ n$, we have

$$\frac{1}{2}||F_n|| \leq ||F_n - F|| \ \forall \ n \qquad (4.17)$$

We can choose $\{F_n\}$ such that

$$\lim_{n \to \infty} ||(F_n - F)|| = 0 \qquad (4.18)$$

because $\{F_n\}$ is a dense subset of S. This implies from (4.17) that

$$||F_n|| = 0 \ \forall \ n$$

Thus, using (4.16)–(4.18), we have

$$1 = ||F|| = ||F - F_n + F_n|| \leq ||F - F_n|| + ||F_n||$$
$$\leq ||F - F_n|| + 2||F - F_n||$$

or

$$1 = ||F|| = 0$$

which is a contradiction. Hence, our assumption is false and $X = M$.

4.3 Topologies on Normed Spaces

4.3.1 Compactness in Normed Spaces

The concept of compactness is introduced in topological and metric spaces in Appendices (Definition A.91 and Theorem A.7). In a metric space, concepts of compactness and sequentially compactness are identical. A subset A of a normed space X is called *compact* if every sequence in A has a convergent subsequence whose limit is an element of A. Properties of compactness in normed spaces can be described by the following theorems.

Theorem 4.8 *A compact subset A of a normed space X is closed and bounded. But the converse of this statement is in general false.*

Theorem 4.9 *In a finite-dimensional normed space X, any subset A of X is compact if and only if A is closed and bounded.*

Theorem 4.10 *If a normed space X has the property that the closed unit ball $M = \{x/||x||\}$ is compact, then X is finite-dimensional.*

Proof (*Proof of Theorem* 4.8) We prove here that every compact subset A of X is closed and bounded. For the converse, we refer to Sect. 4.4 where noncompactness of unit closed ball is shown.

Let A be a compact subset of a normed space X. For every $x \in \bar{A}$, there is a sequence $\{x_n\}$ in A such that $x_n \to x$ by Theorem A.6(3). Since A is compact, $x \in \bar{A}$. Hence, A is closed because $x \in \bar{A}$ was arbitrary. Now we prove that A is bounded. If A were unbounded, it would contain an unbounded sequence $\{y_n\}$ such that $||y_n - b|| > n$, where b is any fixed element. This sequence could not have a convergent subsequence, since a convergent subsequence must be bounded, by Theorem A.6(4).

We need the following lemma in the proof of Theorem 4.9.

Lemma 4.2 *Let* $\{x_1, x_2, \ldots, x_n\}$ *be a linearly set of vectors in a normed space X (of any dimension). Then, there is a number $c > 0$ such that for every choice of scalars* $\alpha_1, \alpha_2, \ldots, \alpha_n$, *we have*

$$||\alpha_1 x_1 + \alpha_2 x_2 + \cdots + \alpha_n x_n|| \geq c(|\alpha_1| + |\alpha_2| + \cdots + |\alpha_n|)$$

See Chap. 11 for proof.

Proof (*Proof of Theorem* 4.9) Compactness implies closedness and boundedness by Theorem 4.8, and we prove the converse. Let A be closed and bounded. Let $dim X = n$ and $\{e_1, e_2, \ldots, e_n\}$ be a basis for X. We consider any sequence $\{x_m\}$ in A. Each x_m has representation $x_m = \alpha_1(m)e_1 + \alpha_2(m)e_2 + \cdots + \alpha_n(m)e_n$. Since A is bounded, so is the sequence $\{x_m\}$, say, $||x_m|| \leq k$ for all m. By Lemma 4.2

$$k \geq ||x_m|| = \left\| \sum_{j=1}^{m} \alpha_j^{(m)} \right\| \geq c \sum_{j=1}^{m} |\alpha_j^{(m)}|$$

where $c > 0$. Hence, the sequence of numbers $\{\alpha_j(m)\}$, j fixed, is bounded and, by Theorem A.10(2), has a limit point α_j, $1 \leq j \leq n$. As in the proof of Theorem 4.8, we conclude that $\{x_m\}$ has subsequence $\{z_m\}$ which converges to $z = \sum \alpha_j e_j$. Since A is closed, $z \in A$. This shows that arbitrary sequence $\{x_m\}$ in A has a subsequence which converges in A. Hence, A is compact.

Proof (*Proof of Theorem* 4.10) Let M be compact but $dim X = \infty$, then we show that this leads to a contradiction. We choose any x_1 of norm 1. This x_1 generates a one-dimensional subspace X_1 of X, which is closed, by Theorem C.10, and is a proper subspace of X since $dim X = \infty$. By Riesz's theorem, there is an $x_2 \in X$ of norm 1 such that

$$||x_2 - x_1|| \geq \alpha = \frac{1}{2}$$

The elements x_1, x_2 generate a two-dimensional proper closed subspace X_2 of X. Again by Riesz's theorem (Theorem 2.3, see proof in Problem 4), there is an $x_3 \in X$ of norm 1 such that for all $x \in X_2$ we have

$$\|x_3 - x\| \geq \frac{1}{2}$$

In particular,

$$\|x_3 - x_1\| \geq \frac{1}{2}$$
$$\|x_3 - x_2\| \geq \frac{1}{2}$$

Proceeding by induction, we obtain a sequence $\{x_n\}$ of elements $x_n \in M$ such that

$$\|x_m - x_n\| \geq \frac{1}{2} \quad m \neq n$$

It is clear that $\{x_n\}$ cannot have a convergent subsequence. This contradicts to the compactness of M. Hence, our assumption $dimX = \infty$ is false and $dimX < \infty$.

4.3.2 Strong and Weak Topologies

A normed space can be a topological space in many ways as one can define different topologies on it. This leads to alternative notions of continuity, compactness, and convergence for a given normed space. As shown in Chap. 2, every normed space X is a metric space and hence it is a topological space. More precisely, a basis for the strong topology is the class of open balls centered at the origin

$$S_r(0) = \{x \in X / \|x\| < r, r > 0\} \tag{4.19}$$

A *weak topology* is generated by the base comprising open sets determined by bounded linear functionals on X; namely

$$B_r(0) = \{x \in X / |F(x)| = |(F, x)| < r, \; for \; all \; finite$$
$$family \; of \; elements \; F \in X^*, \; r > 0\} \tag{4.20}$$

If T_1 is the topology generated by the norm (4.19) and T_2 is the topology induced by linear functionals (Eq. (4.20)), then $T_2 \subset T_1$; that is, T_2 is weaker than T_1.

Let X be the dual of a Banach space Y; that is, $X = Y^*$. Then, the *weak**-topology (weak star topology) is generated by the base at the origin consisting of the open sets

$$\{x \in X = Y^\star / |\langle F_v, x \rangle| = |F_v(x)| < r,$$
$$for\ all\ finite\ family\ of\ elements\ v \in Y,\ r > 0\} \qquad (4.21)$$

where

$$(F_v, u) = (u, v)\ u \in Y^\star,\ v \in Y \qquad (4.22)$$

$\{F_v / v \in Y\}$ defined by (4.22) is a subspace of $X^\star = Y^{\star\star}$.

It may be observed that the notions of weak topology and *weak**M-topology coincide in a reflexive Banach space in general, and Hilbert space in particular.

4.4 Weak Convergence

4.4.1 Weak Convergence in Banach Spaces

Definition 4.1 Let X be a normed space, $X^\star$ and $X^{\star\star}$ the first and second dual spaces of X, respectively. Then,

1. A sequence $\{x_n\}$ in X is called *weakly convergent* in X, in symbols $x_n \overset{w}{\to} x$ or $x_n \rightharpoonup x$, if there exists an element $x \in X$ such that $\lim\limits_{n \to \infty} f(x_n) - f(x) = 0$ for all $f \in X^\star$; i.e., for $\varepsilon > 0$, there exists a natural number N such that $|f(x_n) - f(x)| \leq \varepsilon$ for $n > N$ and $\forall\ f \in X^\star$.
2. A sequence $\{f_n\}$ in $X^\star$ is called *weakly** convergent to f in $X^\star$ if $\lim\limits_{n \to \infty} f_n(x) - f(x) = 0$ for all $x \in X$.

Remark 4.2 1. If X is a Banach space which is the dual of the normed space Y, and if we bear in mind that Y is a subspace of its second dual, then we define *weak** convergence as follows: A sequence $\{x_n\}$ in X is called weakly* convergent to $x \in X$ if $\lim\limits_{n \to \infty} |x_n(y) - x(y)| = \lim\limits_{n \to \infty} |y(x_n) - y(x)| = 0\ \forall\ y \in Y$. It is clear that the elements of Y define the linear functionals on $Y^\star = X$.
2. It may be verified that every convergent sequence in X is weakly convergent but the converse may not be true. However, these two notions are equivalent if X is finite-dimensional.
3. Weak convergence implies *weak** convergence, but the converse is not true in general. However, these two notions are equivalent if X is a reflexive space.
4. A sequence $\{x_n\}$ in X converges to x in the weak topology if and only if it converges weakly.

Definition 4.2 1. A subset A of a normed space X is called *compact in the weak topology* or *weakly compact* if every sequence $\{x_n\}$ in A contains a subsequence which converges weakly in A.

2. A subset A of a normed space X is called *compact in the weak* topology* or *weak* compact* if every sequence in A contains a subsequence which is *weakly** convergent in A.

The statements of the following theorem can be easily verified:

Theorem 4.11 *1. The limit of a weakly convergent sequence is unique.*
2. *Every convergent sequence in a normed space is weakly convergent.*
3. *If $\{x_n\}$ and $\{y_n\}$ converge weakly to x and y, respectively, then $\{x_n\} + \{y_n\}$ converges weakly to $x + y$.*
4. *If $\{x_n\}$ converges weakly to x and a sequence of scalars $\{\alpha_n\}$ converges to α, then $\{\alpha_n x_n\}$ converges weakly to αx.*

Definition 4.3 1. A sequence $\{x_n\}$ in a normed space X is called a weak Cauchy sequence if $\{f(x_n)\}$ is a Cauchy sequence for all $f \in X^*$.
2. A normed space X is called *weakly complete* if every weak Cauchy sequence of elements of X converges weakly to some other member of X.

Theorem 4.12 *In the normed space X, every weak Cauchy sequence is bounded and hence every weakly convergent sequence with limit x is bounded; i.e., the set $\{||x_k||/k = 1, 2, 3, \ldots\}$ is bounded in X and*

$$||x|| \leq \liminf_{k \to \infty} ||x_k||$$

Also, x belongs to the subspace of X generated by the sequence $\{x_n\}$. For proof, see Bachman and Narici [8].

Now, we mention some theorems giving the characterizations of weak convergence in spaces ℓ_p, L_p, and $C[a, b]$. The proofs can be found, for example, in Kantorovich and Akilov [107] and Liusternik and Sobolev [122].

Theorem 4.13 *A sequence $\{x_n\}$, $x_n = (\alpha_1^n, \alpha_2^n, \ldots, \alpha_k^n, \ldots) \in \ell_p$, $1 < p < \infty$ converges weakly to $x = (\alpha_1, \alpha_2, \ldots, \alpha_k, \ldots) \in \ell_p$ if and only if*

1. *$||x_n|| \leq M$ (M is a positive constant) for all n.*
2. *For every i, $\alpha_i^{(n)} \to \alpha_i$ as $n \to \infty$.*

Remark 4.3 In view of the above theorem for bounded sequences, the concept of the weak convergence in ℓ_p is equivalent to the coordinatewise convergence.

Theorem 4.14 *A sequence $\{f_n(t)\}$ in $L_p(0, 1)$, $1 < p < \infty$, is weakly convergent to $f(t) \in L_p(0, 1)$ if and only if*

1. *$||f_n|| \leq M \; \forall \, n$ (M being a positive constant)*
2. *$\int_0^\lambda f_n(t)dt \to \int_0^\lambda f(t)dt$ for arbitrary $\lambda \in [0, 1]$.*

Theorem 4.15 *A sequence $\{f_n(t)\}$ in $C[a, b]$ is weakly convergent to $f(t)$ in $C[a, b]$ if and only if*

1. $|f_n(t)| \leq M$, where M is a positive constant, uniformly in k, $n = 1, 2, 3, \ldots$ and $t \in [a, b]$.
2. $\lim_{n \to \infty} f_n(t) = f(t)$ for every $t \in [a, b]$.

After slight modification of the proof of Theorem 4.15, we get the following theorem (See A.H. Siddiqi [122]).

Theorem 4.16 *A sequence of continuous functions $\{f_n^k(t)\}$ converges weakly (uniformly in k) to a continuous function $f(t)$ if and only if*

1. $|f_n^k(t)| \leq M$ where M is a positive constant, uniformly in k, $n = 1, 2, 3, \ldots$, and $t \in [a, b]$.
2. $\lim_{n \to \infty} f_n^k(t) = f(t)$ uniformly in k for every $t \in [a, b]$.

Remark 4.4 J.A. Siddiqi [168], [Si 60], Mazhar and A.H. Siddiqi [130, 131], and A.H. Siddiqi [168] have applied Theorems 4.5 and 4.6 to prove several interesting results concerning Fourier coefficients, Walsh–Fourier coefficients, and summability of trigonometric sequences.

Example 4.1 Show that weak convergence does not imply convergence in norm.

Solution 4.1 By the application of the Riesz representation theorem to the Hilbert space $L_2(0, 2\pi)$, we find that

$$f(x) = \langle x, g \rangle = \int_0^{2\pi} x(t)g(t)dt \tag{4.23}$$

Consider the sequence $\{x_n(t)\}$ defined below

$$x_n(t) = \frac{\sin(nt)}{\pi} \ for \ n = 1, 2, 3, \ldots$$

We now show that $\{x_n(t)\}$ is weakly convergent in $L_2(0, 2\pi)$ but is not norm-convergent in $L_2(0, 2\pi)$. From Eq. (4.23), we have

$$f(x_n) = \langle x_n, g \rangle = \frac{1}{\pi} \int_0^{2\pi} \frac{\sin(nt)}{\pi} g(t)dt \tag{4.24}$$

The right-hand side of Eq. (4.24) is the trigonometric Fourier coefficient of $g(t) \in L_2(0, \pi)$. By the Riemann–Lebesgue theorem concerning the behavior of trigonometric Fourier coefficients

$$\frac{1}{\pi} \int_0^{2\pi} \frac{\sin(nt)}{\pi} g(t)dt \to 0 \ as \ n \to \infty$$

or $f(x_n) \to 0$ as $n \to \infty$; i.e., $\{x_n\}$ converges weakly to 0.

We have

$$||x_n - 0|| = ||x_n|| = \left(\int_0^{2\pi} |x_n(t)|^2 dt \right)^{1/2}$$

$$= \left(\int_0^{2\pi} \frac{|\sin nt|^2}{\pi^2} dt \right)^{1/2} = \frac{1}{\sqrt{\pi}}$$

Since $\frac{1}{\pi} \left(\int_0^{2\pi} |\sin^2 nt| dt \right)^{1/2} \neq 0 \ \forall \ n$, $||x_n - 0||$ cannot tend to zero, and therefore, $\{x_n\}$ cannot converge in the norm. Thus, a weakly convergent sequence $\{x_n\}$ need not be convergent in the norm.

Example 4.2 Show that *weak** convergence does not imply weak convergence.

Solution 4.2 Let $X = \ell_1$ (we know that ℓ_1 is the dual of c_0) and $Y = c_0$. Then, $Y^* = \ell_1$. Thus, the dual of Y is X. Let $\{x_k\}$ be a sequence in X defined by the relation

$$x_j^k = \begin{cases} 0 \ if \ k \neq j \\ 1 \ if \ k = j \end{cases}$$

For $y = (y_1, y_2, y_k, \ldots) \in c_0$, let $x^k(y) = y_k$ (x^k belongs to the dual of c_0; i.e., it is a bounded linear functional on c_0).

Since $y \in c_0$, $\lim_{k \to \infty} y_k = 0$ and so $\lim_{k \to \infty} y^k(y) = \lim_{k \to \infty} y_k = 0$ Therefore, the sequence $\{x_k\}$ of the dual space of $Y = c_0$ converges *weakly** to zero. Now, if $z \in X^* = \ell_\infty$ with $z = (z_1, z_2, \ldots)$, then $x_k(z) = z_k$. Since $z \in \ell_\infty$, $\{z_k\}$ is bounded with respect to k but need not converge to zero as $k \to \infty$. In fact, if $z = (1, 1, \ldots)$, then $x_k(z) \to 1$ as $k \to \infty$. Therefore, $\{x_k\}$ does not converge weakly.

This example shows that *weak** convergence does not imply weak convergence and, by Example 4.1, it does not imply norm convergence.

Example 4.3 Prove that the notions of weak and strong convergence are equivalent in finite-dimensional normed spaces.

Solution 4.3 Solution See Bachman and Narici [8] and Kreyszig [117].

4.4.2 Weak Convergence in Hilbert Spaces

By virtue of the Riesz representation theorem, a sequence $\{x_n\}$ in a Hilbert space X is weakly convergent to the limit $x \in X$ if and only if

$$\lim_{n \to \infty} \langle x_n, z \rangle = \langle x, z \rangle$$

for all $z \in X$.

Theorem 4.17 (The weak compactness property) *Every bounded sequence in a Hilbert space X contains a weakly convergent subsequence.*

Proof Let $\{x_k\}$ denote the bounded sequence with bound M, $||x_k|| \le M$. Let Y be the closed subspace spanned by the elements $\{x_k\}$. Suppose that $Y^\perp$ denotes the orthogonal complement of Y. Consider the sequence $\langle x_1, x_n \rangle$. Being the bounded sequence of real numbers, we can extract from it a convergent subsequence by the Bolzano–Weierstrass theorem. Denote this subsequence by $\alpha_n^1 = \langle x_1, x_n^1 \rangle$. Similarly, $\langle x_2, x_n^1 \rangle$ contains a convergent subsequence $\alpha_n^2 = \langle x_2, x_n^2 \rangle$. Proceeding in this manner, consider the diagonal sequence x_n^n. For each $m, \langle x_m, x_n^n \rangle$ converges, since for $n > m$, it is the subsequence of the convergent sequence$m$. Define $F(x) = \lim_n \langle x, x_n^n \rangle$ whenever this limit exists. This limit clearly exists for finite sums of the form

$$x = \sum_{k=1}^{n} a_k x_k$$

which are dense in Y. Hence, for any $y \in Y$, we can find a sequence y_n such that $||y_n - y|| \to 0$ and

$$F(y_n) = \lim_m \langle y_n, x_m^m \rangle$$

Now,

$$\langle y, x_m^m \rangle = \langle y_p, x_m \rangle + \langle y - y_p, x_m^m \rangle$$

where

$$|\langle y - y_p, x_m^m \rangle| \le M ||y - y_p|| \to 0 \ as \ p \to \infty \ uniformly \ in \ M$$

Therefore, $\{\langle y, x_m^m \rangle\}$ converges. Since for any z in $Y^\perp$, $\langle z, x_k \rangle = 0$, it follows that $F(\cdot)$ is defined for every element of X in view of Theorem 3.10. It can be seen that F is linear. F is continuous as $|F(y_m - y)| = \lim_n |\langle y_m - y, x_n^n \rangle| \le M ||y_m - y|| \to 0$ for $y_m \to y$. By the Riesz representation theorem

$$F(x) = \langle x, h \rangle \ \text{for some } h \in X, \ \text{in fact} \in Y$$

It is also clear that $||h|| \le M$ as $|F(x)| = |\langle x, h \rangle| \le M \, ||h||$.

Theorem 4.18 *Suppose $x_n \rightharpoonup x$ and $||x_n|| \to ||x||$. Then, x_n converges strongly to x.*

Proof We know that

$$||x_n - x||^2 = \langle x_n - x, x_n - x \rangle$$
$$= ||x_n||^2 + ||x||^2 - \langle x_n, x \rangle - \langle x, x_n \rangle$$

or

$$\lim_{n\to\infty} ||x_n - x||^2 = \lim ||x_n||^2 + ||x||^2 - \lim_{n\to\infty} \langle x_n, x \rangle - \lim_{n\to\infty} \langle x, x_n \rangle$$
$$= 2||x||^2 - \langle x, x \rangle - \langle x, x \rangle = 2||x||^2 - 2||x||^2$$
$$= 0.$$

Theorem 4.19 *Let $\{x_n\}$ converge weakly to x. Then, we can extract a subsequence $\{x_{n_k}\}$ such that its arithmetic mean*

$$\frac{1}{m} \sum_1^m x_{n_k}$$

converges strongly to x.

Proof Without loss of generality, we can assume x to be zero. Let the first term of x_{n_k} be $x_{n_1} = x_1$. By weak convergence, we can choose x_{n_2} such that $|\langle x_{n_1}, x_{n_2} \rangle| < 1$. Having chosen $x_{n_1}, \ldots, x_{n_k}$, we can clearly choose $x_{n_{k+1}}$ so that

$$|\langle x_{n_i}, x_{n_{k+1}} \rangle| < \frac{1}{k}, \quad i = 1, 2, 3 \ldots, k$$

By the uniform boundedness principle (Theorem 4.26), $||x_{n_k}|| \leq M < \infty$. Therefore,

$$\left\| \frac{1}{k} \sum_{i=1}^k x_{n_i} \right\|^2 = \langle \frac{1}{k} \sum_{i=1}^k x_{n_i}, \frac{1}{k} \sum_{i=1}^k x_{n_i} \rangle$$
$$\leq \left(\frac{1}{k}\right)^2 \left(kM + 2 \sum_{i=2}^k \sum_{j=1}^{i-1} |\langle x_{n_j}, x_{n_i} \rangle| \right)$$
$$\leq \left(\frac{1}{k}\right)^2 (kM + 2(k-1)) \to 0 \text{ as } k \to \infty$$

Corollary 4.1 *Let $F(\cdot)$ be a continuous convex functional on a Hilbert space X. Then,*

$$F(x) \leq \liminf F(x_n) \text{ for all } x_n \rightharpoonup x.$$

Proof Let us consider a subsequence and, if necessary, renumber it so that $\liminf F(x_n) = \lim F(x_m)$ and further renumber the sequence so that by Theorem 4.19, $\frac{1}{n} \sum_{1}^{n} x_m$ converges strongly to x. Then, we have by convexity of F

$$\frac{1}{n} \sum_{k=1}^{n} F(x_k) \geq F\left(\frac{1}{n} \sum_{k=1}^{n} (x_k)\right)$$

Hence,

$$F(x_n) = \lim \frac{1}{n} \sum_{k=1}^{n} F(x_k) \geq \lim F\left(\frac{1}{n} \sum_{k=1}^{n} (x_k)\right) = F(x).$$

4.5 Banach–Alaoglu Theorem

In the first place, we observe that the closed unit ball in a Banach space need not be compact in its norm topology. To prove the statement, consider the normed space ℓ_2. The closed unit ball $\overline{S_1}$ of ℓ_2, namely

$$\overline{S_1} = \{x = (x_1, x_2, \ldots, x_n, x_{n+1} \ldots) \in \ell_2 / \sum_{i=1}^{\infty} |x_i|^2 \leq 1\}$$

is both bounded and closed. That is, $\overline{S_1}$ is a bounded closed subset of ℓ_2. We now show that $\overline{S_1}$ is not compact. For this, it is sufficient to show that there exists a sequence in $\overline{S_1}$, every subsequence of which is divergent. Define a sequence $\{x_n\}$ in $\overline{S_1}$ in the following manner.
$x_n = (0, 0, \ldots, 1, 0, \ldots)$ (All coordinates are zero and nth coordinate is 1). Thus,

$$x_1 = (1, 0, 0, \ldots)$$
$$x_2 = (0, 1, 0, \ldots)$$
$$x_3 = (0, 0, 1, \ldots)$$

.

.

.

$$x_p = (0, 0, 0, 1, \ldots) \quad 1 \text{ at pth coordinate}$$
$$x_q = (0, 0, 0, 0, 1, \ldots) \quad 1 \text{ at qth coordinate}$$

For $p \neq q$, $\|x_p - x_q\| = \left(\sum_{1}^{\infty} |x_p - x_q| \right)^{1/2} = (0 + \cdots + 1^2 + \cdots + 1^2 + \cdots)^{1/2}$

$= \sqrt{2}$. Therefore, the sequence $\{x_n\}$ and all its subsequences are divergent.

As seen above, the closed unit sphere of $\ell_2 = \ell_2^*$ is not compact with respect to its norm topology. However, the following theorem proved by Alaoglu in the early forties, often called *Banach–Alaoglu theorem*, shows that the closed unit sphere in the dual of every normed space is compact with respect to the *weak** topology.

Theorem 4.20 *Suppose X is a normed space and X^* is its dual. Then, the closed unit sphere $S_1^* = \{f \in X^* / \|f\| \leq 1\}$ is compact with respect to the weak*topology.*

Proof Let $C_x = [-\|x\|, \|x\|]$ for $x \in X$ and

$$C = \prod_{x \in X} C_x = \text{cross-product of all } C_x.$$

By the Tychonoff theorem (Theorem A.4(1)), C is a compact topological subspace of R^x. If $f \in S_1^*$, then $|f(x)| \leq \|f\| \|x\| \leq \|x\|$ and so $f(x) \in C_x$.

We can consider $S_1^* \subseteq C$ where $f \in S_1^*$ is associated with $(f(x_1), f(x_2), \ldots, f(x_n)$ $\ldots) \in C$ for $x_1, x_2, \ldots, x_n, \ldots \in X$.

Since C is compact, in view of Theorem A.4(5), it is sufficient to prove that S_1^* is a closed subspace of C. For this, we show that if $g \in \overline{S_1^*}$, then $g \in S_1^*$; that is to say, $\overline{S_1^*} = S_1^*$, which implies that S_1^* is closed. Let $g \in \overline{S_1^*}$; then, $g \in C$ [as $S_1^* \subseteq \bar{C} = C$ in view of Th. A.4(2)], and $|g(x)| \leq \|x\|$.

Now, we show that g is linear. Let $\varepsilon > 0$ be given and $x, y \in X$. Since every fundamental neighborhood of g intersects S_1^*, there exists an $f \in S_1^*$ such that

$$|g(x) - f(x)| < \frac{\varepsilon}{3}$$

$$|g(y) - f(y)| < \frac{\varepsilon}{3}$$

$$|g(x + y) - f(x + y)| < \frac{\varepsilon}{3}$$

Since f is linear, $f(x + y) - f(x) - f(y) = 0$ and we have

$$|g(x + y) - g(x) - g(y)|$$
$$= |[g(x + y) - f(x + y)] - [g(x) - f(x)] - [g(y) - f(y)]|$$
$$\leq |g(x + y) - f(x + y)| + |g(x) - f(x)| + |g(y) - f(y)|$$
$$< \frac{\varepsilon}{3} + \frac{\varepsilon}{3} + \frac{\varepsilon}{3}$$

As this relation is true for arbitrary $\varepsilon > 0$, we have $g(x + y) = g(x) + g(y)$. In the same way, we can show that $g(\alpha x) = \alpha g(x)$ for all real α. Thus, g is linear and bounded. Moreover, $x \neq 0$

$$\frac{1}{||x||}|g(x)| \le 1 \ or \ \left|g\left(\frac{x}{||x||}\right)\right| \le 1$$

as g is linear (For $x = 0$, $g(x) = 0$ and so $|g(x)| = 0 < 1$). Therefore, $g \in S_1^*$.

4.6   Principle of Uniform Boundedness and Its Applications

4.6.1   *Principle of Uniform Boundedness*

Theorem 4.21 *Let X be a Banach space, Y a normed space, and $\{Ti\}$ a sequence of bounded linear operators over X into Y such that $\{Ti(x)\}$ is a bounded subset of Y for all $x \in X$. Then, $\{||T_i||\}$ is a bounded subset of real numbers; i.e., $\{T_i\}$ is a bounded sequence in the normed space $\mathscr{B}[X, Y]$.*

Remark 4.5   1. This principle was first discovered by the French mathematician Henri Leon Lebesgue in 1908 during his investigation concerning the Fourier series, but the principle in its general form was given by Stefan Banach and another famous mathematician H. Steinhaus.
2. Some Russian mathematicians call Solved Problem 4.2 as the Banach–Steinhaus theorem (See Kantorovich and Akilov [107]).
3. Completeness of the space X is essential for the validity of this principle (see Solved Problem 4.3)

Proof (*Proof of Theorem 4.21*) In the first step, we shall show that the hypothesis of the theorem implies the following condition. For some $w \in X$ and some positive $\lambda > 0$, there exists a constant $k > 0$ such that

$$||T_n(x)|| < k \tag{4.25}$$

for all n and $x \in \overline{S_\lambda(w)}$. In the second step, we shall show that this condition implies the desired result. Suppose (4.25) is false; i.e., for any sphere S with an arbitrary point as the center and an arbitrary $r > 0$ as the radius, there exists an integer n_0 and $w \in S$ such that

$$||T_{n_0}(w)|| \ge r \tag{4.26}$$

Since the function $f(x) = ||T_{n_0}(x)||$ is a real-valued function of x that is greater than r at x_0, the continuity implies that there must be a neighborhood of w in which $f(x) > r$; furthermore, we can assume that this neighborhood of w is wholly contained in S. Symbolically, we are asserting the existence of a neighborhood of w, S_r such that $S_r \subset S$ and $T_{n_0}(x) > r \ \forall \ x \in \bar{S}_r$.

If we choose $r = 1$, then by the above fact, there exists some closed set S_1 such that $S_1 \subset S$ and an integer n such that $T_{n_1}(x) > 1 \ \forall x \in S_1$. Further, we can assume that diameter of S_1, $\delta(S_1) < 1$. We repeat the above procedure for S_1 instead of S, and $r = 2$. Thus, we get a closed set $S_2 \subset S_1$ such that $\delta(S_2) < 1/2$ such that for some n_2, $T_{n_2}(x) > 2 \; \forall \, x \in S_2$. Continuing this process, we get a sequence of closed sets $\{S_i\}$ satisfying the conditions of Theorem A.6(7) of Appendix A.3, and hence by this theorem, there exists some point $y \in \bigcap_i S$ which implies that $||T_{n_i}(y)|| > i$ for every i.

This will imply that the hypothesis of the theorem that $\{||Ti(x)||\}$ is a bounded subset of $Y \; \forall \, x \in X$ cannot be true, and this is a contradiction. Hence, our assumption is false and (4.25) holds good.

Let (4.6.1) be true and x be such that $||x|| \leq \lambda$. Then,

$$||T_n(x)|| = ||T_n(x + w) - T_n w||$$
$$\leq ||T_n(x + w)|| + ||T_n w|| \qquad (4.27)$$

Since $||(x + w) - w|| = ||x|| \leq \lambda, x + w \in \overline{S_{\lambda(w)}}$. This implies, by Eq. (4.25), that $||T_n(x + w)|| < k$ for every n. Since $||T_n w|| < k$, Eq. (4.27) implies that $\forall \, n ||T_n x|| < 2k$ for all x such that $||x|| \leq \lambda$. Let x be any nonzero vector and consider

$$\frac{\lambda x}{||x||}.$$

Then,

$$||T_n(x)|| = ||\frac{||x||}{\lambda} T_n(\frac{\lambda x}{||x||})|| \leq \frac{2k}{\lambda} ||x|| \; \forall n$$

The above result also holds good for $x = 0$.

From the definition of the norm of an operator, we have $||T_n|| \leq 2k/\lambda$ for all n; i.e., $\{||T_n||\}$ is a bounded sequence in $\mathscr{B}[X, Y]$. This proves the theorem.

Several applications of Theorem 4.21 in the domain of Fourier analysis and summability theory can be found in Kantorovich and Akilov [107], Goffman and Pedrick [86], Dunford and Schwartz [67], and Mazhar and Siddiqi [130, 131, 190, 190].

4.7 Open Mapping and Closed Graph Theorems

4.7.1 Graph of a Linear Operator and Closedness Property

Definition 4.4 Let X and Y be two normed spaces and D a subspace of X. Then, the linear operator T, defined on D into Y, is called *closed* if for every convergent sequence $\{x_n\}$ of points of D with the limit $x \in X$ such that the sequence $\{Tx_n\}$

is a convergent sequence with the limit y, $x \in D$ and $y = Tx$. This means that if $\lim_{n \to \infty} x_n = x \in X$, and $\lim_{n \to \infty} Tx_n = y$, then $x \in D$ and $y = Tx$.

Definition 4.5 Let X and Y be two normed spaces, D a subspace of X, and T a linear operator defined on D into Y. Then, the set of points $G_T = \{(x, Tx) \in X \times Y / x \in D\}$ is called the *graph* of T.

Remark 4.6 If X and Y are two normed spaces, then $X \times Y$ is also a normed space with respect to the norm $||(x, y)|| = ||x||_1 + ||y||_2$, $|| \cdot ||_1$ and $|| \cdot ||_2$ are norms on X and Y, respectively. We see in the next theorem that G_T is a subspace of $X \times Y$ and hence, a normed subspace of $X \times Y$ with respect to the norm

$$||(x, Tx)|| = ||x|| + ||Tx|| \ for \ x \in D$$

The following theorem gives a relationship between the graph of a linear operator and its closedness property.

Theorem 4.22 *A linear operator T is closed if and only if its graph G_T is a closed subspace.*

Proof Suppose that the linear operator T is closed, then we need to show that the graph G_T of T is a closed subspace of $X \times Y$ (D is a subspace of the normed space X, $T : D \to Y$). In other words, we have to verify the following relations:

1. If (x, Tx), $(x', Tx') \in G_T$; $x, x' \in D$, then $(x, Tx) + (x', Tx') \in G_T$.
2. If $(x, Tx) \in G_T$, $x \in D$, then $\lambda(x, Tx) \in G_T \ \forall \ real \lambda$.
3. For $x_n \in D$, if $(x_n, Tx_n) \in G_T$ converges to (x, y), then $(x, y) \in G_T$.

Verification

1. We have $(x, Tx) + (x', Tx') = (x + x', T(x + x'))$ as T is linear. Since D is a subspace, $x + x' \in D$. Therefore, $(x + x', T(x + x')) \in G_T$.
2. $\lambda(x, Tx) = (\lambda x, \lambda Tx) = (\lambda x, T(\lambda x))$ as T is linear. $\lambda x \in D$ as D is a subspace. Therefore, $(\lambda x, T(\lambda x)) \in G_T$.
3. (x_n, Tx_n) converges to (x, y) is equivalent to

$$||(x_n, Tx_n) - (x, y)|| = ||(x_n - x, Tx_n - y)||$$
$$= ||x_n - x|| + ||Tx_n - y|| \to 0$$

or

$$||x_n - x|| \to 0 \ and \ ||Tx_n - y|| \to 0 \ n \to \infty$$

that is

$$\lim_{n \to \infty} x_n = x \ and \ \lim_{n \to \infty} Tx_n = y$$

Since T is closed, $x \in D$ and $y = Tx$. Thus, $(x, y) = (x, Tx)$, where $x \in D$, and so $(x, y) = (x, Tx) \in G_T$.

To prove the converse, assume that G_T is closed and that $x_n \to x$, $x_n \in D$ for all n, and $Tx_n \to y$.

The desired result is proved if we show that $x \in D$ and $y = Tx$. The given condition implies that $(x_n, Tx_n) \to (x, y) \in \overline{G}_T$. Since G_T is closed, $\overline{G}_T = G_T$; i.e., $(x, y) \in G_T$. By the definition of G_T, $y = Tx$, $x \in D$.

Remark 4.7 1. In some situations, the above characterization of the closed linear operator is quite useful.

2. If D is a closed subspace of a normed space X and T is a continuous linear operator defined on D into another normed space Y, then T is closed (Since $x_n \to x$ and T is continuous, $Tx_n \to Tx$ which implies that if $Tx_n \to y$, the $Tx = y$, and moreover, $x \in D$ as D is a closed subspace of X.)

3. A closed linear operator need not be bounded (The condition under which a closed linear operator is continuous is given by the closed graph theorem). For this, we consider $X = Y = C[0, 1]$ with sup norm (see Example 2.9), and $P[0, 1]$ the normed space of polynomials (see Example 2.10).

We know that $P[0, 1] \subset C[0, 1]$. Let $T : D = P[0, 1] \to [0, 1]$ be defined as

$$T f = \frac{df}{dt}$$

  i. T is a linear operator; see Example 2.31.
  ii. T is closed; let $f_n \in P[0, 1]$.

$$\lim_{n \to \infty} f_n = f \quad and \quad \lim_{n \to \infty} \frac{df_n}{dt} = g$$

To prove that T is closed, we need to show that $T f = \frac{df}{dt} = g$ (It may be observed that the convergence is uniform). We have

$$\int_0^t g(s)ds = \int_0^t \lim_{n \to \infty} \frac{df_n}{ds} ds$$

$$= \lim_{n \to \infty} \int_0^t \frac{df_n}{ds} ds$$

$$= \lim_{n \to \infty} [f_n(t) - f(0)]$$

or

$$\int_0^t g(s)ds = f(t) - f(0).$$

Differentiating this relation, we obtain

$$g = \frac{df}{dt} = Tf$$

   iii. T is not bounded (continuous); see Example 2.31.

Definition 4.6 An operator T on a normed space X into a normed space Y is called open if TA is an open subset of Y whenever A is an open subset of X; i.e., T maps open sets into open sets.

4.7.2 Open Mapping Theorem

Theorem 4.23 (Open mapping theorem) *If X and Y are two Banach spaces and T is a continuous linear operator on X onto Y, then T is an open operator (open mapping).*

The proof of this theorem requires the following lemma whose proof can be found in Simmons [174, 235–236].

Lemma 4.3 *If X and Y are Banach spaces, and if T is a continuous linear operator of X onto Y, then the image of each open sphere centered on the origin in X contains an open sphere centered on the origin in Y.*

Proof (*Proof of Theorem* 4.23) In order to prove that T is an open mapping, we need to show that if O is an open set in X, then $T(O)$ is also an open set in Y. This is equivalent to proving that if y is a point in $T(O)$, then there exists an open sphere centered on y and contained in $T(O)$. Let x be a point in O such that $T(x) = y$. Since O is open, x is the center of an open sphere, which can be written in the form $x + S_r$, contained in O. By Lemma 4.3, $T(S_r)$ contains an open sphere, say S'_{r_1}. It is clear that $y + S'_{r_1}$ is an open sphere centered on y, and the fact that it is contained in $T(O)$ follows from the following relation:

$$y + S'_{r_1} \subseteq y + T(S_r) = T(x) + T(S_r) = T(x + S_r) \subseteq T(O)$$

The following result is a special case of Theorem 4.23 which will be used in the proof of the closed graph theorem.

Theorem 4.24 *A one-to-one continuous linear operator of one Banach space onto another is a homeomorphism. In particular, if a one-to-one linear operator T of a Banach space onto itself is continuous, its inverse T^{-1} is automatically continuous.*

4.7.3 The Closed Graph Theorem

Theorem 4.25 (Closed graph theorem) *If X and Y are Banach spaces, and if T is a linear operator of X into Y, then T is continuous if and only its graph G_T is closed.*

Remark 4.8 In view of the following result of topology, the theorem will be proved if we show that T is continuous whenever G_T is closed.

"If f is a continuous mapping of a topological space X into a Hausdorff space Y, then the graph of f is a closed subset of the product $X \times Y$."

Proof (Proof of Theorem 4.25) Let X_1 be the space X renormed by $||x||_1 = ||x|| + ||T(x)||$. X_1 is a normed space. Since $||T(x)|| \leq ||x|| + ||T(x)|| = ||x||_1$, T is continuous as a mapping of X_1 into Y.

We shall obtain the desired result if we show that X and X_1 have the same topology. Since $||x|| \leq ||x|| + ||T(x)|| = ||x||_1$, the identity mapping of X_1 onto Y is continuous. If we can show that X_1 is complete then, by Theorem 4.24, this mapping is a homeomorphism and, in turn, X and X_1 have the same topology.

To show that X_1 is complete, let $\{x_n\}$ be a Cauchy sequence in it. This implies that $\{x_n\}$ and $\{T(x_n)\}$ are also Cauchy sequences in X and Y, respectively. Since X and Y are Banach spaces, there exist vectors x and y in X and Y, respectively, such that $||x_n - x|| \to 0$ and $||T(x_n) - y|| \to 0$ as $n \to \infty$. The assumption that G_T is closed in $X \times Y$ implies that (x, y) lies on G_T and so $T(x) = y$. Furthermore,

$$||x_n - x||_1 = ||x_n - x|| + ||T(x_n - x)||$$
$$= ||x_n - x|| + ||T(x_n) - T(x)||$$
$$= ||x_n - x|| + ||T(x_n) - y|| \to 0 \text{ as } n \to \infty$$

Therefore, X_1 is complete.

4.8 Problems

4.8.1 Solved Problems

Problem 4.1 Let $\{T_n\}$ be a sequence of continuous linear operators of a Banach space X into a Banach space Y such that $\lim_{n\to\infty} T_n(x) = T(x)$ exists for every $x \in X$. Then, prove that T is a continuous linear operator and

$$||T|| \leq \lim_{n\to\infty} \inf ||T_n||$$

Solution

1. T is linear:

   (a) We have $T(x + y) = \lim\limits_{n\to\infty} T_n(x + y)$. Since each T_n is linear

$$T_n(x + y) = T_n(x) + T_n(y)$$
$$\lim_{n\to\infty} T_n(x + y) = \lim_{n\to\infty} [T_n(x) + T_n(y)]$$
$$= \lim_{n\to\infty} T_n(x) + \lim_{n\to\infty} T_n(y)$$
$$= T(x) + T(y)$$

   or

$$T(x + y) = T(x) + T(y)$$

   (b)

$$T(\alpha x) = \lim_{n\to\infty} T_n(\alpha x)$$
$$= \lim_{n\to\infty} \alpha T_n(x) \text{ as each } T_n \text{ is linear}$$
$$T(\alpha x) = \alpha \lim_{n\to\infty} T_n(x) = \alpha T(x)$$

   Thus, T is linear.

2. Since $\lim\limits_{n\to\infty} T_n(x) = T(x)$

$$\left\| \lim_{n\to\infty} T_n(x) \right\| = \|T(x)\|$$

   or

$$\lim_{n\to\infty} \|T_n(x)\| = \|T(x)\|$$

as a norm is a continuous function and T is continuous if and only if $x_n \to x$ implies $Tx_n \to Tx$ (see Proposition 2.1(3)).

Thus, $\|T_n(x)\|$ is a bounded sequence in Y. By Theorem 4.21 (the principle of uniform boundedness), $\|T_n\|$ is a bounded sequence in the space $B[X, Y]$. This implies that

$$\|T(x)\| = \lim_{n\to\infty} \|T_n(x)\| \leq \lim_{n\to\infty} \inf \|T\| \, \|x\| \tag{4.28}$$

By the definition of $\|T\|$, we get $\|T\| \leq \lim\limits_{n\to\infty} \inf \|T_n\|$ (see Appendix A.4 for lim inf and Definition (2.8(6)) for $\|T\|$.

In view of Eq. (4.28), T is bounded and hence continuous as T is linear.

Problem 4.2 Let X and Y be two Banach spaces and $\{T_n\}$ a sequence of continuous linear operators. Then, the limit $Tx = \lim_{n \to \infty} T_n(x)$ exists for every $x \in X$ if and only if

1. $\|T_n\| \leq M$ for $n = 1, 2, 3, \ldots$
2. The limit $Tx = \lim_{n \to \infty} T_n(x)$ exists for every element x belonging to a dense subset of X.

Solution Suppose that the limit $Tx = \lim_{n \to \infty} T_n(x)$ exists for all $x \in X$. Then clearly $Tx = \lim_{n \to \infty} T_n(x)$ exists for x belonging to a dense subset of X.

$$\|T_n\| \leq M \ for \ n = 1, 2, 3, \ldots, \ follows \ from \ Problem \ 4.1$$

Suppose (1) and (2) hold; then, we want to prove that

$$Tx = \lim_{n \to \infty} T_n(x)$$

exists, i.e., for $\varepsilon > 0$, there exists an N such that $\|T_n(x) - T(x)\| < \varepsilon$ for $n > N$. Let A be a dense subset of X, then for arbitrary $x \in X$, we can find $x' \in A$ such that

$$\|x - x'\| < \delta, \ \text{for an arbitrary} \ \delta > 0 \tag{4.29}$$

We have

$$\|T_n(x) - T(x)\| \leq \|T_n(x) - Tn(x')\| + \|T_n(x') - T(x')\|. \tag{4.30}$$

By condition (2)

$$\|T_n(x') - T(x')\| < \varepsilon_1 \ for \ n > N \tag{4.31}$$

as $x' \in A$, a dense subset of X. Since $T_n's$ are linear

$$\|T_n(x) - T_n(x')\| = \|Tn(x - x')\|$$

By condition (1) and Eq. (4.29), we have

$$\|T_n(x - x')\| \leq \|T_n\| \, \|x - x'\|$$
$$\leq M\delta, \ for \ all \ n \tag{4.32}$$

From Eqs. (4.30)–(4.31), we have

$$\|T_n(x) - T(x)\| \leq M\delta + \varepsilon_1 < \varepsilon, \ for \ all \ n > N.$$

This proves the desired result.

Problem 4.3 Show that the principle of uniform boundedness is not valid if X is only a normed space.

Solution To show this, consider the following example. Let $Y = R$ and X be the normed space of all polynomials

$$x = x(t) = \sum_{n=0}^{\infty} a_n t^n \ where \ a_n = 0 \ \forall \ n > N$$

with the norm $||x|| = \sup_n |a_n|$. X is not a Banach space. Define $T_n : X \to Y$ as follows:

$$T_n(x) = \sum_{k=0}^{n-1} a_k$$

one can check that $||T_n||$ is not bounded.

Problem 4.4 Prove Riesz's lemma (Theorem 2.3).

Solution Since Y is a proper subspace, there must exist an element $y_0 \in X - Y$. Suppose $d = \inf_{x \in Y} ||x - y_0||$. We have $d > 0$, otherwise $x = y_0 \in \overline{Y} = Y$, a contradiction of the fact that $y_0 \in X - Y$. For $\beta > 0$, we can find an element $x_0 \in Y$ such that

$$d \leq ||x_0 - y_0|| < d + \beta \tag{4.33}$$

Since $y_0 \notin Y$, the element $y = (x_0 - y_0)/(||x_0 - y_0||) \notin Y$. Moreover

$$||y|| = ||\{(x_0 - y_0)/(||x_0 - y_0||)\}||$$
$$= ||x_0 - y_0||/||x_0 - y_0|| = 1$$

we have

$$||x - y|| = ||x - \{(x_0 - y_0)/(||x_0 - y_0||)\}||$$
$$= ||x(||x_0 - y_0||) - x_0 + y_0||/(||x_0 - y_0||)$$
$$= ||y_0 - x_1||/||x_0 - y_0||$$

where $x_1 = x_0 - x||x_0 - y_0|| \in Y$. Using the definition of d and (4.33), we get

$$||x - y|| > \frac{d}{d + \beta} = 1 - \frac{\beta}{d + \beta}$$

If β is chosen such that $\frac{\beta}{d+\beta} < \alpha$, then $||x - y|| > 1 - \alpha$. This proves the lemma.

4.8.2 Unsolved Problems

Problem 4.5 Prove the Hahn–Banach theorem for a Hilbert space.

Problem 4.6 Show that $p(x) = \lim \sup x_n$, where $x = (x_n) \in \ell_\infty$, x_n real and satisfies conditions (1) and (2) of Theorem 4.1.

Problem 4.7 If p is a functional satisfying conditions (1) and (2) of Theorem 4.1, show that $p(0) = 0$ and $p(-x) \geq -p(x)$.

Problem 4.8 If $F(x) = F(y)$ for every bounded linear functional F on a normed space X, show that $x = y$.

Problem 4.9 A linear function F defined on m, satisfying the following conditions, is called a Banach limit:

1. $F(x) \geq 0$ if $x = (x_1, x_2, x_3, \ldots, x_n, \ldots)$ and $x_n \geq 0 \,\forall\, n$.
2. $F(x) = F(\sigma x)$, where $\sigma x = \sigma(x_1, x_2, x_3, \ldots) = (x_2, x_3, \ldots)$.
3. $F(x) = 1$ if $x = (1, 1, 1, \ldots)$.

Show that $\lim\limits_{n\to\infty} \inf x_n \leq F(x) \leq \lim\limits_{n\to\infty} \sup x_n \,\forall\, x = (x_n) \in m$, where F is a Banach limit.

Problem 4.10 Let $T_n = A_n$, where the operator $A : \ell_2 \to \ell_2$ is defined by $A(x_1, x_2, x_3, x_4, \ldots) = (x_3, x_4, \ldots)$. Find

$$\lim_{n\to\infty} ||T_n(x)||, \quad ||T_n|| \;and\; \lim_{n\to\infty} ||T_n||$$

Problem 4.11 Prove that the normed space $P[0, 1]$ of all polynomials with norm defined by $||x|| = \sup\limits_i |\alpha_i|$, where $\alpha_1, \alpha_2, \ldots$ are the coefficients of x, is not complete.

Problem 4.12 Show that in a Hilbert space X, a sequence $\{x_n\}$ is weakly convergent to x if and only if $\langle x_n, z \rangle$ converges to $\langle x, z \rangle$ for all $z \in X$.

Problem 4.13 Show that weak convergence in ℓ_1 is equivalent to convergence in norm.

Problem 4.14 Show that all finite-dimensional normed spaces are reflexive.

Chapter 5
Differential and Integral Calculus in Banach Spaces

Abstract In this chapter, differentiation and integration of operators defined on a Banach space into another Banach space are introduced. Basic concepts of distribution theory and Sobolev spaces are discussed, both concepts play very significant role in the theory of partial differential equations. A lucid presentation of these two topics is given.

Keywords Gâteaux derivative · Fréchet derivative · Chain rule · Mean value theorem implicit function theorem · Taylor formula · Generalized gradient (subdifferential) · Compact support · Test functions · Distribution · Distributional derivative · Dirac delta distribution · Regular distribution · Singular distribution Distributional convergence · Integral of distribution · Sobolev space · Green's formula for integration by parts · Friedrich's inequality · Poincaré inequality Sobolev spaces of distributions · Sobolev embedding theorems · Bochner integral

5.1 Introduction

As we know, the classical calculus provides foundation for science and technology and without good knowledge of finite-dimensional calculus a systematic study of any branch of human knowledge is not feasible. In many emerging areas of science and technology calculus in infinite-dimensional spaces, particularly function spaces, is required. This chapter is devoted to differentiation and integration of operators and distribution theory including Sobolev spaces. These concepts are quite useful for solutions of partial differential equations modeling very important problems of science and technology.

© Springer Nature Singapore Pte Ltd. 2018
A. H. Siddiqi, *Functional Analysis and Applications*, Industrial and Applied Mathematics, https://doi.org/10.1007/978-981-10-3725-2_5

5.2   The Gâteaux and Fréchet Derivatives

5.2.1   The Gâteaux Derivative

Throughout this section, U and V denote Banach spaces over R, and S denotes an operator on U into V ($S: U \to V$).

Definition 5.1 (*Gâteaux Derivative*) Let x and t be given elements of U and

$$\lim_{\eta \to 0} \left|\left| \frac{S(x + \eta t) - S(x)}{\eta} - DS(x)t \right|\right| = 0 \tag{5.1}$$

for every $t \in X$, where $\eta \to 0$ in R. $DS(x)t \in Y$ is called the value of the Gâteaux derivative of S at x in the direction t, and S is said to be *Gâteaux differentiable* at x in the direction t. Thus, the Gâteaux derivative of an operator S is itself an operator often denoted by $DS(x)$.

Remark 5.1 (a) If S is a linear operator, then

$$DS(x)t = S(t), \; that \; is, \; DS(x) = S \; for \; all \; x \in U$$

(b) If $S = F$ is a real-valued functional on U; that is, $S: U \to R$, and F is Gâteaux differentiable at some $x \in U$, then

$$DS(x)t = \left[\frac{d}{d\eta} F(x + \eta t) \right]_{\eta=0} \tag{5.2}$$

(c) We observe that the Gâteaux derivative is a generalization of the idea of the directional derivative well known in finite dimensions.

Theorem 5.1 *The Gâteaux derivative of an operator S is unique provided it exists.*

Proof Let two operators $S_1(t)$ and $S_2(t)$ satisfy (5.1). Then, for every $t \in U$ and every $\eta > 0$, we have

$$||S_1(t) - S_2(t)|| = ||(\frac{S(x + \eta t) - S(x)}{\eta} - S_1(t))$$

$$- \left((\frac{S(x + \eta t) - S(x)}{\eta} - S_2(t) \right)$$

$$\leq \left|\left| \left(\frac{S(x + \eta t) - S(x)}{\eta} - S_1(t) \right) \right|\right|$$

$$+ ||(\frac{S(x + \eta t) - S(x)}{\eta} - S_2(t))|| \to 0$$

$$as \;\; \eta \to 0$$

Therefore, $||S_1(t) - S_2(t)|| = 0$ for all $t \in X$ which implies that $S_1(t) = S_2(t)$.

Definition 5.2 *(Gradient of a Functional)* Let F be a functional on U. The mapping $x \to DF(x)$ is called the *gradient* of F and is usually denoted by ∇F.

It may be observed that the gradient ∇ is a mapping from U into the dual space U^* of U.

Example 5.1 Let $U = R^n$, $V = R$, $e_1 = (1, 0, 0, \ldots)$, $e_2 = (0, 1, 0, 0, \ldots), \ldots, e_n = (0, 0, \ldots, 0, 1)$. Then

$$DF(x)e_i = \frac{\partial F}{\partial x_i}, \quad i = 1, 2, \ldots n$$

where $F : R^n \to R$ and $\frac{\partial F}{\partial x_i}$ is the partial derivative of F.

Example 5.2 Let $F : R^2 \to R$ be defined by

$$F(x) = \frac{x_1 x_2^2}{x_1^2 + x_2^2}$$

where $x = (x_1, x_2) \in R^2$, $x \notin (0, 0)$ and $F(0) = 0$. Then, for $t = (t_1, t_2)$

$$DF(0)t = \lim_{\eta \to 0} \frac{F(0 + \eta t) - F(0)}{\eta}$$

$$= \lim_{\eta \to 0} \frac{F(\eta t)}{\eta}$$

$$= \lim_{\eta \to 0} \frac{(\eta t_1)(\eta t_2)^2}{\eta[(\eta t_1)^2 + (\eta t_2)^2]}$$

$$= \lim_{\eta \to 0} \frac{t_1 t_2^2}{t_1^2 + t_2^2}$$

$$= \frac{t_1 t_2^2}{t_1^2 + t_2^2}$$

Example 5.3 Let $F : R^2 \to R$ be defined as

$$F(x) = \frac{x_1 x_2}{x_1^2 + x_2^2}, \quad x = (x_1, x_2) \neq 0 \text{ and } F(0) = 0.$$

Then

$$DF(0)t = \lim_{\eta \to 0} \frac{F(\eta t)}{\eta}$$

$$= \lim_{\eta \to 0} \frac{\eta^2 t_1 t_2}{\eta[\eta^2 t_1^2 + \eta^2 t_2^2]}$$

where $t = (t_1, t_2)$.

$DF(0)t$ exists if and only if

$$t = (t_1, 0) \text{ or } t = (0, t_2)$$

It is clear from this example that the existence of the partial derivatives is not a sufficient condition for the existence of the Gâteaux derivative.

Example 5.4 Let $U = R^n$, $F: R^n \to R$, $x = (x_1, \ldots, x_n) \in R^n$ and $t = (t_1, t_2, \ldots, t_n) \in R^n$. If F has continuous partial derivatives of order 1, then the Gâteaux derivative of F is

$$DF(x)t = \sum_{k=1}^{n} \frac{\partial F(x)}{\partial x_k} t_k \qquad (5.3)$$

For a fixed $a \in U$, the Gâteaux derivative

$$DF(x)t = \left[\sum_{k=1}^{n} \frac{\partial F(x)}{\partial x_k} t_k \right] \qquad (5.4)$$

is a bounded linear operator on R^n into R^n (we know that $(R^n)^\star = R^n$). $DF(a)t$ can also be written as the inner product

$$DF(a)t = \langle y, t \rangle \qquad (5.5)$$

where

$$y = \left(\frac{\partial F(a)}{\partial x_1}, \frac{\partial F(a)}{\partial x_2}, \ldots, \frac{\partial F(a)}{\partial x_n} \right).$$

Example 5.5 Let $X = R^n$, $Y = R^m$, and $F = (F_1, F_2, F_3, \ldots, F_m): R^n \to R^m$ be Gâteaux differentiable at some $x \in R^n$. The Gâteaux derivative can be identified with an $m \times n$ matrix (a_{ij}). If t is the jth coordinate vector, $t = e_j = (0, 0, \ldots, 1, 0, \ldots, 0)$, then

$$\lim_{\eta \to 0} \left\| \frac{F(x + \eta t) - F(x)}{\eta} - DF(x)t \right\| = 0$$

implies

$$\lim_{\eta \to 0} \left\| \frac{F(x + \eta e_j) - F_i(x)}{\eta} - a_{ij} \right\| = 0$$

for every $i = 1, 2, 3, \ldots, m$ and $j = 1, 2, 3, \ldots, n$. This shows that F_i's have partial derivatives at x and $\frac{\partial F_i(x)}{\partial x_j} = a_{ij}$, for every $i = 1, 2, 3, \ldots, m$ and $j = 1, 2, 3, \ldots, n$.

The Gâteaux derivative of F at x has the matrix representation

$$
\begin{bmatrix}
\frac{\partial F_1(x)}{\partial x_1} & \cdots & \frac{\partial F_1(x)}{\partial x_n} \\
\cdot & \cdots & \cdots \\
\frac{\partial F_m(x)}{\partial x_1} & \cdots & \frac{\partial F_m(x)}{\partial x_n}
\end{bmatrix} = (a_{ij}) \tag{5.6}
$$

This is called the *Jacobian matrix* of F at x. It is clear that if $m = 1$, then the matrix reduces to a row vector which is discussed in Example 5.4.

Example 5.6 Let $U = C[a, b]$, $K(u, v)$ be a continuous real function on $[a, b] \times [a, b]$ and $g(v, x)$ be a continuous real function on $[a, b] \times R$ with continuous partial derivative $\frac{\partial g}{\partial x}$ on $[a, b] \times R$. Suppose that F is an operator defined on $C[a, b]$ into itself by

$$
F(x)(s) = \int_a^b K(s, v) g(v, x(v)) dv \tag{5.7}
$$

Then

$$
DF(x)h = \int_a^b K(s, v) \left[\frac{\partial}{\partial x} g(v, x(v)) \right] h(v) dv
$$

Theorem 5.2 (Mean Value Theorem) *Suppose that the functional F has a Gâteaux derivative $DF(x)t$ at every point $x \in U$. Then for any points x, $x + t \in U$, there exists a $\xi \in (0, 1)$ such that*

$$
F(x + t) - F(x) = DF(x + \xi t)t \tag{5.8}
$$

Proof Put $\varphi(\alpha) = F(x + \alpha t)$. Then

$$
\begin{aligned}
\varphi'(\alpha) &= \lim_{\beta \to 0} \left[\frac{\varphi(\alpha + \beta) - \varphi(\alpha)}{\beta} \right] \\
&= \lim_{\beta \to 0} \frac{F(x + \alpha t + \beta t) - F(x + \alpha t)}{\beta} \\
&= DF(x + \alpha t)t
\end{aligned}
$$

By the Mean Value Theorem for real function of one variable applied to φ, we get

$$
\varphi(1) - \varphi(0) = \varphi'(\xi) \text{ for some } \xi \in (0, 1).
$$

This implies that

$$
F(x + t) - F(x) = DF(x + \xi t)t
$$

5.2.2   The Fréchet Derivative

Definition 5.3 (*Fréchet Derivative*) Let x be a fixed point in a Banach space U and V be another Banach space. A continuous linear operator $T: U \to V$ is called the *Fréchet derivative* of the operator $S: X \to Y$ at x if

$$S(x + t) - S(x) = T(t) + \varphi(x, t) \tag{5.9}$$

and

$$\lim_{||t|| \to 0} \frac{||\varphi(x, t)||}{||t||} = 0 \tag{5.10}$$

or, equivalently,

$$\lim_{||t|| \to 0} \frac{||S(x + t) - S(x) - T(t)||}{||t||} = 0. \tag{5.11}$$

The Fréchet derivative of T at x is usually denoted by $dT(x)$ or $T'(x)$. T is called *Fréchet differentiable* on its domain if $T'(x)$ exists at every point of the domain.

Remark 5.2 (a) If $U = R$, $V = R$, then the classical derivative $f'(x)$ of real function $f: R \to R$ at x defined by

$$f'(x) = \lim_{t \to 0} \frac{f(x + t) - f(x)}{t} \tag{5.12}$$

is a number representing the slope of the graph of the function f at x. The Fréchet derivative of f is not a number, but a linear operator on R into R. The existence of the classical derivative $f'(x)$ implies the existence of the Fréchet derivative at x, and by comparison of Eqs. (5.9) and (5.12) written in the form

$$f(x + t) - f(x) = f'(x)t + tg(t) \tag{5.13}$$

we find that T is the operator which multiplies every t by the number $f'(x)$.

(b) In classical calculus, the derivative at a point x is a local linear approximation of the given function in the neighborhood of x. Fréchet derivative can be interpreted as the best local linear approximation. One can consider the change in S when the argument changes from x to $x + t$, and then approximate this change by a linear operator T so that

$$S(x + t) = S(x) + T(t) + E \tag{5.14}$$

where E is the error in the linear approximation.

Thus, E has the same order of magnitude as t except when S is equal to the

Fréchet derivative of T. $E = 0(t)$, so that E is much smaller than t as $t \to 0$. Thus, the Fréchet derivative provide the best linear approximation of T near x.

(c) It is clear from the definition (Eq. (5.11)) that if T is linear, then

$$dS(x) = S(x)$$

that is, if S is a linear operator, then the Fréchet derivative (linear approximation) of S is S itself.

Theorem 5.3 *If an operator has the Fréchet derivative at a point, then it has the Gâteaux derivative at that point and both derivatives have equal values.*

Proof Let $T: X \to Y$, and suppose T has the Fréchet derivative at x, then

$$\lim_{||t|| \to 0} \frac{||S(x + t) - S(x) - T(t)||}{||t||} = 0$$

for some bounded linear operator $S: X \to Y$. In particular for any fixed nonzero $t \in X$, we have

$$\lim_{\eta \to 0} ||\frac{S(x + \eta t) - T(x)}{\eta} - S(t)|| = \lim_{\eta \to 0} \frac{||S(x + \eta t) - S(x) - T(\eta t)||}{||\eta t||} ||t||$$

Thus we see that T is the Gâteaux derivative of S at x.

By Theorem 5.1, Gâteaux derivative is unique and hence the Fréchet derivative is also unique. Example 5.7 shows that the converse of Theorem 5.3 does not hold true, in general.

Theorem 5.4 *Let Ω be an open subset of X and $S: \Omega \to Y$ have the Fréchet derivative at an arbitrary point a of Ω. Then S is continuous at a. This means that every Fréchet differentiable operator defined on an open subset of a Banach space is continuous.*

Proof For $a \in \Omega$, let $\varepsilon > 0$ be such that a $+ t \in \Omega$ whenever $||t|| < \varepsilon$. Then

$$||S(a + t) - S(a)|| = ||T(t) + \varphi(a, t)|| \to 0 \text{ as } ||t|| \to 0$$

Therefore, T is continuous at a.

It may be observed that results of classical calculus can be extended to Fréchet derivatives. For example, the usual rules of sum and product in case of functions of two or more variables apply to Fréchet derivatives.

We present now extension of the Chain Theorem, the Mean Value Theorem, the Implicit Function Theorem, and Taylor's Formula to Fréchet differentiable operators.

Theorem 5.5 (Chain Rule) *Let X, Y, Z be real Banach spaces. If $T : X \to Y$ and $S : Y \to Z$ are Fréchet differentiable at x and $y = T(x) \in Y$, respectively, then $U = SoT$ is Fréchet differentiable at x and*

$$U'(x) = S'(T(x))T'(x)$$

Proof For $x, t \in X$, we have

$$
\begin{aligned}
U(x + t) - U(x) &= S(T(x + t)) - S(T(x)) \\
&= S(T(x + t) - T(x) + T(x)) - S(y) \\
&= S(z + y) - S(y)
\end{aligned}
$$

where $z = T(x + t) - T(x)$. Thus

$$||U(x + t) - U(x) - S'(y)z|| = o(||z||)$$

Since $||z - T'(x)t|| = o(||t||)$, we get

$$
\begin{aligned}
||U(x &+ t) - U(x) - S'(y)T'(x)t|| \\
&= ||U(x + t) - U(x) - S'(y)z + S'(y)z - S'(y)T'(x)t|| \\
&= (||t||) + (||z||)
\end{aligned}
$$

In view of the fact that T is continuous at x, by Theorem 5.4, we obtain $||z|| = o(||t||)$ and so

$$U'(x) = T'(S(x))S'(x).$$

We require the following notation in the Mean Value Theorem and Taylor's formula: If a and b are two points of a vector space, the notation

$$[a, b] = \{x = \alpha a + (1 - \alpha)b \in X / \alpha \in [0, 1]\}$$
$$]a, b[= \{x = \alpha a + (1 - \alpha)b \in X / \alpha \in (0, 1)\}$$

are used to denote, respectively, the closed and open segments with end-points a and b.

Theorem 5.6 (Mean Value Theorem) *Let $S : K \to Y$, where K is an open convex set containing a and b, Y is a normed space and $S'(x)$ exists for each $x \in]a, b[$ and $S(x)$ is continuous on $]a, b[$. Then*

$$||S(b) - S(a)|| \leq \sup_{y \in]a,b[} ||S'(y)|| \, ||b - a|| \qquad (5.15)$$

Proof Let $F \in Y^\star$ and the real function h be defined by

$$h(\alpha) = F((T(1 - \alpha)a + \alpha(b))) \text{ for } \alpha \in [0, 1]$$

Applying the Mean Value Theorem of the classical calculus to h, we have, for some $\overline{\alpha} \in [0, 1]$ and $\overline{x} = (1 - \overline{\alpha})a + \overline{\alpha}b$

$$\begin{aligned}
F(S(b) - S(a)) &= F(S(b)) - F(S(a)) \\
&= h(1) - h(0) \\
&= h'(\overline{\alpha}) = F(S'(\overline{x})(b - a))
\end{aligned}$$

where we have used the chain rule and the fact that a bounded linear functional is its own derivative. Therefore, for each $F \in Y^*$

$$||F(S(b) - S(a))|| \le ||F|| \, ||S'(\overline{x})|| \, ||b - a|| \tag{5.16}$$

Now, if we define the function F_0 on the subspace $[S(b) - S(a)]$ of Y as

$$F_0(\lambda(F(b))) - F(a) = \lambda$$

then $||F_0|| = ||T(b) - T(a)||^{-1}$. If F is a Hahn–Banach extension of F_0 to the entire space Y (Theorem 4.2), we find by substituting in (5.16) that

$$1 = ||F(S(b) - S(a))|| \le ||S(b) - S(a)|| - 1||T'(\overline{x})|| \, ||b - a||$$

which gives (5.15).

Theorem 5.7 (Implicit Function Theorem) *Suppose that X, Y, Z are Banach spaces, C is an open subset of $X \times Y$ and $T : C \to Z$ is continuous. Suppose further that for some $(x_1, y_1) \in C$*

(i) *$T(x_1, y_1) = 0$*
(ii) *The Fréchet derivative of $S(\cdot, \cdot)$ when x is fixed is denoted by $S_y(x, y)$, called the partial Fréchet derivative with respect to y, exists at each point in a neighborhood of (x_1, y_1) and is continuous at (x, y)*
(iii) *$[S_y(x_1, y_1)]^{-1} \in \mathscr{B}[Z, Y]$*

Then there is an open subset D of X containing x_1 and a unique continuous mapping $y : D \to Y$ such that $S(x, y(x)) = 0$ and $y(x_1) = y_1$.

Proof For the sake of convenience, we may take $x_1 = 0$, $y_1 = 0$. Let $A = [T_y(0, 0)]^{-1} \in \mathscr{B}[Z, Y]$. Since C is an open set containing $(0, 0)$, we find that

$$0 \in C_x = \{y \in Y / (x, y) \in C\}$$

for all x sufficiently small, say $||x|| \le \delta$. For each x having this property, we define a function

$$T(x, \cdot) : C_x \to Y$$

by

$$S(x, y) = y - AS(x, y).$$

In order to prove the theorem, we must prove (i) the existence of a fixed point for $T(x, \cdot)$ under the condition that $||x||$ is sufficiently small, and (ii) continuity of the mapping $x \to y(x)$ and $y(x_1) = y_1$. Now

$$S_y(x, y)(u) = u - AS_y(x, y)(u)$$

and $AA^{-1} = AS_y(0, 0)$; therefore, assumptions on S guarantee the existence of $T_y(x, y)$ for sufficiently small $||x||$ and $||y||$, and

$$T_y(x, y)(u) = A[T_y(0, 0) - T_y(x, y)](u).$$

Hence

$$||T_y(x, y)|| \le ||A|| \, ||S_y(0, 0) - S_y(x, y)||.$$

Since S_y is continuous at $(0, 0)$, there exists a constant $L > 0$ such that

$$||S_y(x, y)|| \le L \tag{5.17}$$

for sufficiently small $||x||$ and $||y||$, say, $||x|| \le \varepsilon_1 \le \delta$ and $||y|| \le \varepsilon_2$. Since T is continuous at $(0, 0)$, there exists an $\varepsilon \le \varepsilon_1$ such that

$$||T(x, 0)|| = ||AS(x, 0)|| < \varepsilon_2(1 - L) \tag{5.18}$$

for all x with $||x|| \le \varepsilon$.

We now show that $T(x, \cdot)$ maps the closed ball $S\varepsilon(0) = \{y \in Y / ||y|| \le \varepsilon_2\}$ into itself. For this, let $||x|| \le \varepsilon$ and $||y|| \le \varepsilon_2$. Then by (5.15), (5.17), and (5.18), we have

$$||T(x, y)|| \le ||T(x, y) - T(x, 0)|| + ||T(x, 0)||$$
$$\le \sup_{0 \le \alpha < 1} ||T'y(x, y)|| \, ||y|| + ||T(x, 0)||$$
$$\le L\varepsilon_2 + \varepsilon_2(1 - L) = \varepsilon_2$$

Therefore, for $||x|| < \varepsilon$, $S(x, \cdot) : \overline{S}_{\varepsilon_2}(0) \to \overline{S}_{\varepsilon_2}(0)$. Also, for $y_1, y_2 \in S_{\varepsilon_2}(0)$, we obtain by (5.15) and (5.17)

$$||T(x, y_1) - T(x, y_2)|| \le \sup_{||y|| \le \varepsilon_2} ||T_y(x, y)|| ||y_1 - y_2||$$
$$\le L||y_1 - y_2||$$

The Banach Contraction Mapping Theorem (Theorem 1.1) guarantees that for each x with $||x|| < \varepsilon$, there exists a unique $y(x) \in \overline{S}_{\varepsilon_2}(0)$ such that

$$y(x) = T(x, y(x)) = y(x) - AS(x, y(x)).$$

that is, $T(x, y(x)) = 0$. By uniqueness of y, we have $y(0) = 0$ since $T(0, 0) = 0$. Finally, we show that $x \to y(x)$ is continuous. For if $||x_1|| < \varepsilon$ and $||x_2|| < \varepsilon$, then selecting $y_0 = y(x_2)$ and $y_1 = T(x_1, y_0)$, we have by the error bound for fixed point iteration on the mapping $T(x_1, \cdot)$ (Theorem 1.1, Problem 1.17)

$$||y(x_2) - y(x_1)|| \leq \frac{1}{1 - L}||y_0 - y_1||$$

We can write

$$y_0 - y_1 = y(x_2) - S(x_1, y(x_2)) = S(x_2, y(x_2)) - S(x_1, y(x_2))$$
$$= -A[T(x_2, y(x_2)) - T(x_1, y(x_2))]$$

Therefore, by the continuity of T, $||y(x_2) - y(x_1)||$ can be made arbitrarily small for $||x_2 - x_1||$ sufficiently small.

Corollary 5.1 *If, in addition to conditions of Theorem 5.7, $S_x(x, y)$ also exists on the open set and is continuous at (x_1, y_1), then $F : x \to y(x)$ has a Fréchet derivative at x_1 given by*

$$F'(x_1) = -[S_y(x_1, y_1)]^{-1} S_x(x_1, y_1) \tag{5.19}$$

Proof We set $x = x_1 + h$ and $G(h) = F(x) - y_1$. Then $G(0) = 0$, and

$$||G(h) + [S_y(x_1, y_1)]^{-1}h||$$
$$\leq ||[S_y(x_1, y_1)]^{-1}|| \; ||S_y(x_1, y_1)G(h) + S_x(x_1, y_1)h||$$

and

$$S_y(x_1, y_1)G(h) - S_x(x_1, y_1)h = -S(x_1 + h, y_1 + G(h)) + S(x_1, y_1)$$
$$+ S_y(x_1, y_1)G(h) + S_x(x_1, y_1)h$$

If θ_1, θ_2 are numbers in $(0, 1)$, then

$$||S_y(x_1, y_1)G(h) + S_x(x_1, y_1)h|| \leq \sup_{\theta_1, \theta_2} ||S_x(x_1 + \theta_1 h, y_1 + \theta_2 G(h))$$
$$- S_x(x_1, y_1)|| \; ||h|| + \sup_{\theta_1, \theta_2} ||S_y(x_1 + \theta_1 h, y_1 + \theta_2 G(h))$$
$$- S_y(x_1, y_1)|| \; ||G(h)||$$

Thus, applying continuity of S_x, S_y for $\varepsilon > 0$, we find that $\delta = \delta(\varepsilon)$ such that on $||x - x_1|| \leq \delta$, we have

$$||G(h) + [S_y(x_1, y_1)]^{-1} S_x(x_1, y_1)h||$$
$$\leq \frac{[S_y(x_1, y_1)]^{-1}\varepsilon[1 + ||[S_y(x_1, y_1)]^{-1} S_x(x_1, y_1)||\,||h||}{1 - ||[S_y(x_1, y_1)^{-1}||}$$

The coefficient of $||h||$ can be made as small as needed as $||h|| \to 0$. Thus

$$||F(x) - [F(x_1) - [S_y(x_1, y_1)]^{-1} S_x(x_1, y_1)(x - x_1)]|| = (||x - x_1||)$$

Definition 5.4 Let $S: X \to Y$ be Fréchet differentiable on an open set $\Omega \subseteq X$ and the first Fréchet derivative S' at $x \in \Omega$ is Fréchet differentiable at x, then the Fréchet derivative of S' at x is called the second derivative of S at x and is denoted by $S''(x)$.

It may be noted that if $S: X \to Y$ is Fréchet differentiable on an open set $\Omega \subset X$, then S' is a mapping on X into $\mathscr{B}[X, Y]$. Consequently, if $S''(x)$ exists, it is a bounded linear mapping from X into $\mathscr{B}[X, Y]$. If T'' exists at every point of Ω, then $T'': X \to \mathscr{B}[X, \mathscr{B}[X, Y]]$.

Let $S: \Omega \subset X \to Y$ and let $[a, a+h]$ be any closed segment in Ω. If T is Fréchet differentiable at a, then

$$S(a + h) = S(a) + S'(a)h + ||h||\varepsilon(h), \ \lim_{h \to h} \varepsilon(h) = 0$$

Theorem 5.8 (Taylor's Formula for Twice Fréchet Differentiable Functions) $S: \Omega \subset X \to Y$ and let $[a, a + h]$ be any closed segment lying in Ω. If S is differentiable in Ω and twice differentiable at a, then

$$S(a + h) = S(a) + S'(a)h + \frac{1}{2}(S''(a)h)h + ||h||^2\varepsilon(h)$$
$$\lim_{h \to 0} \varepsilon(h) = 0$$

For proofs of these two theorems and other related results, we refer to Cartan [34], Dieudonné [64], Nashed [143].

Example 5.7 Let $S: R^2 \to R$ be defined by

$$S(x.y) = \begin{cases} \frac{x^3 y}{x^4 + y^2}, & \text{if } x \neq 0 \text{ and } y \neq 0 \\ 0, & \text{if } x = 0 \text{ and } y = 0. \end{cases}$$

It can be verified that S has the Gâteau derivative at $(0, 0)$ and its value is 0. However, it is not Fréchet differentiable at $(0, 0)$ as

$$\frac{|S(x, x^2)|}{||(x, x^2)||} = \frac{|x^3 x^2|}{(x^4 + x^4)} \frac{1}{\sqrt{x^2 + x^4}} = \frac{1}{2} \frac{1}{\sqrt[3]{(1 + x^2)}} \to \frac{1}{2} \text{ as } x \to 0$$

Example 5.8 Let $T: H \to R$ be a functional on a Hilbert space H. Let it be Fréchet differentiable at some $x \in H$, then its Fréchet derivative T' must be a bounded linear functional on H; that is, $T' \in H^\star$. By the Riesz Representation Theorem (Theorem 3.19), there exists an element $y \in H$ such that $T'(x)(z) = \langle z, y \rangle$ for every $z \in H$. The Fréchet derivative $T'(x)$ can thus be identified as y which is named the *gradient* of T with x denoted by $\nabla T(x)$; that is, $T'(x)v = \langle \nabla T(x), v \rangle \, \forall \, v \in H$.

Example 5.9 Let $T: H \to H$ be a bounded linear operator on a Hilbert space H into itself and $F: H \to R$ be a functional defined by $F(x) = \langle x, Tx \rangle$. Then the Fréchet derivative of F is written as

$$F'(x)z = \langle z, (T + T^\star)x \rangle$$

where $T^\star$ is adjoint of T.

Example 5.10 Let T be a self-adjoint bounded operator on a Hilbert space H and φ the quadratic functional defined by

$$\varphi(x) = \frac{1}{2} \langle Tx, x \rangle.$$

Then φ is Fréchet differentiable and gradient of φ, $\nabla \varphi = T$. Furthermore, φ is convex if T is strictly positive.

Example 5.11 Let $T : \ell_2^2 \to R$ be defined by

$$T(x) = x_1 x_2 + x_1^2.$$

Then

$$T'(x) = dT(x) = (x_2 + 2x_1, x_1).$$

Example 5.12 Let $T : C[0, 1] \to C[0, 1]$ be defined by

$$T(x(t)) = x^2(t).$$

Then

$$T'(x)z = 2xz \text{ or}$$
$$dT(x) = T'(x) = 2x \text{ and}$$
$$T''(x) = 2I.$$

Example 5.13 Let $F : L_2(0, 1) \to R$ be defined by

$$F(f) = \int_0^1 f^2(t)dt.$$

Then

$$dF(f) = F'(f) = 2f.$$
$$d^2 F(f) = F''(f) = 2I.$$

Example 5.14 Let the Hammerstein operator $T : C[a, b] \to C[a, b]$ be given by

$$(T(u(t)))(x) = \int_a^b K(x, t) f(t, u(t)) dt$$

where $K(\cdot, \cdot) : [a, b] \times [a, b] \to R$ and $f : [a, b] \times R \to R$ are known. The following relation holds for infinitely differentiable functions:

$$dT(u(t))(z) = \int_a^b K(x, t) \frac{\partial f}{\partial u}(t, u(t)) z(t) dt.$$

Equivalently, the Fréchet derivative of the Hammerstein operator is another integral operator with the kernel $K(x, t) \frac{\partial f}{\partial u}(t, u(t))$

Example 5.15 Consider the map $S : R \to X$, defined by

$$S(t) = e^{At} x$$

where $x \in X$, $A \in B(X)$ and e^A is defined as in Example 2.48. Then

$$dS(t) = Ae^{At} x = AS(t).$$

5.3   Generalized Gradient (Subdifferential)

Definition 5.5 (*Lipschitz Continuity*) Let $\Omega \subset X$, S an operator from Ω into Y. We say that T is Lipschitz (with modulus $\alpha \geq 0$) on Ω, if

$$||S(x_1) - S(x_2)|| \leq \alpha ||x_1 - x_2|| \ \forall \ x_1, x_2 \in \Omega.$$

S is called *Lipschitz near x* (with modulus α) if, for some $\varepsilon > 0$, T is Lipschitz with modulus α on $S_\varepsilon(x)$. If S is Lipschitz near $x \in \Omega$, we say that S is locally Lipschitz on Ω. α is called the Lipschitz exponent.

Definition 5.6 (*Monotone Operators*) Let $S : X \to X^*$, then S is called monotone if

$$(Su - Sv, u - v) \geq 0 \ for \ all \ u, v \in X$$

Note $(\cdot, \cdot)$ denotes the duality between X and $X^\star$; that is, also the value of $Su - Sv$ at $u - v$. In Hilbert space setting, $(\cdot, \cdot)$ becomes the inner product. In this chapter, we have used the notation $\langle \cdot, \cdot \rangle$ also for the duality.
S is called *strictly monotone* if

$$\langle Su - Sv, u - v \rangle > 0 \; for all \; u, v \in X$$

T is called *strongly monotone* if there is a constant $k > 0$ such that

$$\langle Tu - Tv, u - v \rangle \geq k||u - v||^2 \; for all \; u, v \in X$$

Definition 5.7 Let $S : H \to 2^{H^*}$ be a *multivalued operator* on H into $H^\star$. The operator T is said to be *monotone* if

$$(\xi - \eta, u - v) \geq 0 \; \forall \, u, v \in H \; and \; \forall \, \xi \in S(u) \; and \; \eta \in S(v)$$

A monotone operator S is called *maximal monotone* if it is monotone and there does not exist $\overline{S}: H \to 2^{H^*}$ such that $\overline{S}$ is monotone and $Gr(S) \nsubseteq Gr(\overline{S})$; that is, the graph of S does not have any proper extension which is the graph of a monotone operator.

Definition 5.8 (*Generalized Gradient*) The generalized gradient (subdifferential) of $F : X \to R$ at $x \in X$, denoted by $\partial F(x)$, is the set

$$\partial F(x) = \{F' \in X^\star / F'(h) \leq F(x + h) - F(x), \; h \in X\}.$$

An element F' of $\partial F(x)$ is called a subgradient or support functional at x.

Theorem 5.9 ([200]) *If $F : X \to R$ is a convex functional on a normed space X, and F has a Gâteaux derivative at $x \in X$, then $F(x)$ has a unique generalized gradient at x and $\partial F(x) = DF(x)$. Conversely, if $F(x) < \infty$ and the generalized gradient $\partial F(x)$ reduces to a unique subgradient, then $F(x)$ has a Gâteaux derivative at x and $\partial F(x) = DF(x)$.*

Theorem 5.10 *Let $F : H \to R \cup \{\infty\}$, where let H be a Hilbert space, then the generalized gradient $\partial F(x)$ is a monotone operator.*

Proof If $\partial F(x)$ or $\partial F(y)$ is empty, then clearly

$$\langle \partial F(x) - \partial F(y), x - y \rangle \geq 0$$

is satisfied. If this is not the case, choose $F_1 \in \partial F(x)$ and $F_2 \in \partial F(y)$. Then

$$\langle F_1, x - y \rangle \geq F(x) - F(y) \; for \; all \; y \in H$$
$$\langle F_2, y - x \rangle \geq F(y) - F(x) \tag{5.20}$$

By changing sign in second inequality of (5.20), we get

$$\langle -F_2, x - y \rangle \geq F(y) - F(x) \tag{5.21}$$

By adding (5.20) and (5.21), we get

$$\langle F_1 - F_2, x - y \rangle \geq 0.$$

Hence, $\partial F(x)$ is a monotone operator.

Example 5.16 Let

$$f(x) = |x|, \; f : R \to R$$
$$\partial f(x) = \{ \text{sgn} x \} \text{ if } x \neq 0, \, \text{sgn} x = \frac{x}{|x|}, \; x \neq 0$$
$$\partial f(x) = [-1, 1] \text{ if } x = 0.$$

For more details of generalized gradient, we refer to Outrata, Koçvara and Zowe [148] and Rockafellar and Wets [163].

5.4   Some Basic Results from Distribution Theory and Sobolev Spaces

Distribution theory was invented by the French Mathematician Laurent Schwartz around 1950 to resolve the discrepancy created by the Dirac delta function having value zero at all points except one point and having Lebesgue integral

1. This function is named after the famous physicist P.M. Dirac who introduced it in 1930. This contradicted the celebrated Lebesgue theory of integral. A class of Lebesgue integrable functions was introduced by the Russian scientist S.L. Sobolev around 1936 which has been found very useful in many areas of current interest and now known as Sobolev space. The theory of Sobolev space provides a solid foundation for modern theory of ordinary and partial differential equations. We present here important features of these two topics, namely distributions and Sobolev spaces.

5.4.1   Distributions

Let n be a positive integer. A vector of n-tuples $\alpha = (\alpha_1, \alpha_2, \ldots, \alpha_n)$, where $\alpha_i, \; i = 1, 2, \ldots, n$ are nonnegative integers, is called multi-index of dimension n. The number $|\alpha| = \sum_{i=1}^{n} \alpha_i$ is called the magnitude or length of the multi-index.

For given α, β, $\alpha + \beta$ means $(\alpha_1 + \beta_1, \alpha_2 + \beta_2, \ldots, \alpha_n + \beta_n)$

$$\alpha! = \alpha_1! \alpha_2! \ldots \alpha_n!$$
$$C_\beta^\alpha = \frac{\alpha}{\beta!(\alpha - \beta)!}$$
$$x^\alpha = x_1^{\alpha_1} x_2^{\alpha_2} \ldots x_n^{\alpha_n} \; where \; x = (x_1, x_2, \ldots x_n) \in R^n$$

We say that multi-index α, β are related by $\alpha \leq \beta$ if $\alpha_i \leq \beta_i$ for all $i = 1, 2, 3, n$.

Calculus of functions of several variables, specially partially differential for a function of n variables $f = f(x)$, $x = (x_1, x_2, \ldots, x_n)$. Laurent Schwartz introduced the concept of multi-index and a new notation of derivatives given below. In the new terminology, D^α will denote the expression

$$\frac{\partial^{|\alpha|}}{\partial_{x_1}^{\alpha_1} \partial_{x_2}^{\alpha_2} \ldots \partial_{x_n}^{\alpha_n}} D^\alpha f$$

will be called derivative of f of order $|\alpha|$. For $n = 1$, $\alpha_1 = 1$, $D^\alpha f = \frac{\partial^1 f}{\partial x_1}$ which is the classical derivative of a function of single variable denoted by $\frac{df}{dx}(x_1 = x)$. For $n = 2$, $\alpha_1 = 1, \alpha_2 = 1$, $\frac{\partial^2}{\partial x_1 \partial x_2} = D^\alpha f$. This is nothing but the partial derivative of the function of two variables $f(x_1, x_2)$ which is denoted by $\frac{\partial^2}{\partial x_1 \partial x_2}$ We have also

$$D^{(1,1)} f = \frac{\partial^2}{\partial x_2 \partial x_1} = \frac{\partial}{\partial x_2} \left(\frac{\partial f}{\partial x_2} \right)$$

which is equal to

$$\frac{\partial^2}{\partial x_1 \partial x_2} = \frac{\partial^2}{\partial x_1} \left(\frac{\partial f}{\partial x_2} \right)$$

for $f \in C^\infty(\Omega)$. In distributional derivatives to be defined later, we shall not distinguish between $\frac{\partial^2}{\partial x_1 \partial x_2}$ and $\frac{\partial^2}{\partial x_2 \partial x_1}$.

For $\alpha_1 = 1, \alpha_2 = 2$
$$D^{(1,2)} f = \frac{\partial^3 f}{\partial x_1 \partial x_2^2}$$
For $\alpha_1 = 1, \; \alpha_2 = 2, \; \alpha_3 = 1$
$$D^{(1,2,1)} f = \frac{\partial^4 f}{\partial x_1^1 \partial x_2^2 \partial x_3^1}$$

We have

$$D^{(0,0)} f = \frac{\partial^0 f}{\partial x_1^0 \partial x_2^0} = f, \; D^{(1,0)} f = \frac{\partial^1 f}{\partial x_1^1 \partial x_2^0} = \frac{\partial f}{\partial x_1}$$

$$D^{(0,1)} f = \frac{\partial^1 f}{\partial x_1^0 \partial x_2^1} = \frac{\partial f}{\partial x_2}$$

$$D^{(0,1,1)} f = \frac{\partial^2 f}{\partial x_1^0 \partial x_2^1 \partial x_3^1} = \frac{\partial^2 f}{\partial x_2 \partial x_3}$$

$$D^{(0,0,2)} f = \frac{\partial^2 f}{\partial x_1^0 \partial x_2^0 \partial x_3^2} = \frac{\partial^2 f}{\partial x_3^2}.$$

All functions are defined on a bounded subset Ω of R^n into R. The boundary of Ω is denoted by Γ or $\partial \Omega$ and $\overline{\Omega} = \Omega + \Gamma$. We say that $f \in L_2$ if $|f|^2$ is Lebesgue integrable on Ω. As we have seen in Chap. 3, $L_2(\Omega)$ is a Hilbert space with respect to $\|f\| = \left(\int_\Omega |f|^2 dx \right)^{1/2}$ and $\langle f, g \rangle = \int_\Omega fg dx$ $\left(where \int_\Omega f dx \; stands \; for \; \int \ldots \int f(x_1, x_2 \ldots x_n) dx_1 \ldots dx_n \right).$

Throughout our discussion, Ω is a bounded subset of R^n with Lipschitz boundary Γ. A technical definition is slightly complicated; however, broadly it means that the boundary will not contain cuspidal points and edges. For example, in two and three dimensions, domains having the boundary satisfying Lipschitz condition are circles, squares, triangles, spheres, cubes, annuli, etc. In one dimension, $\Omega = (a, b)$. It may be observed that the function representing such a boundary will be smooth or piecewise smooth and will have no singularities.

A function f defined on Ω into R is said to satisfy the *Hölder* condition with exponent λ, $0 < \lambda \leq 1$, if there exists a constant $M > 0$ such that

$$|f(x) - f(y)| \leq M\|x - y\|^\lambda \; \forall \, x, y \in \Omega$$

where $\| \cdot \|$ is the Euclidean norm on R^n. $K = supp f = \{x \in \Omega / f(x) \neq 0\}$ is called the *support* of f. If K is compact, f is said to have compact support.

It can be easily seen that $C_0^\infty(\Omega)$, then space of functions with compact support and having continuous derivatives of all order, is a vector space with respect to usual operations. A sequence $\{\varphi_n\}$ in C_0^∞ is said to converge to an element φ in $C_0^\infty(\Omega)$; namely, $\varphi_n \to \varphi$, if

1. there is a fixed compact set $K \subset \Omega$ such that $supp \varphi_n \subset K$ for all n.
2. φ_n and all its derivatives converge uniformly to $\varphi(x)$ and its derivatives, that is, $D^\alpha \varphi_n \to D^\alpha \varphi$ for all α uniformly.

Definition 5.9 (*Test Functions*) $C_0^\infty(\Omega)$ equipped with the topology induced through the convergence is called the space of test functions and often denoted by $D(\Omega)$. Thus, a test function is an infinitely differentiable function on R^n identically zero outside a compact set.

Definition 5.10 A bounded and linear functional defined on $D(\Omega)$ is known as a *distribution* or *Schwartz distribution*.

It may be observed that the space of distributions is nothing but the dual space of $D(\Omega)$ and is denoted by $D^*(\Omega)$ or $D'(\Omega)$. If $\Omega = R^n$, we simply write D^*. A functional F defined on $D(\Omega)$ is a distribution if it satisfies:

(i) $F(\varphi + \psi) = F(\varphi) + F(\psi)$
(ii) $F(\alpha\varphi) = \alpha F(\varphi)$
(iii) If $\varphi_n \to \varphi$, then $F(\varphi_n) F(\varphi)$

Definition 5.11 (*Distributional Derivative*) Distributional derivative of distribution or generalized derivative is a continuous linear functional denoted by $D^\alpha F$ defined as follows: For $F \in D^*(\Omega)$, $\langle D^\alpha F, \varphi \rangle = (-1)^{|\alpha|} \langle F, D^\alpha \varphi \rangle$ for all $\varphi \in D(\Omega)$.

Definition 5.12 A function $f : \Omega \to R$ is called *locally integrable* if for every compact set $K \subset \Omega$, $\int_K |f| dx < \infty$; that is, f is Lebesgue integrable over any compact set K.

It may be noted that all Lebesgue integrable functions and consequently all continuous functions are locally integrable over $[a, b]$. If $\Omega = S_r$ is equal to a ball of radius $r > 0$ and center $(0, 0)$ in R^2, then $f(r) = \frac{1}{r}, r = |x|$ is locally integrable on S_r. Locally integrable functions f can be identified with distributions F_f defined as follows:

$$\langle F_f, \varphi \rangle = \int_\Omega f\varphi dx. \; If \; n = 1, \; then \langle F_f, \varphi \rangle = \int_a^b f(x)\varphi(x)dx. F_f \equiv f.$$

or simply $F \equiv f$.

Example 5.17 (a)

$$\varphi(x) = \begin{cases} \exp[\frac{-1}{(x-1)(x-3)}], & 1 < x < 3 \\ 0, & \text{outside the open interval } (1,3) \end{cases}$$

$\varphi(x)$ is a test function with support $[1, 3]$.
(b) Let

$$\varphi(x) = \exp(-x^2), x > 0$$
$$= 0, x \leq 0$$
$$\varphi(x) \text{ is a test function.}$$

Example 5.18 $F(\varphi) = 0$, F is linear and continuous on $D(\Omega)$, $\Omega = (a, b)$ and so F is a distribution.

Example 5.19 Let $F(\varphi) = \int_a^b f(x)\varphi(x)dx$, f is a locally integrable function. F is linear and continuous on $D(\Omega)$ and therefore it is a distribution.

Example 5.20 Let $F(\varphi) = \int\limits_a^b |\varphi(x)|^2 dx$, F is continuous on $D(\Omega)$, but not linear; therefore F is not a distribution.

Example 5.21 The Dirac delta distribution is defined as

$$\langle \delta, \varphi \rangle = \varphi(0) \ \forall \ \varphi \in \mathscr{D}(\Omega)$$

δ is linear and continuous on $D(\Omega)$ and hence a distribution.

Two distributions F and G are equal if $\langle F, \varphi \rangle = \langle G, \varphi \rangle$ for all $\varphi \in \mathscr{D}(\Omega)$ such that $supp(\varphi) \subset (a, b)$. The Heaviside function H is defined by

$$H(x) = \begin{cases} 0, & x < 0 \\ 1/2, & x = 0 \\ 1, & x > 0 \end{cases}$$

Let H_1 be defined by

$$H_1(x) = \begin{cases} 0, & x \leq 0 \\ 1, & x > 0 \end{cases}$$

H and H_1 generate the same distribution and such functions are identified in distribution theory.

Definition 5.13 (*Regular and Singular Distribution*) A distribution $F \in \mathscr{D}^*$ is said to be a *regular distribution* if there exists a locally integrable function f such that

$$\langle F, \varphi \rangle = \int\limits_\Omega f\varphi dx \tag{5.22}$$

for every $\varphi \in D$. A distribution is called a *singular distribution* if (5.22) is not satisfied. In case $n = 1$

$$\langle F, \varphi \rangle = \int_a^b f(x)\varphi(x)dx \tag{5.23}$$

Remark 5.3 (a) We check below that (5.22) and, in particular, (5.23) defines a distribution. Since any φ in $\mathscr{D}(R)$ has bounded support contained in $[a, b]$, we note $\int\limits_a^b f(x)\varphi(x)dx$ exists (being integral of the product of an integrable and a continuous function). Thus, F is a well-defined functional on $\mathscr{D}(R)$. Linearity is clear as

$$\langle F, \varphi_1 + \varphi_2 \rangle = \int_a^b f(x)(\varphi_1 + \varphi_2)dx = \int_a^b f(x)\varphi_1(x)dx +$$

$$\int_a^b f(x)\varphi_2(x)dx = \langle F, \varphi_1 \rangle + \langle F, \varphi_2 \rangle$$

$$\langle F, \alpha\varphi \rangle = \int_a^b f(x)(\alpha\varphi(x))dx = \alpha\langle F, \varphi \rangle$$

Let $\varphi_n \to \varphi$, then

$$|\langle F, \varphi_n \rangle - \langle F, \varphi \rangle| = \left| \int_a^b f(x)(\varphi_n(x) - \varphi(x))dx \right|$$

$$\leq \sup |\varphi_n(x) - \varphi(x)| \int_a^b |f(x)|dx \to 0 \; as \; n \to \infty$$

This implies that F is continuous.

(b) $\langle F, \varphi \rangle$ is treated as an average value of F and so distributions are objects having average values in the neighborhoods of every point. Very often, distributions do not have values at points which resemble physical interpretation: If quantity is measured, then it does not yield exact value at a single point.

Example 5.22 Let Ω be an open or just a measurable set in R^n. The function F defined by

$$\langle F, \varphi \rangle = \int_\Omega \varphi dx$$

is a distribution. It is a regular distribution since

$$\langle F, \varphi \rangle = \int_{R^n} \chi_\Omega \varphi dx.$$

where χ_Ω is a characteristic function of Ω. In particular, if $\Omega = (0, \infty) \times \ldots (0, \infty)$, we obtain a distribution

$$\langle H, \varphi \rangle = \int_0^\infty \int_0^\infty \ldots \int_0^\infty \varphi dx_1 dx_2 \ldots dx_n$$

which is also called the Heaviside function.

Example 5.23 Let α be a multi-index. The functional F on D defined by

$$\langle F, \varphi \rangle = D^\alpha \varphi(0)$$

is a distribution. In particular, for $n = 1$, F defined by $\langle F, \varphi \rangle = \varphi'(0) =$ the value of the first derivative of φ at '0'is a distribution. In case $n = 2$, F defined by

$$\langle F, \varphi \rangle = \frac{\partial \varphi}{\partial x_k}(0), k = 1, 2 \ldots$$

is a distribution.

Example 5.24 $|x|$ is a locally integrable. It is differentiable in the classical sense at all points except $x = 0$. It is differentiable in the sense of distribution at $x = 0$ and distributional derivative at $x = 0$ is computed below:

$$\langle |x|, \varphi \rangle = (-1)\langle |x|, \varphi \rangle$$

$$= -\int_{-\infty}^{\infty} |x| \varphi'(x) dx$$

$$= \int_{-\infty}^{0} x \varphi'(x) dx + \int_{0}^{\infty} x \varphi'(x) dx$$

$$= -\int_{-\infty}^{0} \varphi(x) dx + \int_{0}^{\infty} \varphi(x) dx$$

integrating by part and using the fact that φ vanishes at infinity. Let a function sgn (read as signum of x) be defined by

$$sgn(x) = \begin{cases} -1, & \text{for } x < 0 \\ 0, & \text{for } x = 0 \\ 1, & \text{for } x > 0 \end{cases}$$

Then

$$\langle |x|', \varphi \rangle = \int_{-\infty}^{\infty} sgn(x) \varphi(x) dx \ \forall \varphi \in D(\Omega).$$

This means that $|x|' = sgn(x)$.

Example 5.25 Let us consider the Heaviside function H defined by

$$H = \begin{cases} 0, & \text{for } x \le 0 \\ 1, & \text{for } x > 0 \end{cases}$$

H is locally integrable and can be identified with a distribution. The generalized derivative of this is the Dirac distribution distributional derivative of Heaviside function

$$\langle H', \varphi \rangle = - \int_0^\infty H(x)\varphi'(x)dx$$

$$= - \int_0^\infty \varphi'(x)dx = \varphi(0)$$

Hence, $H'(x) = \delta(x)$.

Example 5.26 The generalized (distributional) derivative of the Dirac delta distribution is defined by

$$\langle \delta', \varphi \rangle = \langle \delta, \varphi' \rangle = -\varphi'(0)$$

The nth derivative $\delta(n)$ of δ is given by

$$\langle \delta^{(n)}, \varphi \rangle = (-1)^n \varphi^{(n)}(0)$$

Definition 5.14 Let F and G be distributions and α and β be scalars, then $F + G$ is defined by

  i. (Addition) $\langle F + G, \rangle = \langle F, \varphi \rangle + \langle G, \varphi \rangle$ *for all* $\varphi \in D$.
  ii. (Scalar multiplication) αF is defined by $\langle \alpha F, \varphi \rangle = \alpha \langle F, \varphi \rangle$, *for all* $\varphi \in D$ where α is a scalar.
  iii. If F is a distribution and h is an infinitely differentiable (smooth) function, then hF is defined by

$$\langle hF, \varphi \rangle = \langle F, h\varphi \rangle \ \forall \ \varphi \ in \ D.$$

It may be remarked that smoothness of h is essential for defining the operation product.

Theorem 5.11 *Let F and G be distributions and h a smooth function. Then their distribution derivatives satisfy the following relation:*

$$(F + G)' = F' + G'$$
$$(\alpha F)' = \alpha F'$$
$$(hF)' = h'F + hF'.$$

Proof

$$\langle (F+G)', \varphi \rangle = -\langle F+G, \varphi' \rangle = -\langle F, \varphi' \rangle - \langle G, \varphi' \rangle$$
$$= \langle F', \varphi \rangle + \langle G', \varphi \rangle = \langle F' + G', \varphi \rangle$$

This implies $(F+G)' = F' + G'$. Similarly, we get $(\alpha F)' = \alpha F'$.

$$\langle (hF)', \varphi \rangle = -\langle hF, \varphi' \rangle$$
$$= -\langle F, h\varphi' \rangle \text{ by Definition 5.14}$$
$$= -\langle F, (h\varphi)' - h'\varphi \rangle$$
$$= -\langle F, (h)' \rangle + \langle F, h'\varphi \rangle \text{ by Definition 5.11}$$
$$= \langle hF', \varphi \rangle + \langle h'F, \varphi \rangle \text{ by Definition 5.14}$$
$$= \langle hF' + h'F, \varphi \rangle$$

Thus, $(hF)' = hF' + h'F = h'F + hF'$.

It may be observed that the distribution generated by the derivative f' of a function $f : R \rightarrow R$ is the same as the derivative of the distribution f; these two possible ways of interpreting the symbol F' for a differentiable function f are identical. The main advantage of distribution theory over classical calculus is that every distribution has distributional derivative while there are functions which are not differential in the classical sense.

Theorem 5.12 *If F is any distribution, then $D^\alpha F$ (Definition 5.11) is a distribution for any multi-index α. In particular, F' defined by $\langle F', \varphi \rangle = -\langle F, \varphi' \rangle$ is a distribution.*

Proof (i)

$$\langle D^\alpha F, \lambda\varphi \rangle = (-1)^{|\alpha|} \langle F, D^\alpha(\lambda\varphi) \rangle$$
$$= \lambda(-1)^{|\alpha|} \langle F, D^\alpha \varphi \rangle$$

or

$$\langle D^\alpha F, \lambda\varphi \rangle = \lambda \langle D^\alpha F, \varphi \rangle$$

(ii)

$$\langle D^\alpha F, \varphi + \psi \rangle = (-1)^{|\alpha|} \langle F, D^\alpha(\varphi + \psi) \rangle$$
$$= (-1)^{|\alpha|} \langle F, D^\alpha \varphi \rangle + (-1)^{|\alpha|} \langle F, D^\alpha \psi \rangle$$
$$= \langle D^\alpha F, \varphi \rangle + \langle D^\alpha F, \psi \rangle$$

Thus, D^α is linear. We now show that D^α is continuous. Let $\varphi_n \rightarrow \varphi$; that is, $||\varphi_n - \varphi|| \rightarrow 0$ *as* $n\infty$, then $||\varphi_n' - \varphi'|| \rightarrow 0$ as $n \rightarrow \infty$ and, in general,

$||D^\alpha \varphi_n - D^\alpha \varphi|| \to 0$ as $n \to \infty$ by a well-known result on uniform convergence

$$|\langle D^\alpha F, \varphi_n \rangle - \langle D^\alpha F, \varphi \rangle| = |(-1)^{|\alpha|} \langle F, D^\alpha \varphi_n \rangle - (-1)^\alpha \langle F, D^\alpha \varphi \rangle|$$
$$= |(-1)^{|\alpha|} \langle F, D^\alpha \varphi_n - D^\alpha \varphi \rangle|$$
$$\leq ||F|| \, ||D^\alpha \varphi_n - D^\alpha \varphi|| \to 0$$

as $n \to \infty$ by CSB inequality

and the fact that $||F||$ is finite and $||D^\alpha \varphi_n - D^\alpha \varphi|| \to 0$ as $n \to \infty$.

Convergence of Distributions

Definition 5.15 (*Convergence of Distribution*) A sequence of distributions $\{F_n\}$ is called convergent to a distribution F if

$$\langle F_n, \varphi \rangle \to \langle F, \varphi \rangle \text{ for every } \varphi \in \mathscr{D}$$

It may be observed that this definition does not involve the existence of a limiting distribution toward which the sequence tends. It is presented in terms of the sequence itself unlike the usual definitions of convergence in elementary analysis, where one is required to introduce the concept of limit of a sequence before the definition of convergence.

Theorem 5.13 (Distributional Convergence) *Let* $F, F_1, F_2, \ldots, F_n \ldots$ *be locally integrable functions such that* $F_n \to F$ *uniformly in each bounded set, then* $D^\alpha F_n \to D^\alpha F$ *(in the sense of distribution Definition 5.15).*

Proof We have

$$\langle D^\alpha F_n, \varphi \rangle = (-1)^{|\alpha|} \langle F_n, D^\alpha \varphi \rangle \to (-1)^{|\alpha|} \langle F, D^\alpha \varphi \rangle$$
$$= \langle D^\alpha F, \varphi \rangle$$

for every test function φ. This proves the desired result.
For $\alpha = 1$, we obtain that

$$\langle F'n, \varphi \rangle \to \langle F', \varphi \rangle \text{ whenever } \langle F_n, \varphi \rangle \to \langle F, \varphi \rangle \text{ uniformly}$$

Example 5.27 Let

$$f_n(x) = \frac{n}{\pi(1 + n^2 x^2)}$$

We show that the sequence $f_n(x)$ converges in the distributional sense (Definition 5.15) to the Dirac delta function $\delta(x)$. Let φ be a test function. To prove our assertion, we must show that (Fig. 5.1)

Fig. 5.1 Graph of
$f_1(x)$, $f_2(x)$, $f_3(x)$ of
Example 5.27

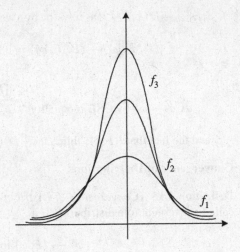

$$\int\limits_{-\infty}^{\infty} f_n(x)\varphi(x)dx \to \varphi(0) \; as \; n \to \infty$$

Or equivalently

$$\int\limits_{-\infty}^{\infty} f_n(x)\varphi(x)dx - \varphi(0) \to 0 \; as \; n \to \infty$$

Since

$$\int\limits_{-\infty}^{\infty} f_n(x)dx = \int\limits_{-\infty}^{\infty} \frac{ndx}{\pi(1+n^2x^2)} = 1 \; \forall \, n \in N$$

we have

$$\int\limits_{-\infty}^{\infty} f_n(x)\varphi(x)dx - \varphi(0) = \int\limits_{-\infty}^{\infty} f_n(x)(\varphi(x) - \varphi(0))dx$$

$$= \int\limits_{-\infty}^{-a} f_n(x)(\varphi(x) - \varphi(0))dx + \int\limits_{-a}^{a} f_n(x)(\varphi(x) - \varphi(0))dx$$

$$+ \int\limits_{a}^{\infty} f_n(x)(\varphi(x)dx - \varphi(0))dx$$

where $[-a, a]$ is an interval containing the support of φ. Consequently

$$\left| \int_{-\infty}^{\infty} f_n(x)\varphi(x)dx - \varphi(0) \right| = \left| \int_{-\infty}^{\infty} f_n(x)(\varphi(x) - \varphi(0))dx \right|$$

$$\leq \left| \varphi(0) \int_{-\infty}^{-a} f_n(x)dx \right| + \left| \int_{-a}^{a} f_n(x)(\varphi(x)dx - \varphi(0)) \right|$$

$$+ \left| \varphi(0) \int_{a}^{\infty} f_n(x)dx \right|$$

By direct integration, we obtain that

$$\lim_{n \to \infty} \left| \varphi(0) \int_{-\infty}^{-a} f_n(x)dx \right| = 0 \text{ and } \lim_{n \to \infty} \left| \varphi(0) \int_{a}^{\infty} f_n(x)dx \right| = 0$$

Now we show that

$$\lim_{n \to \infty} \left| \int_{-a}^{a} f_n(x)(\varphi(x) - \varphi(0))dx \right|$$

We have

$$\left| \int_{-a}^{a} f_n(x)(\varphi(x) - \varphi(0))dx \right| \leq \int_{-a}^{a} |f_n| \, |\varphi(x) - \varphi(0)|dx$$

$$\leq \max |\varphi'(x)| \int_{-a}^{a} |xf_n(x)|dx$$

This holds because $|\varphi(x) - \varphi(0)| \leq \max |\varphi'(x)| \, |x|$, by the mean value theorem. It can be seen that

$$\lim_{n \to \infty} \int_{-a}^{a} |xf_n(x)|dx = 0$$

Therefore, we have the desired result.

Example 5.28 Let $f_n(x)$ be as in Example 5.27. Then

$$f_n'(x) = -\frac{2}{\pi}\frac{n^3 x}{(1+n^2x^2)^2} \to \delta'(x)$$

in the distributional sense. However, $f_n'(x) \to 0$ in the sense of classical convergence (elementary analysis convergence).

Theorem 5.14 (Termwise Differentiation) *If F, F_1, F_2, $\ldots$, F_n $\ldots$ are distributions such that $F_n \to F$ as $n \to \infty$, then $F_n' \to F'$. In fact, this result is true for any multi-index α; that is*

$$D^\alpha F_n \to D^\alpha F$$

Proof $\langle F'n, \varphi \rangle = -\langle F_n, \varphi' \rangle \to -\langle F, \varphi' \rangle = \langle F', \varphi' \rangle$. Also $\langle D^\alpha F_n, \varphi \rangle = (-1)^{|\alpha|}$ $\langle F_n, D^\alpha \varphi \rangle \to (-1)^{|\alpha|} \langle F, D^\alpha \varphi \rangle = \langle D^\alpha F, \varphi \rangle$ for every test function.

It may be noted that for termwise differentiation in the classical case the sequence $\{F_n\}$ must converge uniformly and the differentiated sequence also must converge uniformly. It may be observed that in view of Theorem 5.14 a series of distributions can be differentiated term-by-term without imposing uniform convergence condition.

The Integral of a Distribution

Definition 5.16 (*Integral or Antiderivative of a Distribution*) If F is a distribution. Then a distribution G is called an integral or antiderivative of F provided $G' = F$ (first distributional derivative of G is equal to F).

Theorem 5.15 (Existence of Integral of a Distribution) *Every distribution $F \in D^\star(R)$ has an integral.*

Proof Let $\varphi_0 \in D(R)$ be a fixed test function such that

$$\int_{-\infty}^{\infty} \varphi_0(x)dx = 1$$

Then, for every test function $\varphi \in D(R)$, there exists a test function $\varphi_1 \in D(R)$ such that

$$\varphi = K\varphi_0 + \varphi_1$$

where $K = \int_{-\infty}^{\infty} \varphi(x)dx$ and $\int_{-\infty}^{\infty} \varphi_1(x)dx$ Let $F \in D^\star(R)$. Define a functional G on $D(R)$ by

$$\langle G, \varphi \rangle = \langle G, K\varphi_0 + \varphi_1 \rangle = KC_0 - \langle F, \psi \rangle$$

where C_0 is a constant and ψ is the test function defined by

$$\psi(x) = \int\limits_{-\infty}^{x} \varphi_1(t)dt$$

Thus G is a distribution and $G' = F$.

Theorem 5.16 *If distributional derivative of distribution F is zero then it must be constant. As a consequence any two integrals of a distribution differ by a constant function.*

Proof Let $\varphi \in D(R)$. Using the notation of the proof of Theorem 5.15, we get

$$\langle F, \varphi \rangle = \langle F, K\varphi_0 + \varphi_1 \rangle = \langle F, K\varphi_0 \rangle + \langle F, \varphi_1 \rangle$$

$$= \langle F, \varphi_0 \rangle \int\limits_{-\infty}^{\infty} \varphi(x)dx$$

Since $\langle F, \varphi_1 \rangle = -\langle F, \psi \rangle = 0$. Thus, F is a regular distribution generated by the constant function $C = \langle F, \varphi_0 \rangle$. Let G_1 and G_2 be two integrals of a distribution F. Then $G_1' = F$, $G_2' = F$. This implies that $G_1' - G_2' = 0$ or $(G_1 - G_2)' = 0$. By the first part, $G_1 - G_2$ is a constant function. Every $f \in C_\infty(\Omega)$ having classical derivative has also distributional derivative.

Before the conclusion of this section, we make certain useful observations required in the next subsection.

Remark 5.4 (i) If $f \in C_m(\Omega)$, then all the classical partial derivatives of f to order m are also generalized derivatives (distributional derivatives).

(ii) Let f be locally integrable such that $f \in L_1(K)$ for any compact subset K of Ω. Then

$$F(\varphi) = \int\limits_{\Omega} f(x)\varphi(x)dx, \quad \varphi \in D(\Omega)$$

defines a distribution on Ω denoted by F_f. If $f = g$ a.e. then $F_f = F_g$

(iii) $S(R^n) = T = \{f \in C^\infty; \ \forall \alpha, \beta \in N^n, \ x^\alpha Df \to 0 \text{ as } |x| \to \infty\}$ is the space of functions of C^∞ of rapid decay at infinity. This is not a normed space but a complete metric space with respect to the metric

$$d(f, g) = \sum_{\alpha, \beta \in N^n} a_{\alpha\beta} \frac{d_{\alpha\beta}(f - g)}{1 + d_{\alpha\beta}(f - g)}, \quad f, g \in S$$

where $d_{\alpha\beta}(f) = \sup\limits_{x \in R^n} |x^\alpha Df(x)|$ and $a_{\alpha\beta}$ are chosen such that

$$\sum_{\alpha, \beta} a\alpha, \beta = 1.$$

It can be proved that S is dense in L_p for all p with $1 \leq p < \infty$ but S is not dense in L_∞. S is also a vector space. For $F \in S$, the Fourier transform of F is defined as

$$\hat{F}(y) = \mathscr{F}f(y) = \int_{R^n} e^{-ixy} f(x)dx, \ y \in R^n.$$

Space of tempered distributions is the vector space of all continuous linear functionals defined on S introduced above, denoted by S'. Each element of S' is a distribution and is called a tempered distribution. An interesting characterization of tempered distribution can be found in [$\{DaLi \ 90\}$, $p.508$, see $H^s(R^n)$]. The elements of L_p, $1 \leq p \leq \infty$ can be identified with tempered distributions (and in particular $S \subset S'$).

It may be observed that an arbitrary distribution F will not have a Fourier transform, at least as a distribution. However, we can define a Fourier transform denoted by $\hat{F}$ for each tempered distribution $T \in S'$ as $\langle \hat{T}, \varphi \rangle = \langle T, \hat{\varphi} \rangle$ for all $\varphi \in S$.

For detailed discussion on Fourier transform of tempered distribution, see $\{DaLi90\}$.

5.4.2 Sobolev Space

$H^m(\Omega) = \{f \in L_2(\Omega) | D^\alpha f \in L_2(\Omega), \ |\alpha| \leq m\}$, m being any positive integer, is called the *Sobolev space* of order m. $H^m(\Omega)$ is a Hilbert space with respect to the inner product

$$\langle f, g \rangle_{H^m(\Omega)} = \sum_{|\alpha| \leq m} \langle D^\alpha f, D^\alpha g \rangle_{L_2(\Omega)}$$

$$\text{For } m = 1, \ \Omega = (a, b), \ D^\alpha f = \frac{df}{dx}$$

$$\langle f, g \rangle_{H^1(a,b)} = \langle f, g \rangle_{L_2(a,b)} + \left\langle \frac{df}{dx}, \frac{dg}{dx} \right\rangle_{L_2(a,b)}$$

$$= \int_a^b f(x)g(x)dx + \int_a^b \frac{df}{dx}\frac{dg}{dx}dx$$

It can be checked that $H^1(a, b)$ is a Hilbert space.

For $m = 2$, $\Omega = S_r = \{(x, y)/x^2 + y^2 \leq r\}$ = circle with origin as the center; or $\Omega = \{(x, y)/a < x < b, \ c < y < d\}$ = a rectangle with sides of length $b - a$ and $d - c$

$$\langle f, g \rangle_{H^2(\Omega)} = \sum_{|\alpha| \leq 2} \langle D^\alpha f, D^\alpha g \rangle_{L_2(\Omega)}$$

where

$$\alpha = (0, 0), (0, 1), (1, 0), (1, 1), (0, 2), (2, 0)$$

$$|\alpha| = \alpha_1 + \alpha_2 \leq 2, \sum_{|\alpha| \leq 2} \langle D^\alpha f, D^\alpha g \rangle = \langle D^{(0,0)} f, D^{(0,0)} g \rangle_{L_2(\Omega)}$$

$$+ \langle D^{(0,1)} f, D^{(0,1)} g \rangle_{L_2(\Omega)} + D^{(1,0)} f, D^{(1,0)} g \rangle_{L_2(\Omega)}$$

$$+ \langle D^{(1,1)} f, D^{(1,1)} g \rangle_{L_2(\Omega)} + D^{(2,0)} f, D^{(2,0)} g \rangle_{L_2(\Omega)}$$

$$+ \langle D^{(0,2)} f, D^{(0,2)} g \rangle_{L_2(\Omega)} = \langle f, g \rangle_{L_2(\Omega)} + \left\langle \frac{\partial f}{\partial x_1}, \frac{\partial g}{\partial x_1} \right\rangle_{L_2(\Omega)}$$

$$+ \left\langle \frac{\partial f}{\partial x_2}, \frac{\partial g}{\partial x_2} \right\rangle_{L_2(\Omega)} + \left\langle \frac{\partial^2}{\partial x_1 \partial x_2}, \frac{\partial^2}{\partial x_1 \partial x_2} \right\rangle_{L_2(\Omega)}$$

$$+ \left\langle \frac{\partial^2 f}{\partial x_1^2}, \frac{\partial^2 g}{\partial x_1^2} \right\rangle_{L_2(\Omega)} + \left\langle \frac{\partial^2 f}{\partial x_1^2}, \frac{\partial^2 g}{\partial x_2^2} \right\rangle_{L_2(\Omega)}$$

$$= \int_a^b \int_c^d fg \, dx_1 dx_2 + \int_a^b \int_c^d \frac{\partial f}{\partial x_1} \frac{\partial g}{\partial x_1} dx_1 dx_2$$

$$+ \int_a^b \int_a^b \frac{\partial^2 f}{\partial x_2^2} \frac{\partial^2 g}{\partial x_2^2} dx_1 dx_2 + \int_a^b \int_c^d \frac{\partial^2 f}{\partial x_1 \partial x_2} \frac{\partial^2 g}{\partial x_1 \partial x_2} dx_1 dx_2$$

$$H^m(\Gamma) = \{ f \in L_2(\Gamma) / D^\alpha f \in L_2(\Gamma), \, |\alpha| \leq m \}$$

where $\Gamma = \partial \Omega$ denotes the boundary of Ω. For $L_2(\Gamma)$, see Appendix A.4. $H^m(\Gamma)$ is a Hilbert space with respect to

$$\langle f, g \rangle_{H^m(\Gamma)} = \sum_{|\alpha| \leq m} \langle D^\alpha f, D^\alpha g \rangle_{L_2(\Gamma)}$$

$$\langle f, g \rangle_{H^1(\Gamma)} = \int_\Gamma fg \, dx + \int_\Gamma \frac{df}{dx} \frac{dg}{dx} dx$$

$$\langle f, g \rangle_{H^2(\Gamma)} = \sum_{|\alpha| \leq m} \langle D^\alpha f, D^\alpha g \rangle_{L_2(\Gamma)}$$

$$= \int_\Gamma fg \, d\Gamma + \int_\Gamma \frac{\partial f}{\partial x_1} \frac{\partial g}{\partial x_2} d\Gamma$$

$$+ \int_\Gamma \frac{\partial f}{\partial x_2} \frac{\partial g}{\partial x_2} d\Gamma + \int_\Gamma \frac{\partial^2 f}{\partial x_1^2} \frac{\partial^2 g}{\partial x_1^2} d\Gamma$$

$$+ \int_\Gamma \frac{\partial^2 f}{\partial x_1^2} \frac{\partial^2 g}{\partial x_1^2} d\Gamma.$$

The restriction of a function $f \in H^m(\Omega)$ to the boundary Γ is the *trace* of f and is denoted by Γ_f; that is, $\Gamma_f = f(\Gamma) =$ value of f at Γ.

$$H_0^m(\Omega) = \{f \in H^m(\Omega)/\Gamma_f = 0\} = closure\ of\ D(\Omega)\ in\ H^m(\Omega)$$
$$H^m(R^n) = H_0^m(R^n),\ m \geq 0.$$

The dual space of $H_0^m(\Omega)$, that is, the space of all bounded linear functionals on $H_0^m(\Omega)$ is denoted by $H^{-m}(\Omega)$.

$$f \in H^{-m}(\Omega)\ \text{if and only if}$$
$$f = \sum_{|\alpha| \leq m} D^\alpha g\ for\ some\ g \in L_2(\Omega)$$

Sobolev Space with a Real Index: $H^s(R^n)$

A distribution is a tempered distribution if and only if it is the derivative of a continuous function with slow growth in the usual sense, that is, a function which is the product of $P(x) = (1 + |x|^2)^{k/2}$, $k \in N$ by a bounded continuous function on R^n. For $s \in R$, $H^s(R^n)$ is the space of tempered distributions F, such that

$$(1 + |y|^2)^{s/2} \hat{F} \in L_2(R^n),\ y \in R^n$$

where $\hat{F}$ is the Fourier transform of F. $H^s(R^n)$ equipped with the inner product

$$\langle F, G \rangle_s = \int_{R^n} (1 + |y|^2)^{s/2} \hat{F}(y)\overline{\hat{G}(y)}d\xi$$

$\left(\text{with associated norm} \|F\|_s = \left(\int_{R^n}(1 + |y|^2)^s |\hat{F}(y)|^2 dy\right)^{1/2}\right)$ is a Hilbert space (for proof see Solved Problem 5.7).

If $s_1 \geq s_2$, then $H^{s_1}(R^n) \subset H^{s_2}(R^n)$ and the injection is continuous (Solved Problem 5.7). For $s = m \in N$, the space $H^s(R^n)$ coincides with $H^m(R^n)$ introduced earlier. It is interesting to note that every distribution with compact support in R^n is in $H^s(R^n)$ for a certain $s \in R$. For more details, see Dautray and Lions {*DaLi* 90}.

Theorem 5.17 (Green's Formula for Integration by Parts)

(i) $\int_\Omega v \Delta u dx + \int_\Omega grad\ u.grad\ v dx = \int_\Gamma \frac{\partial u}{\partial n} d\Gamma$

(ii) $\int_\Omega (u \Delta v - v \Delta u)dx = \int_\Gamma (u \frac{\partial v}{\partial n} - v \frac{\partial u}{\partial n})d\Gamma$

It is clear that (i) is a generalization of the integration by parts formula stated below (for $n = 1$, $\Omega = [a, b]$)

$$\int_a^b u''(x)v(x)dx = u'(b)v(b) - u'(a)v(a) - \int_a^b u'(x)v'(x)dx$$

Theorem 5.18 (The Friedrichs Inequality) *Let Ω be a bounded domain of R^n with a Lipschitz boundary. Then there exists a constant $k_1 > 0$, depending on the given domain, such that for every $f \in H_1(\Omega)$*

$$\|f\|_{H^1(\Omega)}^2 \leq k_1 \left\{ \sum_{j=1}^n \int_\Omega \left(\frac{\partial f}{\partial x_j} \right)^2 dx + \int_\Gamma f^2 d\Gamma \right\} \tag{5.24}$$

Theorem 5.19 (The Poincaré Inequality) *Let Ω be a bounded domain of R^n with a Lipschitz boundary. Then there exists a constant, say k_2, depending on Ω such that for every $f \in H^1(\Omega)$*

$$\|f\|_{H^1(\Omega)}^2 \leq k_2 \left\{ \int_\Omega \sum_{j=1}^n \left| \frac{\partial f}{\partial x_j} \right| dx + \left(\int_\Omega f dx \right)^2 \right\}$$

The above two inequalities hold for elements of $L_2(\Omega)$.
For $f \in H^1(\Omega)$ such that $\int_\Omega f(x)dx = 0$

$$\|f\|_{L_2(\Omega)}^2 \leq k_1 \sum_{j=1}^n \left\| \frac{\partial f}{\partial x_j} \right\|_{L_2(\Omega)}^2 , \quad k_1 > 0 \text{ constant depending on } \Omega.$$

Definition 5.17 (*Sobolev Space of Distributions*) Let $H^m(\Omega)$ denote the space of all distributions F such that $D^\alpha F \in L_2(\Omega)$ for all α, $|\alpha| \leq m$ equipped with the norm

$$\|F\|_m = \left(\sum_{|\alpha| \leq m} \|D^\alpha F\|_{L_2(\Omega)}^2 \right)^{1/2} \tag{5.25}$$

Here, the derivatives are taken in the sense of distributions and the precise meaning of the statement $D^\alpha F \in L_2(\Omega)$ is that there is a distribution F_f constructed in Remark 5.4.2 with $f \in L_2(\Omega)$ such that $D^\alpha F = F_f$. Hence

$$\|D^\alpha F\|_{L_2(\Omega)} = \sup \frac{|\langle F, D^\alpha \varphi \rangle_{L_2(\Omega)}|}{\|\varphi\|} \tag{5.26}$$

Theorem 5.20 *$H^m(\Omega)$ is a Hilbert space with respect to the inner product*

$$\langle F, G \rangle = \sum_{|\alpha| \le m} \langle D^\alpha F, D^\alpha G \rangle_{L_2(\Omega)} \tag{5.27}$$

More generally, if $H^{m,p}(\Omega)$ or $W^{m,p}(\Omega)$ denotes the space of all functions $f \in L_p(\Omega)$ such that $D^\alpha f \in L_p$, $1 \le p < \infty$, $|\alpha| \le m$, then this space is a Banach space. $W^{m,p}(\Omega)$ is usually called Sobolev space with index p or generalized Sobolev space.

Proof It is easy to check conditions of inner product. We prove completeness of $H^m(\Omega)$. Let $\{F_k\}$ be a Cauchy sequence in $H^m(\Omega)$. Then for any $|\alpha| \le m$, $\{D^\alpha F_k\}$ is a Cauchy sequence in $L_2(\Omega)$ and since $L_2(\Omega)$ is complete, there exists $F^\alpha \in L_2(\Omega)$ such that

$$\lim_{k \to \infty} \int_\Omega |D^\alpha F_k - F^\alpha(x)|^2 dx = 0$$

Now, since $D^\alpha F_k$ is locally integrable, we may compute

$$F_{D^\alpha F_k}(\varphi) = \int_\Omega D^\alpha F_k(x)\varphi(x)dx = (-1)^{|\alpha|} \int_\Omega F_k(x) D^\alpha \varphi(x)dx$$

$$= (-1)^{|\alpha|}\langle F_k(x), D^\alpha \varphi(x) \rangle_{L_2(\Omega)}, \varphi \in D(\Omega)$$

Also,

$$D^\alpha F_{F^0} = (-1)^{|\alpha|}\langle F^0, D^\alpha \varphi(x) \rangle_{L_2(\Omega)}$$

Thus

$$\| F_{D^\alpha F_k} - D^\alpha F_{F^0} \| = \sup_\varphi \frac{|\langle F_k - F^0, D^\alpha \varphi \rangle_{L_2(\Omega)}|}{\|\varphi\|_{L_2(\Omega)}}$$

and

$$F_{D^\alpha F_k} \to D^\alpha F_{F^0} \text{ as } k \to \infty.$$

We also have

$$\| F_{D^\alpha F_k} - F_{F^\alpha} \| = \sup_\varphi \frac{|\langle F_k - F^0, D^\alpha \varphi \rangle_{L_2(\Omega)}|}{\|\varphi\|_{L_2(\Omega)}} \to 0 \text{ as } k \to \infty$$

so that $D^\alpha F_{F^0} = F_F$; $\alpha \le m$ and the distributional derivative of F^0 is the distribution associated with F^α. Hence, $H^m(\Omega)$ is complete.

Let for $1 \le p < \infty$

$$\|f\|_p = \left(\int_\Omega |f(x)|^p dx \right)^{1/p} \qquad (5.28)$$

and

$$\|f\|_{m,p} = \left(\sum_{0 \le |\alpha| \le m} \int_\Omega |D^\alpha f(x)|^p \right)^{1/p} \qquad (5.29)$$

$$Df(x) = \frac{df}{dx}, \quad D^0 f = f \qquad (5.30)$$

$$\langle f, g \rangle_{L_2(\Omega)} = \int_\Omega fg \, dx \qquad (5.31)$$

$$\langle f, g \rangle_{m,2} = \left(\sum_{0 \le |\alpha| \le m} \int_\Omega D^\alpha f \, D^\alpha g \, dx \right) \qquad (5.32)$$

Note that

$$\|f\|_{m,p} = \left(\sum_{0 \le |\alpha| \le m} \|D^\alpha f\|_p^p \right)^{1/p} \qquad (5.33)$$

$$\langle f, g \rangle_{m,2} = \sum_{0 \le |\alpha| \le m} \langle D^\alpha f, \, D^\alpha g \rangle_{L_2(\Omega)} \qquad (5.34)$$

For $m = 0$, we obtain that $\|f\|_{0,p} = \|f\|_p$ and $\langle f, g \rangle_{0,2} = \langle f, g \rangle_{L_2(\Omega)}$. Conditions of norm for $\|f\|_m$ can be easily checked using Minkowski's inequality. Completeness of $H^{m,p}(\Omega)$ can be proved on the lines of the proof of Theorem 5.20.

Definition 5.18 $H_0^{m,p}(\Omega)$ denotes the closure of $C_0^\infty(\Omega)$ in $H^{m,p}(\Omega)$.

Equivalently $f \in H^{m,p}(\Omega)$ belongs to $H_0^{m,p}(\Omega)$ if and only if there exists a sequence $\{f_n\}$ in $C_0^\infty(\Omega)$ with $\|f_n - f\| \to 0$ as $n \to \infty$.

Theorem 5.21 $H_0^{m,p}(\Omega)$ *is a Banach space*, $1 \le p < \infty$, *with norm* $\|\cdot\|_{m,p}$ *defined by Eq. (5.33) and* $H_0^{m,2}(\Omega)$ *is a Hilbert space with the scalar product* $\langle \cdot, \cdot \rangle$ *given by Eq. (5.34).*

5.4.3 The Sobolev Embedding Theorems

Definition 5.19 (*Embedding Operator*) Let X and Y be Banach spaces with $X \subseteq Y$. The embedding operator $j : X \to Y$ is defined by $j(x) = x$ for all $x \in X$. The embedding $X \subseteq Y$ is called continuous if and only if j is continuous, that is

$$||x||_Y \leq k||x||_X \ \forall \ x \in X \ and \ k \ is \ a \ constant$$

The embedding $X \subseteq Y$ is called *compact* if and only if j is compact, that is, it is continuous and each bounded sequence $\{x_n\}$ in X contains a sub-sequence $\{x_{n_k}\}$ which is convergent in Y.

One may consider an embedding as an injective linear operator $j : X \to Y$. Since j is injective, we can identify u with $j(u)$. In this sense, we write $X \subseteq Y$. Let

$$||f||_{1,2} = \left(\int_\Omega f^2 dx + \sum_{i=1}^n \left(\frac{\partial f}{\partial x_i} \right)^2 \right)^{1/2} \tag{5.35}$$

$$||f||_{1,2,0} = \left(\int_\Omega \sum_{i=1}^n \left(\frac{\partial f}{\partial x_i} \right)^2 dx \right)^{1/2} \tag{5.36}$$

Theorem 5.22 *Let Ω be a bounded region in R^n with $n \geq 1$. Then*

(a) *The norms $||f||_{1,2}$ and $||f||_{1,2,0}$ are equivalent on $H_0^{1,2}(\Omega)$.*
(b) *The embedding $H^{1,2}(\Omega)$ usually denoted by $H(\Omega) \subseteq L_2(\Omega)$ is compact.*
(c) *The embeddings*

$$L_2(\Omega) \supseteq H_0^{1,2}(\Omega) \supseteq H_0^{2,2}(\Omega) \supseteq H_0^{3,2}(\Omega) \supseteq \cdots$$

are compact.

Theorem 5.23 *Let Ω be a bounded region in R^n, $n \geq 1$ and have sufficiently smooth boundary, that is, $\partial\Omega \in C^{0,1}$ (If $n = 1$, then Ω is a bounded open interval). Then*

(a) *(Density). $C^\infty(\Omega)$ is dense in $H^{1,2}(\Omega)$.*
(b) *(Compact Embedding). The embedding $H^{1,2}(\Omega) \subseteq L_2(\Omega)$ is compact.*
(c) *(Equivalent Norms). Each of the following norms*

$$||f||_{1,2}^\star = \left(\int_\Omega \sum_{i=1}^n \left(\frac{\partial f}{\partial x_i} \right)^2 dx + \left| \int_\Omega f dx \right|^2 \right)^{1/2} \tag{5.37}$$

$$||f||_{1,2}^{\star\star} = \left(\int_\Omega \sum_{i=1}^n \left(\frac{\partial f}{\partial x_i} \right)^2 dx + \int_\Omega f^2 d\Omega \right)^{1/2} \tag{5.38}$$

$\left(In \ case \ n = 1, \ we \ take \int\limits_{d\Omega} f^2 d\Omega = (f(a))^2 + (f(b))^2 \right)$ *is an equivalent nor-m on $H^{1,2}(\Omega)$, namely*

$$\|f\|_{1,2} = \left(\int_\Omega |f(x)|^2 dx + \int_\Omega \left| \frac{\partial f}{\partial x} \right|^2 dx \right)^{1/2}$$

$$= \left(\|f\|^2 + \left\| \frac{\partial f}{\partial x} \right\|^2 \right)^{1/2}$$

(d) *(Regularity). For $m - j > n/2$, the embedding*

$$H^{m,2}(\Omega) \subseteq C^j(\overline{\Omega})$$

is continuous, that is, each continuous function $f \in H^{m,2}(\Omega)$ belongs to $C^j(\overline{\Omega})$ after changing the value of f on a set of n-dimensional Lebesgue measure zero, if necessary.

(e) *(Generalized boundary function). There exists only one linear continuous operator $A \colon H_{1,2}(\Omega) \to L_2(\partial \Omega)$ with the property that, for each $f \in C^1(\overline{\Omega})$, the function $Af \colon \partial \Omega \to R$ is the classical boundary function to u; that is, Af is the restriction of $f \colon \overline{\Omega} \to R$ to the boundary $\partial \Omega$.*
For $f \in H^{1,2}(\Omega)$, the function $Af \in L_2(\partial \Omega)$ is called the generalized boundary function of f. The values of Af are uniquely determined on $\partial \Omega$ up to changing the values on a set of surface measure zero. Let $f \in H_0^{1,2}(\Omega)$, then $Af = 0$ in $L_2(\Omega)$; that is

$$Af(x) = 0 \ for \ almost \ all \ x \in \partial \Omega \tag{5.39}$$

Corollary 5.2 *Let Ω be a bounded region in R^n with $n \geq 1$ and $\partial \Omega \in C^{0,1}$. Then*

(a) *$C^\infty(\overline{\Omega})$ is dense in $H^{m,2}(\Omega)$ for $m = 0, 1, 2, \ldots$*
(b) *The embeddings*

$$L_2(\Omega) \supseteq H^{1,2}(\Omega) \supseteq H^{2,2}(\Omega) \supseteq H^{3,2}(\Omega) \supseteq \ldots$$

are compact.
(c) *If $f \in H_0^{m,2}(\Omega)$ with $m \geq 1$, then $A(D_f^\alpha) = 0$ in $L_2(\partial \Omega)$ for $|\partial| \leq m - 1$, that is, $D^\alpha f = 0$ on $\partial \Omega$ for $\alpha \colon |\colon| \leq m - 1$, $f = 0$ on $\partial \Omega$.*

We may observe that embedding theorems provide conditions under which Sobolev spaces are embedded in spaces of continuous functions, $C^k(\overline{\Omega})$. In one dimension $(n = 1) \Omega$ is a subset of the real line and the functions in $H^1(\Omega)$ are continuous. For $\Omega \subset R^2 (n = 2$, subsets of the plane), one require that a function be in $H^2(\Omega)$ in order to ensure its continuity.

Remark 5.5 (i) Equivalent norms $\| \cdot \|_{1,2}$ and $\| \cdot \|_{1,2,0}$ given, respectively, by (5.35) and (5.36) on $H_0^{1,2}(\Omega)$ are applied for solving the first boundary value problem for linear second-order elliptic equations. This corresponds to Poincaré - Friedrichs inequality.

(ii) For solving the second-order (respectively third-order) boundary value problems, we require the equivalence of the norms $|| \cdot ||_{1,2}$ and $|| \cdot ||_{1,2}^{\star}$ (respectively $|| \cdot ||_{1,2}$ and $|| \cdot ||_{2}^{\star\star}$) on $H^{1,2}(\Omega)$. This also corresponds to classical inequalities obtained by Poincaré and Friedrichs. It may be observed that norms $|| \cdot ||_1$, $|| \cdot ||_2$, $|| \cdot ||_3$ and $|| \cdot ||_4$ defined below are equivalent.

$$||f||_1 = \left(\int_a^b (f')^2 dx + \left| \int_a^b f dx \right|^2 \right)^{1/2} , \quad ||f||_2 = \left(\int_a^b (f')^2 dx + f^2(a) + f^2(b) \right)^{1/2}$$

$$||f||_3 = \left(\int_a^b (f^2 + (f')^2) dx \right)^{1/2} \quad and \quad ||f||_4 = \left(\int_a^b ((f')^2) dx \right)^{1/2}$$

where f' denotes the derivative of f in the classical sense.

(iii) Compactness of certain embeddings is used to prove the equivalence of norms on Sobolev spaces.

(iv) For solving eigenvalue problems, we require the compactness of the embedding $H_0^{1,2}(\Omega) \subseteq L_2(\Omega)$.

(v) Regularity statement of Theorem 5.23(d) is applied to prove regularity of generalized solutions. We are required to show that $f \in H^{m,2}(\Omega)$ for sufficiently large m; that is, the generalized solution f has generalized derivatives up to order m. Then, for $j < m - n/2$, we obtain $f \in C^j(\overline{\Omega})$; that is, f has classical derivatives up to order j. In particular, $H^{m,2}(\Omega) \subseteq C^{m1}(\overline{\Omega})$, $m = 1, 2, 3, \ldots$, if $\Omega \subset R$. In R^2 and R^3, we have that

$$H^{m,2}(\Omega) \subseteq C^{m-2}(\overline{\Omega}), \ m = 2, 3 \ldots$$

(vi) The boundary operator A is vital for the formulation of boundary conditions in the generalized sense. For example, let $f \in H^{1,2}(\Omega)$ and $g \in H^{1,2}(\Omega)$. Then the boundary condition

$$f = g \ on \ \partial\Omega$$

is to be understood in the sense

$$Af = Ag \ in \ L_2(\Omega)$$

that is, $Af(x) = Ag(x)$ for almost all $x \in \partial\Omega$. For proofs of the above results, we refer to Zeidler [201].

5.5 Integration in Banach Spaces

We discuss here some definitions and properties of spaces comprising functions on a real interval $[0, T]$ into a Banach space X. These concepts are of vital importance for studying parabolic differential equations, modeling problems of plasticity, sandpile growth, superconductivity, and option pricing.

Definition 5.20 Let (Ω, A, μ) be a finite measure space and X a Banach space. $u : \Omega \to X$ is called *strongly measurable* if there exists a sequence $\{u_n\}$ of simple functions such that $\|u_n(w) - u(w)\|_X \to 0$ for almost all w as $n \to \infty$.

Definition 5.21 (*Bochner Integral*) Let (Ω, A, μ) be a finite measure space, and X a Banach space. Then, we define the Bochner integral of a simple function $u : \Omega \to X$ by

$$\int_E u \, d\mu = \sum_{i=1}^{\infty} c_i \mu(E \cap E_i)$$

for any $E \in \mathscr{A}$, where c_i's are fixed scalars.

The *Bochner integral* of a strongly measurable function $u : \Omega \to X$ is the strong limit (if it exists) of the Bochner integral of an approximating sequence $\{u_n\}$ of simple functions. That is,

$$\int_E u \, d\mu = \lim_{n \to \infty} \int_E u_n \, d\mu.$$

Remark 5.6 (i) The Bochner integral is independent of the approximating sequence.

(ii) If u is strongly measurable, u is Bochner integrable if and only if $\|u(\cdot)\|_X$ is integrable.

Definition 5.22 $L_p(0, T; X)$, $1 \leq p < \infty$ consists of all strongly measurable functions $f : [0, T] \to X$ for which

$$\int_0^T \|f(t)\|_X^p \, dt < \infty.$$

Theorem 5.24 $C^m([0, T], X)$ *consisting of all continuous functions* $f : [0, T] \to X$ *that have continuous derivatives up to order m on* $[0, T]$ *is a Banach space with the norm*

$$\|f\| = \sum_{k=0}^{} \sup_{0 \leq t \leq T} \|f^k(t)\|_X \tag{5.40}$$

Theorem 5.25 $L_p(0, T; X)$ *is a Banach space with the norm*

$$\|f\| = \left(\int_0^T \|f(t)\|_X^p dt \right)^{1/2} \tag{5.41}$$

Let X be a Hilbert space, then $L_2(0, T; X)$ is a Hilbert space with respect to the inner product

$$\langle f, g \rangle_{L_2(0,T;X)} = \int_0^T \langle f, g \rangle_X dt \tag{5.42}$$

Remark 5.7 (a) In $L_p(0, T; X)$, two functions are identically equal if they are equal except on a set of measure zero.

(b) $L_\infty(0, T; X)$ denotes the space of all measurable functions which are essentially bounded. It is a Banach space with the norm

$$\|f\| = ess \sup_{0 \le t \le T} \|f(t)\|_X$$

(c) If the embedding $X \subseteq Y$ is continuous, then the embedding

$$L_p(0, T; X) \subseteq L_q(0, T; Y), \ 1 \le q \le p \le \infty$$

is also continuous.

(d) Let X^* be the dual space of a Banach space X, then $(L_p(0, T; X))^*$, the dual of $L_p(0, T; X)$ can be identified with $L_p(0, T; X^*)$; that is, we can write $(L_p(0, T; X))^* = L_p(0, T; X^*)$.

(e) Proofs of Theorems 5.24 and 5.25 are on the lines of Problems 2.3, 3.4, and 2.22.

Definition 5.23 (*Generalized Derivative*) Let $f \in L_1(0, T; X)$ and $g \in L_1(0, T; Y)$ where X and Y are Banach spaces. The function g is called the nth generalized derivative of the function f on $(0, T)$ if

$$\int_0^T \varphi^{(n)}(t) f(t) dt = (-1)^n \int_0^T \varphi(t) g(t) dt \text{ for all } \varphi \in C_0^\infty(0, T) \tag{5.43}$$

We write $g = f^{(n)}$.

Remark 5.8 (a) (Uniqueness of generalized derivative). The nth generalized derivative is unique, that is, if h is another nth generalized derivative, then $h = g$ almost everywhere on $(0, T)$; that is, $h = g$ in $L_1(0, T; Y)$.

(b) (Relationship between generalized derivative and distributions). Let $f \in L_1(0, T; X)$, then a distribution F is associated with f by the relation

$$F(\varphi) = \int_0^T \varphi(t)f(t)dt \ for \ all \ \varphi \in C_0^\infty(0, T)$$

For each n, this distribution has an nth derivative F(n) defined by

$$\langle F^{(n)}, \varphi \rangle = (-1)^n \langle F, \varphi^{(n)} \rangle \ for \ all \ \varphi \in C_0^\infty(0, T) \tag{5.44}$$

If (5.43) holds, then $F^{(n)}$ can be represented by

$$\langle F^{(n)}, \varphi \rangle = \int_0^T \varphi(t)f^{(n)}(t)dt \ for \ all \ \varphi \in C_0^\infty(0, T) \tag{5.45}$$

As we know, the advantage of the distribution concept is that each function $f \in L_1(0, T; X)$ possesses derivatives of every order in the distributional sense. The generalized derivative (Definition 5.23) singles out the cases in which by (5.44), the nth distributional derivative of f can be represented by a function $g \in L_1(0, T; Y)$. In this case, we set $f^{(n)} = g$ and write briefly $f \in L_1(0, T; X)$, $f^{(n)} \in L_1(0, T; Y)$.

Theorem 5.26 (Generalized Derivative and Weak Convergence) *Let X and Y be Banach spaces and let the embedding $X \subseteq Y$ be continuous. Then it follows from*

$$f_k^{(n)} = g_k \ on \ (0, T) \ \forall k \ and \ fixed \ n \geq 1$$

and $f_k \rightharpoonup f$ in $L_p(0, T; X)$ as $k \longrightarrow \infty$, $g_k \rightharpoonup g$ in $L_q(0, T; Y)$ as $k \to \infty$, $1 \leq p, q < \infty$ that $f^{(n)} = g$ on $(0, T)$. (See Zeidler [201, 419–420], for proof).

Theorem 5.27 *For a Banach space X, let $H^{m,p}(0, T; X)$ denote the space of all functions $f \in L_p(0, T; X)$ such that $f^{(n)} \in L_p(0, T; X)$, where $n \leq m$ and $f^{(n)}$ denote the nth generalized derivative of f. Then $H^{m,p}(0, T; X)$ is a Banach space with the norm*

$$\|f\|_{H^{m,p}(0,T;X)} = \left(\sum_{i=0}^m \|f_{L_p(0,T;X)}^{(i)}\| \right)^{1/p} \quad (f^{(0)} = f) \tag{5.46}$$

If X is a Hilbert space and $p = 2$, then $H^{m,2}(0, T; X)$ is a Hilbert space with the inner product

$$\langle f, g \rangle_{H^{m,2}(0,T;X)} = \int_0^T \sum_{i=0}^m \langle f^i, g^i \rangle_X dt \tag{5.47}$$

Remark 5.9 (a) The Proof of Theorem 5.27 is similar to that of Theorem 5.20.

(b) For $x < y$

$$\|f(y) - f(x)\|_X \leq \int_x^y \|f'(t)\|_X dt \tag{5.48}$$

holds

(c) The embedding $H^{1,2}(0, T; X) \subset C([0, T], H)$, where H is a Hilbert space, is continuous, that is, there exists a constant $k > 0$ such that

$$\|f\|_{C([0,T],H)} \leq k\|f\|_{H^{1,2}(0,T;H)}$$

Example 5.29 Let $X = Y$, $f \in C^n([0, T], Y)$, $n \geq 1$. Then the continuous nth derivative $f^{(n)} : [0, T] \to Y$ is also the generalized nth derivative of f on $(0, T)$. For $f \in C^1([0, T]; Y)$ and $\varphi \in C_0^\infty(0, T)$, $(\varphi f)' = \varphi' f + \varphi f'$. We obtain the classical integration by parts formula

$$\int_0^T \varphi' f dt = -\int_0^T \varphi f' dt$$

Repeated applications of this formula give

$$\int_0^T \varphi^n(t) f(t) dt = (-1)^n \int_0^T \varphi(t) f^n(t) dt$$

For more details we refer to [1, 61, 62, 88, 181, 201], [A 71, Hr 80].

5.6 Problems

5.6.1 Solved Problems

Problem 5.1 If the gradient $\nabla F(x)$ of a function $F : X \to R$ exists and $\|\nabla F(x)\| \leq M$ for all $x \in K$, where K is a convex subset of X, then show that

$$|F(u) - F(v)| \leq M\|v - u\|$$

Solution 5.1 By Theorem 5.2, we have

$$
\begin{aligned}
|F(u) - F(v)| &= |F'(u + \lambda(v - u))(v - u)| \\
&= |\langle \nabla F(u + \lambda(v - u)), (v - u) \rangle| \\
&\leq \|\nabla F(u + \lambda(v - u))\| \, \|v - u\| \\
&\leq M\|v - u\|
\end{aligned}
$$

as $||\nabla F(u)|| \leq M$ for all $u \in K$ [Since K is convex; for all $u, v \in K$, $(1 - \lambda)u + \lambda v \in K$].

Problem 5.2 Let $f : R^3 \to R$ possess continuous second partial derivatives with respect to all three variables, and let $F : C^1[a, b] \to R$ be defined by

$$F(x) = \int_a^b f(x(t), x'(t), t)dt$$

Show that the Fréchet derivative of F, $dF(x)h$, is given by

$$dF(x)h = \int_a^b \left[\frac{\partial f}{\partial x} - \frac{d}{dt}\left(\frac{\partial f}{\partial x'}\right) \right] h \, dt + \left[\frac{\partial f}{\partial x'} h \right]_a^b$$

Solution 5.2

$$F(x + h) - F(x) = \int_a^b f(x(t) + h(t), x'(t) + h(t), t)$$

$$- f(x(t), x'(t), t)dt$$

$$= \int_a^b \left(\left(\frac{\partial f}{\partial x}(x(t), x'(t), t) \right) h(t) \right.$$

$$+ \left(\frac{\partial f}{\partial x'}(x(t), x'(t), t) \right) h(t)dt + r(h, h) \quad (5.49)$$

where $r(h, h) = 0(||h||_{C[a,b]})$, i.e.

$$\frac{r(h, h)}{||h||_{C[a,b]}} \to 0 \text{ as } ||h||_{C[a,b]} \to 0$$

Hence

$$dF(x)h = \int_a^b \left[\frac{\partial f}{\partial x}(x(t), x'(t), t)h(t) + \frac{\partial f}{\partial x'}(x(t), x'(t), t)h(t) \right] dt$$

$$= \int_a^b \left[\frac{\partial f}{\partial x} - \frac{d}{dx}\frac{\partial f}{\partial x'} \right] h dt + \left[\frac{\partial f}{\partial x'} h \right]_a^b$$

after integration by part.

Problem 5.3 Let $a(\cdot, \cdot) \colon X \times X \to R$ be a bounded symmetric bilinear form on a Hilbert space X and J a functional on X, often called "energy functional", defined by

$$J(u) = \frac{1}{2} a(u, u) - F(u), \ where \ F \in X^{\star}$$

Find the Fréchet derivative of J.

Solution 5.3 For an arbitrary $\phi \in X$

$$
\begin{aligned}
J(u + \phi) &= \frac{1}{2} a(u + \phi, u + \phi) - F(u + \phi) \\
&= \frac{1}{2} a(u, u) + \frac{1}{2} a(\phi, u) + \frac{1}{2} a(u, \phi) + \frac{1}{2} a(\phi, \phi) \\
&\quad - F(u) - F(\phi)
\end{aligned}
$$

by the bilinearity of $a(\cdot, \cdot)$. Using the symmetry of $a(\cdot, \cdot)$, $[a(u, \phi) = a(\phi, u)]$, we get

$$
\begin{aligned}
J(u + \phi) &= \left\{ \frac{1}{2} a(u, u) - F(u) \right\} + \left\{ \frac{1}{2} a(u, \phi) - F(\phi) \right\} + \frac{1}{2} a(\phi, \phi) \\
&= J(u) + \{ a(u, \phi) F(\phi) \} + \frac{1}{2} a(\phi, \phi)
\end{aligned}
$$

or

$$
\begin{aligned}
\frac{\|[J(u + \phi) - J(u) - \{ a(u, \phi) - F(\phi) \}]\|}{\|\phi\|_X} &= \frac{1}{2} \frac{|a(\phi, \phi)|}{\|\phi\|_X} \\
\leq \frac{1}{2} \frac{M \|\phi\|_X \|\phi\|_X}{\|\phi\|_X} \ as \ a(\cdot, \cdot) \ is \ bounded.
\end{aligned}
$$

This implies that

$$\lim_{\|\phi\|_X \to 0} \frac{\|[J(u + \phi) - J(u) - \{ a(u, \phi) - F(\phi) \}]\|}{\|\phi\|_X} = \frac{1}{2} \frac{|a(\phi, \phi)|}{\|\phi\|_X} = 0$$

or

$$dJ(u)\phi = a(u, \phi) - F(\phi).$$

Since J defined in this problem is Fréchet differentiable, it is also Gâteaux differentiable and $DJ(u)\phi = dJ(u)\phi$. The derivative of this function is often used in optimal control problems and variational inequalities.

Problem 5.4 Prove that a linear operator T from a Banach space X into a Banach space Y is Fréchet differentiable if and only if T is bounded.

Solution 5.4 Let T be a linear operator and Fréchet differentiable at a point. Then T is continuous (and hence bounded) by Theorem 5.4. Conversely, if T is a bounded linear operator, then $||T(x+t) - Tx - Tt|| = 0$, proving that T is Fréchet differentiable and $T' = T$.

Problem 5.5 Prove that for $f \in C_0^\infty(\Omega)$, $\Omega \subset R^2$, there is a constant K depending on Ω such that

$$K \int_\Omega f^2 dx \leq \int_\Omega \left[\left(\frac{\partial f}{\partial x_1} \right)^2 + \left(\frac{\partial f}{\partial x_2} \right)^2 \right] dx.$$

Solution 5.5 Let $f \in C_0^\infty(\Omega)$. Consider a rectangle $Q = [a, b] \times [c, d]$ as in Fig. 5.2 with $\overline{\Omega} \subset Int\, Q$. Note that f vanishes outside Ω. Then

$$f(x, y) = \int_c^y \frac{\partial}{\partial y} f(x, t) dt \ for \ all \ (x, y) \in Q.$$

By Hölder's inequality, we get

$$|f(x, y)|^2 \leq \left(\int_c^y dt \right) \left(\int_c^y \left(\frac{\partial f}{\partial y} \right)^2 dt \right) \leq (d - c) \int_c^d \left(\frac{\partial f}{\partial y}(x, t) \right)^2 dx.$$

Integrating over Q, we get

$$\int_Q f^2 dx \leq (d - c)^2 \int_Q \left(\frac{\partial f}{\partial y} \right)^2 dx$$

Problem 5.6 Let Q be a closed square in R^2, with side length 1, then show that

$$\int_Q f^2 dx \leq \int_Q \sum_{i=1}^2 \left(\frac{\partial u}{\partial x_1} \right)^2 dx + \left| \int_K f dx \right|^2 \ for \ all \ f \in C^1(K).$$

Solution 5.6 We consider the square

$$Q = \{ (\xi, \eta) \in R^2 : -1/2 \leq \xi, \eta \leq 2 \}.$$

Let $f \in C^1(Q)$, and let $x = (\xi, \eta)$ and $y = (\alpha, \beta)$. Then

Fig. 5.2 Geometrical
explanation in Problem 5.5

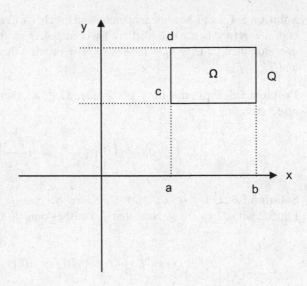

$$f(x) - f(y) = \int\limits_{a}^{\xi} f_\xi(t, \beta)dt + \int\limits_{\beta}^{\eta} f_\eta(\eta, t)dt.$$

By the inequality $(a + b)^2 \leq 2(a^2 + b^2)$ and by the Hölder inequality

$$(f(x) - f(y))^2 = f(x)^2 + f(y)^2 - 2f(x) f(y)$$

$$\leq 2 \int\limits_{-1/2}^{1/2} [(f_\xi(t, \beta))^2 + (f_\eta(\xi, t))^2]dt.$$

Integration over Q with respect to (x, y) yields

$$2 \int\limits_{\Omega} f^2 dx \leq 2 \int\limits_{\Omega} (f_\xi^2 + f_\eta^2)dx + 2 \left(\int\limits_{\Omega} fdx \right)^2.$$

This is the desired result. Here, $f_\xi = \frac{\partial f}{\partial \xi}$; $C^0(\Omega)$ is equal to the space of continuous functions on Ω into R; $C^k(\Omega)$ is equal to the space of all functions defined on Ω into R with compact support whose derivatives up to kth order exist and are continuous; and $C_0^\infty(\Omega) = \bigcap_{k=0}^{\Omega} C^k(\Omega)$ is equal to space of all functions defined on Ω into R having derivatives of all orders.

Problem 5.7 For each real numbers s, the space $H^s(R^n)$ has the following properties:

(i) $H^s(R^n)$ is a Hilbert space.

(ii) If $s_1 \geq s_2$ then $H^{s_1}(R^n) \subset H^{s_2}(R^n)$ and the injection is continuous.

Solution 5.7 (i) We verify the completeness of $H^s(R^n)$ as checking of inner product conditions are straightforward. Let $\{F_j\}$ be a Cauchy sequence in $H^s(R^n)$, then $\{(1 + |y|^2)^{1/2s}\hat{F}j\}$ is a Cauchy sequence in $L_2(R^n)$ which is complete and so $\{(1 + |y|^2)^{1/2s}\hat{F}j\} \to G$ in $L_2(R^n)$. Hence, G is a tempered distribution and since $(1 + |y|^2)^{-1/2s}G$ is of slow growth $(1 + |y|^2)^{-1/2s}G$ is a tempered distribution. Thus, there exists a tempered distribution F such that $(1 + |y|^2)^{-1/2s}G = \hat{F}$. Then $F(1 + |y|^2)^{1/2s}\hat{F} \in L_2(R^n)$; that is, $F \in H^s(R^n)$ and $F_j \to F$ in $H^s(R^n)$.

(ii) This follows from the fact that $(1 + |y|^2)^{s_2} \leq (1 + |y|^2)^{s_1}$ if $s_1 \geq s_2$.

5.6.2 Unsolved Problems

Problem 5.8 Let $F: C[0, 1] \to R$ defined by

$$F(f) = \int_0^1 (tf(t) + (f(t)^2)dt$$

Then find the Gâteaux derivative of F.

Problem 5.9 Let $F : \ell_2^2 \to R$ be defined by

$$F(x) = x_1^2 + x_1 x_2^2, \text{ where } x = (x_1, x_2)$$

Find the Fréchet derivative of F.

Problem 5.10 Let $T : R \to \ell_\infty$ be defined by

$$T(t) = \left(1, t, \frac{t^2}{2!}, \ldots, \frac{t^n}{n!}, \ldots\right)$$

Find the Fréchet derivative of T.

Problem 5.11 Let H_1 and H_2 be two Hilbert spaces and $T \in B(H_1, H_2)$. For $y \in H_2$, let $F : H_1 \to R$ be defined by

$$F(x) = ||Tx - y||^2/2$$

Prove that

$$\nabla F(x) = T^\star T x - T^\star y$$

Problem 5.12 (a) Let $f_1, f_2 : R^n \to R$ be Lipschitz near x, then show that $\partial(f_1 + f_2)(x) \subset \partial f_1(x) + \partial f_2(x)$.

(b) If f_1 is Lipschitz near x and f_2 is continuously differentiable on a neighborhood of x, then show that

$$\partial(f_1 + f_2)(x) = \partial f_1(x) + \nabla f_2(x).$$

Problem 5.13 Let $f : R^n \to R$ be Lipschitz on an open set containing the line segment $[x, y]$. Then show that there is a point $u \in (x, y)$ such that $f(y) - f(x) \in \langle \partial f(u), y - x \rangle$.

Problem 5.14 Let $F : X \to R$ be a functional. If F has a subgradient at x, then show that F is weakly lower semicontinuous at x. If F has a subgradient at all x in some open convex set $K \subset X$, then show that F is convex and weakly lower semicontinuous in K.

Problem 5.15 Let $F : X \to R$ be a convex functional on an open convex set K of X. Then prove that at every $x \in K$, there exists at least one subgradient.

Problem 5.16 (a) Justify that

$$\varphi(x) = \begin{cases} e^{(x^2 - 1)^{-1}}, & if |x| < 1 \\ 0, & otherwise \end{cases}$$

is a test function. Is the function

$$\psi(x) = \begin{cases} (x_1^2 + \cdots + x_n^2 - 1)^{-1} \\ 0. \end{cases}$$

a test function on R^n?

(b) Show that $\varphi(\alpha x + \beta x)$, $f(x)\varphi(x)$, and $\varphi^{(k)}(x)$; where α and β are constants, φ is as in (a), $f(x)$ is an arbitrary smooth function and k is a positive integer, are test functions.

(c) Let $\varphi_1, \varphi_2, \ldots, \varphi_n$ be test functions on R. Then

$$\varphi(x) = \varphi(x_1)\varphi(x_2), \ldots, \varphi_n(x_n)$$

is a test function on R^n.

Problem 5.17 Show that $\|f(t) - f(s)\|_X \leq \int_s^t \|f'(u)\|_X du$, where X is a Banach space.

Problem 5.18 Define the concept of convolution for distributions and give examples.

Problem 5.19 Show that $D^\alpha \delta(\lambda t - \mu)(\varphi) = \frac{1}{|\lambda|\lambda^\alpha} D^\alpha \delta\left(t - \frac{\mu}{\lambda}\right)$, where $\delta(\cdot)$ denotes the Dirac delta function.

Problem 5.20 Show that the sequence of regular distribution $\{sinnt\}$ converges to the zero distribution.

Problem 5.21 Let $H^1(\Omega)$, $\Omega \subseteq R^2$ denote the set of all $f \in L_2(\Omega)$ such that first partial derivative $\frac{\partial f}{\partial x_i} \in L_2(\Omega)$, $i = 1, 2$. Then show that $H^1(\Omega)$ is a Hilbert space with respect to the inner product $\langle f, g \rangle_1 = \int_\Omega (fg + \nabla f \cdot \nabla g)dx$ where

$$\nabla f \cdot \nabla g = \sum_{i=1}^{2} \frac{\partial g}{\partial x_i} \frac{\partial f}{\partial x_i}.$$

Show further that

$$\int_\Omega f\nabla^2\varphi d\Omega = -\int_\Omega \nabla f \cdot \nabla\varphi d\Omega$$

Problem 5.22 (*Rellich's Lemma*) If Ω is a bounded region in R^n, $n \geq 1$, then prove that embedding $H_0^{1,2}(\Omega) \subseteq L_2(\Omega)$ is compact.

Problem 5.23 Prove Green's formula, namely, the equation

$$\int_\Omega v\Delta u dx + \int_\Omega \langle grad\ u, grad\ v \rangle dx = \int_{d\Omega} v/\Gamma \frac{\partial u}{\partial v} d\Omega$$

Problem 5.24 Show that the embedding $L_\infty(0, T; X) \subseteq L_p(0, T; X)$ is continuous for all $1 \leq p \leq \infty$.

Problem 5.25 Let $f : R \to R$ be defined by

$$f(u) = \begin{cases} |u|^{p-2}u, & if\ u \neq 0 \\ 0, & if\ u = 0 \end{cases}$$

Show that

(a) if $p > 1$, then f is strictly monotone,
(b) if $p = 2$, then f is strongly monotone.

Problem 5.26 Show that for $u \in L_2(0, T; H)$ and $\frac{\partial u}{\partial t} \in L_2(0, T; H^*)$

$$\frac{d}{dt} \langle u, v \rangle = \left\langle \frac{\partial u}{\partial t}, v \right\rangle.$$

Chapter 6
Optimization Problems

Abstract Notion of optimization of a functional defined on a normed space by Banach space and Hilbert space is discussed in this chapter. All classical results concerning functions defined on R^n are obtained as special cases. Well-known algorithms for optimization are presented.

Keywords Optimization in Hilbert space · Convex programming · Quadratic programming · Linear programming · Calculus of variation · Minimization of energy functional · Newton algorithm · Conjugate gradient methods

6.1 Introduction

Finding maximum and minimum (extremum) of a real-valued function defined on a subset of real line is an optimization problem. It is an important topic of calculus. There is a close connection between this problem and derivative of the associated function. Celebrated mathematician Pierre de Fermat has obtained a simple result that if extremum of real-valued function exists at a point, then its derivative at that point is zero. In eighties, several books appeared devoted to optimization problem in the setting of vector spaces involving functionals particularly in the setting of R^n, Hilbert spaces, C[a, b] etc. Applications of optimization problems to diverse fields have been studied. We discuss in this chapter main results of this theme. Current developments concerning algorithmic optimization and related software may be seen in [46, 52, 71, 87, 99, 109, 115, 124, 137, 150, 151].

6.2 General Results on Optimization

Definition 6.1 Let U be a normed space and f a real-valued function defined on a nonempty closed convex subset K of X. The *general optimization* problem denoted by (P) is finding an element $u \in K$ such that $f(u) \leq f(v)$ for all $v \in K$. If such an element u exists, we write

© Springer Nature Singapore Pte Ltd. 2018   227
A. H. Siddiqi, *Functional Analysis and Applications*, Industrial and Applied Mathematics,
https://doi.org/10.1007/978-981-10-3725-2_6

$$f(u) = \inf_{v \in K} f(v)$$

If right-hand side exists, we say that f attains minimum at u. If $K \neq U$, this problem is referred to as the *constrained optimization problem*, while the case $K = U$ is called the *unconstrained optimization problem*.

Definition 6.2 Suppose A is a subset of a normed space X and f a real-valued function on A. f is called to have a *local* or *relative minimum (maximum)* at a point $x_0 \in A$ if there is an open sphere $S_r(x_0)$ of U such that $f(x_0) \leq f(x)$ $(f(x) \leq f(x_0))$ holds for all $x \in S_r(x_0) \cap A$. If f has either a relative minimum or relative maximum at x_0, then f is called to have a *relative extremum*. The set A on which an extremum problem is defined is often called the *admissible set*.

Theorem 6.1 *Suppose* $f : U \to R$ *is a Gâteaux differentiable functional at* $x_0 \in U$ *and* f *has a local extremum at* x_0, *then* $Df(x_0)t = 0$ *for all* $t \in U$.

Proof For every $t \in U$, the function $f(x_0 + \alpha t)$ (of the real variable) has a local extremum at $\alpha = 0$. Since it is differentiable at 0, it follows from classical calculus that

$$\left[\frac{d}{d\alpha} f(x_0 + \alpha t) \right]_{\alpha=0} = 0$$

This implies that $Df(x_0)t = 0$ for all $t \in U$, that is, the proof of the theorem.

Remark 6.1 (i) Theorem 6.1 implies that if a functional $f : X \to R$ is Fréchet differentiable at $x_0 \in X$ and has a relative extremum at x_0, then $dT(x_0) = 0$.
(ii) Suppose f is a real-valued functional on a normed space U and x_0 a solution of (P) on a convex set K. If f is a Gâteaux differentiable at x_0, then

$$Df(x_0)(x - x_0) \geq 0 \forall\, x \in K$$

Verification Since K is a convex set, $x_0 + \alpha(x - x_0) \in K$ for all $\alpha \in (0, 1)$ and $x \in K$. Hence

$$Df(x_0)(x - x_0) = \left[\frac{d}{d\alpha} f(x_0 + \alpha(x - x_0)) \right]_{\alpha=0} \geq 0$$

Theorem 6.2 *Suppose K is a convex subset of a normed space U.*

i. *If $J : K \to R$ is a convex function, then existence of a local minima of J implies existence of solution of problem (P).*
ii. *Let $J : O \subset U \to R$ is a convex function defined over an open subset of X containing K and let J be Fréchet differentiable at a point $u \in K$. Then, J has a minimum at u (u is a solution of (P) on K) if and only if*

$$J'(u)(v - u) \geq 0 \text{ for every } v \in K \tag{6.1}$$

If K is open, then (6.1) is equivalent to

$$J'(u) = 0 \tag{6.2}$$

Equation (6.2) is known as the Euler's equation.

Proof   1. Let $v = u + w$ be any element of K. By the convexity of J

$$J(u + \alpha w) \leq (1 - \alpha)J(u) + \alpha J(v), 0 \leq \alpha \leq 1$$

which can also be written as

$$J(u + \alpha w) - J(u) \leq \alpha(J(v) - J(u)), 0 \leq \alpha \leq 1.$$

As J has a local minimum at u, there exists α_0 such that $\alpha_0 > 0$ and $0 \leq J(u + \alpha_0 w) - J(u)$, implying $J(v) \geq J(u)$.

2. The necessity of (6.1) holds even without convexity assumption on J by Remark 6.1(ii). For the converse, let

$$J(v) - J(u) \geq J'(u)(v - u) \text{ for every } v \in K.$$

Since J is convex

$$J((1 - \alpha)u + \alpha v) \leq (1 - \alpha)J(u) + \alpha J(v) \text{ for all } \alpha \in [0, 1]$$

or

$$J(v) - J(u) \geq \frac{J(u + \alpha(v - u))}{\alpha}$$

or

$$J(v) - J(u) \geq \lim_{\alpha \to 0} \frac{J(u + \alpha(v - u))}{\alpha} = J'(u)(v - u) \geq 0.$$

This proves that for $J'(u)(v - u) \geq 0$, J has a minimum at u.

A functional J defined on a normed space U is called *coercive* if $\lim\limits_{||x|| \to \infty} J(x) = \infty$.

Theorem 6.3   (Existence of Solution in R^n) *Suppose K is a nonempty, closed convex subset of R^n and $J : R^n \to R$ a continuous function which is coercive if K is unbounded. Then, (P) has at least one solution.*

Proof Let $\{u_k\}$ be a minimizing sequence of J, that is, a sequence satisfying conditions $u_k \in K$ for every integer k and $\lim\limits_{k \to \infty} J(u_k) = \inf_{v \in K} J(v)$. This sequence is necessarily bounded, since the functional J is coercive, so that it is possible to find a subsequence $\{u_{k'}\}$ which converges to an element $u \in K$ (K being closed). Since J is continuous, $J(u) = \lim\limits_{k' \to \infty} J(u_{k'}) = \inf_{v \in K} J(v)$. This proves the theorem.

Theorem 6.4 (Existence of Solution in Infinite-Dimensional Hilbert Space) *Suppose K is a nonempty, convex, closed subset of a separable Hilbert space H and $J : H \to R$ is a convex, continuous functional which is coercive if K is unbounded. Then, the optimization problem (P) has at least one solution.*

Proof As in the previous theorem, K must be bounded under the hypotheses of the theorem. Let $\{u_k\}$ be a minimizing sequence in K. Then, by Theorem 4.17, $\{u_k\}$ has a weakly convergent subsequence $u_{k'} \rightharpoonup u$. By Corollary 4.1, $J(u) \leq$ $\liminf J(u_{k'})$, $u_{k'} \rightharpoonup u$ which, in turn, shows that u is a solution of (P).

It only remains to show that the weak limit u of the sequence $\{u_{k'}\}$ belongs to the set K. For this, let P denote the projection operator associated with the closed, convex set K; by Theorem 3.12

$$w_\ell \in K \text{ implies } \langle Pu - u, w_\ell - Pu \rangle \geq 0 \text{ for every integer } \ell$$

The weak convergence of the sequence $\{w_\ell\}$ to the element u implies that

$$0 \leq \lim_{\ell \to \infty} \langle Pu - u, w_\ell - Pu \rangle = \langle Pu - u, u - Pu \rangle = -||u - Pu||^2 \leq 0$$

Thus, $Pu = u$ and $u \in K$.

Remark 6.2 (i) Theorem 6.4 remains valid for reflexive Banach space and continuity replaced by weaker condition, namely weak lower semicontinuity. For proof, see Ekeland and Temam [71] or Siddiqi [169].
(ii) The set S of all solutions of (P) is closed and convex.
   Verification
   Let u_1, u_2 be two solutions of (P); that is, $u_1, u_2 \in S$. $\alpha u_1 + (1 - \alpha)u_2 \in K$, $\alpha \in$ $(0, 1)$ as K is convex. Since J is convex

$$J(\alpha u_1 + (1 - \alpha)u2) \leq \alpha J(u_1) + (1 - \alpha)J(u_2)$$

Let $\lambda = \inf_{v \in K} J(v) = J(u_1)$ and $\lambda = \inf_{v \in K} J(v) = J(u_2)$, then

$$\lambda \leq J(\alpha u_1 + (1 - \alpha)u_2) \leq \alpha\lambda + (1 - \alpha)\lambda = \lambda$$

that is, $\lambda = J(\alpha u_1 + (1 - \alpha)u_2)$ implying $\alpha u_1 + (1 - \alpha)u_2 \in S$. Therefore, S is convex.
Let $\{u_n\}$ be a sequence in S such that $u_n \to u$. For proving closedness, we need to show that $u \in S$. Since J is continuous

$$J(u) = \liminf_{n \to \infty} J(u_n) \leq \lambda \leq J(u)$$

This gives

$$J(u) = \lambda \text{ and so } u \in S$$

(iii) The solution of Theorem 6.4 is unique if J is strictly convex.
    Verification
    Let $u_1, u_2 \in S$ and $u_1 \neq u_2$. Then $\frac{u_1 + u_2}{2} \in S$ as S is convex.
    Therefore, $J\left(\frac{u_1 + u_2}{2}\right) = \lambda$. Since J is strictly convex.

$$J\left(\frac{u_1 + u_2}{2}\right) < \frac{1}{2}J(u_1) + \frac{1}{2}J(u_2) = \frac{1}{2}\lambda + \frac{1}{2}\lambda = \lambda$$

This is a contradiction. Hence, $u_1 \neq u_2$ is false and $u_1 = u_2$.

6.3 Special Classes of Optimization Problems

6.3.1 Convex, Quadratic, and Linear Programming

For $K = \{v \in X / \varphi_i(v) \leq 0, 1 \leq i \leq m', \varphi_i(v) = 0, m' + 1 \leq i \leq m\}$, (P) is called a *nonlinear programming problem*. If φ_i and J are convex functionals, then (P) is called a *convex programming problem*. For $X = R^n, K = \{v \in R^n / \varphi_i(v) \leq d_i, 1 \leq i \leq m\}, J(v) = \frac{1}{2}\langle Av, v \rangle - \langle b, v \rangle, A = (a_{ij})$, an $n \times n$ positive definite matrix, and $\varphi_i(v) = \sum_{j=1}^{n} a_{ij}v_j$; (P) is called a *quadratic programming problem*. If

$$J(v) = \sum_{j=1}^{n} \alpha_i v_i, X = R^n, K = \left\{v \in R^n / \sum_{j=1}^{n} a_{ij}v_j \leq d_i, 1 \leq i \leq m\right\},$$

$A = (a_{ij}), n \times n$ positive definite matrix, then (P) is called a *linear programming problem*.

6.3.2 Calculus of Variations and Euler–Lagrange Equation

The classical calculus of variation is a special case of (P) where we look for the extremum of functionals of the type

$$J(u) = \int_a^b F(x, u, u')dx \left(u'(x) = \frac{du}{dx}\right) \tag{6.3}$$

u is a twice continuously differentiable function on $[a, b]$; F is continuous in x, u and u' and has continuous partial derivatives with respect to u and u'.

Theorem 6.5 *In order that functional $J : U \rightarrow R$, where U is a normed space has a minimum or maximum (extremum) at a point $u \in U$ must satisfy the Euler–Lagrange equation*

$$\frac{\partial F}{\partial u} - \frac{d}{dx}\left(\frac{\partial F}{\partial u'}\right) = 0 \tag{6.4}$$

in $[a, b]$ with the boundary condition $u(a) = \alpha$ and $u(b) = \beta$. J and F are related by (6.3).

Proof Let $u(a) = 0$ and $u(b) = 0$, then

$$J(u + \alpha v) - J(u) = \int_a^b [F(x, u + \alpha v, u' + \alpha v') - F(x, u, u')]dx \tag{6.5}$$

Using the Taylor series expansion

$$F(x, u + \alpha v, u' + \alpha v') = F(x, u, u') + \alpha \left(v\frac{\partial F}{\partial u} + v'\frac{\partial F}{\partial u'}\right) + \frac{\alpha^2}{2!}\left(v\frac{\partial F}{\partial u} + v'\frac{\partial F}{\partial u'}\right)^2 + \cdots$$

it follows from (6.5) that

$$J(u + \alpha v) = J(u) + \alpha dJ(u)(v) + \frac{\alpha^2}{2!}d^2 J(u)(v) + \cdots \tag{6.6}$$

where the first and the second Fréchet differentials are given by

$$dJ(u)v = \int_a^b \left(v\frac{\partial F}{\partial u} + v'\frac{\partial F}{\partial u'}\right) dx \tag{6.7}$$

$$d^2 J(u)v = \int_a^b \left(v\frac{\partial F}{\partial u} + v'\frac{\partial F}{\partial u'}\right)^2 dx \tag{6.8}$$

The necessary condition for the functional J to have an extremum at u is that $dJ(u)v = 0$ for all $v \in C^2[a, b]$ such that $v(a) = v(b) = 0$; that is

$$0 = dJ(u)v = \int_a^b \left(v\frac{\partial F}{\partial u} + v'\frac{\partial F}{\partial u'}\right) dx \tag{6.9}$$

Integrating the second term in the integrand in (6.9) by parts, we get

$$\int_a^b \left[\frac{\partial F}{\partial u} - \frac{d}{dx}\left(\frac{\partial F}{\partial u'}\right)\right] vdx + \left[v\frac{\partial F}{\partial u'}\right]_a^b = 0 \tag{6.10}$$

Since $v(a) = v(b) = 0$, the boundary terms vanish and the necessary condition becomes

$$\int_a^b \left[\frac{\partial F}{\partial u} - \frac{d}{dx} \left(\frac{\partial F}{\partial u'} \right) \right] v \, dx = 0 \; for \; all \; v \in C^2[a, b] \qquad (6.11)$$

for all functions $v \in C^2[a, b]$ vanishing at a and b. This is possible only if

$$\frac{\partial F}{\partial u} - \frac{d}{dx} \left(\frac{\partial F}{\partial u'} \right) = 0 \; (see \; Problem \; 6.1)$$

Thus, we have the desired result.

6.3.3 Minimization of Energy Functional (Quadratic Functional)

A functional of the type

$$J(v) = \frac{1}{2} a(v, v) - F(v) \qquad (6.12)$$

where $a(\cdot, \cdot)$ is a bilinear and continuous form on a Hilbert space H and F is an element of the dual space H^* of H, which is called an *energy functional* or a *quadratic functional*.

Theorem 6.6 *Suppose $a(\cdot, \cdot)$ is coercive and symmetric, and K is a nonempty closed convex subset of H. Then (P) for J in (6.12) has only one solution on K.*

Proof The bilinear form induces an inner product over the Hilbert space H equivalent to the norm induced by the inner product of H. In fact, the assumptions imply that

$$\sqrt{\alpha}\|v\| \le (a(v, v))^{1/2} \le \sqrt{\|a\|}\|v\| \qquad (6.13)$$

where $\|a\|$ is given in Remark 3.23. Since F is a linear continuous form under this new norm, the Riesz representation theorem (Theorem 3.19), there exists a unique element $u \in X$ such that

$$F(v) = a(u, v) \; for \; every \; v \in K \qquad (6.14)$$

In view of the remark made above and (6.14), (6.12) can be rewritten as

$$J(v) = \frac{1}{2} a(v, v) - a(u, v) = \frac{1}{2} a(v - u, v - u) - \frac{1}{2} a(u, u)$$

$$= \frac{1}{2} \langle v - u, v - u \rangle - \frac{1}{2} \langle u, u \rangle \; for \; all \; v \in K \; and \; unique \; u.$$

Therefore, $\inf_{v \in K} J(v)$ is equivalent to $\inf_{v \in K} ||v - u||$. Thus, in the present situation, (P) amounts to looking for the projection x of the element u on to the set K. By Theorem 3.11, (P) has a unique solution.

Example 6.1 Let $X = K = R^n, J : v \in R^n \to J(v)$, where

$$J(v) = \frac{1}{2}||Bv - c||_m^2 - \frac{1}{2}||c||_m^2, \tag{6.15}$$

B is an $m \times n$ matrix, and $|| \cdot ||_m$ denotes the norm in R^m. Since

$$J(v) = \frac{1}{2}\langle B^t Bv, v \rangle_n - \langle B^t u, v \rangle_n$$

the problem is one of quadratic programming if and only if the symmetric matrix is positive definite. Theorem 6.6 yields the existence of the solution.

We will examine in Chap. 7 existence of solutions of boundary value problems representing interesting physical phenomena formulating in terms of minimization of the energy functional.

6.4 Algorithmic Optimization

6.4.1 Newton Algorithm and Its Generalization

The Newton method deals with the search of zeros of the equation $F(x) = 0$, $F :$ $U \subset X \to Y, X$ and Y are normed spaces, in particular for $X = Y = R$, $F : R \to R$ or $X = R^n$ and $Y = R^n$, $F : R^n \to R^n$ and U an open subset of X (open interval of R or open ball of R^n). Once we have this method, the functional F can be replaced by F' or ∇F to obtain the algorithm for finding the extrema of F, that is, zeros of F' or ∇F which are extremum points of F. One can easily check that if $F : [a, b] \to R$ and $|F'(x)| < 1$, then $F(x) = 0$ has a unique solution; that is, F has a unique zero. For the function $F : U \subset R \to R$, U the open subset of R, the Newton method is defined by the sequence

$$u_{k+1} = u_k - \frac{F(u_k)}{F'(u_k)}, \ k \geq 0 \tag{6.16}$$

u_0 is an arbitrary point of open set U. The geometric meaning of (6.16) is that each point u_{k+1} is the intersection of the axis with the tangent at the point u_k. This particular case suggests the following generalization for the functional $F : U \subset X \to Y :$ For an arbitrary point $u_0 \in U$, the sequence $\{u_k\}$ is defined by

$$u_{k+1} = u_k - \{F'(u_k)\}^{-1} F(u_k) \tag{6.17}$$

under the condition that all the points u_k lie in U. If $X = R^n, Y = R^n, F(u) = 0$ is equivalent to

$$F_1(u) = 0, \ u = (u_1, u_2, \ldots, u_n) \in R^n$$
$$F_2(u) = 0$$
$$F_3(u) = 0$$
$$\cdot$$
$$\cdot$$
$$\cdot$$
$$F_n(u) = 0$$

where $F_i : R^n \to R$, $i = 1, 2, \ldots, n$. A single iteration of the Newton method consists in solving the linear system

$$\left. \begin{array}{c} F'(u_k)\Delta u_k = -F(u_k), \text{ with matrices} \\ F'(u_k) = \frac{\partial F_i(u_k)}{\partial x_j} \end{array} \right\}$$

and then setting

$$u_{k+1} = u_k + \Delta u_k.$$

It may be noted that if F is an affine function, that is, $F(x) = A(x) - b, A = (a_{ij})$ is a square matrix of size n; that is, $A \in \mathcal{A}_n(R)$ and $b \in R^n$, then the iteration described above reduces to the solution of the linear system $Au_{k+1} = b$. In that case, the method converges in a single iteration.

We now look for

(i) sufficient conditions which guarantee the existence of a zero of the function F, and
(ii) an algorithm for approximating such an element u, that is, for constructing a sequence $\{u_k\}$ of points of U such that

$$\lim_{k \to \infty} u_k = u.$$

We state below two theorems concerning the existence of a unique zero of F.

Theorem 6.7 *Let X be a Banach space, U an open subset of X, Y a normed linear space and $F : U \subset X \to Y$ differentiable over U. Suppose that there exist three constants α, β and γ such that $\alpha > 0$ and $S_\alpha(u_0) = \{u \in X \,/\, \|u - u_0\| \le \alpha\} \subseteq U$*

(i) $\sup_{k \ge 0} \sup_{u \in S_\alpha(u_0)} \|A_k^{-1}(u)\|_{B[X,Y]} \le \beta$, $Ak(u) = A_k \in B[X, Y]$ *is bijective.*
(ii) $\sup_{k \ge 0} \sup_{x' \in S_\alpha(u_0)} \|F(x') - A_k(x')\|_{B[X,Y]} \le \frac{\gamma}{\beta}$, *and $\gamma < 1$.*
(iii) $\|F(u_0)\| \le \frac{\alpha}{\beta}(1 - \gamma)$.

Then, the sequence defined by

$$u_{k+1} = u_k - A_k^{-1}(u_{k'})F(u_k), \ k \geq k' \geq 0 \tag{6.18}$$

is entirely contained within the ball and converges to a zero of F in $S_\alpha(u_0)$ which is unique. Furthermore

$$\|u_k - u\| \leq \frac{\|u_1 - u_0\|}{1 - \gamma}\gamma^k. \tag{6.19}$$

Theorem 6.8 *Suppose X is a Banach space, U is an open subset of X, $F : U \subset X \rightarrow Y$, and Y a normed space. Moreover, suppose that F is continuously differentiable over U. Suppose that u is a point of U such that*

$$\left\{ \begin{array}{l} F(u) = 0, A = F'(u) : X \rightarrow Y, \ bounded \ linear \ and \ bijective \\ \sup_{k \geq 0} \|A_k - A\|_{B[X,Y]} \leq \frac{\lambda}{\|A^{-1}\|_{B[X,Y]}}, \ and \ \lambda < \frac{1}{2}. \end{array} \right\}$$

Then, there exists a closed ball, $S_r(u)$, with center u and radius r such that for every point $u_0 \in S_r(u)$, the sequence $\{u_k\}$ defined by

$$u_{k+1} = u_k - A_k^{-1}F(u_k), \ k \geq 0 \tag{6.20}$$

is contained in $S_r(u)$ and converges to a point u, which is the only zero of F in the ball $S_r(u)$. Furthermore, there exists a number γ such that

$$\gamma < 1 \ and \ \|u_k - u\| \leq \gamma^k\|u_0 - u\|, \ k \geq 0 \tag{6.21}$$

Theorem 6.7 yields the following result:

Corollary 6.1 *Let U be an open subset of a Banach space X and $F : U \subset X \rightarrow R$ which is twice differentiable in the open set U. Suppose that there are three constants: α, β, γ such that $\alpha > 0$ and $S_\alpha(u_0) = \{v \in X| \ \|v - u_0\| \leq \alpha\} \subset U$, $A_k(v) \in B[X, X^\star]$ and bijective for every $v \in S_\alpha(u)$ and*

$$\left. \begin{array}{l} \sup_{k \geq 0} \sup_{u \in S_\alpha(u_0)} \|A_k^{-1}(u)\|_{B[X^\star,X]} \leq \beta \\ \sup_{k \geq 0} \sup_{v' \in S_\alpha(u_0)} \|F''(v) - A_k(v)\|_{B[X,X^\star]} \leq \frac{\gamma}{\beta}, \ and \\ \gamma < 1, \ \|F(u_0)\|_{X^\star} \leq \frac{\alpha}{\beta}(1 - \gamma). \end{array} \right\}$$

Then, the sequence $\{u_k\}$ defined by

$$u_{k+1} = u_k - A_k^{-1}(u_{k'})F'(u_k), \ k \geq k' \geq 0$$

is contained in the ball $S_\alpha(u_0)$ and converges to a zero of F', say u, which is the only zero in this ball. Furthermore

$$||u_k - u|| \leq \frac{||u - u_0||}{1 - \gamma} \gamma^k.$$

Theorem 6.8 yields the following result.

Corollary 6.2 *Let U be an open subset of a Banach space X and $F : U \subset X \to R$ a function which is twice differentiable in U. Moreover, let u be a point of U such that*

$$F'(u) = 0, \ F''(u) \in B[X, X^\star] \ and \ bijective$$

$$\sup_k ||A_k - F''(u)||_{B[X,X^\star]} \leq \frac{\lambda}{||(F''(u))^{-1}||_{B[X^\star,X]}} \ and \ \lambda < \frac{1}{2}$$

Then, there exists a closed ball $S_r(u)$ with center u and radius $r > 0$ such that, for every point $u_0 \in S_r(u)$, the sequence $\{u_k\}$ defined by $u_{k+1} = u_k - A_k^1 F(u_k)$ is contained in $S_r(u)$ and converges to the point u, which is the only zero of F' in the ball. Furthermore, $u_{k+1} = u_k - A_k^{-1}(u_k)F'(u_k)$ converges geometrically; namely, there exists a γ such that $\gamma < 1$ and $||u_k - u|| \leq \gamma^k ||u_0 - u||, \ k \geq 0$.

Remark 6.3 (i) Let $X = R^n$, the generalized Newton method of Corollary 6.2 takes the form

$$u_{k+1} = u_k - A_k^{-1}(u_{k'})\nabla F(u_k), \ k \geq k' \geq 0 \tag{6.22}$$

where $A_k(u_k)$ are invertible matrices of order n, and $\nabla F(u_k)$ denotes the gradient vector of the function F at the point u_k ($(R^n)^\star$ is identified with R^n). In particular, the original Newton method corresponds to

$$u_{k+1} = u_k - \{\nabla^2 F(u_k)\}^{-1} \nabla F(u_k), \ k \geq 0 \tag{6.22a}$$

where the matrix $\nabla^2 F(u_k)$ is Hessian of the function F at the point u.

(ii) The special case, $A_k(u_k) = \varphi^1 I$, is known *as the gradient method with fixed parameter.*

(iii) The special case, $A_k(u_{k'}) = -\varphi_k^{-1} I$, is called the *gradient method with variable parameter.*

(iv) The special case, $A_k(u_{k'}) = -(\varphi(u_k))^{-1} I$, is called the *gradient method with optimal parameter*, where the number $\varphi(u_k)$ (provided it exists) is determined from the condition

$$F(u_k - \varphi(u_k))\nabla F(u_k) = \inf_{\varphi \in R} F(u_k - \varphi \nabla F(u_k)). \tag{6.23}$$

General Definition of the Gradient Method

Every iterative method for which the point u_{k+1} is of the form

$$u_{k+1} = u_k - \varphi_k \nabla F(u_k), \ \varphi_k > 0$$

is called a *gradient method*. If φ_k is fixed, it is called a *gradient method with fixed parameter*, whereas if φ_k is variable, it is called a *gradient method with variable parameters*.

Theorem 6.9 *Suppose $X = R^n$ and suppose that the functional $F : X \to R$ is elliptic, that is, there is a positive constant α such that $F(x) \geq \alpha||x||^2$ for all $x \in X$. Then, the gradient method with optimal parameter converges.*

Remark 6.4 (a) The following properties of elliptic functionals are quite useful (For details, we refer to Ciarlet [46]):

(a) Suppose $F : H \to R$ (H is a Hilbert space, in particular $X = R^n$) is strictly convex and coercive, then it satisfies the inequality

$$F(v) - F(u) \geq \langle \nabla F(u), v - u \rangle + \frac{\alpha}{2}||v - u||^2 \text{ for every } u, v \in X \quad (6.24)$$

(b) If F is twice differentiable, then it is elliptic if and only if

$$\langle \nabla^2 F(u)w, w \rangle \geq \alpha||w||^2 \text{ for every } w \in H \quad (6.25)$$

(c) A quadratic functional F over R^n

$$\left.\begin{array}{c} F(v) = \frac{1}{2}\langle Av, v \rangle - \langle y, v \rangle, \text{ A is the } n \times n \text{ matrix and} \\ A = A^t, \text{ is elliptic if and only if} \\ \langle \nabla^2 F(u)w, w \rangle = \langle Aw, w \rangle \geq \lambda_1||w||^2, \ \forall\, u, w \in R^n \end{array}\right\} \quad (6.26)$$

where λ_1 denotes the smallest eigenvalue of A.

(d) Let $J(v) = \frac{1}{2}\langle Av, v \rangle - \langle y, v \rangle, A \colon R^n \to (R^n)^* = R^n$. Since $\nabla J(u_k)$ and $\nabla J(u_{k+1})$ are orthogonal and $\nabla J(v) = Av - y$, we have

$$\langle \nabla J(u_{k+1}), \nabla J(u_k) \rangle = \langle A(u_k - \varphi(u_k)(Au_k - y)) - y, Au_k - y \rangle = 0$$

This implies that $\varphi(u_k) = \frac{||w_k||^2}{\langle Aw_k, w_k \rangle}$ where $w_k = Au_k - y = \nabla J(u_k)$. A single iteration of the method is done as follows:

(i) calculate vector $w_k = Au_k - y$

(ii) calculate the number

$$\varphi(u_k) = \frac{||w_k||^2}{\langle Aw_k, w_k \rangle}$$

(iii) calculate the vector

$$u_{k+1} = u_k - \varphi(u_k)w_k$$

Theorem 6.10 *Suppose $F : R^n \to R$ is a differentiable functional. Let there be two positive constants α and β such that*

(i) $\langle \nabla F(v) - \nabla F(u), v - u \rangle \geq \alpha ||v - u||^2$ for all $v, u \in R^n$ and $\alpha > 0$
(ii) $||\nabla F(v) - \nabla F(u)|| \leq \beta ||v - u||$ for every $u, v \in R^n$.

Moreover, there are two numbers a and b such that

$$0 < a \leq \varphi_k \leq b < \frac{2\alpha}{\beta^2} \text{ for every } k$$

Then, the gradient method with variable parameter converges and the convergence is geometric in the sense that there exists a constant γ depending on α, β, a, b such that $\gamma < 1$ and $||u_k - u|| \leq \gamma_k ||u_0 - u||$.

Remark 6.5 (i) If F is twice differentiable, then condition (ii) can also be written in the form $\sup ||\nabla^2 F(u)|| \leq \beta$.

(ii) In the case of an elliptic quadratic functional $F(v) = \frac{1}{2}\langle Av, v \rangle - \langle y, v \rangle$, one iteration of the method takes the form

$$u_{k+1} = u_k \varphi_k (Au_k - y), \ k \geq 0$$

and by Theorem 6.10 that the method is convergent if $0 < \alpha \leq \varphi_k \leq b \leq 2\lambda_1/\lambda_n^2$, where λ_1 and λ_n are, respectively, the least and the largest eigenvalues of the symmetric positive definite matrix A.

Proof (*Proof of Theorem 6.7*) First, we prove that for every integer $k \geq 1$

$$||u_k - u_{k-1}|| \leq \beta ||F(u_{k-1})||$$

$$||u_k - u_0|| \leq \alpha \text{ equivalently } u_k \in S_\alpha(u_0)$$

$$||F(u_k)|| \leq \frac{\gamma}{\beta} ||u_k - u_{k1}||$$

We apply the finite induction principle for the proof. Let us show that the results are true for $k = 1$; that is

$$||u_1 - u_0|| \leq \beta ||F(u_0)||$$

$$||u_1 - u_0|| \leq \alpha, \ ||F(u_1)|| \leq \frac{\gamma}{\beta} ||u_1 - u_0||$$

Putting $k = 0$ in relation (6.18), we get

$$u_1 - u_0 = -A_0^{-1}(u_0)F(u_0) \tag{6.27}$$

which implies that $||u_1 - u_0|| \leq \beta ||F(u_0)|| \leq \alpha(1 - \gamma) \leq \alpha$ by the hypotheses of the theorem. Further, from (6.27), we can write

$$F(u_1) = F(u_1) - F(u_0) - A_0(u_0)(u_1 - u_0)$$

By the Mean Value Theorem applied to the function $u \to F(u) - A_0(u_0)u$, we have

$$||F(u_1)|| \leq \sup_{u \in S_\alpha(u_0)} ||F'(u) - A_0(u_0)|| \, ||(u_1 - u_0)|| \leq \frac{\gamma}{\beta}||(u_1 - u_0)||$$

by condition (ii) of the theorem.

Let us assume that the desired results are true for the integer $k = n - 1$. Since $u_n - u_{n1} = -A_{n-1}^{-1}(u_{(n-1)\gamma})F(u_{n-1})$, it follows that $||u_n - u_{n-1}|| \leq \beta||F(u_{n-1})||$ which gives the first relation for $k = n$. Then, we have

$$||u_n - u_{n-1}||^{-1} = ||A_{n-1}^{-1}(u_{(n-1)\gamma})F(u_{n-1})|| \leq \beta||F(u_{n-1})||$$

$$\leq \beta\frac{\gamma}{\beta}||u_{n-1} - u_{n-2}||$$

$$\cdots$$
$$\cdots$$
$$\cdots$$

$$\leq \gamma^{n-1}||u_1 - u_0||$$

This implies that

$$||u_n - u_0|| \leq \sum_{i=1}^{n} ||u_i - u_{i-1}|| \leq \left\{\sum_{i=1}^{n} \gamma_i\right\}||u_1 - u_0||$$

$$\leq \frac{||u_1 - u_0||}{1 - \gamma} \leq \alpha, \text{ by condition(iii)}$$

which means that $u_n \in S_\alpha(u_0)$.

For proof of the last relation, we write

$$F(u_n) = F(u_n) - F(u_{n-1}) - A_{n-1}(u_{(n-1)\gamma})(u_n - u_{n-1})$$

By applying the Mean Value Theorem to the function $u \to F(u) - A_{(n-1)} \times (u_{(n-1)\gamma})u$, we get

$$||F(u_k)|| \leq \sup_{u \in S_\alpha(u_0)} ||F'(u) - A_{n1}(u_{(n-1)\gamma})|| \, ||u_n - u_{n-1}||$$

$$\leq \frac{\gamma}{\beta}||u_n - u_{n-1}||$$

and the last relation is established for n. Hence, these three relations are true for all integral values of k.

We now prove the existence of a zero of the functional F in the ball $S_\alpha(u_0)$. Since

$$\left.\begin{array}{l} ||u_{k+m} - u_k|| \leq \sum_{i=1}^{m-1} ||u_{k+i+1} - u_{k+i}|| \\ \leq \gamma^k \sum_{i=0}^{m-1} \gamma^i||u_1 - u_0|| \leq \frac{\gamma^k}{1-\gamma}||u_1 - u_0|| \to 0 \text{ as } k \to \infty \end{array}\right\} \tag{6.28}$$

where $\{u_k\}$ is a Cauchy sequence of points in the ball $S_\alpha(u_0)$ which is a closed subspace of a complete metric space X (X, a Banach space). This implies that there exists a point $u \in S_\alpha(u_0)$ such that

$$\lim_{k \to \infty} u_k = u$$

Since F is differentiable and therefore continuous, we get

$$||F(u)|| = \lim_{k \to \infty} ||F(u_k)|| \le \frac{\gamma}{\beta} \lim_{k \to \infty} ||u_k - u_{k-1}|| = 0$$

which, in turn, implies $F(u) = 0$ by the first axiom of the norm. By taking the limit $m \to \infty$ in (6.28), we find that

$$||u_k - u|| \le \frac{\gamma^k}{1 - \gamma} ||u_1 - u_0||$$

is the desired result concerning geometric convergence. Finally, we show that u is unique. Let v be another zero of F; that is, $F(v) = 0$. Since $F(u) = F(v) = 0$

$$v - u = -A_0^{-1}(F(u) - F(v) - A_0(u_0)(v - u))$$

from which it follows that

$$||v - u|| = ||A_0^{-1}(u_0)|| \sup_{u \in S_\alpha(u_0)} ||F'(v) - A_0(u_0)|| \, ||(v - u)|| \le \gamma ||v - u||$$

which implies that $u = v$ as $\gamma < 1$.

Proof (*Proof of Theorem* 6.8)

(i) First, we show the existence of constants and β such that

$$\alpha > 0, \ S_\alpha(u) = \{x \in X / ||x - u|| \le \alpha\} \subset U \tag{6.29}$$

and

$$\sup_{k \ge 0} \sup_{x \in S_\alpha(u_0)} ||I - A_k^{-1}F'(x)|| \le \gamma \le 1 \tag{6.30}$$

For every integer k, we can write $A_k = A(I + A^{-1}(A_k - A))$ with $||A^{-1}(A_k - A)||\lambda < 1$ in view of a condition of the theorem. Thus, A_k are isomorphisms from X onto Y, and moreover

$$||A_k^{-1}|| = ||(A(I + A^{-1}(A_k - A)))^{-1}||$$

$$\le ||(I + A^{-1}(A_k - A)))^{-1}|| \, ||A^{-1}|| \le \frac{||A^{-1}||}{1 - \lambda}$$

by Theorem 2.11 and $||(I + B)^{-1}|| \leq \frac{1}{1-||B||}$. This implies that

$$||I - A_k^{-1}A|| = ||A_k^{-1}A_k - A_k^{-1}A|| \leq ||A_k^{-1}||\,||A_k - A||$$

$$\leq ||A_k^{-1}||\frac{\lambda}{||A_k^{-1}||}\ for\ \lambda < \frac{1}{2}$$

$$\leq \frac{||A_k^{-1}||}{1 - \lambda}\frac{\lambda}{||A_k^{-1}||}$$

or

$$||I - A_k^{-1}A|| \leq \frac{\lambda}{1 - \lambda} = \beta' < 1$$

Let be such that $\beta' < \beta' + \delta = \gamma < 1$. This implies that

$$||I - A_k^{-1}F'(u)|| \leq ||I - A_k^{-1}A|| + ||A_k^{-1}(A - F'(u))||$$

from which (6.29) and (6.30) follow immediately keeping in mind the continuity of the derivative F' and the fact that $A = F'(u)$.

(ii) Let u_0 be any point of the ball $S_\alpha(u)$ and $\{u_k\}$ be the sequence defined by $u_{k+1} = u_k - A_k^{-1}F(u_k)$; each of these elements lies in $S_\alpha(u)$. This implies that $\{u_k\}$ is well defined. Since $F(u) = 0$, we have

$$u_{k+1} - u = u_k - A_k^{-1}F(u_k) - (u - A_k^{-1}F(u))$$

By the Mean Value Theorem applied to the function, $x \rightarrow x - A_k^{-1}F(x)$ shows that

$$||u_{k+1} - u|| \leq \sup_{x \in S_\alpha(u)} ||I - A_k^{-1}F'(x)||\,||u_k - u|| \leq \gamma ||u_k - u||$$

By (6.30) and continuing in this way, we get

$$||u_{k+1} - u|| \leq \gamma^{k-1}||u_1 - u||$$

which is the geometric convergence. This relation also implies that $u_k \rightarrow u$ as $k \rightarrow \infty$ as $\gamma < 1$.

(iii) The zero of F, point u, is unique. For this, let v be another point such that $F(v) = 0$. The sequence $\{u_k\}$ corresponding to $u_0 = v$ is a stationary sequence, since $u_1 = u_0 - A_k^{-1}F(u_0) = u_0$, and on the other hand, it converges to the point u by the above discussion. This implies $u = v$.

We cite Ciarlet [46, pp. 300–301] for the proof of Theorem 6.9; here, we prove Theorem 6.10.

Proof (*Proof of Theorem* 6.10) In a gradient method with variable parameter, we have $u_{k+1} = u_k - \varphi_k \nabla F(u_k)$. Since $\nabla F(u) = 0$ for a minima at u, we can write $u_{k+1} - u = (u_k - u) - \varphi_k \{\nabla F(u_k) - \nabla F(u)\}$. This implies that $||u_{k+1} - u||^2 = ||u_k - u||^2 - 2\varphi_k < \nabla F(u_k) - \nabla F(u), u_k - u > +\varphi_k^2 ||\nabla F(u_k) - \nabla F(u)||^2 \leq \{1 - 2\alpha\varphi_k + \beta^2 \varphi_k^2\} ||u_k - u||^2$, under the condition that $\varphi_k > 0$. If

$$0 \leq \alpha \leq \varphi_k \leq b \leq \frac{2\alpha}{\beta^2}$$

then

$$1 - 2\alpha\varphi_k + \beta^2 \varphi_k^2 < 1$$

and so

$$||u_{k+1} - u|| \leq \gamma ||u_k - u|| \leq \gamma^{k+1} ||u - u_0||$$

where $\gamma < 1$ which depends on α, a, b and β. This also implies the geometric convergence of $\{u_k\}$.

6.4.2 Conjugate Gradient Method

The conjugate gradient method deals with the minimization of the quadratic functional on $X = R^n$; that is

$$J : v \in R^n \rightarrow \frac{1}{2} \langle Av, v \rangle - \langle b, v \rangle$$

or

$$J(v) = \frac{1}{2} \langle Av, v \rangle - \langle b, v \rangle$$

where A is the $n \times n$ matrix. Starting with an initial arbitrary vector u_0, we set $d_0 = \nabla J(u_0)$. If $\nabla J(u_0) = 0$, the algorithm terminates. Otherwise, we define the number

$$r_0 = \frac{\langle \nabla J(u_0), d_0 \rangle}{\langle Ad_0, d_0 \rangle}$$

then the numbers u_1 are given by

$$u_1 = u_0 - r_0 d_0$$

Assuming that the vectors $u_1, d_1, \ldots, u_{k-1}, d_{k-1}, u_k$ have been constructed which assumes that the gradient vectors $\nabla J(u_\ell)$, $0 \le \ell \le k-1$ are all nonzero, one of the two situations will prevail: $\nabla J(u_k) = 0$ and the process terminates, or $\nabla J(u_k) \ne 0$, in which case we define the vector

$$d_k = \nabla J(u_k) + \frac{||\nabla J(u_k)||^2}{||\nabla J(u_{k-1})||^2 d_{k-1}}$$

then the numbers r_k and u_{k+1} are given by

$$r_k = \frac{\langle \nabla J(u_k), d_k \rangle}{\langle Ad_k, d_k \rangle}$$

and

$$u_{k+1} = u_k - r_k d_k$$

respectively.

This beautiful algorithm was invented by Hestenes and Stiefel in 1952. The method converges in at most n iterations.

The study of the conjugate gradient method for nonquadratic function on R^n into R began in sixties. Details of these methods and their comparative merits may be found in Polak [150] and other Refs. [151, 153]. We present here the essential ingredients of these two best methods, namely Fletcher–Reeves (FR) and Polak–Ribiére (PR).

Let $F : R^n \to R$; we look for $\inf_{v \in R^n} F(v)$ where F is twice differentiable. The point at which $\inf_{v \in R^n} F(v)$ is attained will be denoted by $arg \inf(x)$. Starting with an arbitrary vector u_0, one assumes the vectors $u_1, u_2, \ldots, u_k$ to have been constructed, which means that the gradient vectors $\nabla F(u_i)$, $0 \le i \le n-1$, are nonzero. In such situations, either $\nabla F(u_n) = 0$ and the algorithm terminates, or $\nabla F(u_n) \ne 0$, in which case, vectors u_{n+1} are defined (if they exist and are unique) by the relations

$$u_{n+1} = u_n - r_n d_n \text{ and } F(u_{n+1}) = \inf_{\rho \in R} F(u_n - \rho d_n)$$

the successive descent directions d_i being defined by the recurrence relation

$$d_0 = \nabla F(u_0)$$
$$d_i = \nabla F(u_i) + \frac{\langle \nabla F(u_i), \nabla F(u_i) - \nabla F(u_{i-1}) \rangle}{||\nabla F(u_{i-1})||^2} d_{i-1}, \ 1 \le i \le n$$
$$r_i = \frac{\langle \nabla F(u_i), \nabla F(u_i) - \nabla F(u_{i-1}) \rangle}{||\nabla F(u_{i-1})||^2}.$$

is called the Polak–Ribiére formula, and in this case, the conjugate gradient method is called the *Polak–Ribiére conjugate gradient method*, and one denotes r_i by r_i^{PR}.

The case

$$r_i = \frac{||\nabla F(u_i)||^2}{||\nabla F(u_{i-1})||^2}$$

is called the Fletcher–Reeves formula, and the corresponding method is called the *Fletcher–Reeves conjugate gradient method.* Such r_i is denoted by r_i^{FR}. It may be noted that the Polak–Ribiére conjugate gradient method is more efficient in practice.

Polak–Ribiére Conjugate Gradient Algorithm

Data $u_0 \in R^n$
Step 0. Set $i = 0$, $d_0 = \nabla F(u_0)$, and $h_0 = -d_0$.
Step 1. Compute the step size

$$\lambda_i = arg \inf_{\lambda \geq 0} F(u_i + \lambda h_i)$$

Step 2. update: Set $u_{i+1} = u_i + \lambda_i h_i$.

$$d_{i+1} = \nabla F(u_{i+1})$$
$$r_i^{PR} = \frac{\langle d_{i+1}, d_{i+1} - d_i \rangle}{||d_i||^2}$$
$$h_{i+1} = -d_{i+1} + r_i^{PR} h_i.$$

Step 3. Replace i by $i + 1$ and go to Step 1.

Polak–Reeves Conjugate Gradient Algorithm

Data. $u_0 \in R^n$.
Step 0. Set $i = 0$, $d_0 = \nabla F(u_0)$, and $h_0 = -d_0$.
Step 1. Compute the step size

$$\lambda_i = arg \inf_{\lambda \geq 0} F(u_i + \lambda h_i)$$

Step 2. Update: Set $u_{i+1} = u_i + \lambda_i h_i$

$$d_{i+1} = \nabla F(u_{i+1})$$
$$r_i^{FR} = \frac{||d_{i+1}||^2}{||d_i||^2}$$
$$h_{i+1} = -d_{i+1} + r_i^{FR}.$$

Step 3. Replace i by $i + 1$ and go to Step 1.

6.5   Problems

Problem 6.1 If $f \in C[a, b]$ and $\int_a^b f(t)h(t)dt = 0$ for all $h \in C^1[a, b]$ with $h(a) = h(b) = 0$, then prove that $f = 0$.

Problem 6.2 Let H be a Hilbert space, K a convex subset of H, and $\{x_n\}$ a sequence in K such that $\lim\limits_{n \to \infty} ||x_n|| = \inf\limits_{x \in H} ||x||$. Show that $\{x_n\}$ converges in X. Give an illustrative example in R^2.

Problem 6.3 Let $K = \left\{ x = (x_1, x_2, \ldots x_n) / \sum\limits_{i=1}^{n} x_i = 1 \right\}$ be a subset of R^n. Find a vector of minimum norm in K.

Problem 6.4 Let H be a normed space and K a subset of H. An element $\bar{x} \in K$ is called a best approximation to an arbitrary element $x \in H$ if

$$d(x, K) = \inf\limits_{y \in K} ||y - x|| = ||\bar{x} - x||$$

The approximation problem is a special type of optimization problem which deals with minimizing a translate of the norm function, or if $x = 0$, it deals with minimizing the norm function itself. Let K be a finite-dimensional closed subset of a normed space U. Show that every point of U has a best approximation.

Problem 6.5 Let $X = C[-\pi, \pi]$ be an inner product space with the inner product

$$\langle f, g \rangle = \int_{-\pi}^{\pi} f(x)g(x)dx$$

and K be the subspace spanned by the orthonormal set

$$\left\{ \frac{1}{\sqrt{2\pi}}, \frac{1}{\sqrt{\pi}} \cos(x), \frac{1}{\sqrt{\pi}} \sin(x), \ldots, \frac{1}{\sqrt{\pi}} \cos(nx), \frac{1}{\sqrt{\pi}} \sin(nx) \right\}$$

Find the best approximation to:

1. $f(x) = x$
2. $f(x) = |x|$.

Problem 6.6 Show that for each $f \in C[a, b]$, there exists a polynomial $P_n(t)$ of maximum degree n such that for every

$$g \in Y = span\{f_0(t), f_1(t), \ldots, f_n(t)\}, \ f_j(t) = t_j,$$
$$\max\limits_{a \le t \le b} |f(t) - P_n(t)| \le \max\limits_{a \le t \le b} |f(t) - g(t)|.$$

Problem 6.7 Let $m \geq n > 0$ and $A = (a_{ij})$, $i = 1, 2, \ldots, m$, $j = 1, 2, \ldots, n$ and $y \in R^m$. Then, write down the solution of optimization problem for the function $F(x) = ||Ax - y||$ over R^n.

Problem 6.8 Let $Ax = y$, where A is an $n \times n$ matrix and x and y are elements of R^n. Write down a sequence of approximate solutions of this equation and examine its convergence. Under what condition on A, this equation has necessarily the unique solution?

Problem 6.9 Explain the concept of the steepest descent and apply it to study the optimization of the functional F defined on a Hilbert space H as follows:

$$F(x) = \langle Ax, x \rangle - 2\langle y, x \rangle$$

where A is a self-adjoint positive definite operate on H, $x, y \in H$. For any $x_1 \in H$, construct a sequence $\{x_n\}$ where

$$x_{n+1} = x_n + \frac{\langle z_n, z_n \rangle}{\langle Az_n, z_n \rangle} z_n$$

for appropriately chosen z_n and show that $\{x_n\}$ converges to x_0 in H which is the unique solution of $Ax = y$. Furthermore, show by defining

$$F(x) = \langle A(x - x_0), x - x_0 \rangle$$

that the rate of convergence satisfies

$$\langle x_n, x_n \rangle \leq \frac{1}{m} F(x_n) \leq \frac{1}{m} \left(1 - \frac{m}{M}\right)^{n-1} F(x_1).$$

Problem 6.10 Write a short note on nonsmooth optimization problem.

Problem 6.11 Develop Newton's method for nonsmooth optimization.

Problem 6.12 Verify Euler's Equation (6.2).

Chapter 7
Operator Equations and Variational Methods

Abstract In this chapter, existence of solution of some well-known partial differential equations with boundary conditions is studied.

Keywords Neumann–Dirichlet boundary value problem · Galerkin method · Ritz method · Eigenvalue problems · Laplace equation · Poisson equation · Stoke problem · Navier–Stokes equation · Heat equation · Telegrapher's equation Helmholtz equation · Wave equation · Schrödinger equation

7.1 Introduction

The chapter deals with representation of real-world problems in terms of operator equations. Existence and uniqueness of solutions of such problems explored. Approximation methods like Galerkin and Ritz are presented.

7.2 Boundary Value Problems

Let Ω be a bounded or unbounded region of R^n (in application, $n = 1, 2, 3$), Γ or $\partial\Omega$ be its boundary, L and S be linear differential operators, $u(x)$, $x = (x_1, x_2, \ldots x_n) \in R^n$ be a function on R^n belonging to a Hilbert or Banach space of functions, and $f(x)$ and $g(x)$ be given functions of x. Then,

$$Lu(x) = f(x) \ in \ \Omega \qquad (7.1)$$

$$Su(x) = g(x) \ on \ \Gamma \qquad (7.2)$$

is known as linear *boundary value problem* (BVP) for u. u is an unknown function which is called a *solution*. The boundary value problems of the type

$$Lu = f\left(x_i, u, \frac{\partial u}{\partial x_j}, \dots\right) \; in \; \Omega \tag{7.3}$$

$$Su = g(x_j) \; on \, \Gamma, \; j = 1, 2, 3, \dots \tag{7.4}$$

is known as *nonlinear boundary value problems*.

We shall see that these problems can be expressed in the form of finding solution of the abstract variation problem, namely: Find $u \in H$ (H is a Hilbert space) such that

$$a(u, v) = F(v) \; \forall \, v \in H \tag{7.5}$$

where $a(\cdot, \cdot)$ is a bilinear form with appropriate conditions and $F \in H^\star$.

A boundary value problem can also be expressed in the form of the operator equation

$$Au = v \tag{7.6}$$

where A is a linear or nonlinear operator on a Hilbert or Banach space into another one of such spaces. Existence and uniqueness of solutions of (7.5) and (7.6) is given by the Lax–Milgram Lemma. We present a general form of this lemma in Theorem 7.5.

Example 7.1 Consider the BVP

$$- \Delta u = -\left(\frac{\partial^2 u}{\partial x^2} + \frac{\partial^2 u}{\partial y^2}\right) = f \; in \Omega \tag{7.7}$$

$$u = 0 \; on \; \Gamma \tag{7.8}$$

where $\Omega \subset R^2$.
In this example,

$$L = \Delta = \frac{\partial^2(\cdot)}{\partial x^2} + \frac{\partial^2(\cdot)}{\partial y^2} = \text{Laplacian operator}$$

$$S = I, g = 0$$

This is called the *Dirichlet boundary value problem*. The equation in (7.7) is known as *Poisson's equation*.

By the classical solution of this BVP, we mean the function $u(x, y)$ which is continuous in the closed domain $\overline{\Omega}$, satisfies (7.7) in the open domain Ω and is equal to zero on the boundary Γ. By assumption $f \in C(\overline{\Omega})$, the solution $u \in C^2(\overline{\Omega})$, the space of continuous functions with continuous partial derivatives up to second-order inclusive, and equals zero on Γ. The set D_A of these admissible functions

$$D_A = \{u(x) \in C^2(\overline{\Omega}), \ x \in \Omega \subset R^2, \ u = 0 \ on \ \Gamma\}$$

forms a vector space. If the boundary condition (7.8) is nonhomogeneous; that is, $u = g(x), g(x) \neq 0$, then D_A is not a vector space.

In one dimension, the boundary value problem (7.7)–(7.8) takes the form

$$-\frac{d^2u}{dx^2} = f(x) \ in \ (a, b) \tag{7.9}$$

$$u(a) = u(b) = 0 \tag{7.10}$$

Example 7.2 The BVP of the type

$$-\Delta u = f \ on \ \Omega \tag{7.11}$$

$$\frac{\partial u}{\partial n} = g \ on \ \Gamma \tag{7.12}$$

where $\frac{\partial u}{\partial n}$ denotes the derivative of u in the direction of outward normal to the boundary, Γ, is called the *linear non-homogeneous Neumann BVP*.

Example 7.3 The BVP

$$-\Delta u = f \ on \ \Omega \tag{7.13}$$

$$u = 0 \ on \ \Gamma_1 \tag{7.14}$$

$$\frac{\partial u}{\partial n} = g \ on \ \Gamma_2 \tag{7.15}$$

where $\Gamma = \Gamma_1 \cup \Gamma_2$ is known as the *mixed* BVP of *Dirichlet* and *Neumann*. Dirichlet and Neumann BVPs are of elliptic type.

Example 7.4 The BVP of the type

$$\frac{\partial u}{\partial t} + Lu = f \ on \ \Omega_T = (0, T) \times \Omega \tag{7.16}$$

$$Su = g \ on \ \Gamma_T = (0, T) \times \Gamma \tag{7.17}$$

$$u(0, x) = u_0 \ on \ \Omega \tag{7.18}$$

where $f = f(t, x)$, $g = g(t, x)$, and $u_0 = u_0(x)$ is known as the *initial-boundary value problem of parabolic type*.

If we take $L = -\Delta$, $S = I$, $g = 0$, and $f = 0$, then we obtain the heat equation

$$\frac{\partial u}{\partial t} - \Delta u = 0 \ on \ [0, T] \times \Omega \tag{7.19}$$

$$u = 0 \ on \ [0, T] \times \partial \Omega \tag{7.20}$$

$$u(0, x) = u_0 \tag{7.21}$$

The one-dimensional heat equation with initial-boundary conditions is given as

$$\frac{\partial u}{\partial t} = \frac{\partial^2 u}{\partial x^2}, \quad 0 < x < b$$

$$u(0, t) = \frac{\partial u}{\partial x}(b, x) = 0, \text{ for } t > 0$$

$$u(x, 0) = K, \text{ for } 0 < x < b$$

Example 7.5 Hyperbolic equations with initial-boundary conditions:

(i)

$$\frac{\partial u}{\partial t} + a \frac{\partial u}{\partial x} = 0, t > 0, \; x \in R \; where \; a \in R/\{0\}.$$

   (a) $u(0, x) = u_0(x), \; x \in R.$
   (b) $u(t, 0) = \varphi_1(t) \; if \; a > 0, \; u(t, 0) = \varphi_2(t) \; if \; a < 0.$
        where φ_1 and φ_2 are given functions.

(ii)

$$\frac{\partial^2 u}{\partial t^2} = c^2 \frac{\partial^2 u}{\partial x^2}, t > 0, \; x \in R$$

   (*wave equation*) with initial data

$$u(0, x) = u_0 \; and \; \frac{\partial u}{\partial t}(0, x) = u_1(x)$$

(iii)

$$\frac{\partial u}{\partial t} + \frac{\partial F(u)}{\partial x} = 0, \; t > 0, \; x \in R$$

   where $F(u)$ is a nonlinear function of u, and $u(0, x) = u_0(x)$.
   A special case

$$\frac{\partial u}{\partial t} + u \frac{\partial u}{\partial x} = 0$$

is Burger's equation.

7.3 Operator Equations and Solvability Conditions

7.3.1 Equivalence of Operator Equation and Minimization Problem

Let T be a positive operator on a Hilbert space H and let D_T denote domain of T. $D_T \subset H$. U is called a solution of the operator equation

$$Tx = y \tag{7.22}$$

if $Tu = y$. It will be that if u is a solution, then energy functional J (Sect. 6.3.3) attains its minimum at u. The converse also holds.

Theorem 7.1 *Let $T : D_T \subset H \to H$, where H is a Hilbert space, be shown self-adjoint and positive operator on D_T and let $y \in H$. Then, the energy functional*

$$J(u) = \frac{1}{2}\langle Tu, u \rangle - \langle y, u \rangle \tag{7.23}$$

attains its minimum value at $x \in D_T$ if and only if x is a solution of (7.22).

Proof Suppose x is the solution of Eq. (7.22). Then, $Tx = y$ and we have

$$J(u) = \frac{1}{2}\langle Tu, u \rangle - \langle Tx, u \rangle$$
$$= \frac{1}{2}[\langle T(u - x), u \rangle - \langle Tx, u \rangle]$$

Using properties of self-adjointness.

$$J(u) = \frac{1}{2}[\langle T(u - x), u \rangle - \langle Tu, x \rangle + \langle Tx, x \rangle - \langle Tx, x \rangle]$$
$$= \frac{1}{2}[\langle T(u - x), u \rangle - \langle Tu - Tx, x \rangle - \langle Tx, x \rangle]$$
$$= \frac{1}{2}[\langle T(u - x), u - x \rangle - \langle Tx, x \rangle]$$
$$= J(x) + \frac{1}{2}\langle T(u - x), u - x \rangle.$$

We have

$$\langle T(u - x), u - x \rangle > 0 \text{ for every } u \in D_T$$
$$= 0 \text{ if and only if } u - x = 0 \text{ in } D_T \text{(by positiveness of T)}.$$

from which we conclude that

$$J(u) \geq J(x) \tag{7.24}$$

where the equality holds if and only if $u = x$.

Inequality (7.23) implies that the quadratic functional $J(u)$ assumes its minimum value at the solution $x \in D_T$ of Eq. (7.22); any other $u \in D_T$ makes $J(u)$ larger than $J(x)$.

To prove the converse, suppose that $J(u)$ assume its minimum value at $x \in D_T$. This implies $J(u) \geq J(x)$ $for\ u \in D_T$. In particular, for $u = x + \alpha v$, $v \in D_T$, and α a real number, we get

$$J(x + \alpha v) \geq J(x).$$

We have

$$2J(x + \alpha v) = \langle T(x + \alpha v), x + \alpha v \rangle - 2\langle y, x + \alpha v \rangle$$
$$\text{(keeping in mind definition of J)}$$

or

$$2J(x + \alpha v) = \langle Tx, x \rangle + 2\alpha \langle Tx, v \rangle$$
$$+ \alpha^2 \langle Tv, v \rangle - 2\alpha \langle y, v \rangle - 2\langle y, x \rangle \tag{7.25}$$

Since $x \in D_T$ and $y \in H$ are fixed elements, it is clear from Eq. (7.25) that for fixed $v \in D_T$, $J(x + \alpha v)$ is a quadratic function of α. Then, $J(x + \alpha v)$ has a minimum D_T at $\alpha = 0$ (i.e., J has a minimum at x) if the first derivative of $J(x + \alpha v)$ with respect to α is zero at $\alpha = 0$. We have

$$\left[\frac{d}{d\alpha} J(x + \alpha v) \right]_{\alpha=0} = 0$$

or

$$2\langle Tx, v \rangle - 2\langle y, v \rangle = 0$$

or

$$\langle Tx - y, v \rangle = 0 \ \forall\, v \in D_T.$$

For $v = Tx - y$, $\langle Tx - y, Tx - y \rangle = 0$. This implies that $Tx - y = 0$ or $Tx = y$. Thus, we have proved that the element $x \in D_T$ that minimizes the energy functional is a solution of Eq. (7.22).

Remark 7.1 (i) Theorem 7.1 is important as it provides an equivalence between the solution of operator equation $Au = y$ in D_T with the element at which minimum of the energy (quadratic) functional is attained. This variational formulation, that is, expressing the operator equation as the problem of minimizing an energy functional, helps in establishing existence and uniqueness results and yields the approximate solutions. Theorem 6.6 yields existence and uniqueness of the minimization problem of the energy functionals where T is associated with a coercive and symmetric bilinear form $a(u, v)$ via Theorem 3.37. Therefore, for such a T, operator equation (7.22) has a unique solution.

(ii) Let T be associated with bilinear form through Theorem 3.37. Then, operator equation (7.22) has a unique solution by the Lax–Milgram Lemma (Theorem 3.39).

(iii) Finding the solution of (7.22), one can use algorithms of Chap. 6.

7.3.2 Solvability Conditions

Suppose T is a linear operator with domain D_T, where D_T is a dense in a Hilbert space H. Determine y for which Eq. (7.22), namely

$$Tu = y \quad u \in D_T,$$

has solution. The existence of solution of non-homogeneous equation (7.22) depends on the homogeneous adjoint equation

$$T^\star v = 0 \tag{7.26}$$

where $T^\star$ is the adjoint operator of T.

Theorem 7.2 *Suppose that T is a closed bounded below linear operator defined on a Hilbert space H. Then, Eq. (7.22) has solutions if and only if*

$$N(T^\star)^\perp = R(T), \tag{7.27}$$

namely the orthogonal complement of the null space of the adjoint operator $T^\star$ of T (Definitions 2.8(8), (3.3(3)) and (3.8) is equal to the range of T.

Proof We first prove that $\overline{R(T)} = R(T)$; that is, the range of closed and bounded below operator is closed. Let $y_n \in R(T)$ with $y_n \to y$ and $Tv_n = y_n$. Our goal is to show that $y \in R(T)$. Let $u_n = v_n - Pv_n$, where P is the projection operator from D_T onto $N(T)$. Then, $u_n \in D_T \cap N(T)^\perp$ and $Tu_n = y_n$. Since T is bounded below, we have

$$\|u_n - u_m\| \le \frac{1}{c}\|T(u_n - u_m)\| = \frac{1}{c}\|y_n - y_m\|$$

which shows that $\{u_n\}$ is a Cauchy sequence in D_T, and therefore $u_n \to u$ in H. $Tu_n \to y$ and $u_n \to u$ implies that $u \in D_T$ and $Tu = y$ by virtue of closedness of T. Therefore, $R(T)$ is closed. Now, we show that (7.27) is equivalent to

$$\langle y, v \rangle = 0 \text{ for all } v \in N(T^\star) \tag{7.28}$$

where y is data in (7.22). $R(T)^{\perp\perp} = R(T) = N(T^\star)^\perp$ or $R(T)^\perp = N(T^\star)$ as R(T) is closed. From this, we obtain $\langle Tu, v \rangle = \langle u, T^\star v \rangle = 0$ for all $u \in D_T$ and $v \in N(T^\star)$. Thus, (7.27) and (7.28) are equivalent. Let $y \in R(T)$ implying $u \in D_T$ such that $Tu = y$. Since T is bounded below, T has a continuous inverse T^{-1} defined on its range, and therefore, $u = T^{-1}y$. Moreover, we have

$$||u|| = ||T^{-1}y|| \le c||y||$$

To prove the converse, let $u = T^{-1}y$. Hence, $Tu = y$ and $\langle Tu, v \rangle = \langle y, v \rangle$ for any $v \in R(T)$. By the definition of $T^\star$

$$\langle Tu, v \rangle = \langle u, T^\star v \rangle = 0 \text{ for any } v \in N(T^\star).$$

In other words, $u \in N(T^\star)^\perp$.

The Lax–Milgram Lemma (Theorem 3.39) can be given as:

Theorem 7.3 *Suppose $T : H \to H$ is a linear bounded operator on a Hilbert space H into itself. Let there be a constant $\alpha > 0$ such that*

$$|\langle Tu, u \rangle| > \alpha ||u||^2 \ \forall \, u \in H,$$

that is, T is strongly monotone (elliptic or coercive). Then for each given $y \in H$, the operator equation

$$Tu = y, u \in H$$

has a unique solution.

The following generalization of the Lax–Milgram Lemma is known:

Theorem 7.4 *Suppose T is a linear bounded operator defined on a Hilbert space H into its adjoint space $H^\star$. Furthermore, let T be strongly monotone; that is,*

$$(Tu, u) \ge \alpha ||u||^2 \ \forall \, u \in H \ (\alpha > 0)$$

Then,

$$Tu = f, \ u \in H$$

for each given $f \in H^\star$, has a unique solution.

Proof By the Riesz Representation Theorem (Theorem 3.19), there exists an element $w \in H$ such that

$$\langle Tu, v \rangle = \langle w, v \rangle \; \forall \, v \in H$$

and $\|w\| = \|Tu\|$. (Tu, v), $Tu \in H^*$, denotes the value of the functional $Tu \in H^*$ at the point v. Let $Su = w$. Then

$$\|Su\| = \|Tu\| \leq \|T\| \, \|u\|.$$

$S : H \to H$ is linear and bounded. Moreover,

$$\langle Su, u \rangle = (Tu, u) \geq \alpha \|u\|^2 \; \forall \, u \in H.$$

$\langle Su, u \rangle$ is real as H is a real Hilbert space. By the Riesz Representation Theorem, there exists an element $y \in H$ such that

$$(f, v) = \langle y, v \rangle \; \forall \, v \in H.$$

The operator equation $Tu = f$ can be written as

$$(Tu, v) = (f, v) \; \forall \, v \in H$$

or equivalently

$$\langle Su, v \rangle = \cdots (f, v) \; \forall \, v \in H$$

or

$$Su = f, \; f \in H$$

which has a unique solution by Theorem 7.3.

The Lax–Milgram Lemma has been generalized [201, pp. 174–175] in the following form:

Theorem 7.5 *Suppose* $T : D_T \subseteq H \to H$ *be a linear operator on the Hilbert space H over the field of real or complex numbers. Then for each* $y \in H$, *the operator equation*

$$Tu = y \; u \in H$$

has at most one solution in case one of the following five conditions is satisfied:

(a) Strict monotonicity (positivity):

$$Re\langle Tu, u \rangle > 0 \; \forall \, u \in D_T \text{ with } u \neq 0.$$

(b) A priori estimate (stability):

$$\beta||u|| \leq ||Tu|| \ \forall \, u \in D_T \ and \ fixed \ \beta > 0$$

(c) Contractivity of (T-I):

$$||Tu - u|| < ||u|| \ \forall \, u \in D_T \ with \ u \in 0$$

(d) Monotone type: With respect to an order cone on H

$$Tu \leq Tv \ implies \ u \leq v$$

(e) Duality: There exists an operator $S : D_S \subseteq H \rightarrow H$ with $\overline{R(S)} = H$ and $\langle Tu, v \rangle = \langle u, Sv \rangle$ for all $u \in D_T$, $v \in D_S$.

Corollary 7.1 *Suppose $T : D_T \subseteq H \rightarrow H$ is a linear operator on the Hilbert space H, where D_T and D_{T^*} are dense in H. Then we have the following two statements are equivalent:*

(a) For each $y \in H$, the equation $Tu = y$ has at most one solution.
(b) $R(T^\star)$ is dense in H, that is, $\overline{R(T^\star)} = H$.

Theorem 7.6 *Equation $Tu = y$, where $T : D_T \subseteq H \rightarrow H$ is an operator on the Hilbert space H, has at most one solution if the operator $T : D_T \subseteq H \rightarrow H$ is strictly monotone (elliptic) and $Re\langle Tu - Tv, u - v \rangle > 0$ for all $u, v \in D_T$ with $u \neq v$.*

7.3.3 Existence Theorem for Nonlinear Operators

Theorem 7.7 *Let H be a Hilbert space and T an operator on H into itself satisfying the conditions*

$$||Tu - Tv||H \leq \lambda ||u - v||_H (T \ Lipschitz \ continuous) \tag{7.29}$$

and

$$\langle Tu - Tv, u - v \rangle H \geq \mu ||u - v||_H^2 (strongly \ monotone) \tag{7.30}$$

where $\lambda > 0$, $\mu > 0$ and $\lambda > \mu$. Then, the equation

$$Tu = y \tag{7.31}$$

has exactly one solution for every $y \in H$.

Proof Let $y \in H$, and $\varepsilon > 0$. Let the operator S_ε be defined as $S_\varepsilon u = u - \varepsilon(Tu - y)$. For any $u, v \in H$, we have

$$||S_\varepsilon u - S_\varepsilon v||^2 = ||u - v||_H^2 - 2\varepsilon \langle Tu - Tu, u - v \rangle + \varepsilon^2 ||Tu - Tv||_H^2$$
$$\leq (1 - 2\varepsilon\mu + \varepsilon^2\lambda^2)||u - v||_H^2$$

Thus, if we choose $\varepsilon > 0$ so that $\varepsilon < \frac{2\mu}{\lambda^2}$ and if we take $\alpha = (1 - 2\varepsilon\mu + \varepsilon^2\lambda^2)^{1/2}$, then the operator S_ε is a contraction with constant $\alpha < 1$. The operator equation (7.31) has a unique solution by the Banach contraction mapping theorem.

It may be observed that Theorem 7.4 has been proved in a general form by F.E. Browder (see, for example, [78, p. 243]) where Hilbert space is replaced by reflexive Banach space, Lipschitz continuity is replaced by a weaker condition, namely, demicontinuity, and monotonicity coercivity is replaced by strong monotonicity.

7.4 Existence of Solutions of Dirichlet and Neumann Boundary Value Problems

One-dimensional Dirichlet Problem

The Dirichlet problem in one dimension is as follows: Find u such that

$$u'' = -f \text{ or } -\frac{d^2u}{dx^2} = f \quad 0 < x < 1$$
$$u(0) = u(1) = 0$$

A variational formulation Eq. (7.5) of this problem is

$$a(u, v) = F(v) \text{ for } v \in H_0^1(0, 1)$$

where $a(u, v) = \int\limits_{-1}^{1} u'(x)v'(x)dx$ and $F(v) = \int\limits_{0}^{1} v(x)f(x)dx$

$H = H_0^1(0, 1)$ (See Sect. 5.4.2, $m = 1$, $\Omega = (0, 1)$, $\partial\Omega = \{0, 1\}$. "denotes the first derivative, while " denotes the second derivative.

Verification Multiply both sides of the given equation by $v(x)$ and apply integration by parts and the boundary condition $u(0) = u(1) = 0$ to get

$$\int\limits_{0}^{1} u'(x)v'(x)dx = \int\limits_{0}^{1} f(x)v(x)dx$$

which is the desired equation for $v \in H_0^1(0, 1)$. This equation has unique solution if $a(\cdot, \cdot)$ is bilinear, bounded, and coercive, and F is linear and bounded.

$$a(u + w, v) = \int_0^1 (u' + w')v' dx = \int_0^1 u'v' dx + \int_0^1 w'v' dx$$

$$= a(u, v) + a(w, v)$$

$$a(\alpha u, v) = \int_0^1 \alpha u' v' dx = \alpha \int_0^1 u'v' dx = \alpha a(u, v)$$

$$|a(u, v)| \leq \int_0^1 |u'(x)v'(x)| dx \leq \left(\int_0^1 |u'(x)|^2 dx \right)^{1/2} \left(\int_0^1 |v'(x)|^2 dx \right) \quad (By\ the\ CSB)$$

$$\leq M||u||\ ||v|| \text{ by Remark 5.5(i).}$$

Thus $a(\cdot, \cdot)$ is bilinear and bounded.

$$|a(u, u)| = \int_0^1 |u'(x)|^2 dx \geq \alpha ||u||_{H_0^1}$$

$$\alpha > 0\ by\ the\ Poincaré\ Inequality\ (see\ Remark\ 5.5(i)).$$

Therefore, $a(\cdot, \cdot)$ is coercive, and by the Lax Milgram Lemma, the one-dimensional Dirichlet problem has a unique solution. Since $a(u, v) = a(v, u)$; that is, $a(u, v)$ is symmetric, the solution of this is equivalent to the solution of the minimization problem for the energy functional

$$J(u) = \frac{1}{2} a(u, u) - F(u) = \frac{1}{2} \int_0^1 (u(x))^2 dx - \int_0^1 f(x)u(x)\ dx$$

Algorithms of Chap. 6 can also be applied, if necessary, to find a solution to the minimization problem and consequently to the original problem.

The n-dimensional Dirichlet Problem

Finding u defined on Ω such that

$$\Delta u = f\ in\ \Omega \tag{7.32}$$

$$u = 0\ on\ \Gamma \tag{7.33}$$

i.e., finding the solution of the linear elliptic BVP or Dirichlet problem is equivalent to determining the solution u of the variational problem

$$a(u, v) = L(v) \tag{7.34}$$

for all $v \in H_0^1(\Omega)$

$$a(u, v) = \sum_{i=1}^{n} \int_{\Omega} \frac{\partial u}{\partial x_i} \frac{\partial v}{\partial x_i} dx \qquad (7.35)$$

and

$$L(v) = \int_{\Omega} f v \, dx \qquad (7.36)$$

Verification Let $f \in L^2(\Omega)$, $u \in H^2(\Omega)$. Multiplying Eq. (7.32) by a function $v \in H_0^1(\Omega)$ and by integrating the resulting equation, we have

$$-\int_{\Omega} \Delta u v \, dx = \int_{\Omega} f v \, dx \qquad (7.37)$$

By Theorem 5.17, we have

$$-\int_{\Omega} \Delta u v \, dx = -\sum_{i=1}^{n} \int_{\Omega} \frac{\partial^2 u}{\partial x_i^2} v \, dx$$

$$= \sum_{i=1}^{n} \int_{\Omega} \frac{\partial u}{\partial x_i} \frac{\partial v}{\partial x_i} dx - \sum_{i=1}^{n} \int_{\Gamma} v \frac{\partial u}{\partial n} ds$$

Since $v \in H_0^1(\Omega)$, $v/\Gamma = \gamma_v = 0$ and consequently

$$\sum_{i=1}^{n} \int_{\Gamma} v \frac{\partial u}{\partial n} ds = 0$$

Therefore, we get

$$-\int_{\Omega} \Delta u v \, dx = -\sum_{i=1}^{n} \int_{\Omega} \frac{\partial u}{\partial x_i} \frac{\partial v}{\partial x_i} dx \qquad (7.38)$$

By Eqs. (7.37) and (7.38), we obtain Eq. (7.34) where $a(\cdot, \cdot)$ and L are given by Eqs. (7.35) and (7.36), respectively.

Existence of the solution of Eq. (7.34): Since

$$a(u, v) = \sum_{i=1}^{n} \int_{\Omega} \frac{\partial u}{\partial x_i} \frac{\partial v}{\partial x_i} dx$$

$$= \sum_{i=1}^{n} \int_{\Omega} \frac{\partial v}{\partial x_i} \frac{\partial u}{\partial x_i} dx = a(v, u),$$

$a(\cdot, \cdot)$ is symmetric.

$$a(u, u) = \sum_{i=1}^{n} \int_{\Omega} \frac{\partial u}{\partial x_i} \frac{\partial u}{\partial x_i} dx$$

$$\geq k_1 ||u||_{H_0^1} \; (by \; Theorem \; 5.18, \; see \; also \; Remark \; 5.5)$$

Thus, $a(\cdot, \cdot)$ is coercive. We have

$$L(v_1 + v_2) = \int_{\Omega} f(v_1 + v_2) \, dx$$

$$= \int_{\Omega} f v_1 \, dx + \int_{\Omega} f v_2 \, dx = L(v_1) + L(v_2)$$

$$L(\lambda v) = \int_{\Omega} f(\lambda v) = \int_{\Omega} f(\lambda v) \, dx = \lambda \int_{\Omega} f v \, dx$$

where λ is a scalar. Also

$$|L(v)| = |\int_{\Omega} f v \, dx| \leq \left(\int_{\Omega} |f|^2 \, dx \right)^{1/2} \left(\int_{\Omega} |v|^2 \, dx \right)^{1/2}$$

by the CSB inequality. Since

$$f \in L_2(\Omega) ||f|| = \left(\int_{\Omega} |f|^2 \, dx \right)^{1/2} \leq k, \; k > 0$$

and

$$||v||_{H_0^1}(\Omega) = \left(\int_{\Omega} |v|^2 \, dx \right)^{1/2}$$

we get

$$|L(v)| \leq k ||v||_{H_0^1}(\Omega).$$

Thus, L is a bounded linear functional on H_0^1. By Theorem 3.39, Eq. (7.34) has a unique solution.

On the lines of Dirichlet boundary value problem, it can be shown that Neumann BVP of Example 7.2 is the solution of the variation problem (7.39)

$$a(u, v) = L(v) \tag{7.39}$$

for $v \in H_0^1(\Omega)$

$$a(u, v) = \sum_{i=1}^{n} \int_{\Omega} \frac{\partial u}{\partial x_i} \frac{\partial u}{\partial x_i} dx$$

$$L(v) = \int_{\Omega} fv \, dx + \int_{\Gamma} gv \, d\Gamma \tag{7.40}$$

and vice versa.

7.5 Approximation Method for Operator Equations

7.5.1 Galerkin Method

Let us consider the operator equation

$$Tu = y \quad u \in H \tag{7.41}$$

together with the Galerkin equations

$$P_n T u_n = P_n y \quad u_n \in H, \ n = 1, 2, \ldots \tag{7.42}$$

where H is a real separable infinite-dimensional Hilbert space, $H_n = \text{span}\{w_1, w_2, \ldots w_n\}$, $\{w_n\}$ is a basis in H, and $P_n : H \rightarrow H_n$ is the orthogonal projection operator from H into H_n. Since P_n is self-adjoint, Eq. (7.42) is equivalent to

$$\langle Tu_n, w_j \rangle = \langle y, wj \rangle u_n \in H_n, \ j = 1, 2, 3, \ldots n \tag{7.43}$$

Definition 7.1 For given $y \in H$, Eq. (7.41) is called *uniquely approximation-solvable* if the following conditions hold:

(a) Equation (7.41) has a unique solution u.
(b) There exists a number m such that for all $n \geq m$, the Galerkin Equation (7.42) or (7.43) has a unique solution u_n.

(c) The Galerkin method converges; that is, $||u_n - u||_H \to 0$ as $n \to \infty$.

Theorem 7.8 *Under the given conditions, for each $y \in H$, Eq. (7.41) is uniquely approximation-solvable in the case where the linear operator $T : H \to H$ satisfies one of the following four properties:*

(a) $T = I + S$, *where $S : H \to H$ is linear and k-contractive, namely $||S|| < 1$. For $n \geq m$*

$$||u - u_n|| \geq (1 - ||S||)^{-1} d(u, H_n) \tag{7.44}$$

holds.

(b) $T = I + U$, *where $U : H \to H$ is linear and compact (image of a bounded set A under U is precompact; that is, $\overline{U(A)}$ is compact), and $Tu = 0$ implies $u = 0$. Under these conditions,*

$$||uu_n|| \leq const. d(u, H_n) \tag{7.45}$$

(c) *T is linear, bounded, and strongly monotone (or coercive); that is, $\langle Tu, u \rangle \geq \alpha ||u||^2$ for all $u \in H$ and fixed $\alpha > 0$. Under the hypotheses for $\alpha > 0$,*

$$\alpha ||u - u_n|| \leq ||Tu_n - y|| \tag{7.46}$$

(d) $T = S + U$, *where $S : H \to H$ is linear continuous and coercive and U is linear and compact, and $u = 0$ whenever $Tu = 0$.*

In cases (a) and (c), $m = 1$. In case (b), m is independent of y. We prove here cases (a) and (c) and refer to Zeidler [201] for the other cases.

Proof A. Since $P_n u_n = u_n$ for $u_n \in H_n$ and $||P_n S|| \leq ||S|| < 1$, ($||P_n|| = 1$ by Theorem 3.9(3)). The following equations

$$u + Su = y, \ u \in H \tag{7.47}$$

and

$$u_n + P_n Su_n = P_n y, \ u_n \in H_n \tag{7.48}$$

have unique solutions by the Banach contraction mapping theorem (Theorem 1.1). Furthermore,

$$||(I + P_n S)^{-1}|| = \left\|\sum_{k=0}^{\infty} (P_n S)^k\right\| \leq \sum_{k=0}^{\infty} ||S||^k$$
$$= (1 - ||S||^{-1})$$

By (7.47) and (7.48)

$$(1 + P_n S)(u - u_n) = u - P_n u$$

Hence,

$$||u - u_n|| \leq (1 - ||S||^{-1})||u - P_n u|| = (1 - ||S||^{-1})d(u, P_n u)$$

B. We have $P_n u = u$ for all $u \in H$, and so

$$\langle P_n T u, u \rangle = \langle T u, P_n u \rangle \geq \alpha ||u||^2 \tag{7.49}$$

Thus, the operator $P_n T : H_n \to H_n$ is strongly monotone. The two operator Eqs. (7.42) and (7.43) have unique solutions by the Lax–Milgram Lemma. If $n \geq j$, then it follows from (7.43) that

$$\langle T u_n, w_j \rangle = \langle y, w_j \rangle \tag{7.50}$$

$$\langle T u_n, u_n \rangle = \langle y, u_n \rangle \tag{7.51}$$

By (7.51)

$$\alpha ||u_n||^2 \leq \langle y, u_n \rangle \leq ||y|| \, ||u_n||$$

This yields a priori estimate

$$c||u_n|| \leq ||y||; \; that \; is, \; \{u_n\} \; is \; bounded$$

Let $\{u_n\}$ be a weakly convergent subsequence with $u_{n'} \rightharpoonup v$ as $n \to \infty$ (Theorem 4.17). By (7.50), $\langle T u_{n'}, w \rangle \to \langle y, w \rangle$ as $n \to \infty$ for all $w \in \bigcup_n H_n$. Since $\bigcup_n H_n$ is dense in H and $\{T u_n\}$ is bounded, we obtain

$$T u_n \rightharpoonup y \; as \; n \to \infty \; (Theorem \; 4.17)$$

Since T is linear and continuous,

$$T u_{n'} \rightharpoonup T v \; as \; n \to \infty \; (Theorem \; 4.11(i))$$

Hence, $T v = y$; that is, $v = u$. Since the weak limit u is the same for all weakly convergent subsequences of $\{u_n\}$, we get $u_n \rightharpoonup u$ as $n \to \infty$. It follows from

$$\alpha ||u_n - u||^2 \leq \langle T(u_n - u), u_n - u \rangle$$
$$= \langle y, u_n \rangle - \langle T u_n, u \rangle - \langle T u, u_n - u \rangle \to 0 \; as \; n \to \infty;$$

that is, $u_n \to u$ as $n \to \infty$. Therefore,

$$\alpha ||u_n - u||^2 \leq ||Tu_n - Tu|| \, ||u_n - u||$$

and $Tu = y$. Therefore, $\alpha ||u_n - u|| \leq ||Tu_n - y||$.

7.5.2   Rayleigh–Ritz–Galerkin Method

The Rayleigh–Ritz–Galerkin method deals with the approximate solution of (7.41) in the form of a finite series

$$u_m = \sum_{j=1}^{m} c_j \phi_j + \phi_0 \tag{7.52}$$

and its weak formulation (variation formulation), $a(u, v) = F(v)$, where the coefficients c_j, named the Rayleigh–Ritz–Galerkin coefficients, are chosen such that the abstract variational formulation $a(v, w) = F(v)$ holds for $v = \phi_i$, $i = 1, 2, \ldots m$; that is,

$$a\left(\phi_i, \sum_{j=1}^{m} c_j \phi_j + \phi_0\right) = F(\phi_i), \; i = 1, 2 \ldots m \tag{7.53}$$

Since $a(\cdot, \cdot)$ is bilinear, (7.53) becomes

$$\sum_{j=1}^{m} a(\phi_i, \phi_i) c_j = F(\phi_i) - a(\phi_i, \phi_0) \tag{7.54}$$

or

$$Ac = b \tag{7.55}$$

where

$$A = (a(\phi_i, \phi_0))_{ij}$$

is a matrix

$$b^T = [b_1, b_2, \ldots, b_m] \; with \; b_i = F(\phi_i)$$

and

$$c = [c_1, c_2, \ldots_m]$$

which gives a system of m linear algebraic equations in m unknowns c_i. The columns (and rows) of coefficient matrix A must be linearly independent in order that the coefficient in (7.55) can be inverted. Thus, for symmetric bilinear forms, the Rayleigh–Ritz–Galerkin method can be viewed as one that seeks a solution of the form in Eq. (7.52) in which the parameters are determined by minimizing the quadratic functional (energy functional) given by Eq. (6.12). After substituting um of Eq. (7.52) for u into $J(u) = \frac{1}{2}a(u, u) - F(u)$ and integrating the functional over its domain, $J(u)$ becomes an ordinary function of the parameters $c_1, c_2, \ldots$. The necessary condition for the minimum of $J(c_1, c_2, \ldots, c_m)$ is that

$$\frac{\partial J(\ldots)}{\partial c_1} = \frac{\partial J(\ldots)}{\partial c_2} = \cdots = \frac{\partial J(\ldots)}{\partial c_m} = 0 \tag{7.56}$$

This leads to m linear algebraic equations in c_j, $j = 1, 2, \ldots, m$. It may be noted that (7.54) and (7.56) are the same in the symmetric case while they differ in the non-symmetric case. Thus, we obtain the same c_i's by solving (7.54) and (7.56) separately. In the non-symmetric case, we get the m unknowns by solving the linear algebraic equations (matrix equations) (7.55). The selection of $\{\phi_j\}$, $j = 1, 2, \ldots, m$ is crucial, and this should be the basis of the Hilbert space under consideration.

7.6 Eigenvalue Problems

7.6.1 Eigenvalue of Bilinear Form

Suppose $a(u, v)$ is a symmetric coercive and bounded bilinear form defined on the Hilbert space H associated with the operator equation

$$T : H \subset L_2(\Omega) \to H, \ Tu = y \text{ in } \Omega, \Omega \subset R^n \tag{7.57}$$

$H = H^m(\Omega)$.

The problem of finding a number λ and a nonzero function $u \in H$ such that

$$a(u, v) = \lambda \langle u, v \rangle_{L_2(\Omega)} \tag{7.58}$$

is called the *eigenvalue problem of the bilinear form* $a(u, v)$. The weak problem associated with the operator equation

$$Tu - \lambda u = y \tag{7.59}$$

comprises finding $u \in H$ such that

$$a(u, v) - \lambda \langle u, v \rangle_{L_2(\Omega)} = \langle y, v \rangle_{L_2(\Omega)} \tag{7.60}$$

7.6.2　Existence and Uniqueness

Theorem 7.9 *An element $u \in H$ is the weak solution of Problem (7.59) if and only if*

$$u - \lambda T u = T y \tag{7.61}$$

holds in H. The number λ is an eigenvalue of the bilinear form $a(u, v)$, and the function $u(x)$ is the corresponding eigenfunction if and only if

$$u - \lambda T u = 0, \ u \neq 0 \tag{7.62}$$

holds in H.

Proof Suppose $u(x)$ is a weak solution of Problem (7.60); namely,

$$a(u, v) = \langle y + \lambda u, v \rangle_{L_2(\Omega)}$$

Then by the definition of operator T, we get

$$u = T(y + \lambda u) \ or \ u - \lambda T u = T y$$

To see the converse, let $u - \lambda T u = T y$, and then it is clear that u is the weak solution of (7.59). If Problem (7.58) has a nontrivial solution, then by given T we get

$$a(u, v) = \langle \lambda u, v \rangle_{L_2(\Omega)} \Rightarrow u = T(\lambda u) = \lambda T u$$

Thus, Eq. (7.62) holds in H and vice versa.

In view of a well-known result (see Rektorys [159] or Reddy [158] or Zeidler [201]), the symmetric, coercive, and bounded bilinear form $a(u, v)$ has a countable set of eigenvalues, each of which being positive

$$\lambda_1 \leq \lambda_2 \leq \lambda_3 \leq ..., \ \lim_{n \to \infty} \lambda_n = \infty \tag{7.63}$$

The corresponding orthogonal system of eigenfunctions $\{\phi_i\}$ constitutes a basis in the space $H_0^m(\Omega)$, hence in H. Furthermore, we have

$$\lambda_1 = \min_{v \in H, \ v \neq 0} \frac{\langle v, v \rangle_H}{||v||_{L_2(\Omega)}^2} = \frac{\langle u_1, u_2 \rangle_H}{\langle u_1, u_2 \rangle_{L_2(\Omega)}} \tag{7.64}$$

For $\lambda \neq \lambda_n$, $n = 1, 2, 3, \ldots$, Problem (7.58) has exactly one weak solution for every $y \in L_2(\Omega)$. If the eigenfunctions $\phi_i(x)$ belong to the domain D_T of the operator T, then the relation

$$\langle T\phi_j, v \rangle = a(\phi_j, v) = \lambda_j \langle \phi_j, v \rangle \text{ for every } v \in H$$

gives the classical eigenvalue problem

$$T\phi_j = \lambda_j \phi_j \tag{7.65}$$

If we choose $\{\phi_j\}$ as the basis of $H_0^m(\Omega)$, then for the N-parametric Ritz solution of $Tu = y$, we write

$$u_N(x) = \sum_{j=1}^{N} c_j \phi_j$$

where $c_j = \langle y, j \rangle$.

Thus, eigenfunctions associated with the eigenvalue problem $Tu = \lambda u$ can be used to advantage in the Ritz method to find the solution of the equation

$$Tu = y$$

7.7 Boundary Value Problems in Science and Technology

Example 7.6 (The Laplace Equation)

$$\Delta u = 0 \tag{7.66}$$

where

$$\Delta u = \frac{\partial^2 u}{\partial x^2} + \frac{\partial^2 u}{\partial y^2} + \frac{\partial^2 u}{\partial z^2} \tag{7.67}$$

The operator Δ is known as the Laplace operator in dimension three. Equation (7.66) represents the electrostatic potential without the charges, the gravitational potential without the mass, the equilibrium displacement of a membrane with a given displacement of its boundary, the velocity potential for an inviscid, incompressible, irrotational homogeneous fluid in the absence of sources and sinks, the temperature in steady-state flow without sources and sinks, etc.

Example 7.7 (*The Poisson Equation*)

$$\Delta u = -f(x, y, z) \tag{7.68}$$

where f is known.

Equation (7.68) is the model of the electrostatic potential in the presence of charge, the gravitational potential in the presence of distributed matter, the equilibrium displacement of a membrane under distributed forces, the velocity potential for an inviscid, incompressible, irrotational homogeneous fluid in the presence of distributed sources or sinks, the steady-state temperature with thermal sources or sinks, etc.

Example 7.8 (*The Nonhomogeneous Wave Equation*)

$$\frac{\partial^2 u}{\partial t^2} - \Delta u = -f(x, y, z) \tag{7.69}$$

This equation represents many interesting physical situations. Some of these are the vibrating string, vibrating membrane, acoustic problems for the velocity potential for the fluid flow through which sound can be transmitted, longitudinal vibrations of an elastic rod or beam, and both electric and magnetic fields without charge and dielectric.

Example 7.9 (*Stokes Problem*) Find u such that

$$-\mu \Delta u + \nabla p = f \; in \; \Omega \tag{7.70}$$
$$div \; u = 0 \; in \; \Omega \tag{7.71}$$
$$u = 0 \; on \; \Gamma \tag{7.72}$$

This BVP represents the phenomenon of the motion of an incompressible viscous fluid in a domain Ω, where $u = (u_1, u_2, \ldots, u_n)$ is the velocity of the fluid, p denotes the pressure, $f = (f_1, f_2, \ldots, f_n) \in (L_2(\Omega))^n$ represents the body force per unit volume, and μ is the viscosity

$$\nabla p = grad \, p = \sum_{i=1}^n \frac{\partial p}{\partial x_i} e_i, \; e_i = (0, 0, \ldots, 1, 0 \ldots)$$

i.e., the ith coordinate is 1

$$div \; u = \sum_{i=1}^n \frac{\partial u_i}{\partial x_i}.$$

Example 7.10 (*The Navier–Stokes Equation*) The stationary flow of a viscous Newtonian fluid subjected to gravity loads in a bounded domain Ω of R^3 is governed by the BVP

$$-r\Delta u + \sum_{i=1}^{3} \frac{\partial u_i}{\partial x_i} + \nabla p = f \; in \; \Omega \tag{7.73}$$

$$div \, u = 0 \; in \Omega \tag{7.74}$$

$$u = 0 \; on \; \ell = \partial\Omega \tag{7.75}$$

where u represents the velocity, p the pressure, f the body force per unit volume and ∇p, $div u$ have the same meaning as in Example 7.9 for $n = 3$. $r = \frac{\mu}{dv\rho} = 1/R$, where R is called the Reynolds number. Here μ is the viscosity of the fluid, d, a length characterizing the domain Ω, v a characteristic velocity of the flow, and ρ the density of the fluid.

Example 7.11 (*Heat Equation*) The following equation governs the diffusion process or heat conduction to a reasonable approximation:

$$\frac{\partial u}{\partial t} = \frac{\partial^2 u}{\partial x^2}, \; x \in (-1, 1), \; t \in (0, \infty) \tag{7.76}$$

This equation is called the heat equation.

The boundary conditions for this equation may take a variety of forms. For example, if the temperatures at the end points of a rod are given, we would have the BVP

$$\frac{\partial u}{\partial t} = \frac{\partial^2 u}{\partial x^2} \; on \; \Omega, \; \Omega = \{(x, t)/-1 < x < 1, \; 0 < t < \infty\}$$

$$u(\pm 1, t) = e^{-t} \; 0 \le t < \infty$$

$$u(x, 0) = 1 \; |x| \le 1$$

Let $u(x, 0) = u_0(x) = 1$.

The solution of this BVP, $u(x, t)$, will give the temperature distribution in the rod at time t. The heat equation has recently been applied for predicting appropriate bet for shares (for details, see [145] and references therein).

Example 7.12 (*The Telegrapher's Equation*)

$$\frac{\partial^2 \phi}{\partial t^2} + \alpha \frac{\partial \phi}{\partial t} + b\phi = \frac{\partial^2 \phi}{\partial x^2} \tag{7.77}$$

where α and β are constants. This equation arises in the study of propagation of electrical signals in a cable transmission line. Both the current I and the voltage V satisfy an equation of the form (7.77). We also find such equation in the propagation of pressure waves in the study of pulsatile blood flow in arteries, and in one-dimensional random motion of bugs along a hedge.

Example 7.13 (*The Inhomogeneous Helmholtz Equation*)

$$\Delta\psi + \lambda\psi = -f(x, y, z) \tag{7.78}$$

where λ is a constant.

Example 7.14 (*The Biharmonic Wave Equation*)

$$\Delta\psi - \frac{1}{c^2}\frac{\partial^2\psi}{\partial t^2} = 0 \tag{7.79}$$

In elasticity theory, the displacement of a thin elastic plate in small vibrations satisfies this equation. When ψ is independent of time t, then

$$\Delta^2\psi = 0 \tag{7.80}$$

This is the equilibrium equation for the distribution of stress in an elastic medium satisfied by Airy's stress function ψ. In fluid dynamics, the equation is satisfied by the stream function ψ in an incompressible viscous fluid flow. Equation (7.80) is called the Biharmonic equation.

Example 7.15 (*The Time-Independent Schrödinger Equation in Quantum Mechanics*)

$$\frac{h^2}{2m}\Delta\psi + (E - V)\psi = 0 \tag{7.81}$$

where m is the mass of the particle whose wave function is ψ, ℓ is the universal Planck's constant, V is the potential energy, and E is a constant. If $V = 0$, in (7.81) we obtain the Helmholtz equation.

We prove here the existence of the solution of Stokes equation and refer to Chipot [39], Debnath and Mikusinski [63], Quarteroni and Valli [155], Reddy [156], Reddy [158] and Siddiqi [169] for variational formulation and existence of solutions of other boundary value problems. A detailed presentation of variational formulation and study of solution of parabolic equations including classical heat equation is given in Chipot [39], Chaps. 11 and 12.

Existence of the Solution of Stokes Equations We have

$$-\mu\Delta u + grad\, p - f = 0 \; in \; u \in \Omega \tag{7.82}$$
$$div\, u = 0 \; in \; u \in \Omega \tag{7.83}$$
$$u = 0 \; on \; \Gamma \tag{7.84}$$

where $f \in L_2(\Omega)$ is the body force vector, $u = (u_1, u_2, u_3)$ is the velocity vector, p is the pressure, and μ is the viscosity. We introduce the following spaces:

$$\left. \begin{array}{l} D = \{u \in C_0^\infty(\Omega)/div \\ H = \{u \in H_0^1(\Omega) \times H_0^1(\Omega)/div\; u = 0\} \\ Q = \left\{p \in L_2(\Omega)/ \int\limits_\Omega p\, dx = 0\right\} \end{array} \right\} \qquad (7.85)$$

The space H is equipped with the inner product

$$\langle v, u \rangle_H = \sum_{i=1}^n \int\limits_\Omega grad\; v_i . grad\; u_i dx \qquad (7.86)$$

where n is the dimension of the domain $\Omega \subset R^n$.

The weak formulation of Eqs. (7.82)–(7.84) is obtained using the familiar procedure (i.e., multiply each equation with a test function and applying Greens formula for integration by parts (Theorem 5.17). We obtain for $v \in D$

$$\langle -\mu \Delta u + grad p - f, v \rangle = 0$$

$$or \quad \mu \sum_{i=1}^n \int\limits_\Omega grad\; u_i . grad\; v_i dx = \langle grad p, v \rangle + \langle f, v \rangle$$

$$= \langle f, v \rangle \text{ for every } v \in D$$

As for $v \in D$, we have $div\; v = 0$ and $v = 0$ on Γ, giving

$$a(v, u) = (v, f) \text{ for every } v \in D$$

where

$$a(v, u) = \sum_{i=1}^n \int\limits_\Omega \mu grad\; v_i . grad\; u_i dx \qquad (7.87)$$

We now have the following weak problem: find $u \in H$ such that

$$a(v, u) = \langle v, f \rangle \qquad (7.88)$$

holds for every $v \in H$.

The proof that the weak solution of Eq. (7.88) is the classical solution of Eqs. (7.82)–(7.84) follows from the argument (see Temam [183])

$$\langle v, -\mu \Delta u - f \rangle_{L_2(\Omega)} = 0 \text{ for every } v \in V \qquad (7.89)$$

This does not imply that $-\mu \Delta u - f = 0$ because v is subjected to the constraint (because $v \in H$) $div v = 0$. Instead, Eq. (7.89) implies that

$$- \mu \Delta u - f = -grad\ p \qquad (7.90)$$

because (necessary and sufficient)

$$\langle v, grad p \rangle = \langle div\ v, p \rangle = 0 \text{ for every } p \in Q \qquad (7.91)$$

The bilinear form in Eq. (7.88) satisfies the conditions of the Lax–Milgram Lemma. The continuity follows from Eq. (7.87) using the CBS inequality

$$|a(v, u)| = \left| \sum_{i=1}^{n} \int_{\Omega} \mu grad\ v_i . grad/u_i dx \right|$$

$$\leq \mu \left[\sum_{i=1}^{n} \int_{\Omega} |grad\ v_i|^2 dx \right]^{1/2} \left[\sum_{i=1}^{n} \int_{\Omega} |grad\ u_i|^2 dx \right]^{1/2}$$

$$= \mu ||v||_H\ ||u||_H$$

The V-ellipticity of $a(\cdot, \cdot)$ follows from

$$|a(v, v)| = \mu \sum_{i=1}^{n} \int_{\Omega} grad\ v_i . grad/v_i dx$$

$$\leq \mu ||v||_H^2 \geq \alpha ||v||_H^2$$

for $\alpha \geq \mu$. Thus, (7.88) has one and only solution in space H.

7.8　Problems

Problem 7.1 Prove the existence of solution of the following boundary value problem

$$-\frac{d^2 u}{dx^2} + \frac{du}{dx} + u(x) = f \ in \ [0, 1]$$

$$\left(\frac{du}{dx}\right)_{x=0} = \left(\frac{du}{dx}\right)_{x=1} = 0$$

Problem 7.2 Suppose $a(u, v)$ is a coercive and bounded bilinear form on $H = H^m(\Omega)$ (Sobolev space of order m). Then prove the continuous dependence of the weak solution of the operator equation $Su = y$, where S is induced by the given bilinear form, on the given data of the problem; that is, if the function changes slightly in the norm of $L_2(\Omega)$ to $\tilde{y}$, then the weak solution u also changes slightly in the norm of $H^m(\Omega)$ to $\tilde{u}$. Prove that

$$||u - \tilde{u}||_{H^m(\Omega)} \leq \beta ||y - \tilde{y}||_{L_2(\Omega)}$$

Problem 7.3 Find the solution of the following equation using the Raleigh–Ritz method.

$$-\frac{d^2 u}{dx^2} = \cos(\pi x), \ 0 < x < 1$$

with (a) the Dirichlet boundary condition

$$u(0) = u(1) = 0$$

and (b) the Neumann boundary condition

$$u'(0) = u'(1) = 0$$

Problem 7.4 Prove the existence of the following boundary value problem

$$Tu = f \ in \ \Omega \subset R^n$$
$$u = 0 \ on \ \partial\Omega$$

where

$$Tu = -\sum_{i,j=1}^{n} \frac{\partial}{\partial x_i} \left[a_{ij} \frac{\partial u}{\partial x_j} \right] + a_0 u$$

$a_{ij} \in C^1(\overline{\Omega})$, $1 \leq i, j, \leq n$, $a_0 \in C^1[\overline{\Omega}]$, $x = (x_1, x_2, \ldots, x_n) \in R^n$.

Problem 7.5 Let $\Omega \subset R^n$. Show that the Robin boundary condition value problem

$$-\Delta u + u = f \ in \ \Omega, \ f \in L_2(\Omega)$$
$$\frac{\partial u}{\partial n} + \alpha u = 0 \ on \ \partial\Omega, \alpha > 0$$

has a unique solution.

Problem 7.6 Prove that $\sup_{u \in M} ||u - P_n u|| \to 0$ as $n \to \infty$ provided the nonempty subset M of H is compact, where $\{P_n\}$ is a projection operator on a Hilbert space H into H_n, a finite-dimensional subspace of H, $\dim H_n = n$.

Problem 7.7 Suppose T is a bounded linear operator on a Banach space X into another Banach space Y. Suppose that the equation

$$Tu = y, \ u \in X \tag{7.92}$$

has an approximate solution, namely $\exists$ constants $K > 0$ and $\mu \in (0, 1)$ such that for each $a \in Y$, there exists a $u(a) \in X$ with

$$||Tu(a) - a|| \leq \mu||a||, \ ||u(a)|| \leq K||a||$$

Show that, for each $b \in Y$, the Eq. (7.92) has a solution u with $||u|| \leq K(1-\mu)||b||$.

Problem 7.8 Let us consider the operator equation

$$u + Su = y, \ u \in X \tag{7.93}$$

together with the corresponding approximation equation

$$u_n + S_n u_n = P_n y \ u_n \in X_n, \ n = 1, 2, 3, \ldots \tag{7.94}$$

Let us make the following assumptions:

(i) X_n is a subspace of the Banach space X and dim $X_n = n$ and $P_n : X \to X_n$ is a projection operator onto X_n.
(ii) The operators $S : X \to X$ and $S_n : X_n \to X_n$ are linear and bounded and $I + S : X \to X$ is onto.
(iii) There exists a constant β_n with $d(Su, X_n) \leq \beta_n ||u||$ for all $u \in X$.
(iv) As $n \to \infty$, $||P_n S - S_n||_{X_n} \to 0$, $||P_n||\beta_n \to 0$ and $||P_n||d(y, X_n) \to 0$ for all $y \in X$ hold. Show that Eq. (7.94) is uniquely approximation-solvable and

$$||u_n - u|| \leq K \left(||P_n S - S_n||_{X_n} + ||P_n||(\beta_n + d(y, X_n)) \right)$$
$$||u_n - u|| \leq K \left(||P_n S - S_n||_{X_n} + ||P_n||(d(y, X_n)) \right)$$

where K is an appropriate constant. (This problem was studied by Kontorovich).

Problem 7.9 Show that $u\Delta v = \nabla \cdot (u\nabla v) - (\nabla u) \cdot (\nabla v)$.

Problem 7.10 Suppose that H is a real Hilbert space and S is a linear operator on H. Prove that S is continuous and S^{-1} exists if

$$\inf_{||x||=1} (\langle Sx, x \rangle + ||Sx||) > 0.$$

(This is a generalized version of the Lax–Milgram Lemma proved by Jean Saint Raymond in 1997).

Chapter 8
Finite Element and Boundary Element Methods

Abstract In this chapter, finite element and boundary element methods are introduced. Functional analysis plays important role to reduce the problem in discrete form amenable to computer analysis. The finite element method is a general technique to construct finite-dimensional spaces of a Hilbert space of some classes of functions such as Sobolev spaces of different orders and their subspaces in order to apply the Ritz and Galerkin methods to a variational problem. The boundary element method comprises transformation of the partial differential equation describing the behavior of an unknown inside and the boundary of the domain into an integral equation relating to any boundary values, and their finding out numerical solution.

Keywords Abstract error estimation · Céa lemma · Strange first lemma · Strange second lemma · Internal approximations · Finite elements · Finite element method for boundary value problem · Weighted residual methods · Boundary element method

8.1 Introduction

Finite and boundary element methods are well-known numerical methods to solve different types of boundary value problems (BVPs) representing real-world systems. Concepts of functional analysis play very significant role in formulation of boundary value problems amenable to computer simulation. The finite element method mainly deals with approximation of a Hilbert space as a Sobolev space with a finite-dimensional subspace. It also encompasses error estimation solution on the function space with a solution on the finite-dimensional space. This method is based on procedure like partition of domain Ω in which the problem is posed into a set of simple subdomains, known as elements; often these elements are triangles, quadrilaterals, tetrahedra etc.

A Sobolev space defined on Ω is approximated by functions defined on subdomains on Ω with appropriate matching conditions at interfaces. It may be observed that finite element method is nothing but approximation in Sobolev spaces.

A. H. Siddiqi, *Functional Analysis and Applications*, Industrial and Applied Mathematics,
https://doi.org/10.1007/978-981-10-3725-2_8

A systematic study of variational formulation of the boundary value problems and their discretization began in the early seventies. In early 1950s, engineer Argyris started the study of certain techniques for structural analysis which are now known as the primitive finite element method. The work representing the beginning of finite element was contained in a paper of Turner, Clough, Martin, and Topp [187], where endeavor was made for a local approximation of the partial differential equations of linear elasticity by the usage of assembly strategies, an essential ingredient of finite element method. In 1960, Clough termed these techniques as "finite element method." Between 1960 and 1980, several conferences were organized in different parts of the world, mainly by engineers, to understand the intricacies of the method. A paper by Zlamal [203] is considered as the first most significant mathematical contribution in which analysis of interpolation properties of a class of triangular elements and their application to the second- and fourth-order linear elliptic boundary value problems is carried out. Valuable contributions of Ciarlet, Strang, Fix, Schultz, Birkhoff, Bramble and Zlamal, Babuska, Aziz, Varga, Raviart, Lions, Glowinski, Nitsche, Brezzi have enriched the field. Proceedings of the conferences edited by Whiteman [196] and the book by Zienkiewicz and Cheung [202] have popularized the method among engineers and mathematicians alike. The Finite Element Handbook edited by Kardestuncer and Norrie [108, 112] and the Handbook of Numerical Analysis edited by Ciarlet and Lions [47] provide contemporary literature. Wahlbin [189] presents some of the current research work in this field. References [28, 29] are also interesting references for learning the finite element. In short, there is no other approximation method which has had such a significant impact on the theory and applications of numerical methods. It has been practically applied in every conceivable area of engineering, such as structural analysis, semiconductor devices, meteorology, flow through porous media, heat conduction, wave propagation, electromagnetism, environmental studies, safing sensors, geomechanics, biomechanics, aeromechanics, and acoustics.

The finite element method is popular and attractive due to the following reasons: The method is based on weak formulation (variational formulation) of boundary and initial value problems. This is a critical property because it provides a proper setting for the existence of even discontinuous solution to differential equations, for example, distributions, and also because the solution appears in the integral of a quantity over a domain. The fact that the integral of a measurable function over an arbitrary domain can be expressed as the sum of integrals over an arbitrary collection of almost disjoint subdomains whose union is the original domain and is a very important point in this method. Due to this fact, the analysis of a problem can be carried out locally over a subdomain, and by making the subdomain sufficiently small, polynomial functions of various degrees are sufficient for representing the local behavior of the solution. This property can be exploited in every finite element program which allows us to focus the attention on a typical finite element domain and to find an approximation independent of the ultimate location of that element in the final mesh. The property stated above has important implications in physics and continuum mechanics, and consequently, the physical laws will hold for every finite portion of the material.

Some important features of the finite element methods are

1. Arbitrary geometries,
2. Unstructured meshes,
3. Robustness,
4. Sound mathematical foundation.

Arbitrary geometries means that, in principle, the method can be applied to domains of arbitrary shapes with different boundary conditions. By unstructured meshes, we mean that, in principle, one can place finite elements anywhere from the complex cross-sections of biological tissues to the exterior of aircraft, to internal flows in turbomachinery, without the use of a globally fixed coordinate frame. Robustness means that the scheme developed for assemblage after local approximation over individual elements is stable in appropriate norms and insensitive to singularities or distortions of the meshes (this property is not available in classical difference methods).

The method has a sound mathematical basis as convergence of an approximate solution of the abstract variational problem (a more general form is variational inequality problem) and error estimation of the abstract form in a fairly general situation, and their special cases have been systematically studied during the last two decades. These studies make it possible to lift the analysis of important engineering and physical problems above the traditional empiricism prevalent in many numerical and experimental studies.

The main objective is to discuss tools and techniques of functional analysis required in finite element and boundary element methods.

Boundary Element Method

The classical theory of integral equations is well known; see, for example, Tricomi [186], Kupradze [118], Mikhlin [135, 136]. Attention of engineers was drawn toward boundary element methods by systematic work of researchers at Southampton University in the eighties and nineties. Real-world problems were modeled by integral equations on boundary of a region. Appropriate methods were developed to solve it. This method has been applied in a variety of physical phenomena like transient heat conduction, thermo-elasticity, contact problems, free boundary problems, water waves, aerodynamics, elastoplastic material behavior, electromagnetism, soil mechanics. There is a vast literature published on these topics in the last fifteen years which can be found in books by Brebbia [22–27], Antes and Panagiotopoulos [3], Chen and Zhou [37], and Hackbusch [92]. The boundary element method is a rapidly expanding area of practical importance. We illustrate here the basic ingredients of this method with examples.

8.2   Finite Element Method

8.2.1   *Abstract Problem and Error Estimation*

Suppose H is a Hilbert space and $a(\cdot, \cdot)$ is a bounded bilinear form on $H \times H$ into R. For each $F \in H^{\star}$, the dual space of H (the space of all bounded linear functionals on H). The variational problem is to find $u \in H$ such that

$$a(u, v) = F(v) \; \forall \, v \in H \tag{8.1}$$

Equation (8.1) has a unique solution in view of the Lax–Milgram Lemma (Theorem 3.39) provided $a(\cdot, \cdot)$ is coercive or elliptic. Finding a finite-dimensional subspace H_h of H such that $\exists \, u_h \in H_h$

$$a(u_h, v_h) = F(v_h) \; \forall \, v_h \in H_h \tag{8.2}$$

is known as *finite element method*. Equation (8.1) is known as the *abstract variational problem*, and Eq. (8.2) is called the *approximate problem*. If H_h is not a subspace of H, the above method is known as the *nonconformal finite method*.

Equation (8.2) can be written as

$$AU = B \tag{8.3}$$

where $U = (\alpha_1, \alpha_2, \alpha_3, ..., \alpha_{N(h)})$, $N(h) =$ dimension of H_h

$$A^t = (a(w_i, w_j))i, \; j \tag{8.4}$$

$$B = (F(w_1), F(w_2), ..., F(w_{N(h)}) \tag{8.5}$$

$$u_h = \sum_{i=1}^{N(h)} \alpha_i w_i \tag{8.6}$$

$$v_h = \sum_{i=1}^{N(h)} \beta_i w_i \tag{8.7}$$

α_i and β_j are real numbers, $i, j = 1, 2, 3, ..., N(h)$. The choice of the basis $\{w_i\}_i$ of $H_h, i = 1, 2, ..., N(h)$ is of vital importance; namely, choose a basis of H_h which makes A a sparse matrix so that the computing time is reasonably small. In the terminology of structural engineers, A and $F(w_j)$ are called the *stiffness matrix* and the *load vector*, respectively.

If $a(\cdot, \cdot)$ is symmetric, then finding the solution of (8.1) is equivalent to finding a solution of the optimization problem

$$J(u) = \inf_{v \in H} J(v) \tag{8.8}$$

where $J(v) = \frac{1}{2}a(v, v) - F(v)$ is called the *energy functional*. Here, finite element method is known as Rayleigh–Ritz–Galerkin method. Equation (8.2) and the approximate problems of (8.8), namely

$$J(u_h) = \inf_{v_h \in H_h} J(v_h) \tag{8.9}$$

where $J(vh) = \frac{1}{2}a(v_h, v_h) - F(v_h)$ have unique solutions.

Finding $\|u - u_h\|$, where u and u_h are the solutions of (8.1) and (8.2), respectively, is known as the *error estimation*. The problem $u_h \to u$ as $h \to 0$; that is, $\|u_h - u\| \to 0$ as $h \to 0$ or $n = \frac{1}{h} \to \infty$ is known as the *convergence problem*.

Error Estimation

Theorem 8.1 (Céa's Lemma) *There exists a constant C independent of the subspace H_h such that*

$$\|u - u_h\|_H \leq C \inf_{v_h \in H_h} \|u - v_h\|_{H_h} = Cd(u, H_h) \tag{8.10}$$

where $C = \frac{M}{\alpha}$ is independent of H_h, M is the constant associated with the continuity (boundedness) of $a(\cdot, \cdot)$, and α is the coercivity constant. If $a(\cdot, \cdot)$ is symmetric, then the degree of approximation is improved; that is, we get $C = \sqrt{\frac{M}{\alpha}}$ which is less than the constant in the nonsymmetric case.

Proof By (8.1) and (8.2), we get $a(u, v) - a(u_h, v_h) = F(v) - F(v_h)$ and this gives $a(u, v_h) - a(u_h, v_h) = 0$, for $v = v_h$. By bilinearity of $a(\cdot, \cdot)$, we get

$$a(u - u_h, v_h) = 0 \ \forall \ v_h \in H_h \Rightarrow a(u - u_h, v_h - u_h) = 0 \tag{8.11}$$

by replacing v_h by $v_h - u_h$. Since $a(\cdot, \cdot)$ is elliptic

$$a(u - u_h, u - u_h) \geq \alpha \|u - u_h\|^2$$

or

$$\frac{1}{a}(u - u_h, u - v_h + v_h - u_h) \geq \|u - u_h\|.$$

or

$$\frac{1}{a}[(u - u_h, u - v_h) + a(u - u_h, v_h - u_h)] \geq u - u_h.$$

Using (8.11), this becomes

$$\frac{1}{a}[(u - u_h, u - v_h)] \geq \|u - u_h\|^2$$

or

$$\frac{1}{a}M||u - u_h||\,||u - v_h|| \geq ||u - u_h||^2$$

using boundedness of $a(\cdot, \cdot)$; namely

$$||a(u, v)|| \leq M||u||\,||v||$$

This gives us

$$||u - u_h|| \leq \frac{M}{\alpha}||u - v_h|| \, \forall \, v \in H$$

or

$$||u - u_h|| \leq C \inf_{v_h \in H_h} ||u - v_h||.$$

When the bilinear form $a(\cdot, \cdot)$ is symmetric, it leads to a remarkable interpretation of the approximate solution; namely, the approximate solution u_h is the projection of the exact solution u over the subspace H_h with respect to the inner product $a(\cdot, \cdot)$ (induced by the bilinear form which is denoted by the bilinear form itself.) as $a(u - u_h, v_h) = 0$ for all $v_h \in H_h$. Thus, we get

$$a(u - u_h, u - u_h) = \inf_{v_h \in H_h} a(u - v_h, u - v_h).$$

By the properties of ellipticity and boundedness of $a(\cdot, \cdot)$, we get

$$\alpha||u - u_h|| \leq M||u - v_h||\,||u - v_h||$$

or

$$||u - u_h|| \leq \sqrt{\frac{M}{\alpha}}||u - v_h|| \, \forall \, v_h \in H_h$$

Thus

$$||u - u_h|| \leq C \inf_{v_h \in H_h} ||u - v_h||, \; where \; C = \sqrt{\frac{M}{\alpha}},$$

where we have a smaller constant $\sqrt{\frac{M}{\alpha}}$ as $M \geq \alpha$.

Remark 8.1 Inequality (8.10) indicates that the problem of estimating the error $||u - u_h||$ is equivalent to a problem in approximation theory, namely, the determination of the distance $d(u, H_h) = \inf_{v_h \in H_h} ||u - v_h||$ between a function $u \in H$ and a subspace

H_h of H. Under appropriate condition on u, one can show that $d(u, H_h) \leq C(u)h^\beta$ for $\beta > 0$, where $C(u)$ is independent of h and dependent on u, and so $||u - u_h|| \leq C(u)h^\beta$. In this case, we say that the order of convergence is β or, equivalently, we have an order $O(h^\beta)$, and we write $||u - u_h|| = O(h^\beta)$.

We prove below a theorem for the abstract error estimation, called as the first Strang lemma due to Gilbert Strang, [180]. Here, the forms $a_h(\cdot, \cdot)$ and $F_h(\cdot)$ are not defined on the space H, since the point values are not defined in general for functions in the space $H^1(\Omega)$.

Theorem 8.2 (First Strang Lemma) *Suppose H is a Hilbert space and $H_h(\cdot)$ its finite-dimensional subspace. Further, let $a(\cdot, \cdot)$ be a bilinear bounded and elliptic form on H and $F \in H^*$. Assume that u_h is the solution of the following approximate problem: Find $u_h \in H_h$ such that*

$$a_h(u_h, v_h) = F_h(v_h) \text{ for all } v_h \in H_h \tag{8.12}$$

where $a_h(\cdot, \cdot)$ is a bilinear and bounded form defined on H_h and $F_h(\cdot)$ is a bounded linear functional defined on H_h. Then there exists a constant C independent of H_h such that

$$||u - u_h|| \leq C(u) \left(\inf_{v_h \in H_h} \left\{ ||u - v_h|| + \sup \frac{|a(v_h, w_h)|}{||w_h||} \right\} + \sup_{w_h \in H_h} \frac{F(w_h) - F_h(w_h)}{||w_h||} \right)$$

provided $a_h(\cdot, \cdot)$ is uniformly H_h-elliptic, that is, $\exists \beta > 0$ such that $a_h(v_h, v_h) \geq \beta ||v_h||^2$ for all $v_h \in H_h$ and all h.

It may be observed that although $a_h(\cdot, \cdot)$ and $F_h(\cdot)$ are not defined for all the elements of H, Eq.(8.12) has a unique solution under the given conditions.

Proof We have

$$||u - u_h|| \leq ||u - v_h|| + ||u_h - v_h||$$

by the triangular inequality of the norm and

$$\beta ||u_h - v_h||^2 \leq a_h(u_h - v_h, u_h - v_h) \text{ by coercivity} \tag{8.13}$$

By continuity of the bilinear form $a(\cdot, \cdot)$, (8.13) takes the form

$$\beta ||u_h - v_h||^2 \leq a(u - v_h, u_h - v_h) + \{a(v_h, u_h - v_h) - a_h(v_h, u_h - v_h)\} + \{F_h(u_h - v_h) - F(u_h - v_h)\}$$

or

$$\beta ||u_h - v_h|| \leq M ||u - v_h|| + \frac{|a(v_h, u_h - v_h) - a_h(v_h, u_h - v_h)|}{||u_h - v_h||}$$

$$+ \frac{|F_h(u_h - v_h) - F(u_h - v_h)|}{||u_h - v_h||}$$

$$\leq M||u - v_h|| + \sup \frac{|a(v_h, w_h) - a_h(v_h, w_h)|}{||w_h||}$$

$$+ \sup_{w_h \in H_h} \frac{|F_h(w_h) - F(w_h)|}{||w_h||}$$

By putting the value of $||u_h - v_h||$ in the first inequality and taking the infimum over H_h, the desired result is proved.

Now, we present a theorem on abstract error estimation known as the Second Strang Lemma due to Gilbert Strang [180]. In this lemma, H_h is not contained in the space H. The violation of the inclusion $H_h \subset H$ results from the use of finite elements that are not of class C^0; that is, that are not continuous across adjacent finite elements.

Theorem 8.3 (Second Strang Lemma) *Suppose u_h is a solution of the following approximate problem (discrete problem): Find $u_h \in H_h$ (a finite-dimensional space having the norm equivalent to the norm of $H^1(\Omega)$) such that*

$$a_h(u_h, v_h) = F(v_h) \text{ for all } v_h \in H_h \tag{8.14}$$

where $a_h(\cdot, \cdot)$ is as in Theorem 8.2 and $F \in H^\star$. Then there exists a constant C independent of the subspace H_h such that

$$||u - u_h||_{H^1} \leq C \left(\inf_{v_h \in H_h} ||u - v_h|| + \sup_{w_h \in H_h} \frac{||a_h(u, w_h) - F(w_h)||}{||w||_{H^1}} \right)$$

where H_h need not be a subspace of $H = H^1(\Omega)$.

Proof Let v_h be an arbitrary element in the space H_h. Then in view of the uniform H_h-ellipticity and continuity of the bilinear forms ah and of the definition of the discrete problem, we may write

$$\beta ||u_h - v_h||_H^2 \leq a_h(u_h - v_h, u_h - v_h)$$

$$= a_h(u - v_h, u_h - v_h)\{F(u_h - v_h) - a_h(u, u_h - v_h)\}$$

from which we deduce

$$\beta ||u_h - v_h||_H \leq M||u - v_h||_H + \frac{|F(u_h - v_h) - a_h(u, u_h - v_h)|}{||u_h - v_h||_H}$$

$$\leq M||u - v_h||_H + \sup_{w_h \in V_h} \frac{|F(w_h) - a_h(u, w_h)|}{||w_h||_H}$$

We obtain the desired result from the above inequality and the triangular inequality

$$\|u - u_h\|H \leq \|u - v_h\|_H + \|u_h - v_h\|_H$$

Remark 8.2 (i) Theorem 8.2 is a generalization of *Céa 's* lemma as $a_h(\cdot, \cdot) = a(\cdot, \cdot)$ and $F_h(\cdot) = F(\cdot)$ in the case of conformal finite element method (the case when $H_h \subset H$).

(ii) Problem (8.1) can be expressed in the form

$$Au = f \tag{8.15}$$

where $A : H \to H$ bounded and linear.

By Theorem 3.37, there exists a bounded linear operator A on H into itself ($H^* = H$) if H is a real Hilbert space such that $\langle Au, v \rangle = a(u, v)$. By the Riesz theorem for each v, there exists a unique $f \in H$ such that $F(v) = \langle f, v \rangle$. Therefore, (8.1) can be expressed as $\langle Au, v \rangle = \langle f, v \rangle$ implying (8.15).

Convergence Results As a consequence of Theorem 8.1, we find that $\|u_h - u\| \to 0$ as $h \to 0$; equivalently, the approximate solution u_h converges to the exact solution of (8.1) subject to the existence of a family $\{H_h\}$ of subspaces of the space H such that for each $u \in H$

$$\inf_{v_h \in H_h} \|u - v_h\| = 0 \ as \ h \to 0 \tag{8.16}$$

If $\|u - u_h\| \leq Ch^\alpha$ for $\alpha > 0$ where C is a positive constant independent of u and u_h and h is the characteristic length of an element, then α is called the rate of convergence. It may be noted that the convergence is related to the norm under consideration, say, L_1-norm, energy norm (L_2-norm), or L_∞-norm (or sup norm).

Corollary 8.1 *Let there exist a dense subspace U of H and a mapping $r_h : H \to H_h$ such that $\lim_{h \to 0} \|v - r_h v\| = 0 \ \forall u \in U$. This implies*

$$\lim_{h \to 0} \|u - u_h\| = 0.$$

Proof Let $\varepsilon > 0$. Let $v \in U$ such that for $C > 0$, $\|u - v\| \leq \varepsilon/2C$ (U is dense in H) and h sufficiently small such that $\|v - r_h v\| \leq \varepsilon/2C$. Then by Theorem 8.1, and in view of these relations

$$\begin{aligned}
\|u - u_h\| &\leq C \inf_{v_h \in H_h} \|u - v_h\| \\
&\leq C\|u - r_h v\|, \ because \ r_h v \in H_h \\
&\leq C\|u - v\| + \|v - r_h v\| \\
&\leq \frac{C\varepsilon}{2C} + \frac{C\varepsilon}{2C} \\
&= \varepsilon.
\end{aligned}$$

Therefore

$$\lim_{h \to 0} ||u - u_h|| = 0.$$

We refer to Wahlbin [189], Ciarlet [45], Ciarlet and Lions [47], Kardestuncer and Norrie [108] for further study.

8.2.2 Internal Approximation of $H^1(\Omega)$

This is related to finding a finite-dimensional subspace H_h of $H^1(\Omega)$. To achieve this goal, first, we define a triangulation.

Definition 8.1 Let $\Omega \subset R^2$ be a polygonal domain. A finite collection of triangles T_h satisfying the following conditions is called a triangulation:

A. $\overline{\Omega} = \bigcup_{K \in T_h} \overline{K}$, $\overline{K}$ denotes a triangle with boundary sides.
B. $K \cap K_1 = \phi$ for $K, K_1 \in T_h$, $K \neq K_1$.
C. $K \cap K_1 =$ a vertex or a side; i.e., if we consider two different triangles, their boundaries may have one vertex, common or one side common.

Remark 8.3 Let $P(K)$ be a function space defined on $K \in T_h$ such that $P(K) \subset H^1(K)$ (Sobolev space of order 1 on K). Generally, P(K) will be a space of polynomials or functions close to polynomials of some degree

Theorem 8.4 *Let $C^0(\overline{\Omega})$ be the space of continuous real-valued functions on $\overline{\Omega}$ and $H_h = \{v_h \in C^0(\overline{\Omega})/v_h/K \in P(K), K \in T_h\}$, where v_h/K denotes the restriction of v_h on K and $P(K) \subset H^1(K)$. Then $H_h \subset H^1(\Omega)$.*

Proof Let $u \in H_h$ and v_i be a function defined on Ω such that $v_i/K = \frac{\partial}{\partial x_i}(u/K)$. v_i/K is well defined as $u/K \in H^1(K)$. Moreover $v_i \in L_2(\Omega)$ as $v_i/K = \frac{\partial}{\partial x_i}(u/K) \in L_2(K)$. The theorem will be proved if we show that

$$v_i = \frac{\partial u}{\partial x_i} \in D^\star(\Omega)$$

$[\frac{\partial u}{\partial x_i} \in D^\star(\Omega)$ implies that $u \in H^1(\Omega)$ which, in turn, implies $H_h \subset H^1(\Omega)]$. For all $\phi \in D(\Omega)$, we have

$$(v_i, \phi) = \int_\Omega v_i \phi dx = \sum_{K \in T_h} \int_K v_i \phi dx$$

$$= \sum_{K \in T_h} \int_K \frac{\partial}{\partial x_i}(u/K)\phi dx$$

$$= \sum_{K \in T_h} \left[-\int_K (u/K)\frac{\partial \phi}{\partial x_i}dx + \int_{\Gamma = \partial K} (u/K)\phi \eta_i^k d\Gamma \right]$$

by the generalized Green's formula, where η_i^k denotes the ith component of the outer normal at Γ. Therefore

$$(v_i, \phi) = \int_\Omega u \frac{\partial \phi}{\partial x_i} dx + \sum_{K \in T_h} \int_\Gamma (u/K) \phi \eta_i^k d\Gamma$$

The second term on the right-hand side of the above relation is zero as u is continuous in Ω, and if K_1 and K_2 are two adjacent triangles, then $\eta_i^{K_1} = -\eta_i^{K_1}$. Therefore

$$(v_i, \phi) = -\int_\Omega u \frac{\partial \phi}{\partial x_i} dx = \left(\frac{\partial u}{\partial x_i}, \phi \right)$$

which implies that

$$v_i = \frac{\partial u}{\partial x_i} \in D^\star(\Omega)$$

Remark 8.4 Let $h = \max_{K \in T_h} (\text{diameter of K})$, $N(h)$ = the number of nodes of the triangulation, $P(K) = P_1(K)$ = space of polynomials of degree less than or equal to 1 in x and y

$$H_h = \{v_h / K \in P(K), \ K \in T_h\}$$

A. It can be seen that $H_h \subset C^0(\overline{\Omega})$.
B. The functions w_i, $i = 1, 2, \ldots, N(h)$, defined by

$$w_i = 1 \ \ at \ the \ i\text{th} \ node$$
$$0 \ \ at \ other \ nodes$$

form a basis of H_h.
C. In view of (2) and Theorem 8.4, H_h defined in this remark is a subspace of $H^1(\Omega)$ of dimension $N(h)$.

8.2.3 Finite Elements

Definition 8.2 In R^n, a (nondegenerate) n-*simplex* is the convex hull K of $(n + 1)$ points $a_j = (a_{ij})_{i=1}^n \in R^n$, which are called the *vertices* of the n-simplex, and which are such that the matrix

$$A = \begin{vmatrix} a_{11} & a_{12} & \cdots & a_{1,n+1} \\ a_{21} & a_{22} & \cdots & a_{2,n+1} \\ a_{31} & a_{32} & \cdots & a_{3,n+1} \\ \cdot & \cdot & \cdots & \cdot \\ \cdot & \cdot & \cdots & \cdot \\ \cdot & \cdot & \cdots & \cdot \\ a_{n1} & a_{n2} & \cdots & a_{n,n+1} \\ 1 & 1 & \cdots & 1 \end{vmatrix} \qquad (8.17)$$

is regular; i.e., $(n + 1)$ points a_j are not contained in a hyperplane. In other words, K is n-simplex if

$$K = \left\{ x = \sum_{j=1}^{n+1} \lambda_j a_j / 0 \le \lambda_j \le 1, 1 \le j \le n + 1, \sum_{j=1}^{n+1} \lambda_j = 1 \right\}$$

Remark 8.5 A. 2-simplex is a triangle.
B. 3-simplex is a tetrahedron.

Definition 8.3 The barycentric coordinates $\lambda_j = \lambda_j(x)$, $1 \le j \le n + 1$, of any point $x \in R^n$, with respect to the $(n + 1)$ points a_j, are defined to be the unique solution of the linear system

$$\sum_{j=1}^{n+1} a_{ij} \lambda_j = x_i, \ 1 \le i \le n \qquad (8.18)$$

$$\sum_{j=1}^{n+1} \lambda_j = 1 \qquad (8.19)$$

whose matrix is precisely the matrix A given in (8.17).

Remark 8.6 (i) If K is a triangle with vertices a_1, a_2, a_3, and a_{ij}, $j = 1, 2$, are the coordinates of a_i, $i = 1, 2, 3$, then for any $x \in R^2$, the *barycentric coordinates* $\lambda_i(x)$, $i = 1, 2, 3$, of x will be the unique solution of the linear system

$$\sum_{i=1}^{3} \lambda_j a_{ij} = x_j, \ j = 1, 2$$

$$\sum_{j=1}^{3} \lambda_i = 1 \qquad (8.20)$$

(ii) The barycentric coordinates are affine functions of $x_1, x_2, \ldots, x_n$; i.e., they belong to the space P_1 (the space of all polynomials of degree 1).

$$\lambda_i = \sum_{j=1}^{n} b_{ij} x_j + b_{in+1}, \ 1 \le i \le n+1 \qquad (8.21)$$

where the matrix $B = (b_{ij})$ is the inverse of matrix A.

(iii) The barycentric or center of gravity of an n-simplex K is that point of K all of whose barycentric coordinates are equal to $1/(n+1)$.

Example 8.1 Let $n = 2$, then K is a triangle. Let a_1, a_2, a_3 be its vertices. The barycentric coordinates of a_1, a_2, a_3 are $\lambda_1 = (1, 0, 0), \lambda_2 = (0, 1, 0)$, and $\lambda_3 = (0, 0, 1)$, respectively. The barycentric coordinates of the centroid G of K are $(1/3, 1/3, 1/3)$.

Remark 8.7 Using Cramers rule, we determine from Eq. (8.20) that

$$\lambda_1 = \frac{\begin{vmatrix} x_1 & a_{21} & a_{31} \\ x_2 & a_{22} & a_{32} \\ 1 & 1 & .1 \end{vmatrix}}{\begin{vmatrix} a_{11} & a_{21} & a_{31} \\ a_{12} & a_{22} & a_{32} \\ 1 & 1 & 1 \end{vmatrix}}$$

$$= \frac{area \ of \ the \ triangle \ xa_2a_3}{area \ of \ the \ triangle \ a_1a_2a_3}$$

Similarly,

$$\lambda_2 = \frac{area \ of \ the \ triangle \ a_1xa_3}{area \ of \ the \ triangle \ a_1a_2a_3}$$

$$\lambda_3 = \frac{area \ of \ the \ triangle \ a_1a_2x}{area \ of \ the \ triangle \ a_1a_2a_3}$$

Remark 8.8 A. The equation of the side a_2a_3 in the barycentric coordinates is $\lambda_1 = 0$.

B. The equation of the side a_1a_3 in the barycentric coordinates is $\lambda_2 = 0$.

C. The equation of the side a_1a_2 in the barycentric coordinates is $\lambda_3 = 0$.

Definition 8.4 (*Ciarlet, 1975*) Suppose K is a polyhedron in R^n; P_K is the space of polynomials of dimension m, and $\sum_K$ is a set of distributions with cardinality m. Then the triplex $(K, P_K, \sum_K)$ is called a finite element if

$$\sum_k \{L_i \in D^*/i = 1, 2, \ldots, m\}$$

is such that for a given $\alpha_i \in R, \ 1 \le i \le m$, the system of equations

$$L_i(p) = \alpha_i \ for \ 1 \le i \le m$$

has a unique solution $p \in P_K$. The elements L_i, $i = 1, 2, ..., n$ are called degrees of freedom of P_K.

Remark 8.9   1. If K is n-simplex, $(K, P_K, \sum_K)$ is called the *simplicial finite element*.

2. If K is 2-simplex, i.e., a triangle, then $(K, P_K, \sum_K)$ is called a *triangular finite element*.

3. If K is 3-simplex, i.e., a *tetrahedron*, the finite element $(K, P_K, \sum_K)$ is called *tetrahedral*.

4. If K is a rectangle, then $(K, P_K, \sum_K)$ is called a *rectangular finite element*.

Remark 8.10   1. Very often, $\sum_K$ is considered as the set of values of $p \in P_K$ at the vertices and middle points of the triangle and rectangle in the case of triangular and rectangular elements, respectively.

2. Generally, L_i may be considered the Dirac mass concentrated at the vertices and the middle point and Σ_K may comprise the Dirac mass or Dirac delta distribution and its derivative (for Dirac mass, see Examples 5.21 and 5.26).

Remark 8.11   1. Very often, K itself is called a *finite element*.

2. The triplex $(K, P_1, \{a_i\}_{i=1}^3)$, where K is a triangle, P_1 a space of polynomials of degree ≤ 1, and a_1, a_2, a_3 the vertices of K, is called the *triangular finite element of type (I)*.

3. The triplex $(K, P_1, \{a_i\}_{i=1}^3)$, where K is a rectangle with sides parallel to the axes and a_1, a_2, a_3, and a_4 are corners, is called a *rectangular finite element of type (I)*.

Example 8.2 (Finite Element of Degree 1) Let K be a triangle, $P_K = P_1(K)$ is equal to the space of polynomials of degree less than or equal to 1 and the space generated by $1, x, y = [1, x, y]$.

$$dim\, P_K = 3$$
$$\sum_K = \{\delta_{a_i}/a_i, i = 1, 2, 3 \text{ are vertices}\}$$

where δ_{a_i} is the Dirac mass concentrated at the point a_i. Then

$$\left(K, P_K, \sum_K\right) \text{ is a finite element of degree } 1$$

Remark 8.12   1. $L_i = \delta_{a_i}$ is defined as $\delta_{a_i}(p) = p(a_i)$, $i = 1, 2, 3$. It is a distribution. In order to show that $L_1(p) = \alpha_i$, $1 \leq i \leq 3$, has a unique solution, we are required to show that $p(a_i) = \alpha_i$, $1 \leq i \leq 3$, and $p \in P_1(K)$ has a unique solution. This is equivalent to showing that $p(a_i) = 0$ has a unique solution. It suffices to show that it is unique existence of solution is assured. We know that a polynomial $p \in P_1(K)$ is completely determined if its values at three noncollinear points are given; so $p(a_1) = 0$ has a unique solution.

2. If λ_i are the barycentric coordinates

$$\delta_{a_i}(\lambda_j) = \lambda_j(a_i) = \delta_{ij}$$

Hence, λ_j, $j = 1, 2, 3$, forms a basis for $P_1(K)$, and if $p \in P_1(K)$, then

$$p = \sum_{i=1}^{3} p(a_i)\lambda_i$$

Example 8.3 (*Finite Element of Degree 2*) Let K be a triangle. Then

$$P_K = P_2(K) = [1, x, y, x^2, xy, y^2]$$
$$\sum_K = \{\delta_{a_i}, \delta_{a_{ij}}/1 \leq i \leq 3, 1 \leq i \leq j \leq 3\}$$

where a_i denote the vertices of K and a_{ij} the middle points of the side $a_i a_j$. (K, Σ_K, P_K) is a finite element of degree 2.

Example 8.4 (*Finite Element of Degree 3*) Let K be a triangle

$$P_K = P_3(K) = Span\{1, x, y, x^2, xy, y^2, x^3, x^2 y, xy^2, y^2\}$$
$$dim P_K = 10$$
$$\sum_K = \{\delta_{a_{ii}}, \delta_{a_{ij}}, \delta_{a_{123}}/1 \leq i \leq 3, 1 \leq i \leq j \leq 3\}$$

where a_i denote the vertices of K, $\delta_{a_{123}}$ is the centroid of K and $a_{iij} = \frac{2}{3}a_i + \frac{1}{3}a_j$. Then (K, P_K, Σ_K) is a finite element of degree 3.

Definition 8.5 If $p_j \in P_K$, $1 \leq j \leq m$, is such that

$$L_i(p_j) = \delta_{ij}, \ 1 \leq i \leq m, 1 \leq j \leq m, L_i \in \sum_K$$

then $\{p_j\}$ forms a basis of P_K and any $p \in P_K$ can be written as

$$p = \sum_{i=1}^{m} L_i(p)(p_i)$$

$\{p_j\}$ is called the sequence of *basis functions* of the finite element.

Definition 8.6 Let (K, P_K, Σ_K) be a finite element for each $K \in T_h$, where T_h is a triangulation of a polygonal domain Ω. Let $\sum_K = \bigcup_{K \in T_h} \sum_K$ and

$$H_h = \{v_h/v_h/K \in P_K, \ K \in T_h\}$$

We say that the finite element method is *conforming* if $H_h \subset H^1(\Omega)$. Otherwise, it is called *nonconforming*.

8.3   Applications of the Finite Method in Solving Boundary Value Problems

We explain here how the finite element method can be used to solve boundary value problems. This can be also used to solve problems of Chap. 7.

Example 8.5 Let us illustrate different steps in the finite element method solution of the following one-dimensional two-boundary problem:

$$-\frac{d^2 u}{dx^2} + u = f(x), \quad 0 < x < 1 \tag{8.22}$$

$$u(0) = u(1) = 0 \tag{8.23}$$

where $f(x)$ is a continuous function on $[0, 1]$.

We further assume that f is such that Eq. (8.22) with (8.23) has a unique solution.
Let $H = \{v | v$ is a continuous function on $[0, 1]$ and v' is piecewise continuous and bounded on $[0, 1]$, and $v(0) = v(1) = 0\}$. Multiplying both sides of (8.22) by an arbitrary function $v \in H$ and integrating the left-hand side by parts, we get

$$\int_0^1 \left(\frac{du}{dx} \frac{dv}{dx} + uv \right) dx = \int_0^1 f(x) v(x) dx \tag{8.24}$$

We can write (8.24) as

$$a(u, v) = F(v) \text{ for every } v \in H \tag{8.25}$$

where

$$a(u, v) = \int_0^1 \left(\frac{du}{dx} \frac{dv}{dx} + uv \right) dx \tag{8.26}$$

and

$$F(v) = \int_0^1 f(x) v(x) \, dx \tag{8.27}$$

It can be seen that $a(\cdot, \cdot)$ given by (8.26) is symmetric and bilinear form. It can be shown that finding the solution of (8.25) is equivalent to finding the solution of (8.22) and (8.23).

Now, we discretize the problem in (8.25). We consider here $H_n = \{v_h | v_h$ is continuous piecewise linear function$\}, h = \frac{1}{n}$. Let $0 = x_0 < x_1 < x_2 < \cdots < x_n < x_{n+1} = 1$ be a partition of the interval $[0, 1]$ into subintervals $I_j = [x_{j-1}, x_j]$ of length $h_j = x_j - x_{j-1}$, $j = 1, 2, \ldots, n+1$. With this partition, Hn is associated with the set of all functions $v(x)$ that are continuous on $[0, 1]$ linear on each subinterval I_j, $j = 1, 2, \ldots, n+1$, and satisfy the boundary conditions $v(0) = v(1) = 0$. We define the basis function $\{\varphi_1, \varphi_2, \ldots, \varphi_n\}$ of H_n as follows:

(i) $\varphi_j(x_i) = 1 \ \ if \ i = j$
    $0 \ \ if \ i \neq j$

(ii) $\varphi_j(x)$ is a continuous piecewise linear function.
    $\varphi_j(x)$ can be computed explicitly to yield

$$\varphi_j(x) = \frac{x - x_{j-1}}{h_j} \ \ if \ x_{j-1} \leq x \leq x_j$$
$$\frac{x_{j+1} - x}{h_{j+1}} \ \ if \ x_j \leq x \leq x_{j+1}$$

See Fig. 8.1
    Since $\varphi_1, \varphi_2, \ldots, \varphi_n$ are the basis functions, any $v \in H_n$ can be written as

$$v(x) = \sum_{i=1}^{n} v_i \varphi_i(x), \ where \ v_i = v(x_i)$$

It is clear that $H_n \subset H$. The discrete analogue (8.25) reads: Find $u_n \in H_n$ such that

$$a(u_n, v) = F(v) \ for \ every \ v \in H_n \tag{8.28}$$

Fig. 8.1 Basis function in Example 8.5

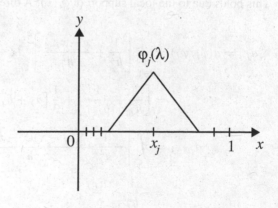

Now, if we choose $u_n = \sum_{i=1}^{n} \alpha_i \varphi_i(x)$ and observe that Eq. (8.28) holds for every function $\varphi_j(x)$, $j = 1, 2, 3, ..., n$, we get n equations, namely

$$a\left(\sum_{i=1}^{n} \alpha_i \varphi_i(x), \varphi_j(x)\right) = F(\varphi_j), \, for \, every \, j = 1, 2, 3, ..., n$$

By the linearity of $a(\cdot, \cdot)$, we get

$$\sum_{i=1}^{n} \alpha_i a(\varphi_i(x), \varphi_j(x)) = F(\varphi_j) \, for \, every \, j = 1, 2, 3, ..., n$$

This can be written in the matrix form

$$AS = (F_n) \qquad\qquad (8.29)$$

where $A = (a_{ij})$ is a symmetric matrix given by

$$a_{ij} = a_{ji} = a(\varphi_i, \varphi_j)$$

$$S = (\alpha_1, \alpha_2, ..., \alpha_n)^T = \begin{bmatrix} \alpha_1 \\ \alpha_2 \\ \cdot \\ \cdot \\ \cdot \\ \alpha_n \end{bmatrix}$$

and $(F_n)_i = F(\varphi_i)$. The entries of the matrix A can be computed explicitly. We first notice that

$$a_{ij} = a_{ji} = a(\varphi_i, \varphi_j) = 0 \text{ if } |i - j| \geq 2$$

This holds due to the local support of $\varphi_i(x)$. A direct computation gives us

$$a_{j,j} = a(\varphi_j, \varphi_j) = \int_{j-1}^{j} \left[\frac{1}{h_j^2} + \frac{x - x_{j-1}}{h_j^2}\right]^2 dx + \int_{j}^{j+1} \left[\frac{1}{h_{j+1}^2} + \frac{x_{j+1} - x}{h_{j+1}^2}\right]^2 dx$$

$$= \left[\frac{1}{h_j} + \frac{1}{h_{j+1}}\right] + \frac{1}{3}\left[h_j + h_{j+1}\right]$$

$$a_{j,j-1} = \int_{j-1}^{j} \left[-\frac{1}{h_j^2} + \frac{(x_j - x)(x - x_{j-1})}{h_j}\right] dx$$

$$= -\frac{1}{h_j} + \frac{h_j}{6}$$

Thus, system (8.29) can be written as

$$
\begin{pmatrix} a_1 & b_1 & \cdots & 0 \\ b_1 & \cdots & \cdots & \cdots \\ \cdots & \cdots & \cdots & \cdots \\ 0 & \cdots & b_{n-1} & a_n \end{pmatrix}
\begin{bmatrix} \alpha_1 \\ \alpha_2 \\ . \\ . \\ . \\ \alpha_n \end{bmatrix}
=
\begin{bmatrix} (F_n)_1 \\ (F_n)_2 \\ . \\ . \\ . \\ (F_n)_n \end{bmatrix}
\tag{8.30}
$$

where $a_j = \frac{1}{h_j} + \frac{1}{h_{j+1}} + \frac{1}{3}[h_j + h_{j+1}]$ and $b_j = -\frac{1}{h_j} + \frac{h_j}{6}$. In the special case of uniform grid $h_j = h = 1/n + 1$, the matrix takes the form

$$
A = \frac{1}{h}
\begin{pmatrix} 2 & -1 & \cdots & \cdots & 0 \\ -1 & 2 & \cdots & \cdots & \cdots \\ \cdots & \cdots & \cdots & \cdots & \cdots \\ \cdots & \cdots & \cdots & \cdots & \cdots \\ 0 & \cdots & \cdots & -1 & 2 \end{pmatrix}
+ \frac{h}{6}
\begin{pmatrix} 4 & 1 & \cdots & 0 \\ 1 & \cdots & \cdots & \cdots \\ \cdots & \cdots & \cdots & 1 \\ 0 & \cdots & 1 & 4 \end{pmatrix}
$$

Example 8.6   Let us consider the Neumann homogeneous boundary problem

$$
\Delta u + u = f \ in \ \Omega, \ \Omega \subset R^2
$$
$$
\frac{\partial u}{\partial n} = 0 \ on \ \Gamma
\tag{8.31}
$$

The variational formulation of Eq. (8.31) is as follows:

$$
H = H^1(\Omega)
$$
$$
a(u, v) = \int_\Omega \left\{ \sum_{i=1}^{2} \frac{\partial u}{\partial x_i} \frac{\partial v}{\partial x_i} + uv \right\} dx
$$
$$
L(v) = \int_\Omega fv \, dx
\tag{8.32}
$$

An internal approximation of this problem is as follows:

$$
H_h = v_h \in C^0(\Omega)/\forall K \in T_h, \ v_h/K \in P_1(K)
\tag{8.33}
$$

This is the result of Theorem 8.4. A basis of H_h is given by the following relation:

$$
w_i(a_j) = \delta_{ij}, \ 1 \le i \le N(h), \ 1 \le j \le N(h)
$$

where $a_1, a_2, ..., a_{N(h)}$ are the vertices of the triangulation T_h. We have

$$\forall v_h \in H_h, \ v_h = \sum_{i=1}^{N(h)} v_h(a_i) w_i$$

The finite element solution of Eq. (8.32) is equivalent to finding u_h such that

$$a(u_h, v_h) = L(v_h) \ \forall \ v_h \in H_h \tag{8.34}$$

Since $\{w_i\}_{i=1}^{N(h)}$ is a basis of H_h, the solution

$$u_h = \sum_{k=1}^{N(h)} \gamma_k w_k \tag{8.35}$$

of Eq. (8.34) is such that the coefficients γ_k are the solutions of the following linear system:

$$\sum_{k=1}^{N(h)} a(w_k, w_\ell) \gamma_k = L(w_\ell) \ \ for \ 1 \leq \ell \leq N(h) \tag{8.36}$$

where

$$a(w_k, w_\ell) = \sum_{K \in T_h} \int_K \left(\sum_{i=1}^{2} \frac{\partial w_k}{\partial x_i} \frac{\partial w_\ell}{\partial x_i} + w_k w_\ell \right) dx \tag{8.37}$$

and

$$L(w_\ell) = \sum_{K \in T_h} \int_K f(w_\ell) \, dx \tag{8.38}$$

Thus, if we know the stiffness matrix $a(w_k, w_\ell)$ and the load vector $L(w_i)$, γ_k can be determined from Eq. (8.36), and putting these values in Eq. (8.35), the solution of Equation (8.34) can be calculated.

Practical Method to Compute
$a(w_j, w_i)$. We have

$$a(u, v) = \int_\Omega (\nabla u . \nabla v + uv)/dx$$

$$= \sum_{K \in T_h} \int_\Omega (\nabla u . \nabla v + uv)/dx \tag{8.39}$$

In the element K, we can write

$$u(x) = \sum_{\alpha=1}^{3} u_{m_{\alpha K}} \lambda_{\alpha}^{K}(x)$$

$$v(x) = \sum_{\beta=1}^{3} v_{m_{\beta K}} \lambda_{\beta}^{K}(x)$$

where $(m_{\alpha K})_{\alpha=1,2,3}$ denotes the three vertices of the element K and $\lambda_{\alpha}^{K}(x)$ the associated barycentric coordinates. Then Eq. (8.39) can be rewritten as

$$a(u, v) = \sum_{K \in T_h} \sum_{\alpha, \beta=1}^{3} u_{m_{\alpha K}} v_{m_{\beta K}} a_{\alpha \beta}^{K}$$

where

$$a_{\alpha \beta}^{K} = \int_{K} (\nabla \lambda_{\alpha}^{K} \cdot \nabla \lambda_{\beta}^{K} + \lambda_{\beta}^{K} \lambda_{\alpha}^{K}) dx$$

The matrix $A^{K} = (a_{\alpha \beta}^{K})_{1 \leq \alpha \leq 3, 1 \leq \beta \leq 3}$ is called the *element stiffness matrix* of K.

8.4 Introduction of Boundary Element Method

In this section, we discuss weighted residual methods, inverse problem, and boundary solutions. We explain here the boundary element method through Laplace equation with Dirichlet and Neumann boundary conditions. We refer to [3, 13, 23, 25, 26, 26, 27, 37, 61, 92, 165, 178] for further details of error estimation, coupling of finite element and boundary element methods, and applications to parabolic and hyperbolic partial differential equations.

8.4.1 Weighted Residuals Method

Let us consider a boundary value problem in terms of operators:

$$Tu = -f \text{ in } \Omega \tag{8.40}$$

$$Su = g \text{ on } \Gamma_1 \tag{8.41}$$

$$Lu = h \text{ on } \Gamma_2 \tag{8.42}$$

where $\Omega \subset R^2$, $\ell = \Gamma_1 + \Gamma_2 = $ boundary of Ω, $T : D_T \subseteq H \rightarrow H$ is a bounded linear self-adjoint operator, and H is a Sobolev space of order 1 or 2 ($H^1(\Omega)$ or $H^2(\Omega)$). Equations (8.40)–(8.42) can be written as

$$Au = f \ in \ \Omega \tag{8.43}$$

If $v \in D_T$ is such that

$$\langle Av - f, \psi_k \rangle = 0 \quad for \ every \ k = 1, 2, 3, \ldots \tag{8.44}$$

where $\{\psi_k\}$ is a basis in H, then $Av - f = 0$ in H; that is, v is a solution of (8.43). Thus, finding a solution of (8.43) is equivalent to finding a solution of (8.44).

This is the essence of the weighted residuals method. We note that an element w is not necessarily represented by the basis $\{\psi_i\}$. Any basis $\{\varphi_i\}$ in D_T can be used to represent w, while the residual

$$R = Aw_N - f \tag{8.45}$$

with

$$w_N = \sum_{i=1}^{N} \alpha_i \varphi_i \tag{8.46}$$

and N is an arbitrary but fixed positive integer and is made orthogonal to the subspace spanned by ψ_k; that is,

$$\langle Aw_N - f, \psi_k \rangle = 0 \quad k = 1, 2, 3, \ldots. \tag{8.47}$$

Equation (8.47) is recognized by different names for different choices of ψ_k. The general method is called *weighted residuals method*. This method is related to finding solution (8.43) in terms of Eq. (8.46) where α_i are determined by (8.47). This leads to N equations for determining N unknowns $\alpha_1, \alpha_2, \ldots, \alpha_N$. For linear A, (8.47) takes the form.

$$\sum_{i=1}^{N} \alpha_i \langle A\varphi_i, \psi_k \rangle = \langle f, \psi_k \rangle, \quad k = 1, 2, 3, \ldots, N \tag{8.48}$$

$\varphi_i \in D_A$ means that φ_i are differentiable $2m$ times provided A is a differentiable operator of order $2m$, and satisfies the specified boundary conditions. For more details, see Finlayson [75].

Let ψ_i be Dirac delta functions. Then $\langle R, w \rangle = \int_{\Omega} Rw \, d\Omega = 0$, where

$$w = \beta_1 \delta_1 + \beta_1 \delta_2 + \beta_3 \delta_3 + \cdots + \beta_n \delta_n, \ \delta_i, \ i = 1, 2, \ldots, n$$

are Dirac delta functions.

In this framework, the weighted residuals method is called the *collocation method* (see, for example, [27, 61, 157]).

8.4.2 *Boundary Solutions and Inverse Problem*

We look for functions satisfying the boundary conditions in the weighted residual method but do not satisfy exactly the governing equations. On the other hand, namely, one looks for functions satisfying exactly the governing equation and approximately satisfying the boundary conditions. Let us explain the ideas with Laplace equation.

$$T = \Delta = \text{ Laplace's operator} \tag{8.49}$$

$$Su = u \text{ on } \Gamma_1 \tag{8.50}$$

$$Lu = \frac{\partial u}{\partial n} \text{ on } \Gamma_2 \tag{8.51}$$

$$\int_\Omega (\Delta u - b) w \, d\Omega = 0 \tag{8.52}$$

We have

$$\int_\Omega \sum_{k=1}^n \left(\frac{\partial u}{\partial x_k} \frac{\partial w}{\partial x_k} + bw \right) d\Omega = \int_\Gamma qw \, d\Gamma \tag{8.53}$$

where $q = \frac{\partial u}{\partial n}$ by the integration by parts.
Integrating (8.53), we get

$$\int_\Omega (\Delta w) u \, d\Omega = \int_\Gamma u \frac{\partial w}{\partial n} d\Gamma - \int_\Gamma qw \, d\Gamma + \int_\Omega bw \, d\Omega. \tag{8.54}$$

By introducing the boundary conditions

$$Su = u = \bar{u} \quad \text{on } \Gamma_1$$

$$Lu = q = \bar{q} \quad \text{on } \Gamma_2$$

Equation (8.54) can be written as

$$\int_\Omega (\Delta w) u \, d\Omega = \int_{\Gamma_1} \bar{u} \frac{\partial w}{\partial n} d\Gamma + \int_{\Gamma_2} u \frac{\partial w}{\partial n} d\Gamma$$

$$- \int_{\Gamma_1} qw \, d\Gamma + \int_{\Gamma_2} \bar{q} w \, d\Gamma + \int_\Omega bw \, d\Omega \tag{8.55}$$

By the boundary conditions and integrating (8.55) twice, we obtain

$$\int_\Omega (\Delta u - b) w \, d\Omega = \int_{\Gamma_2} (q - \bar{q}) w \, d\Gamma - \int_{\Gamma_1} (u - \bar{u}) \frac{\partial w}{\partial n} d\Gamma \tag{8.56}$$

Equation (8.56) can be written in terms of three errors or residuals as

$$\int_\Omega Rw \, d\Omega = \int_{\Gamma_2} R_2 w \, d\Gamma - \int_{\Gamma_1} R_1 \frac{\partial w}{\partial n} \, d\Gamma, \tag{8.57}$$

where

$$R = \Delta u - b, \quad R_1 = u - \bar{u}, \quad R_2 = q - \bar{q}.$$

For homogeneous Laplace equation $\Delta u = 0$ on Ω, Eq. (8.56) takes the form

$$\int_\Omega (\Delta u)w \, d\Omega = \int_{\Gamma_2} (q - \bar{q})w \, d\Gamma - \int_{\Gamma_1} (u - \bar{u}) \frac{\partial w}{\partial n} \, d\Gamma \tag{8.58}$$

as $b = 0$. As $\Delta u = 0$, we have to satisfy

$$\int_{\Gamma_2} (q - \bar{q})w \, d\Gamma = \int_{\Gamma_1} (u - \bar{u}) \frac{\partial w}{\partial n} \, d\Gamma \tag{8.59}$$

We obtain for $u = w$ a method known as the *method of Trefftz*.

By Green's theorem

$$\int_\Omega (u\Delta w - w(\Delta u)) \, d\Omega = \int_\Gamma \left(\frac{\partial u}{\partial n}w - u\frac{\partial w}{\partial n} \right) d\Gamma \tag{8.60}$$

where $\Gamma = \Gamma_1 + \Gamma_2$.

By choosing u and w as the same function, we have $\Delta u = \Delta w = 0$ and can write $w = \delta u$. In view of this, (8.60) becomes

$$\int_\Gamma \frac{\partial u}{\partial n} \delta u \, d\Gamma = \int_\Gamma u\frac{\partial \delta u}{\partial n} \, d\Gamma \tag{8.61}$$

Equation (8.61) reduces to

$$\int_{\Gamma_1} q\delta u \, d\Gamma + \int_{\Gamma_2} \bar{q}\delta u \, d\Gamma = \int_{\Gamma_1} \bar{u}\frac{\partial \delta u}{\partial n}d\Gamma + \int_{\Gamma_2} u\frac{\partial \delta u}{\partial n} \, d\Gamma \tag{8.62}$$

by applying boundary conditions.

8.4.3 Boundary Element Method

Boundary element schemes are related to inverse relationship (8.55). For the weighting function, one uses a set of basis functions which eliminates the domain integrals and reduces the problem to evaluating integrals on boundary. Steps involved in the boundary element method:

(i) Converting boundary value problem into the boundary integral equation.
(ii) Discretization of the boundary into a series of elements over which the potential and its normal derivatives are supposed to vary according to interpolation functions. Elements could be straight line, circular areas, parabolas, etc.
(iii) By the collocation method, the discretized equation is applied to a number of particular nodes within each element where values of the potential and its normal derivatives are associated.
(iv) Evaluation of the integrals over each element by normally using a numerical quadrature scheme.
(v) To derive a system of linear algebraic equations imposing the prescribed boundary conditions and to find its solution by direct or iterative methods.
(vi) Finding u at the internal points of given domain.

We illustrate these steps with the help of the following example: Let us consider the boundary value problem

$$\begin{cases} \Delta u(x) = 0 \; in \; \Omega \\ u(x) = \bar{u} \; on \; \Gamma_1 \\ \frac{\partial u}{\partial n} = \overline{q(x)} \; on \; \Gamma_2 \end{cases} \tag{8.63}$$

where $\Gamma = \Gamma_1 + \Gamma_2$, $\Gamma =$ boundary of the region $\Omega \subset R^2$. The weighted residual Equation (8.56) takes the following form for this boundary value problem:

$$\int_\Omega \Delta u(x) \, u^\star(x, y) \, d\Omega(x) = \\ \left. \begin{array}{l} \int_{\Gamma_2} (q(x) - \bar{q}) u^\star(x, y) \, d\Gamma \\ - \int_{\Gamma_1} (u(x) - \bar{u}) q^\star(x, y) \, d\Gamma \end{array} \right\} \tag{8.64}$$

where $u^\star(\cdot, \cdot)$ is interpreted as the weighting function and

$$q^\star(x, y) = \frac{\partial u^\star(x, y)}{\partial n} \tag{8.65}$$

Integrating by parts (8.64) with respect to x_i, we have

$$-\int_{\Omega} \frac{\partial u(x)}{\partial x_i} \frac{\partial u^\star(x, y)}{\partial x_i} \, d\Omega(x) =$$

$$\left.\begin{array}{c} \int_{\Gamma_1} q(x)u^\star(x, y)d\Gamma(x) \\ -\int_{\Gamma_2} \bar{q}(x)u^\star(x, y)\, d\Gamma(x) \\ -\int_{\Gamma_1} [u(x) - \bar{u}(x)]q^\star(x, y)d\Gamma(x) \end{array}\right\} \quad (8.66)$$

where $i = 1, 2$ and Einstein's summation convention is followed for repeated indices. Using integration by parts once more, we get

$$\int_{\Omega} \Delta u^\star(x, y)u(x) \, d\Omega(x) = -\int_{\Gamma} q(u)u^\star(x, y) \, d\Gamma(x)$$

$$+ \int_{\Gamma} u(x)q^\star(x, y)d\Gamma(x) \quad (8.67)$$

keeping in mind that $\Gamma = \Gamma_1 + \Gamma_2$.

Recalling the following properties of the Dirac delta function $\delta(x, y)$:

$$\left.\begin{array}{ll} \delta(x, y) = 0, & \text{if } x \neq y \\ \quad\quad\; = \infty, & \text{if } x = y \\ \int_{\Omega} u(x)\delta(x, y) \, d\Omega(x) = u(y) \end{array}\right\} \quad (8.68)$$

assuming $u^\star(x, y)$ to be the fundamental solution of two-dimensional Poisson's equation; namely

$$\Delta u^\star(x, y) = -2\pi\delta(x, y) \quad (8.69)$$

and putting the value of $\Delta u^\star(x, y)$ from (8.69) into (8.67), we have

$$2\pi u(y) + \int_{\Gamma} u(x)q^\star(x, y) \, d\Gamma(x) = \int_{\Gamma} q(x)u^\star(x, y) \, d\Gamma(x) \quad (8.70)$$

Considering the point y to be on the boundary and accounting for the jump of the left hand, (8.70) gives the integral equation on the boundary of the given domain Ω (boundary integral equation)

$$c(y)u(y) + \int_{\Gamma} u(x)q^\star(x, y) \, d\Gamma(x) = \int_{\Gamma} q(x)u^\star(x, y) \, d\Gamma(x) \quad (8.71)$$

Fig. 8.2 Constant boundary
elements

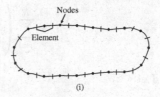

Fig. 8.3 Linear boundary
elements

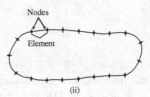

Remark 8.13 (i) For a Neumann boundary problem, we are required to solve a
Fredholm equation of the second kind.

(ii) For a Dirichlet boundary problem, we are required to solve a Fredholm equation
of first kind in unknown $q(x) = \frac{\partial u}{\partial n_i}$ (normal derivative).

(iii) For Cauchy boundary problem, we are required to solve a mixed integral
equation for the unknown boundary data. Equation (8.71) can be discretized
into a large number of elements; see Figs. 8.2 and 8.3.

For the constant element case, the boundary is discretized into N elements. Let
N_1 belong to Γ_1 and N_2 to Γ_2, where the values of u and q are taken to be constant
on each element and equal to the value at the mid node of the element. We observe
that in each element the value of one of the two variables u or q is known. Equation
(8.71) can be converted to (8.72).

$$c_i u_i + \int_\Gamma u q^\star \, d\Gamma = \int_\Gamma q u^\star \, d\Gamma \tag{8.72}$$

where $u^\star = \frac{1}{2\pi} \log\left(\frac{1}{r}\right)$. Here, we have chosen $u_i(y) = u_i$ and $c(y) = c_i$.

Equation (8.72) can be discretized as follows:

$$c_i u_i + \sum_{j=1}^{N} \int_{\Gamma_j} u q^\star \, d\Gamma = \sum_{j=1}^{N} \int_{\Gamma_j} u^\star q \, d\Gamma \tag{8.73}$$

It can be checked that c_i is $\frac{1}{2}$ as for a constant element, and the boundary element
is always smooth. (8.73) is the discrete form of the relationship between node i at
which the fundamental solution is applied and all the j elements, including the case
$i = j$, on the boundary. The values of u and q inside the integrals in Eq. (8.73) are
constants at one element and, consequently,

$$\frac{1}{2}u_i + \sum_{j=1}^{N}\left(\int_{\Gamma_j} q^\star \, d\Gamma\right)u_j = \sum_{j=1}^{N}\left(\int_{\Gamma_j} u^\star \, d\Gamma\right)q_j \tag{8.74}$$

Let

$$\bar{H}_{ij} = \int_{\Gamma_j} q^\star \, d\Gamma, \text{ and } G_{ij} = \int_{\Gamma_j} u^\star \, d\Gamma \tag{8.75}$$

(A symbol involving i and j is used to indicate that the integrals $\int_{\Gamma_j} q^\star \, d\Gamma$ relate the ith node with the element j over which the integral is taken).

Equation (8.74) takes the form

$$\frac{1}{2}u_i + \sum_{j=1}^{N}\bar{H}_{ij}u_j = \sum_{j=1}^{N}G_{ij}q_j \tag{8.76}$$

Here, the integrals in (8.75) are simple and can be evaluated analytically but, in general, numerical techniques will be employed. Let us define

$$H_{ij} = \begin{cases} \bar{H}_{ij}, & \text{if } i \neq j \\ \bar{H}_{ij} + \frac{1}{2}, & \text{if } i = j \end{cases} \tag{8.77}$$

then (8.76) can be written as

$$\sum_{j=1}^{N}H_{ij}u_j = \sum_{j=1}^{N}G_{ij}q_j$$

which can be expressed in the form of matrix equation as

$$AX = F \tag{8.78}$$

where X is the vector of unknown u's and q's, and A is a matrix of order N. Potentials and fluxes at any point are given by

$$u_i = \int_{\Gamma} qu^\star \, d\Gamma - \int_{\Gamma} uq^\star \, d\Gamma \tag{8.79}$$

Equation (8.79) represents the integral relationship between an internal point, the boundary values of u and q and its discretized form is

$$u_i = \sum_{j=1}^{N} q_j G_{ij} - \sum_{j=1}^{N} u_i \overline{H}_{ij} \tag{8.80}$$

The values of internal fluxes can be determined by Eq.(8.79) which gives us

$$\frac{\partial u}{\partial x_i} = \int_\Gamma \int_\Gamma q \frac{\partial u^\star}{\partial x_l} d\Gamma - \int_\Gamma \int_\Gamma u \frac{\partial q^\star}{\partial x_1} d\Gamma \tag{8.81}$$

where x_l are the coordinates, $l = 1, 2$.

Remark 8.14 (i) The values of $u(x)$ at any internal point of the domain are determined by Eq. (8.71).

(ii) The values of derivatives of u, at any internal point y with Cartesian coordinates $x_i(y)$, $i = 1, 2$, can be determined from the equation

$$\frac{\partial u(y)}{\partial x_i(y)} = \frac{1}{2\alpha\pi} \left\{ \begin{array}{l} \int_\Gamma q(x) \frac{\partial u^\star(x,y)}{\partial x_i(y)} d\Gamma(x) \\ -\int_\Gamma u(x) \frac{\partial q^\star(x,y)}{\partial x_i(y)} d\Gamma(x) \end{array} \right\} \tag{8.82}$$

Differentiating (8.70), we get (8.82) by applying appropriate conditions.

(iii) A significant advantage of the boundary element method is the relaxation of the condition that the boundary surface is smooth (Lyapunov); that is, it can be used for surfaces having corners or edges.

Remark 8.15 $\overline{H}_{ij}$ and G_{ij} can be determined by simple Gauss quadrature values for all elements (except one node under consideration) as follows:

$$\overline{H}_{ij} = \int_{\Gamma_j} q^\star d\Gamma = \frac{l_j}{2} \sum_{k=1}^{n} q_k^\star w_k \tag{8.83}$$

and

$$G_{ij} = \int_{\Gamma_j} u^\star d\Gamma = \frac{l_j}{2} \sum_{k=1}^{n} u_k^\star w_k \tag{8.84}$$

where l_j is the element length and w_k is the weight associated with the numerical integration point k.

Example 8.7 Solve

$$\left. \begin{array}{l} \frac{d^2 u}{dx^2} + u = -x \ in \ \Omega, \ \Omega = (0, 1) \\ u(0) = u(1) = 0 \end{array} \right\} \tag{8.85}$$

(by the boundary value problem)

Since $\Delta w = \frac{d^2 w}{dx^2} + w$ and $b = -x$, by Eq. (8.55), we get

$$\int_0^1 \left(\frac{d^2 w}{dx^2} + w \right) u \, dx + \int_0^1 xw \, dx + [qw]_{x=0}^{x=1} - \left[\bar{u} \frac{dw}{dx} \right]_{x=0}^{x=1} = 0 \quad (8.86)$$

In boundary element method, one can choose the fundamental solution as the weighting function w. Denote this solution by $w^\star$, to emphasize its special character and is the solution of

$$\frac{d^2 w^\star}{dx^2} + w^\star = \delta_i \quad (8.87)$$

where δ_i indicates a Dirac delta function which is different from zero at the point i of coordinate ξ. The Dirac delta function is such that

$$\int_0^1 \left(\frac{d^2 w^\star}{dx^2} + w^\star \right) u \, dx = \int_0^1 \delta_i u \, dx = u_i \quad (8.88)$$

In view of this and taking into account $\bar{u} = 0$, we get

$$u_i = -\int_0^1 xw^\star \, dx - [qw^\star]_{x=0}^{x=1} \quad (8.89)$$

The function $w^\star$ which satisfies

$$\int_0^1 \left(\frac{d^2 u}{dx^2} + u + x \right) w \, dx + \left[(u - \bar{u}) \frac{dw}{dx} \right]_{x=0}^{x=1} = 0 \quad (8.90)$$

or (8.88) is $w^\star = \frac{1}{2} \sin r$, where $r = |x - \xi|$.

Putting the value of $w^\star$ into Eq. (8.89) to obtain a system of equations (one at $x = 0$; the other at $x = 1$) from which the two values of q at $x = 0$, $x = 1$ can be found. That is,

$$q_0 = \frac{1}{\sin 1} - 1, \quad q_1 = \frac{\cos 1}{\sin 1} - 1$$

Equation (8.89) can be used to calculate the values of the u function at any internal point. If we choose ξ to be at the midpoint of our internal domain, we obtain the value of u at $x = \frac{1}{2}$ as

$$u\left(\frac{1}{2}\right) = -\frac{1}{2}\int_0^1 x \sin\left(\frac{1}{2} - x\right) dx - \frac{1}{2}\int_{1/2}^1 x \sin\left(x - \frac{1}{2}\right) dx$$

$$- q_1 \sin\frac{1}{2} + q_0 \sin\frac{1}{2}$$

in which

$$r = \begin{cases} \frac{1}{2} - x & \text{for } 0 < x < \frac{1}{2} \\ x - \frac{1}{2} & \text{for } \frac{1}{2} < x < 1 \end{cases}$$

Thus, we obtain

$$u\left(\frac{1}{2}\right) = -\frac{1}{2}[(\cos 1 - 1)/\sin 1] \sin\frac{1}{2} - \frac{1}{2}\left(\frac{1}{2} - \sin\frac{1}{2}\right) - \frac{1}{2}\left(\frac{1}{2} + \sin\frac{1}{2} - \cos\frac{1}{2}\right)$$

$$= 0.06974694$$

which is the same as the exact solution; that is,

$$u\left(\frac{1}{2}\right) = \frac{\sin\frac{1}{2}}{\sin 1} - 1 = 0.06974694$$

8.5 Problems

Problem 8.1 Illustrate application of the finite element method to solve the following boundary value problem

$$-\frac{d^2y}{dx^2} = 1, \ 0 \le x \le 1$$

with $y(0) = 0$, $y'(0) = 0$.

Problem 8.2 Find numerical solution using the finite element method:

$$-\frac{d^2u}{dx^2} + u(x) = f(x), \ 0 < x < 1$$

$$u(0) = u(1) = 0$$

Problem 8.3 Discuss the error estimation between the exact and finite element solution of the following differential equation:

$$-\frac{d^2u}{dx^2} = 4, \ 0 < x < 1$$

$$u(0) = u(1) = 0$$

Problem 8.4 Let $0 = x_0 < x_1 < x_2 < ... < x_n = 1$ be a partition of $[0, 1]$ and V the vector space of functions v such that

(i) $v \in C^0([0, 1])$.
(ii) $v/[x_{i-1}, x_i]$ is a linear polynomial, $i = 1, ..., n$.
(iii) $v(0) = 0$.

Let

$$H = \left\{ v \in L_2(0, 1) / \int_0^1 \left(\frac{dv}{dx} \right)^2 dx < \infty \text{ and } v(0) = 0 \right\}.$$

Let for each $i = 1, 2, ..., n$, φ_i be defined as $\varphi_i(x_j) = \delta_{ij} =$ Kronecker delta. Let interpolant $v_I \in V$ for $v \in C^0[0, 1]$ be defined as

$$V_I = \sum_{i=1}^n v(x_i) \varphi_i.$$

Show that $||u - u_I|| \leq Ch \left|\left| \frac{d^2 u}{dx^2} \right|\right|$ for all $u \in H$, where $h = \max_{1 \leq i \leq n} (x_i - x_{i-1})$, u_I is the interpolant of u and C is independent of h and u.

Problem 8.5 Let H and V be two Hilbert spaces. Let P denote the problem of finding $u \in H$ such that

$$A(u, v) = F(v) \ \forall \ v \in V$$

where $A : H \times V \to R$ and $F \in V^*$, dual space of V. Prove that P has a unique solution $u \in H$ provided there exist constants α and β such that

(a) $|A(u, v)| \leq \beta ||u|| \ ||v|| \ for \ all \ u \in H$ and $v \in V$.
(b) $\sup_{u \in V, \ u \neq 0} \frac{A(u,v)}{||v||_V} \geq \alpha ||u||_H \ for \ all \ u \in H$.
(c) $\sup_{u \in H} A(u, v) > 0 \ for \ all \ v \in V, v \neq 0$.

Furthermore, $||u||_H \leq \frac{||F||_V}{\alpha}$.

Problem 8.6 Formulate a theorem analogous to Theorem 8.1 for the bilinear form considered in Problem 8.5.

Problem 8.7 Find the error between the exact solution and a finite element solution of the following boundary value problem:

$$-\frac{d}{dx} \left[(1+x)\frac{du}{dx} \right] = 0, \ 0 < x < 1$$

$$u(0) = 0, \ \left[(1+x)\frac{du}{dx} \right]_{x=1} = 1$$

Problem 8.8 Solve the following BVP:

$$-\Delta u = 2(x + y) - 4$$

in the square with vertices $(0, 0)$, $(1, 0)$, $(1, 1)$, $(0, 1)$, where the boundary conditions are

$$u(0, y) = y^2, \; u(x, 0) = x^2, \; u(1, y) = 1 - y, \; u(x, 1) = 1 - x$$

Problem 8.9 Solve the Poisson equation

$$\frac{d^2 u}{dx^2} + \frac{d^2 u}{dy^2} = 4$$

with boundary conditions $u = 0$ on $\Gamma (x = \pm 1, y = \pm 1)$ by the boundary element method.

Chapter 9
Variational Inequalities and Applications

Abstract In this chapter, we discuss mathematical models of real-world problems known as variational inequalities introduced systematically by J.L. Lions and S. Stampachia in early seventies. Modeling, discretization, algorithms, and visualization of solutions along with updated references are presented.

Keywords Signorini contact problems · Complementarily problem · Rigid punch problem · Lions–Stampacchia theorem · Minti lemma · Ritz–Galerkin method · Finite element methods for variational inequalities · Parallel algorithm · Obstacle problem · Membrane problem

9.1 Motivation and Historical Remarks

The study of variational inequalities was systematically initiated in the early sixties by J.L. Lions and G. Stampacchia. We have seen in Chap. 7 that boundary value problems can be expressed in terms of operator equations. However, in many situations, boundary conditions are uncertain and such problems are expressed in terms of operator inequalities. These problems occur in well-known branches of physics, engineering, economics, business, and trade.

This development provided solutions to several problems of science, technology, and finance earlier thought to be inaccessible. The beauty of this study lies in the fact that an unknown part of an elastic body that is in contact with a rigid body becomes an intrinsic part of the solution and no special technique is required to locate it. In short variational inequalities model free boundary value problems. The techniques of variational inequalities are based on tools, and techniques of functional analysis discussed in this book. Methods of variational inequalities have been applied to modeling, error estimation, and visualization of industrial problems.

9.1.1 Contact Problem (Signorini Problem)

In practically every structural and mechanical system, there is a situation in which a deformable body contact with another body comes in. The contact of two

© Springer Nature Singapore Pte Ltd. 2018

A. H. Siddiqi, *Functional Analysis and Applications*, Industrial and Applied Mathematics, https://doi.org/10.1007/978-981-10-3725-2_9

bodies tells how loads are distributed into a structure where structures are supported to sustain loads. Thus, properties of contact may play a significant role in the behavior of the structure related to deformation, motion, and the distribution of stresses, etc.

In 1933, Signorini studied the general equilibrium problem of a linear elastic body in contact with a rigid frictionless body. He himself refined his results in 1959 by considering a rigid body a base on which an elastic body is placed. In 1963, Fichera presented a systematic study by formulating the problem in the form of an inequality involving an operator or bilinear form and a functional. He proved an existence theorem for a convex subset of a Hilbert space. In 1967, Lions and Stampacchia presented an existence theorem of variational inequalities and proved an existence theorem for nonsymmetric bilinear form. Modeling important phenomena of physics and engineering formulated in terms of variational inequalities was presented in a book form by Duvaut and Lions [DuLi 72]. This is one of the best source for applied research in this area.

Mosco and Bensoussan introduced several generalization of variational inequalities in seventies. Their results and other related results are presented in Mosco [139], Bensoussan and Lions [17, 18], Baiocchi and Capelo [9], and Kinderlehrer and Stampacchia [112]. The book of Kinderlehrer and Stampacchia presents interesting fundamental results and applications to free boundary problems of lubrication, filtration of a liquid through porous media, deflection of an elastic beam, and Stefan problem. Applications of finite element methods for numerical solutions of variational inequalities were presented in book form by Glowinski, Lions and Trémoliéres [85], and Glowinski [83]. A comprehensive study of solid mechanics through variational inequalities was presented by Jayme Mason [105]. Applications of the theory of variational inequalities to the dam problem were systematically presented by Baiocchi and Capelo [9]. During the same period, the book of Chipot [38] entitled variational inequalities and Flow in Porous Media appeared which treated the obstacle and the dam problems. A monograph on the study of contact problems in elasticity through variational inequalities and the finite element methods (discretization, error estimation, and convergence of approximate solution to the variational solution of the problem) was written by Kikuchi and Oden [111]. See also [53, 68].

9.1.2   Modeling in Social, Financial and Management Sciences

Complementarity problem is a well-known topic of operations research. This includes as special cases linear and quadratic programming. Cottle, Karmardian, and Lions noted in 1980 that the problems of variational inequality are a generalization of the complementarity problem. Interesting results have been obtained on finite-dimensional spaces like R^n; see [81, 82]. It was also noted that the equilibrium conditions of the traffic management problem had the structure of a variational inequality. This provided a study of more general equilibrium systems. Stella Deformas identified network equilibrium conditions with a variational inequality problem. Since the equilibrium theory is very well established in economics, game theory, and other areas of social sciences, the link of variational inequality could be established with all those themes.

Stella and Anna Nagurney jointly developed a general multinetwork equilibrium model with elastic demands and a asymmetric interactions formulating the equilibrium conditions as a variational inequality problem. They studied a projection method for computing the equilibrium pattern. Their work includes a number of traffic network equilibrium models, spatial price equilibrium models, and general economic equilibrium problems with production. There are series of papers on successful application of variational inequality theory in the setting of finite-dimensional spaces to qualitative analysis and computation of perfectly and imperfectly competitive problems, for example [68, 141, 142]. Studied problems are: (i) dynamical evolution of competitive systems underlying many equilibrium problems and (ii) existence, uniqueness, and iterative methods. It has been shown that American option pricing of financial mathematics and engineering is modelled by an evolution variational inequality [198]. A refined numerical method is presented in the Ref. [SiMaKo 00].

9.2 Variational Inequalities and Their Relationship with Other Problems

9.2.1 Classes of Variational Inequalities

Let H be a real Hilbert space with the inner product $\langle \cdot, \cdot \rangle$. Let $|| \cdot ||$ denote the norm induced by $\langle \cdot, \cdot \rangle$ and let $(\cdot, \cdot)$ denote the duality between H and H^*, where H^* is the dual of H. In case of the real Hilbert space, $\langle \cdot, \cdot \rangle = (\cdot, \cdot)$. Let K be a closed convex subset of H. Assume that T, S, A are nonlinear operators on H into H^* and $a(\cdot, \cdot)$ is bilinear form on H. Let F be a fixed element of H^*. The following forms of variational inequalities are used to model problems of different areas:

Model 1 [Stampacchia–Lions] Find $u \in K$ such that

$$a(u, v - u) \geq \langle F, v - u \rangle \ \ for \ all \ v \in K \tag{9.1}$$

where $a(\cdot, \cdot)$ is a bilinear form on $H \times H$ into R.

Model 2 [Stampacchia–Hartman–Browder] Find $u \in K$ such that

$$\langle Tu, v - u \rangle \geq \langle F, v - u \rangle \ for \ all \ v \in K \tag{9.2}$$

Model 3 [Duvaut–Lions] Find $u \in K$ such that

$$a(u, v - u) + j(v) - j(u) \geq \langle F, v - u \rangle \ for \ all \ v \in K \tag{9.3}$$

where $j(\cdot)$ is an appropriate real-valued function.

Model 4 Find $u \in K$ such that

$$\langle Tu, v - u \rangle \geq \langle Au, v - u \rangle \ \ for \ all \ v \in K \tag{9.4}$$

Model 5 Find $u \in K$ such that

$$(Tu, v - u) + \phi(u, v) - \phi(u, u) \geq (Au, v - u) \ \ for \ all \ v \in K \tag{9.5}$$

where $\phi(\cdot, \cdot) : H \times H \to R$ is an appropriate function.

Model 6 [Bensoussan–Lions] Find $u \in K(u)$, $K : H \to 2H$ such that

$$(Tu, v - u) \geq \langle Fu, v - u \rangle \ \forall \, v \in K(u) \tag{9.6}$$

This model is known as the *quasi-variational inequality*. In particular, T can be chosen as $\langle Tu, v \rangle = a(u, v)$ and $K(u)$ as $K(u) = m(u) + K$, $m : H \to H$ is a nonlinear map, or $K(u) = \{v \in H / u \leq Mv, M : H \to H$, where $\leq$ is an ordering of H$\}$.

Model 7 Find $u \in K(u)$ such that

$$(Tu, v - u) + \phi(u, v) - \phi(u, u) \geq (Au, v - u) \ \ for \ all \ v \in K(u) \ . \tag{9.7}$$

where $\phi(\cdot, \cdot)$ and $K(u)$ are the same as in Model 5 and Model 6, respectively.

Model 8 Find $u \in K(u)$ such that

$$(Tu, Sv - Su) \geq \langle Au, Su - Sv \rangle \, for \, all \, Sv \in K, \ Su \in K \tag{9.8}$$

It may be observed that the variational inequalities in Models 1 to 8 are called the elliptic variational inequalities.

Model 9 (Parabolic Variational Inequalities) Find $u(t)$, where $u : [0, T] \to H$ and $[0, T]$ is a time interval, $0 < T < \infty$ such that

$$\left\langle \frac{\partial u}{\partial t}, v - u \right\rangle + a(u, v - u) \geq \langle F, v - u \rangle = F(v - u) \forall \, v \in K, and$$

$$t \in (0, T), \ u(t) \in K \text{ for almost all } t \tag{9.9}$$

Here, $F \in L_2(0, T, H^\star) = \{F : [0, T] \to H^\star | F(t) \in L_2\}$, a Hilbert space. For $\frac{\partial u}{\partial t} = 0$, we get the elliptic variational inequality.

Model 10 (Rate-independent Evolution Variational Inequality) Find $u : [0, T] \to H$ such that

$$\left\langle \frac{\partial u}{\partial t}, v - u \right\rangle + a(u, v - u') + j(v) - j(u')$$

$$\geq F(v - u') \, for \, v \in H \tag{9.10}$$

where $u' = \frac{\partial u}{\partial t}$.

For information concerning more general form of inequalities like vectorvalued variational inequalities, simultaneous variational inequalities, generalized quasi-variational inequalities, and implicit variational problems, we refer to Siddiqi, Manchanda, and Kocvara [SiMa 00], [167], Brokate and Siddiqi [31], and Giannessi [82].

9.2.2 Formulation of a Few Problems in Terms of Variational Inequalities

(1) **Minimization of a Single-valued Real Function**

Let $f : [a, b] \rightarrow R$ be a differentiable real-valued function defined on the closed interval of R. We indicate here that finding the point $x_0 \in I = [a, b]$ such that

$$f(x_0) \leq f(x) \ for \ all \ x \in I; \ that \ is$$
$$f(x_0) = inf f(x), \ x \in I$$

is equivalent to finding the solution of a variational inequality. We know that

(a) if $a < x_0 < b$, then $f'(x_0) = 0$
(b) if $x_0 = a$, then $f'(x_0) \geq 0$
(c) if $x_0 = b$, then $f'(x_0) \leq 0$

From this, we see that

$$\langle f'(x_0), x - x_0 \rangle \geq 0 \ for \ all \ x \in [a, b], \ x_0 \in [a, b] \qquad (9.11)$$

that is, x_0 is a solution of Model 1, where $F = 0$, $a(u, v - u) = (f'(u), v - u) = \langle Bu, v - u \rangle$, $B = f' : R \rightarrow R$ linear or x_0 is solution of Model 2, where $F = 0$ and $T = f'$.

(4) **Minimization of a Function of n Variables**

Let f be a differentiable real-valued function defined on the closed convex set K of R^n. We shall show that finding the minima of f, that is, searching the point $x_0 \in K$ such that $f(x_0) = \inf_{x \in X} f(x)$, that is, $f(x_0) \leq f(x)$, for all $x \in K$ is equivalent to finding the point $x_0 \in K$, which is a solution of Model 1 or Model 2. Let $x_0 \in K$ be a point where the minimum is achieved and let $x \in K$. Since K is convex, the segment $(1 - t)x_0 + tx = x_0 + t(x - x_0)$, $0 \leq t \leq 1$, lies in K. $\phi'(t) = f(x_0 + t(x - x_0))$, $0 \leq t \leq 1$, attains its minimum at $t = 0$ as in (1).

$\phi'(0) = \langle grad f(x_0), (x - x_0) \rangle \geq 0$ for any $x \in K$, where $grad f = \nabla f(x) = \left(\frac{\partial f}{\partial x_1}, \frac{\partial f}{\partial x_2}, \ldots, \frac{\partial f}{\partial x_n} \right)$, $x = (x_1, x_2, \ldots, x_n) \in K$. This shows that, $x_0 \in K$ is a solution of the variational inequality, $\langle grad f(x_0), (x - x_0) \rangle \geq 0$ for any $x \in K$, or x_0 solves Model 2 where $u = x_0, v = x$, $T = grad f = \nabla f$, $F = 0$. Thus, if f is a convex function and $u \in K$ is a solution of variational inequality

(Model 2) for $T = grad f = \nabla f$, then

$$f(x) = \inf_{v \in K} f(v)$$

(5) Consider the Following Boundary Value Problem (BVP)

$$-\frac{d^2u}{dx^2} = f(x) \text{ on } (0, 1) \quad (i)$$

$$u(0) \geq 0; \; u(1) \geq 0 \quad (ii)$$

$$\left(\frac{du}{dx}\right)_{x=0} \leq 0 \quad (iii)$$

$$\left(\frac{du}{dx}\right)_{x=1} \geq 0 \quad (iv)$$

$$u(x)\left(\frac{du}{dx}\right)_{x=0} = 0, \; x \in \{0, 1\} \quad (v) \tag{9.12}$$

Let $H = H^1(0, 1)$, $K = \{u \in H^1(0, 1)/u(0) \geq 0, \; u(1) \geq 0\}$. It can be observed that if K is a cone with vertex 0 in Model 1 (the convex set K is called cone with vertex at 0, if for every pair of elements $v, w \in K$, $v+w \in K$, $tv \in K$ holds for every $t \geq 0$), then (9.1) is equivalent to the pair of relations

$$a(u, v) = \langle Bu, v \rangle \geq \langle F, v \rangle a(u, u) = \langle Bu, u \rangle = \langle F, u \rangle \tag{9.13}$$

(B is a bounded linear operator associated with $a(\cdot, \cdot)$ given in the beginning of Sect. 7.3). To verify this, let $Q(u) = B(u) - F$; so (9.12) and (9.13) reduce to

$$\langle Qu, v \rangle \geq 0; \; \langle Qu, u \rangle = 0 \tag{9.14}$$

From Equation (9.14), subtracting the second from the first, $\langle Qu, v - u \rangle \geq 0$ for all $v \in K$. Now if we choose $v = 2u$, which is permissible since K is a cone, we obtain $\langle Qu, u \rangle \geq 0$. By choosing $v = 0$, we have $\langle Qu, u \rangle \leq 0$. By combining these two inequalities, we obtain $\langle Qu, u \rangle = 0$ and, finally adding this to $\langle Qu, v - u \rangle \geq 0$, we get the desired result. We show now that Model 1, where H and K are as chosen above

$$a(u, v) = \langle Bu, v \rangle = \int_0^1 \frac{du}{dx}\frac{dv}{dx}dx \text{ and } \langle F, v \rangle = \int_0^1 f(x)v(x)dx$$

is equivalent to (BVP) (9.11). Model 1 takes the form

$$\int_0^1 \frac{du}{dx}\left[\frac{dv}{dx} - \frac{du}{dx}\right] \geq \int_0^1 f(x)(v(x) - u(x))dx \tag{9.15}$$

for all $v(x) \in K$. It can be checked that K is a cone with vertex 0 and so by the above remark, (9.15) can be rewritten as

$$\int_0^1 \frac{du}{dx}\frac{dv}{dx}dx \geq \int_0^1 f(x)v(x)dx \tag{9.16}$$

$$\int_0^1 \left(\frac{du}{dx}\right)^2 dx = \int_0^1 f(x)u(x)dx \tag{9.17}$$

Assume that the solution u of (9.16) is smoother than strictly required for it to be in $H^1(0, 1)$, for example $u \in C^2(0, 1)$ and $v \in C_0^\infty(0, 1)$. Integrating by parts the left-hand side of (9.16), we find

$$\int_0^1 -\frac{du}{dx}v(x)dx + \left[\left(\frac{du}{dx}\right)_{x=1} v(1) - \left(\frac{du}{dx}\right)_{x=0} v(0)\right]dx$$

$$\geq \int_0^1 f(x)v(x)dx \tag{9.18}$$

Since $v \in K \cap C_0^\infty(0, 1)$ and (9.16) also holds for $(-v)$, we conclude that

$$\int_0^1 \frac{du}{dx}\frac{dv}{dx}dx \leq \int_0^1 f(x)v(x)dx \tag{9.19}$$

This together with (9.16) implies

$$\int_0^1 \frac{du}{dx}\frac{dv}{dx}dx = \int_0^1 f(x)v(x)dx \tag{9.20}$$

An integration by parts here yields (9.17) and, as $v \in C_0^\infty(0, 1)$; that is, $v(0) = v(1)$, the boundary term vanishes and we obtain

$$-\frac{d^2u}{dx^2} = f(x) \; in \; (0, 1)$$

In view of (9.17), (9.18) gives us

$$\left(\frac{du}{dx}\right)_{x=1} v(1) - \left(\frac{du}{dx}\right)_{x=0} v(0) \geq 0, \; for \; u, v \in K \tag{9.21}$$

If we choose $v(1) = 1$, and $v(0) = 0$ in (9.21), we get $\left(\frac{du}{dx}\right)_{x=1} \geq 0$ (BVP) 9.11 (iv)). On the other hand, by choosing $v(1) = 0$, and $v(0) = 1$ in (9.21), we obtain $-\left(\frac{du}{dx}\right)_{x=0} \geq 0$ or $\left(\frac{du}{dx}\right)_{x=0} \leq 0$ which is (BVP) 9.11 (iii).
Since $u \in K$, we have (BVP) 9.11 (ii). (BVP) 9.11 (v) is obtained in a similar way from (9.17). Thus if u is a solution of (9.15), then it is the solution of (BVP) 9.11$(i - v)$. If u is a solution of (BVP) 9.11$(i - v)$, then multiplying both sides

of (BVP) 9.11(i) by $v(x)$ and integrating by parts and using other conditions, we get (9.16) and (9.18). This implies that u is a solution of (9.15).

(6) **Complementarity Problem and Variational Inequality**

The following problem is known as the nonlinear complementarity problem (NCP): Find $u \in R^n$ such that $u \geq 0$, $F(u) \geq 0$ and u, and $F(u)$ are orthogonal, that is, $\langle u, F(u) \rangle = 0$, where $F : R^n \to R^n$. The NCP has been used as a general framework for quadratic programming, linear complementarity problems, mathematical programming, and some equilibrium problems. For details, we refer to Cottle, Pang and Store [53]. It can be easily checked that u is a solution to the NCP if and only if it is a solution of the variational inequality: Find $u \in R^n_+$ such that $\langle F(u), v - u \rangle \geq 0$ for all $v \in R^n_+$, where $R^n_+ = \{v = (v_1, v_2, \ldots, v_n) \in R^n / v_i \geq 0, i = 1, 2, \ldots, n\}$.

Proof Let u be a solution to the NCP. Then
$\langle F(u), v \rangle \geq 0 \; \forall v \in R^n_+$, so

$$\langle F(u), v - u \rangle = \langle F(u), v \rangle - \langle F(u), u \rangle$$
$$= \langle F(u), v \rangle \geq 0 \; because \; \langle F(u), u \rangle = 0$$

Thus, u is a solution of the variational inequality.

To prove the converse, let u be a solution of the variational inequality. Then $v = u + e_i$, $e_i = (0, 0, \ldots, 1, 0, \ldots)$ (1 in the ith place) is an element of R^n_+, so $0 \leq \langle F(u), u + e_i - u \rangle = \langle F(u), e_i \rangle = F_i(u)$ or $F(u) \in R^n_+$. Thus, since $v = 0 \in R^n_+$, $\langle F(u), u \rangle \leq 0$. But $u, F(u) \in R^n_+$ implies that $\langle F(u), u \rangle \geq 0$. Hence, $\langle F(u), u \rangle = 0$; that is u is a solution of the NCP.

(7) **Rigid Punch Problem**

The variational inequality approach for studying the contact problems is well established in the methodology see, for example, Kikuchi and Oden [111], Duvaut and Lions [69], and Brokate and Siddiqi [31, 32] for an extended discussion of the rigid punch problem. In the rigid punch problem, one considers a material body deformed by another body (the punch rigid body of definite shape). If we consider a specific situation as shown in (9.1) for the initial and equilibrium configurations, then the deformation u in the elastic body is a solution of the variational inequality: Find $u \in K$ such that

$$a(u, v - u) \geq \langle F, v - u \rangle \; \forall v \in K$$

where $a(\cdot, \cdot)$ is the usual bilinear form of linear elasticity on a suitable Hilbert space H

$$a(u, v) = \int_\Omega a_{ijkl} \xi_{ij}(\tilde{u}) \xi_{kl}(\tilde{v}) dx$$
$$\xi_{ij}(\tilde{u}) = \frac{1}{2}(\delta_i u_j + \delta_j u_i)$$
$$u = (\tilde{u}, y) = (u_1, u_2, y) \in H$$

The linear elastic material body occupies at rest the domain $\Omega \subset R^2$ and is kept fixed at a part ℓ_0 of its boundary. The punch has one degree of freedom, namely the motion in the vertical direction, and presses from above with a fixed force resulting from its own weight onto the material body. Frictionless contact is assumed. Here, the unknown (the deformation) is $u = (u_1, u_2, y)$, where $u_i : \Omega \to R^2$ and $y \in R$ is the vertical displacement in the downward direction of the rigid punch and

$$\langle F, u \rangle = \int_\Omega f_i u_i dx + Py$$

where $P > 0$ corresponds to the weight of the punch. The constraint set K represents a linearized form of the nonpenetration conditions (i.e., the condition that the punch is always above the elastic body). A general form is

$$K = \{(v_1, v_2, z) | \alpha_i(x)v_i(x) + \beta(x)z \leq \phi(x), \ x \in \ell_c, \ i = 1, 2\}$$

Here, ℓ_c is the part of the upper boundary of Ω, where the contact may possibly occur. The function ϕ denotes the vertical distance (or some modification of it) of the punch from the elastic body in the initial configuration, and the functions α_i, β are connected to the unit normal of the boundary of the punch or of the elastic body. Various forms are possible; for example

$$\eta_i(x)v_i(x) + z \leq \phi(x)$$
$$\eta_i(x)v_i(x) + \eta_2(x)z \leq \eta_2(x)\phi(x)$$
$$v_2(x) + z \leq (x)$$

(8) **Reasonable Price System**

Let $M = \{p = (p_1, p_2, \ldots, p_n) \in R^n : 0 \leq p_i \leq 1\}$ and let N be a compact, convex, nonempty set in R^n. Suppose there are n commodities $C_1, C_2, C_3, \ldots, C_n$. Let p_i be the price of C_i with $0 < p_i < 1$. We assign a set $K(p) \subseteq N$ to each price vector $p \in M$. $q \in K(p)$ with $q = (q_1, q_2, \ldots, q_n)$ means that the difference between the supply and demand for C_i is equal to q_i. The number

$$\langle p, q \rangle = \sum_{i=1}^n p_i q_i$$

is, therefore, equal to the difference between the value of the commodities which are supplied to the market and the value of the commodities which are demanded by the market. Thus,

$$\langle p, q \rangle = 0, \ q \in K(p)$$

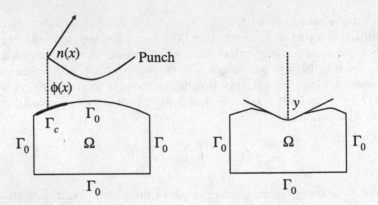

Fig. 9.1 Rigid punch

is the mathematical formulation of the situation "supply equals demand." This ideal situation cannot always be realized. We are therefore satisfied with a weaker condition. First, it is reasonable to assume that

$$\langle p, q \rangle > 0, \text{ for all } p \in M, \ q \in K(p).$$

This is the so-called the law of Walras. Roughly, it means that we only consider economical situations with a supply excess.

We call $(\bar{p}, \bar{q})$ with $\bar{q} \in K(p)$ a Walras equilibrium if and only if the following holds. $0 \leq \langle \bar{p}, \bar{q} \rangle \leq \langle p, \bar{q} \rangle$ for all $p \in M$. Broadly speaking, this means that the value difference between the supply and demand becomes minimal. The vector $\bar{p}$ is called the equilibrium price system. The fundamental problem of mathematical economics is to find conditions for the supply excess map K which ensures the existence of the Walras equilibrium. This is equivalent to finding the solution of Model 6, where $H = R^n$, M = unit cube, $\langle Tu, v \rangle = \langle u, v \rangle$ inner product in R^n. Mosco [139] has given a theorem dealing with the solution of this model (Fig 9.1).

9.3   Elliptic Variational Inequalities

Let H be a Hilbert space and K a nonempty closed convex subset of H. Further, let $a(\cdot, \cdot)$ be a bilinear form on $H \times H$ into R and $F \in H^*$ which is identified with element, say Fu or y of H, by the Reisz representation theorem. Then the Model 1 (variational inequality problem or VIP for short) takes the form: Find $u \in K$ such that

$$a(u, v - u) \geq F(v - u) = \langle y, v - u \rangle \quad for \ all \ v \in K \qquad (9.22)$$

or

$$\langle Au, v - u \rangle \geq F(v - u) = \langle F, v - u \rangle = \langle y, v - u \rangle \qquad (9.23)$$

where $A : H \rightarrow H^\star$, $\langle Au, v \rangle = a(u, v)$, $||A|| = ||a||$. In Model 2, we find $u \in K$ such that

$$\langle Tu, v - u \rangle \geq \langle F, v - u \rangle \ \forall v \in K \qquad (9.24)$$

where $T : H \rightarrow H^\star$ or into itself in the case of real Hilbert space.

In Sect. 9.3.1, we shall study the existence of solution of (9.1) where $a(\cdot, \cdot)$ is bounded and coercive bilinear form (not necessarily symmetric). This is the Lions–Stampacchia theorem. Prior to this theorem, we prove that, for the bilinear bounded coercive and symmetric bilinear form, the optimization problem for the energy functional is equivalent to a variational inequality problem, that is, "Find $u \in K$ such that $J(u) \leq J(v) \forall v \in K$, where $J(v) = \frac{1}{2}a(v, v) - F(v)$ is equivalent to (9.1)."

Section 9.3.2 is devoted to the existence theorem for the VIP (9.1)–(9.3) along with the Ritz–Galerkin and Penalty methods.

9.3.1 Lions–Stampacchia Theorem

The Lions–Stampacchia theorem for symmetric bilinear form follows from Theorem 6.6 and Theorem 9.1 given below.

Theorem 9.1 *Let $J(v) = \frac{1}{2}a(v, v) - F(v)$. If $a(\cdot, \cdot)$ is bilinear bounded coercive and symmetric then the problem of optimization, (P) for $J(\cdot)$ and (VIP) are equivalent. That is to say "Find $u \in K$ such that $J(u) \leq J(v) \forall v \in K$" holds if and only if (9.1) holds, namely there exists $u \in K$ such that $a(u, v - u) \geq F(v - u) \forall v \in K$.*

Proof Let (9.1) hold for all $v \in K$, then we have

$$J(v) = J(u + v - u) = \frac{1}{2}a(u + v - u, u + v - u) - F(u + v - u)$$

$$\Rightarrow J(u + v - u) = \frac{1}{2}a(u, u) + \frac{1}{2}a(v - u, v - u) + \frac{1}{2}a(u, v - u)$$
$$+ a(v - u, u) - F(u) - F(v - u)$$

$$\Rightarrow J(u + v - u) = J(u) + [a(u, v - u) - F(v - u)]$$
$$+ \frac{1}{2}a(v - u, v - u).$$

Since $a(\cdot, \cdot)$ is coercive, the third term is greater than or equal to 0 and the second term is also nonnegative by the assumption that (9.1) holds. Therefore, $J(u) \leq J(v) \forall v \in K$. For the converse, suppose that there exists $u \in K$ such that $J(u) \leq J(v) \forall v \in K$. Therefore,

$$\lim_{t \to 0} \frac{J(u + t(v - u)) - J(u)}{t} \geq 0$$

or the Gâuteax derivative of J in the direction of $(v - u)$ is greater than or equal to 0, that is, $DJ(u)(v - u) \geq 0$.

Applying Theorem 5.3 and Problem 5.3, we get

$$DJ(u)(v - u) = DJ(u)(v - u) = a(u, v - u) - F(v - u) \geq 0$$

$$or \quad a(u, v - u) \geq F(v - u) \, \forall \, v \in K$$

Hence, (9.1) holds.

Theorem 9.2 *If $a(\cdot, \cdot)$ is bilinear bounded and coercive, then VIP (9.1) has a unique solution. Furthermore, the mapping $F \to u$ of H^* into H is Lipschitz continuous, that is, if u_1 and u_2 are solutions corresponding to F_1 and F_2 of (9.1), then $\|u_1 - u_2\| \leq \frac{1}{\alpha}\|F_1 - F_2\|_{H^*}$.*

Proof (1) **Lipschitz Continuity**

Let u_1 and u_2 be two solutions. Then

(a) $a(u_1, v - u_1) \geq F_1(v - u_1)$.
(b) $a(u_2, v - u_2) \geq F_2(v - u_2) = \langle F_2, v - u_2 \rangle$.

Put $v = u_2$ in (a), then we get $a(u_1, u_2 - u_1) \geq F_1(u_2 - u_1)$ or $a(u_1, u_2 - u_1) \leq -F_1(u_2 - u_1)$ by linearity of $a(\cdot, \cdot)$ and $F_1(\cdot)$. Put $v = u_1$ in (b), then we get $a(u_2, u_1 - u_2) \geq F_2(u_1 - u_2)$ or $a(u_2, u_2 - u_1) \leq F_2(u_2 - u_1)$ by linearity of $a(\cdot, \cdot)$ and $F_2(\cdot)$. These two inequalities imply $a(u_1 - u_2, u_1 - u_2) \leq \langle F_1 - F_2, u_1 - u_2 \rangle$. Since $a(\cdot, \cdot)$ is coercive and $F_1 - F_2$ is bounded

$$\alpha \|u_1 - u_2\|^2 \leq \|F_1 - F_2\| \, \|u_1 - u_2\|$$

$$or \quad \|u_1 - u_2\| \leq \frac{1}{\alpha}\|F_1 - F_2\|$$

(3) **Uniqueness of Solution**

Solution is unique for $F_1 = F_2$, $\|u_1 - u_2\| = 0 \Rightarrow u_1 = u_2$, $\Rightarrow$ solution is unique.

(4) **Existence and Uniqueness**

We have $\langle Au, v \rangle = a(u, v)$ by Corollary 3.1, and $F(v) = \langle y, v \rangle$ by the Riesz representation theorem. VIP $\Leftrightarrow \langle Au, v - u \rangle \geq \langle y, v - u \rangle \, \forall \, v \in K \Leftrightarrow \langle \rho Au, v - u \rangle \leq \rho \langle -y, v - u \rangle$ for any $\rho > 0$ or $\langle u \rho (Au - y) - u, v - u \rangle \leq 0 \, \forall \, v \in K$ or find $u \in K$ such that $P_K(u - \rho(Au - y)) = u$ (see Theorem 3.11). Let $T_\rho : H \to H$ defined by $T_\rho(v) = P_K(v - \rho(Av - y))$.

We show that T_ρ has a unique fixed point, say u, which is the solution of VIP (9.1).

$$
\begin{aligned}
||T_\rho(v_1) - T_\rho(v_2)||^2 &= ||P_K(v_1 - \rho(Av_1 - y)) - P_K(v_2 - \rho(Av_2 - y))||^2 \\
&\leq ||(v_1 - \rho(Av_1 - y)) - (v_2 - \rho(Av_2 - y))||^2 \ (By\ Theorem\ 3.12) \\
&= ||(v_1 - v_2) - \rho A(v_1 - v_2)||^2 = ||v_1 - v_2||^2 - 2\rho a(v_2 - v_1, v_2 - v_1) \\
&\quad + \rho^2||A(v_2 - v_1)||^2 \\
&\leq (1 - 2\rho\alpha + \rho^2||A||^2)||(v_2 - v_1)||^2
\end{aligned}
$$

If $0 < \rho < \frac{2\alpha}{||A||^2}$, then $||T_\rho(v_1) - T_\rho(v_2)||^2 \leq \beta||(v_1 - v_2)||^2$, $0 \leq \beta < 1$. By Theorem 1.1, T_ρ has a unique fixed point which is the solution of (9.1).

Remark 9.1 The proof of the above theorem gives a natural algorithm for solving the variational inequality since $v \to P_K(v - \rho(Av - y))$ is a contraction mapping for $0 < \rho < \frac{2\alpha}{||A||^2}$. Hence, we can use the following algorithm to find u :

Let $u^0 \in H$, arbitrarily given, then for $n \geq 0$ assuming that u_n is known, define u^{n+1} by $u^{n+1} = P_K(u^n - \rho(Au^n - y))$. Then $u^n \to u$ strongly in H, where u is a solution of the VIP. In practice, it is not easy to calculate y and A except the case $H = H^\star$. If $a(\cdot, \cdot)$ is symmetric, then $dJ(u) = J'(u) = Au - y$ and hence $u_{n+1} = P_K(u^n - \rho(J'(u^n)))$, where J is the energy functional (Theorem 9.1). This method is known as the *gradient-projection* method with constant step ρ.

9.3.2 Variational Inequalities for Monotone Operators

An operator $T : K \to H^\star$ or, in particular H is called monotone if $\langle Tu - Tv, u - v \rangle \geq 0 \ \forall u, v \in K$. The monotone mapping T is called strictly monotone if $\langle Tu - Tv, u - v \rangle = 0$ implies $u = v$. T is called *coercive* if there exists $v_0 \in K$ such that

$$
\frac{\langle Tv, v - v_0 \rangle}{||v||_H} \to \infty \ as \ ||v||_H \to \infty
$$

Lemma 9.1 (Minty lemma) *Let K be a nonempty closed and convex subset of a Hilbert space H and let $T : K \to H^\star$ be monotone and continuous and $F \in H^\star$. Then the element $u \in K$ is the solution of the variational inequality*

$$
\langle Tu, v - u \rangle \geq \langle F, v - u \rangle \ \forall v \in K \tag{9.25}
$$

if and only if

$$
\langle Tv, v - u \rangle \geq \langle F, v - u \rangle \ \forall v \in K \tag{9.26}
$$

Proof Let (9.25) hold. We have

$$
\langle Tv, v - u \rangle = \langle Tu, v - u \rangle + \langle Tv - Tu, v - u \rangle \ \forall v \in K \tag{9.27}
$$

by keeping in mind $Tv = Tu - Tu + Tv = Tu + Tv - Tu$. Since T is monotone $\langle Tv - Tu, v - u \rangle \geq 0$, (9.27) gives us $\langle Tv, v - u \rangle \geq \langle Tu, v - u \rangle$. This inequality together with (9.25) implies (9.26). To prove the converse, let (9.26) hold for every $v \in K$. In particular, choose $v = (1 - t)u + tw$ where $w \in K$ is arbitrary and $t \in [0, 1]$. Then $v - u = t(w - u)$ and $v = u + t(w - u)$. By (9.26), we have

$$\langle Tv, v - u \rangle = \langle T(u + t(w - u)), t(w - u) \rangle = t\langle T(u + t(w - u)), (w - u) \rangle$$
$$\geq \langle F, v - u \rangle = \langle F, t(w - u) \rangle = t\langle F, w - u \rangle$$

For $t > 0$, we have

$$\langle T(u + t(w - u)), w - u \rangle \geq \langle F, w - u \rangle \tag{9.28}$$

By taking limit as $t \to 0$ in (9.28) we get (9.25), by the continuity of T and the fact that w is arbitrary. We get the following result applying Minty lemma.

Theorem 9.3 *Under the conditions of the Minty lemma, the set of solutions of the variational inequality (9.24) is convex and closed.*

See Problem 9.6.

Theorem 9.4 *Let K be a closed and convex subset of a Hilbert space H and T a continuous, monotone, bounded, and coercive operator defined on K into $H^\star$.*

Then (9.24) has a solution for every $F \in H^\star$. Furthermore, if in addition, T is strictly monotone, then VIP (9.24) has a unique solution.

We will obtain Theorem 9.4 as a consequence of Theorem 9.7.

As we have seen in Sect. 3.4, the projection of an element $x \in H$ onto the set K is characterized by the variational inequality

$$\langle x - P_K(x), y - P_K(x) \rangle \leq 0 \text{ holds } \forall \, y \in K \tag{9.29}$$

This is evident from (9.2).

The left-hand side of (9.29) is just the cosine of the angle α between lines connecting the point $P_K(x)$ with the point x and with the arbitrary point $y \in K$ (see Fig. 9.2), respectively, and we have that $\cos \alpha \leq 0$ as $\alpha \geq \frac{\pi}{2}$ necessarily holds.

Conversely, for every other point $x' \in K$, $x' \neq P_K(x)$, there exists a point $y' \in K$ such that the angle between the lines connecting the point x' with the points x and y' is less then $\dfrac{\pi}{2}$.

Theorem 9.5 *Let K be a nonempty closed convex subset of a Hilbert space H and T an operator on K into H, $F \in H^\star$, $\alpha > 0$. Then u is a solution of the variational inequality (9.24) if and only if*

$$u = P_K(u - \alpha(Tu - F)) \tag{9.30}$$

Fig. 9.2 Geometrical
interpretation of
characterization of
projection in terms of
variational inequality

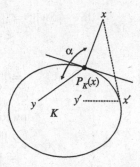

that is, u is the projection of the point $(u - \alpha(Tu - F))$ *or u is the fixed point of the
operator S given by the equation*

$$Su = P_K(u - \alpha(Tu - F)) \tag{9.31}$$

Proof Let $w = u - \alpha(Tu - F)$. By (9.29), (9.30) is equivalent to the inequality: For
$u \in K$, $\langle u - \alpha(Tu - F) - \acute{u}, v - u \rangle \leq 0 \,\forall v \in K$, or $-\alpha \langle Tu - F, v - u \rangle \leq 0 \,\forall v \in K$,
or $\langle Tu, v - u \rangle \geq \langle F, v - u \rangle \,\forall v \in K$ as $\alpha > 0$. Thus, u is the solution of the variational
inequality (9.24).

Theorem 9.6 *Suppose the hypothesis of Theorem (9.5) satisfied besides the following conditions: There exist constants* $\mu > 0$ *and* $\lambda > 0$ *such that*

$$\|Tv - Tw\| \leq \mu \|v - w\| \tag{9.32}$$

and

$$\langle Tv - Tw, v - w \rangle \geq \lambda \|v - w\|^2, \ \forall \, v, w \in K \tag{9.33}$$

Then, (9.24) for every $F \in H$ *has a unique solution. If* α *is so chosen that* $0 < \alpha < \frac{2\lambda}{\mu^2}$
and if u_0 *is an arbitrary element of K, then*

$$u = \lim_{n \to \infty} u_n$$

where

$$u_{n+1} = P_K(u_n - \alpha(Tu_n - F)), \ n = 0, 1, 2, \ldots \tag{9.34}$$

Proof By Theorem 9.5, it is sufficient to show that the operator S (defined by (9.31))
is a contraction operator. Equations (9.32) and (9.33) give

$$||Sv - Sw||^2 = ||P_K(v - \alpha(Tv - F)) - P_K(w - \alpha(Tw - F))||^2$$
$$\leq ||(v - \alpha(Tv - F)) - (w - \alpha(Tw - F))||^2$$

by Theorem 3.12(*a*)

$$= ||(v - w) - \alpha(Tv - Tw)|| = ||v - w||^2$$
$$-2\alpha \langle v - w, Tv - Tw \rangle + \alpha^2 ||Tv - Tw||^2$$

by Remark 3.4(1) and properties of the inner product   (9.35)

$$\leq ||v - w||^2 - 2\alpha\lambda||v - w||^2 + \alpha^2\mu^2||v - w||^2$$
$$= (1 - 2\alpha\lambda + \alpha^2\mu^2)||v - w||^2$$

If $0 < \alpha < \frac{2\lambda}{\mu^2}$, the operator S is a contraction operator with constant $\sqrt{1 - 2\alpha\lambda + \alpha^2\mu^2} = \xi \leq 1$. Thus, we have the desired result by Theorem 1.1 and Problem 1.17.

The Ritz–Galerkin Methods

In the Ritz–Galerkin methods, we find a finite-dimensional subspace H_n of separable Hilbert space H and an element $u_n \in K \cap H_n$, where $K \neq \phi$ is closed convex subset of H such that

$$\langle Tu_n, v - u_n \rangle \geq \langle F, v - u_n \rangle \; \forall \, v \in H \text{ and each } F \in H^*.$$

The solution of the VI (9.24) is obtained as the weak limit of the sequence $\{u_n\}$.

The Penalty Method

Suppose K is a nonempty closed convex subset of a separable Hilbert space H and J is a functional on H. The *penalty method* is to find the conditions under which optimization of J on K is equivalent to solving (9.24). Instead of finding $u \in K$ such that $J(u) \leq J(v) \; \forall \, v \in K$, we look for $u_\varepsilon \in H$ at which a new functional $J : H \to R$ attains a minimum, that is

$$J_\varepsilon(u_\varepsilon) \leq J_\varepsilon(v) \; \forall \, v \in K$$

or

$$J_\varepsilon(u_\varepsilon) = \inf_{v \in K} J_\varepsilon(v)$$

We write here

$$J_\varepsilon(v) = J(v) + \frac{1}{\varepsilon}\phi(v), \; \varepsilon > 0,$$

where ϕ is a functional on H for which $\phi(v) = 0, \; \forall \, v \in K, \; \phi(v) > 0 \; \forall \, v \notin K$. Under appropriate conditions, $u_\varepsilon \rightharpoonup u \in K$ as $\varepsilon \to 0$. ϕ is called the *penalty functional*.

Let T be an operator on H into H^* and a bounded continuous monotone operator β exists which also maps H into H^* such that

$$\beta v = 0 \; if \; and \; only \; if \; v \in K \tag{9.36}$$

Then β is called the *penalty operator* with respect to K. Let $\varepsilon > 0$ be arbitrary and let us look for the point $u_\varepsilon \in H$ such that

$$\left\langle T u_\varepsilon + \frac{1}{\varepsilon} \beta u_\varepsilon, v \right\rangle = \langle F, v \rangle \tag{9.37}$$

or

$$T u_\varepsilon + \frac{1}{\varepsilon} \beta u_\varepsilon = F$$

Thus, variational inequality (9.24) on K is replaced by variational Eq. (9.36) on H. The basic idea is to convert a problem of variational inequality into one of variational equations.

Theorem 9.7 *Let H, T, K be as in Theorem 9.5 and let β be the penalty operator corresponding to the set K and $0 \in K$. Then there exists a solution u_ε of (9.36) for $\varepsilon > 0$ and a sequence $\{\varepsilon_n\}$ such that $\varepsilon_n \to 0$ and u_{ε_n} converges weakly to the solution u of variational inequality (9.24). If, in addition, T is strictly monotone, then VIP (9.24) has a unique solution.*

Proof (1) **Uniqueness**

Let u_1 and u_2 be two solutions of (9.24), that is

$$\langle T u_1, v - u_1 \rangle \geq \langle F, v - u_1 \rangle \; \forall \, v \in K \tag{9.38}$$

$$\langle T u_2, v - u_2 \rangle \geq \langle F, v - u_2 \rangle \; \forall \, v \in K \tag{9.39}$$

Putting $v = u_2$ in (9.37) and $v = u_1$ in (9.38) and adding the resulting inequalities we get $\langle T u_1 - T u_2, u_2 - u_1 \rangle \geq 0$. Consequently, $\langle T u_1 - T u_2, u_1 - u_2 \rangle \leq 0$. Suppose T is strictly monotone $u_1 = u_2$ must hold.

(2) **Existence of Solution of** (9.36)

Since T and β are continuous and monotone for a given $\varepsilon > 0$, the operator $S_\varepsilon = T + \frac{1}{\varepsilon} \beta$ is continuous and monotone. The operator S_ε is also coercive because (9.35) implies that $\beta 0 = 0$, and furthermore, we have $\forall \, v \in K$

$$\langle S_\varepsilon v, v \rangle = \langle T v, v \rangle + \langle \beta v, v \rangle$$

$$= \langle T v, v \rangle + \frac{1}{\varepsilon} \langle \beta v - \beta 0, v - 0 \rangle \geq \langle T v, v \rangle$$

as $\frac{1}{\varepsilon} \langle \beta v - \beta 0, v - 0 \rangle \geq 0$. This implies that

$$\frac{\langle S_\varepsilon v, v \rangle}{||v||_H} \geq \frac{\langle Tv, v \rangle}{||v||_H} \to \infty \; as \; ||v||_H \to \infty$$

Since S_ε satisfies all the conditions of Theorem 7.7, we find that there exists $u_\varepsilon \in H$ such that $S_\varepsilon u_\varepsilon = F$, for each $F \in H^\star$. This means that (9.36) holds for all $v \in H$, and u_ε is a solution of (9.36).

(3) $u_\varepsilon \rightharpoonup u_0 \in K$: We now show that $\{u_{\varepsilon_n}\}$ is a bounded sequence. For $\varepsilon > 0$, we have

$$\langle Tu_\varepsilon, u_\varepsilon \rangle \leq S_\varepsilon u_\varepsilon, u_\varepsilon = \langle F, u_\varepsilon \rangle \leq ||F|| \, ||u_\varepsilon||$$

$$or \qquad \frac{\langle Tu_\varepsilon, u_\varepsilon \rangle}{||u_\varepsilon||_H} \leq ||F||_{H^\star}.$$

This implies that $\{u_\varepsilon, \; \varepsilon > 0\}$ is bounded; otherwise, T is noncoercive, a contradiction of the hypothesis. By Theorem 4.17, there exists a sequence $\{\varepsilon_n\}$, $\varepsilon_n > 0, \varepsilon_n \to 0$, such that the sequence of elements u_{ε_n}, $n \in N$ converges weakly to some element $u_0 \in H$.

Now, we show that $u_0 \in K$. Since β is monotone

$$\langle \beta v - \beta u_\varepsilon, v - u_\varepsilon \rangle \geq 0, \; \forall v \in H \tag{9.40}$$

By (9.36), $\beta u_\varepsilon = \varepsilon_n (F - Tu_{\varepsilon_n}$. Since T is bounded $\{Tu_{\varepsilon_n}\}$ is bounded and thus $\beta u_{\varepsilon_n} \to 0$ for $n \to \infty$. By taking limit in (9.39), we get

$$\langle \beta v, v - u_0 \rangle \geq 0 \; \forall v \in H$$

By choosing $v = u_0 + tw$, where $t > 0$ and $w \in H$; we get $v - u_0 = tw$ and $\langle \beta(u_0 + tw), w \rangle \geq 0$. By the continuity of the operator β, it follows that

$$\langle \beta u_0, w \rangle \geq 0 \text{ for every } w \in H \, (by \; letting \; t \to 0+) \tag{9.41}$$

The inequality also holds for $-w$ and so

$$\langle \beta u_0, w \rangle \leq 0 \; for \; every \; w \in H. \tag{9.42}$$

(9.40) and (9.41) imply that $\langle \beta u_0, w \rangle = 0$ for each $w \in H$. This implies that $\beta u_0 = 0$ and so by (9.35) $u_0 \in K$.

(4) u_0 is the Solution of the Variational Inequality (9.24)

(9.35) and (9.36) and the monotonicity of the operator T and β imply that

$$\langle Tv - F, v - u_{\varepsilon_n} \rangle = \langle Tv - Tu_{\varepsilon_n}, v - u_{\varepsilon_n} \rangle + \langle Tu_{\varepsilon_n} - F, v - u_{\varepsilon_n} \rangle$$

$$= \langle Tv - Tu_{\varepsilon_n}, v - u_{\varepsilon_n} \rangle + \frac{1}{\varepsilon} \langle \beta v - \beta u_{\varepsilon_n}, v - u_{\varepsilon_n} \rangle$$

$$\geq 0 \, \forall v \in K, \text{ by (9.39) } and \; monotonicity \; of \; T$$

or

$$\langle Tv, v - u_0 \rangle \geq \langle F, v - u_0 \rangle \qquad (9.43)$$

u_0 is a solution of (9.24) by virtue of Minty lemma (Lemma 9.1) and (9.42).

9.4 Finite Element Methods for Variational Inequalities

In this section, we present results concerning convergence of the solution of the discretized variational inequality to the solution of continuous form and the error estimation between these two solutions. In Sect. 9.3.1, results concerning Model 1 are presented, whereas a concrete case is discussed in Sect. 9.3.2.

9.4.1 Convergence and Error Estimation

Let H be a real Hilbert space, K a nonempty closed and convex subset of H and $a(\cdot, \cdot)$ bilinear bounded and coercive form on H and $F \in H^\star$. We are interested in the approximation or discretization of the variational inequality (9.22), namely

$$u \in K : a(u, v - u) \geq F(v - u) \forall\, v \in K \qquad (9.44)$$

As shown earlier, there exists a bounded linear operator $A : H \rightarrow H$ such that $\langle Au, v \rangle = a(u, v)$ and $||A|| = ||a||$. Let $\{H_h\}_h$ be a sequence of finite-dimensional subspaces of H, where h is a given parameter converging to 0. Furthermore, let $\{K_h\}_h \ \forall\, h$ be a sequence of closed convex subsets of H (K_h may not be a subset of K) satisfying the following conditions:

If $\{v_h\}_h$ is such that $v_h \in K_h \ \forall\, h$ and $\{v_h\}_h$ is bounded in H, then the weak limit points of $\{v_h\}$ belong to K.
There exist $\chi \subseteq H$, $\bar{\chi} = K$ and $r_h : \chi \rightarrow K_h$ such that

$$\lim_{h \to 0} ||r_h - v||_H = 0 \, v \in \chi$$

Remark 9.2 (a) If $K_h \subseteq K$, then (1) is trivially satisfied as K is weakly closed.
(b) $\cap_h K_h \subseteq K$.
(c) A variant of condition (2) is as follows: $\exists$ a subset $\chi \subseteq H$ such that $\bar{\chi} = K$ and $r_h : \chi \rightarrow K_h$ with the property that for each $v \in \chi$, $\exists\, h_0 = h_0(v)$ with $r_h v \in K_h$ for all $h \leq h_0(v)$, and $\lim_{h \to 0} ||r_h - v||_H = 0$.

Approximate Problem P_h Find $u_h \in K_h$ such that

$$a(u_h, v_h - u_h) \geq F(v_h - u_h) \ \forall\, v_h \in K_h \qquad (9.45)$$

It is clear that variational inequalities (9.43) and (9.44) have a unique solution under the hypothesis. Here, we present two theorems, one related to the convergence of $\{u_h\}$ as $h \to 0$ with respect to the norm of H; that is, $||u_h - u||_H \to 0$ as $h \to 0$, where u and u_h are the solutions of (9.43) and (9.44), respectively; the other provides the order or the upper bound of $||u_h - u||_H$. It may also be observed that (9.44) will be a system of matrix inequalities which are to be evaluated in practical situations.

It may be noted that if $a(\cdot, \cdot)$ is also symmetric, then the solution of (P_h) is equivalent to solving the following programming problem:

Problem P_h'   Find $u_h \in K_h$ such that

$$J(u_k) = \inf_{v_h \in K_h} J(v_h) \tag{9.46}$$

where $J(v_h) = \frac{1}{2}a(v_h, v_h) - \langle F, v_h \rangle$

Theorem 9.8 *Theorem 9.8 We have $\lim_{h \to 0} ||u - u_h|| = 0$, where the above conditions are satisfied.*

Proof The proof is divided into three parts:

1. Priori estimates of $\{u_h\}$
2. Weak convergence of u_h
3. Strong convergence

1. **Estimates for $\{u_h\}$** We show that $\exists$ constants C_1 and C_2 such that

$$||u_h||^2 \le C_1 ||u_h|| + C_2 \; \forall h$$

Since $\{u_h\}$ is the solution of P_h (9.44), we have

$$
\begin{aligned}
a(u_h, v_h - u_h) &\ge F(v_h - u_h) \; \forall \, v_h \in K_h \\
\text{or} \quad a(u_h, v_h) &\ge a(u_h, u_h) + F(v_h) - F(u_h) \\
\text{or} \quad a(u_h, u_h) &\le a(u_h, v_h) - F(v_h - u_h) \\
\text{by linearity of } a(\cdot, \cdot) & \text{ and } F(\cdot) \\
\text{or} \quad \alpha ||u_h||^2 &\le a(u_h, u_h) \le ||A|| \, ||u_h|| \, ||v_h|| \\
&+ ||F||(||v_h|| + ||u_h||) \; \forall \, v_h \in K_h,
\end{aligned}
$$

by applying coercivity of $a(\cdot, \cdot)$, boundedness of F

, and A induced by $a(\cdot, \cdot)$.

Let $v_0 \in \chi$ and $v_h = r_h v_0 \in K_h$. By condition 2 on K_h, we have $r_h v_0 \to v_0$ strongly in H and hence $||v_h||$ uniformly bounded by constant m.

$$||u_h||^2 \le \frac{1}{\alpha}\{(m||A|| + ||F||)||u_h|| + m||F||\} = c_1||u_h|| + c_2$$

where $c_1 = \frac{1}{\alpha}(m||A|| + ||F||)$ and $c_2 = \frac{m}{\alpha}||F|| \Rightarrow ||u_h|| \le c \; \forall \, h$.

(2) **Weak Convergence of** $\{u_h\}$

$\{u_h\}$ is uniformly bounded $\Rightarrow \{u_h\}$ has a weakly convergent subsequence, say $\{u_{h_i}\}$, such that $u_{h_i} \rightharpoonup u^*$ in H (by condition 1 on $\{K_h\}_h$, $u^* \in K$) by Theorem 4.17. Now, we show that u^* is a solution of (VIP). We have, $a(u_{h_i}, u_{h_i}) \le a(u_{h_i, v_{h_i}}) - F(v_{h_i} - u_{h_i}) \; \forall \; v_{h_i} \in K_{h_i}$. Let $v \in \chi$ and $v_{h_i} = r_{h_i} v$. Then this equation takes the form $a(u_{h_i}, u_{h_i}) \le a(u_{h_i}, r_{h_i} v) - F(r_{h_i} v - u_{h_i})$. Since $u_{h_i} \rightharpoonup u^*$ and $r_{h_i} v \to v$ as $h_i \to 0$, taking the limit in this inequality, we get

$$\liminf_{h_i \to 0} a(u_{h_i}, u_{h_i}) \le a(u^*, v) - F(v - u^*) \; \forall \; v \in \chi \tag{9.47}$$

On the other hand, $0 \le a(u_{h_i} - \dot{u}^*, u_{h_i} - u^*) \le a(u_{h_i}, u_{h_i}) - a(u_{h_i}, u^*) - a(u^*, u_{h_i}) + a(u^*, u^*)$ or $a(u_{h_i}, u^*) + a(u^*, u_{h_i}) - a(u^*, u^*) \le a(u_{h_i}, u_{h_i})$. By taking the limit, we get

$$a(u^*, u^*) \le \liminf_{h_i \to 0} a(u_{h_i}, u_{h_i}) \tag{9.48}$$

From (9.46) and (9.47), we get

$$a(u^*, u^*) \le \liminf_{h_i \to 0} a(u_{h_i}, u_{h_i}) \le a(u^*, v) - F(v - u^*) \forall \; v \in \chi \tag{9.49}$$

Therefore, $a(u^*, v - u^*) \ge F(v - u^*) \; \forall \; v \in \chi, u^* \in K$. Since χ is dense in K and $a(\cdot, \cdot)$, F are continuous by the hypotheses; from this inequality, we obtain

$$a(u^*, v - u^*) \ge F(v - u^*) \; \forall \; v \; and \; u^* \in K$$

Since conditions of Theorem 9.2 are satisfied, solution u^* must be equal to u. Hence, u is the only limit point of $\{u_h\}_h$ in the weak topology of H, and therefore, $\{u_h\}_h$ converges weakly to u.

(3) **Strong Convergence**

Since $a(\cdot, \cdot)$ is coercive, we get

$$0 \le \alpha \|u_h - u\|^2 \le a(u_h - u, u_h - u)$$
$$= a(u_h, u_h) - a(u_h, u) - a(u, u_h) + a(u, u) \tag{9.50}$$

Since $r_h v \in K_h$ for any $v \in \chi$, we have

$$a(u_h, u_h) \le a(u_h, r_h v) - F(r_h v - u_h) \; \forall \; v \in \chi \tag{9.51}$$

From (9.49) and (9.50) and keeping in mind $\lim_{h \to 0} u_h = u$ weakly in H and $\lim_{h \to 0} r_h v = v$ strongly in H, we get

$$0 \leq \alpha \liminf \|u_h - u\|^2 \leq \alpha \limsup \|u_h - u\|^2$$
$$\leq a(u, v - u) - F(v - u) \tag{9.52}$$
$$\Rightarrow \lim_{h \to 0} \|u_h - u\|^2 = 0$$

by density and continuity and letting $v = u$ in (9.51). Therefore, u_h converges strongly to u.

Theorem 9.9 *Assume that $(Au - Fy) \in H$, where A is defined through $a(\cdot, \cdot)$ and Fy by F (Theorem 3.19). Then there exists C independent of the subspace H_h of H and a nonempty closed convex subset K_h of H such that*

$$\|u_h - u\| \leq C(\inf_{v_h \in K_h} \{\|u - v_h\|^2 + \|Au - Fy\| \|u - v_h\|\}$$
$$+ \|Au - Fy\| \inf_{v \in K} \|u_h - v\|)^{1/2} \tag{9.53}$$

Corollary 9.1 *If $K = H$, then $Au - Fy = 0$, so that with the choice of $K_h = H_h$ the error estimate (9.52) reduces to the well-known estimate known as Céa's Lemma (Theorem 8.1).*

It may be noted that K_h need not be a subset of K.

Proof

$$\alpha \|u - u_h\|^2 \leq a(u - u_h, u - u_h) = a(u, u) + a(u_h, u_h)$$
$$-a(u, u_h) - a(u_h, u) \tag{9.54}$$
$$\forall v \in K, \ a(u, u) \leq a(u, v) + F(u - v) \text{ by (9.43)} \tag{9.55}$$
$$\forall v_h \in K_h, \ a(u_h, u_h) \leq a(u_h, v_h) + F(u_h - v_h) \text{ by (9.44)} \tag{9.56}$$

Therefore we have, for all $v \in K$ and for all $v_h \in K_h$ by (9.53)–(9.55)

$$\alpha \|u - u_h\|^2 \leq a(u, v - u_h) + a(u_h, v_h - u) + F(u - v) + F(u_h - v_h)$$
$$= a(u, v - u_h) - F(v - u_h) + a(u, v_h - u) - F(v_h - u)$$
$$+ a(u_h - u, v_h - u)$$
$$= \langle F - Au, u - v_h \rangle + \langle F - Au, u_h - v \rangle + a(u - u_h, u - v_h)$$

Thus, we have, for all $v \in K$ and for all $v_h \in K_h$

$$\alpha \|u - u_h\|^2 \leq \|F - Au\| \|u - v_h\| + \|F - Au\| \|u_h - v\| + M \|u - u_h\| \|u - v_h\|$$

Since $\|u - u_h\| \|u - v_h\| \leq \frac{1}{2} \left(\frac{\alpha}{M} \|u - u_h\|^2 + \frac{M}{\alpha} \|u - v_h\|^2 \right)$, we get

$$\frac{\alpha}{2} \|u - u_h\|^2 \leq \|F - Au\|(\|u - v_h\| + \|u_h - v\|) + \frac{M^2}{2\alpha} \|u - v_h\|^2$$

This implies the error estimation (9.52).

9.4.2 Error Estimation in Concrete Cases

Approximation in One Dimension

Let us consider the variational inequality: Find $u \in K$

$$\int_0^1 \frac{du}{dx} \left(\frac{dv}{dx} - \frac{du}{dx} \right) dx \geq \int_0^1 f(v - u) dx \ \forall v \in K \qquad (9.57)$$

where $H = H_0^1(0, 1) = \{v \in H^1(0, 1) / v(0) = v(1) = 0\}$

$$K = \left\{ v \in H / \left| \frac{dv}{dx} \right| \leq 1 \ a.e. \ in \ (0, 1) \right\}$$

$$a(u, v) = \int_0^1 \frac{du}{dx} \frac{dv}{dx}, \ F(v) = \int_0^1 f(x) v(x) dx, \ f \in L_2(0, 1)$$

Let N be a positive integer and $h = \frac{1}{N}$. Let $x_i = ih$ for $i = 0, 1, \ldots, N$ and $e_i = [x_{i-1}, x_i]$, $i = 1, 2, \ldots, N$. Let $H_h = \{v_h \in C[0, 1] / v_h(0) = v_h(1) = 0, v_h / e_i \in P_1, i = 1, 2, \ldots, N\}$. Then

$$K_h = K \cap H_h = \{v_h \in H_h / |v_h(x_i) - v_h(x_{i-1})| \\ \leq h, \ \text{for } i = 1, 2, \ldots, N\}$$

The approximation problem (P_h) takes the form

$$\int_0^1 \frac{du_h}{dx} \left(\frac{dv_h}{dx} - \frac{du_h}{dx} \right) dx \geq \int_0^1 f(v_h - u_h) dx \forall v_h \in K_h, \ u_h \in K_h \qquad (9.58)$$

or, $a(u_h, v_h - u_h) \geq F(v_h - u_h)$.

Theorem 9.10 *Let u and u_h be the solutions of (9.56) and (9.57), respectively, then*

$$\|u - u_h\|_H = O(h) \ for \ f \in L_2(0, 1)$$

Proof Since $u_h \in K_h \subset K$, from (9.56) we have

$$a(u_h, v_h - u_h) \geq \int_0^1 f(u_h - u) dx \qquad (9.59)$$

Adding (9.57) and (9.58), we obtain

$$a(u_h - u, u_h - u) \geq a(v_h - u, u_h - u) + a(u, v_h - u)$$

$$- \int_0^1 f(v_h - u)dx, \forall v_h \in K_h$$

which, in turn, implies that

$$\frac{1}{2}||u_h - u||^2 \leq \frac{1}{2}||v_h - u||^2 + \int_0^1 \frac{du}{dx}\left(\frac{dv_h}{dx} - \frac{du}{dx}\right)dx$$

$$- \int_0^1 f(v_h - u)dx, \ \forall v_h \in K_h \tag{9.60}$$

Since $u \in K \cap H^2(0, 1)$, we obtain

$$\int_0^1 \frac{du}{dx}\frac{d}{dx}(v_h - u)dx = \int_0^1 \frac{d^2u}{dx^2}(v_h - u)dx \leq \left\|\frac{d^2u}{dx^2}\right\|_{L_2} ||v_h - u||_{L_2}$$

But we have

$$\left\|\frac{d^2u}{dx^2}\right\|_{L_2} \leq ||f||_{L_2} \tag{9.61}$$

Therefore, (9.59) becomes

$$\frac{1}{2}||u_h - u||_H^2 \leq \frac{1}{2}||v_h - u||_H^2 + 2||f||_{L_2}||v_h - u||_{L_2} \ \forall v_h \in K_h \tag{9.62}$$

Let $v \in K$, then the linear interpolation is defined by

$$r_h v \in H_h, \ (r_h v)(x_i) = v(x_i), \ i = 0, 1, 2, \ldots, N$$

and we have

$$\frac{d}{dx}(r_h v)/e_i = \frac{v(x_i) - v(x_{i-1})}{h} = \frac{1}{h}\int_{x_{i-1}}^{x_i} \frac{dv}{dx}dx \tag{9.63}$$

Hence, we obtain

$$\left|\frac{d}{dx}(r_h v)\right|_{e_i} \leq 1, \ since \ \left|\frac{dv}{dx}\right| \leq 1, a.e., \ in \ (0, 1)$$

Thus, $r_h v \in K_h$. Let us replace v_h by $r_h u$ in (9.61), then

$$\frac{1}{2}\|u_h - u\|_H^2 \leq \frac{1}{2}\|r_h u - u\|_H^2 + 2\|f\|_{L_2}\|r_h u - u\|_{L_2} \tag{9.64}$$

From (9.60) and by well-known approximation results, we get

$$\|r_h u - u\|_H \leq Ch\|u\|_{H^2(0,1)} \leq Ch\|f\|_{L_2} \tag{9.65}$$

$$\|r_h u - u\|_{L_2(\Omega)} \leq Ch^2\|u\|_{H^2(0,1)} \leq Ch^2\|f\|_{L_2} \tag{9.66}$$

where C denotes constants independent of u and h. By (9.63)–(9.65), we get $\|u_h - u\|_H = O(h)$.

9.5 Evolution Variational Inequalities and Parallel Algorithms

9.5.1 Solution of Evolution Variational Inequalities

Variational inequalities of evolution were introduced by Lions and Stampacchia (see, e.g., Lions [121] for updated references). Let V and H be Hilbert spaces such that

$$V \subset H, \ V \ dense \ in \ H, \ V \hookrightarrow H \ being \ continuous \tag{9.67}$$

We identify H with its dual so that if V^* denotes the dual of V, then

$$V \subset H \subset V^* \tag{9.68}$$

We shall use the following notations: Here, $L_2(0, T, H)$ denote the space of all measurable functions $u : [0, T] \rightarrow H$, which is a Hilbert space with respect to the inner product

$$\langle u, v \rangle_{L_2(0,T,H)} = \int_0^T \langle u(t), v(t) \rangle_H dt \tag{9.69}$$

If H^* is the topological dual of H, then the dual of $L_2(0, T, H) = L_2(0, T, H^*)$ for any Hilbert space H. $H^{1,2}(0, T; H) (H^{1,1}(0, T; H))$ will denote the space of all those elements of $L_2(0, T; H)$ $((L_1(0, T; H))$ such that their distributional derivatives Df also belong to $L_2(0, T; H)$ $(L_1(0, T; H))$.

We also have a set $K \subset V$ such that

$$K \text{ is a closed convex subset of V} \tag{9.70}$$

We do not restrict ourselves generally (it suffices to make a translation) by assuming that

$$0 \in K \tag{9.71}$$

Let f be given such that

$$f \in L_2(0, T; V^*) \tag{9.72}$$

We now consider a bilinear form

$$(u, \hat{u}) \to a(u, \hat{u}) \text{which is continuous on } V \times V$$
$$a(u, \hat{u}) \; is \; \text{symmetric or not}$$
$$a(u, u) \geq \alpha ||u||_V^2, \; \alpha > 0 \; \forall u \in V \tag{9.73}$$

where we denote by $||u||_V$ the norm of u in V. We look for u such that

$$u \in L_2(0, T; V) \cap L^\infty(0, T; H), \; u(t) \in K \text{ a.e.}$$
$$\left\langle \frac{\partial u}{\partial t}, \hat{u} - u \right\rangle + a(u, \hat{u} - u) \geq (f, \hat{u} - u) \; \forall \hat{u} \in K$$
$$u(0) = 0 \tag{9.74}$$

The solution has to be thought of as being a weak solution of (9.73); otherwise, the condition $u(0) = 0$ in (9.75) is somewhat ambiguous. This condition becomes precise if we add the condition

$$\frac{\partial u}{\partial t} \in L_2(0, T; V^*) \tag{9.75}$$

but this condition can be too restrictive. We can introduce weak solutions in the following form: We consider smooth functions $\hat{u}$ such that

$$\hat{u} \in L_2(0, T; V), \; \frac{\partial u}{\partial t} \in L_2(0, T; V^*)$$
$$\hat{u} \in L_2(0, T; V), \; \frac{\partial u}{\partial t} \in L_2(0, T; V^*)$$
$$\hat{u} \in K \; for \; a.e., \; \hat{u}(0) = 0 \tag{9.76}$$

Then, if u satisfies (9.72) and is supposed to be smooth enough, we have (we write $\langle u, \hat{u} \rangle$ instead of $\langle u, \hat{u} \rangle_H$

$$\int_0^T \left[\left\langle \frac{\partial u}{\partial t}, \hat{u} - u \right\rangle + a(u, \hat{u} - u) - (f, \hat{u} - u) \right] dt$$

$$= \int_0^T \left[\left\langle \frac{\partial u}{\partial t}, \hat{u} - u \right\rangle + a(u, \hat{u} - u) - (f, \hat{u} - u) \right] dt + \int_0^T \left\langle \left(\frac{\partial (\hat{u} - u)}{\partial t} \right), \hat{u} - u \right\rangle$$

The last term equals $\frac{1}{2} \|\hat{u}(T) - u(T)\|_H^2$ (since $u(0) = 0$, $\hat{u}(0) = 0$, so that

$$\int_0^T \left[\left\langle \frac{\partial u}{\partial t}, \hat{u} - u \right\rangle + a(u, \hat{u} - u) - (f, \hat{u} - u) \right] dt \geq 0 \qquad (9.77)$$

for all $\hat{u}$ satisfying (9.75).

We then define a weak solution of (9.73) as a function u such that

$$u \in L_2(0, T; V), \ u(t) \in K \ a.e. \qquad (9.78)$$

and which satisfies (9.76) for all $\hat{u}$ satisfying (9.75). The following existence and approximation results are well known:

Theorem 9.11 *If the bilinear form is coercive, symmetric, and bounded, then the evolution variational inequality given by (9.73) has a unique solution.*

Proof (Uniqueness of Solution) Let u_1 and u_2 be two solutions. By putting $u = u_1$ and u_2 in (9.73), adding these inequalities and setting $w = u_1 - u_2$ in the resultant inequality, we obtain

$$-\langle w'(t), w(t) \rangle - a(w(t), w(t)) \geq 0$$

By coercivity, this inequality gives us

$$\frac{1}{2} \frac{d}{dt} |w(t)|^2 + \alpha \|w(t)\|^2 \leq 0$$

$$\frac{1}{2} \frac{d}{dt} |w(t)|^2 \leq 0$$

This implies that $w(t) = 0$ or $u_1 = u_2$. For existence see, for example, Duvaut and Lions [69].

9.5.2   Decomposition Method and Parallel Algorithms

We introduce N couples of Hilbert spaces V_i and H_i, and N convex sets K_i:

$$V_i \subset H_i \subset V_i^* \ i = 1, 2, \ldots, N \tag{9.79}$$

$$K_i \subset V_i, \ K_i \ closed \ convex \ subset \ of \ V_i, \ nonempty \tag{9.80}$$

We are given linear operators r_i such that

$$r_i \ \in \ L(H, H_i) \cap L(V, V_i)$$
$$r_i \ maps \ K \ into \ K_i \ i = 1, 2, \ldots, N \tag{9.81}$$

We are also given a family of Hilbert spaces H_{ij} such that

$$H_{ij} = H_{ji} \ \forall \, i, j \in [1, 2, \ldots, N] \tag{9.82}$$

and a family of operators r_{ij} such that

$$r_{ij} \in L(H_j, H_{ij}) \tag{9.83}$$

The following hypotheses are made:

$$r_j r_{ji} \varphi = r_i r_{ij} \varphi \ \forall \varphi \in V \tag{9.84}$$

If N elements u_i are given such that

$$u_i \in K_i \ \forall i, \ r_{ij} u_j = r_{ji} u_i \ \forall i, j \tag{9.85}$$

then there exists $u \in K$ such that

$$u_i = r_i u, \ and \ moreover \ ||u||_V^2 \le c \left(\sum_{i=1}^N ||u_i||_{V_i}^2 \right) \tag{9.86}$$

The hypothesis

$$K_i = V_i \ for \ a \ subset \ of \ [1, 2, 3, \ldots, N] \tag{9.87}$$

is perfectly acceptable!

We now proceed with the decomposition of the problem. We introduce the following bilinear forms: $c_i(u_i, \hat{u}_i)$ is continuous, symmetric on $H_i \times H_i$, and it satisfies

$$c_i(u_i, u_i) \ge \alpha_i ||u_i||_{H_i}^2, \ \alpha_i > 0, \ \forall u_i \in H_i \tag{9.88}$$

$a_i(u_i, \hat{u}_i)$ is continuous, symmetric or not, on $V_i \times V_i$, and it satisfies

$$a_i(u_i, u_i) \geq \beta_i \|u_i\|_{V_i}^2, \ \beta_i > 0, \ \forall u_i \in V_i \tag{9.89}$$

We assume that

$$\sum_{i=1}^{N} c_i(r_i u, r_i \hat{u}) = \langle u, \hat{u} \rangle_H, \ u, \hat{u} \in H \tag{9.90}$$

$$\sum_{i=1}^{N} a_i(r_i u, r_i \hat{u}) = a(u, \hat{u}), \ u, \hat{u} \in V \tag{9.91}$$

Finally, we assume that the function f is also decomposed as follows: We are given functions $f_i \in L_2(0, T_i; V_i^\star)$ such that

$$\sum_{i=1}^{N} (f_i, r_i \hat{u}) = \langle f, \hat{u} \rangle = \langle f, \hat{u} \rangle \ \hat{u} \in V \tag{9.92}$$

We are now ready to introduce the decomposed approximation. We look for functions $u_i (i = 1, 2, 3, \ldots, N)$ such that

$$c_i \left(\frac{\partial u_i}{\partial t}, \hat{u}_i - u_i \right) + a(u_i, \hat{u}_i - u_i) + \frac{1}{\varepsilon} \sum_{j} \langle r_{ji} u_i - r_{ij} u_j, r_{ji}(\hat{u}_i - u_i) \rangle_{H_{ij}}$$

$$\geq (f_i, \hat{u}_i - u) \ \forall \hat{u}_i \in K_i \tag{9.93}$$

$$u_i \in L_2(0, T; V_i), \ u_i(t) \in K_i \ a.e., \ u_i(0) = 0 \tag{9.94}$$

Remark 9.3 1. It may be remarked that each of the variational inequalities in (9.92) has to be thought of in its weak formulation as introduced above.
2. In (9.92), ε is positive and small. The corresponding term in this equation is a penalty term.
3. In the examples, $\|r_{ji}\|$ is a sparse matrix.

Theorem 9.12 *The set of (decomposed) variational inequalities (9.92)–(9.93) admits a unique solution $u_i = u_i^\varepsilon (i = 1, 2, 3, \ldots, N)$. Further, as $\varepsilon \to 0$, one has*

$$u_i^\varepsilon \to u_i \text{ in } L_2(0, T; V_i) \text{ weakly} \tag{9.95}$$

and

$$u_i = r_i u$$

where u is the solution of (9.73) (weak form (9.76)).

Proof **Step 1 A Priori Estimates** We can assume, without loss of generality, that $0 \in K_i$. Therefore, taking $\hat{u}_i = 0$ in (9.92) is allowed (for a complete proof, the technical details are much more complicated. One has to work first on approximations of (9.92), by using (other) penalty arguments; see the bibliographical references. This simplification gives (we write u_i instead of u_i^ε for the time being)

$$c_i \left(\frac{\partial u_i}{\partial t}, u_i \right) + a_i(u_i, u_i) + \frac{1}{\varepsilon} X_i \le (f_i, u_i) \ (i = 1, \dots, N) \qquad (9.96)$$

where

$$X_i = \sum_j \langle r_{ji} u_i - r_{ij} u_j \rangle_{H_{ij}} \qquad (9.97)$$

We can write

$$X_i = \frac{1}{2} \sum_j ||r_{ji} u_i - r_{ij} u_j||^2_{H_{ij}} + \frac{1}{2} \sum_j ||r_{ji} u_i||^2_{H_{ij}} - \frac{1}{2} \sum_j ||r_{ij} u_j||^2_{H_{ij}} \quad (9.98)$$

But one easily verifies that

$$\sum_i X_i = \frac{1}{2} \sum_j ||r_{ji} u_i - r_{ij} u_j||^2_{H_{ij}} \qquad (9.99)$$

Therefore, by integration in t, in the interval $(0, t)$, of (9.95), and by summing in i, using (9.98), we obtain

$$\frac{1}{2} \sum_i c_i(u_i(t)) + \sum \int_0^t a_i(u_i(s))ds$$

$$+ \frac{1}{2\varepsilon} \sum_{i,j} \int_0^t ||r_{ji} u_s - r_{ij} u_s||^2_{H_{ij}} ds \le \sum_i \int_0^t (f_i, u_i)ds \qquad (9.100)$$

Step 2 It follows from (9.87), (9.88) and (9.99) that, as $\varepsilon \to 0$ (and we now use the notation u_i^ε), u_i^ε remains in a bounded set of

$$L_2(0, T; V_i) \cap L_\infty(0, T; H_i), \ u_i^\varepsilon(t) \in K_i \qquad (9.101)$$

$$\frac{1}{\sqrt{\varepsilon}} (r_{ji} u_i^\varepsilon - r_{ij} u_j^\varepsilon) \ remains \ in \ a \ bounded \ set \ of \ (0, T; H_{ij}) \qquad (9.102)$$

Therefore, we can extract a subsequence, still denoted by u_i^ε, such that

$$u_i \to u_i \ in \ L_2(0, T; V_i) \ weakly, \ u_i(t) \in K_k \qquad (9.103)$$

and, by virtue of (9.101), we have

$$r_{ji}u_i = r_{ij}u_j \ \forall \, i, j \tag{9.104}$$

Notice that we have not used the fact that u_i^ε remains in a bounded set of $L_\infty(0, T; H_i)$. It follows from (9.102), (9.102), and the hypothesis (9.85) that

$$u_i = r_i u^\star, \ u^\star(t) \in K \text{ a.e. } u^\star \in L_2(0, T; V) \tag{9.105}$$

It remains to show that $u^\star = u$ is the solution of (9.73) or (9.76).

Step 3 We use the weak formulation. To avoid slight technical difficulties, we further *weaken* (9.76), by writing it

$$\int_0^T \left[\left(\frac{\partial \hat{u}}{\partial t}, \hat{u} - u \right) + a(\hat{u}, \hat{u} - u) - (f, \hat{u} - u) \right] dt \geq 0$$

$$for \ all \ \hat{u} \ satisfying \ (5.75) \tag{9.106}$$

We introduce $\hat{u}_i$ such that

$$\hat{u}_i \in L_2(0, T; V_i), \ \frac{\partial \hat{u}_i}{\partial t} \in L_2(0, T; V_i^\star)$$
$$\hat{u}_i(t) \in K_i \ for \ a.e. \ t, \ \hat{u}(0) = 0 \tag{9.107}$$

and we replace (9.92) by its (very) weak form

$$\int_0^T [c_i \left(\frac{\partial \hat{u}_i}{\partial t}, \hat{u} - u_i^\varepsilon \right) + a_i(\hat{u}_i, \hat{u}_i - u_i^\varepsilon)$$
$$\frac{1}{\varepsilon} \sum_j \langle r_{ji}\hat{u}_i - r_{ij}\hat{u}_j, r_{ji}(\hat{u}_i - u_i^\varepsilon) \rangle H_{ij}] dt$$

$$\geq \int_0^T (f_i, \hat{u}_i - u_i^\varepsilon) \tag{9.108}$$

Let us now assume that

$$\hat{u}_i = r_i \varphi$$
$$\varphi \in L_2(0, T; V^\star), \ \frac{\partial \varphi}{\partial t} \in L_2(0, T; V^\star), \ \varphi(0) = 0, \ \varphi(t) \in K \tag{9.109}$$

Since $r_{ji}r_i\varphi = r_{ij}r_j\varphi$, the $\frac{1}{\varepsilon}$ terms in (9.107) drop out so that

$$\int_0^T \left[c_i \left(\frac{\partial}{\partial t}(r_i\varphi), r_i\varphi - u_i^\varepsilon \right) + a_i(r_i\varphi, r_i\varphi - u_i^\varepsilon) \right] dt$$

$$\geq \int_0^T (f_i, r_i\varphi - u_i^\varepsilon) dt \qquad (9.110)$$

We can pass to the limit in ε in (9.109). Because of (9.104), we obtain

$$\int_0^T \left[c_i \left(\frac{\partial}{\partial t}(r_i\varphi), r_i\varphi - u_i^\star \right) + a_i(r_i\varphi, r_i\varphi - u_i^\star) \right] dt$$

$$\geq \int_0^T (f_i, r_i\varphi - r_i^\star) dt \qquad (9.111)$$

Summing (9.110) in i and using (9.89)–(9.91), we obtain

$$\int_0^T \left[\left(\frac{\partial \varphi}{\partial t}, \varphi - u_\star \right) + a(\varphi, \varphi - u_\star) - (f, \varphi - u_\star) \right] dt \geq 0$$

so that (by uniqueness) $u^\star = u$.

Parallel Algorithm

We introduce the time step Δt and a semidiscretization. We denote by u_i^n (what we hope is) an approximation of $u_i(n\Delta t)$. We then define u_i^n by

$$c_i \left\langle \frac{u_i^n - u_i^{n-1}}{\Delta t}, \hat{u} - u_i^n \right\rangle + a_i(u_i^n, \hat{u} - u_i^n)$$

$$+ \frac{1}{\varepsilon} \sum_j \left\langle r_{ji}u_i^n - r_{ij}u_j^{n-1}, r_{ji}(\hat{u}_i - u_i^n) \right\rangle H_{ij} \geq (f_i^n, \hat{u}_i - u_i^n) \; \forall \hat{u}_i \in K_i$$

$$u_i^n \in K_i \; (n = 1, 2, \ldots) \; (9.112)$$

where

$$f_i^n = \frac{1}{\Delta t} \int_{(n-1)\Delta t}^{n\Delta t} f_i(t) dt, \; u_i^0 = 0 \qquad (9.113)$$

Remark 9.4 The algorithm (9.111) is parallel. Each u_i^n is computed through the solution of stationary variational inequality. Once the u_j^{n-1} are computed, in the computation of u_i^n, only those j such that $r_{ij} \neq 0$ are used.

Let us now show the stability of the algorithm. Replacing $\hat{u}_i$ by 0 in (9.111), we obtain

$$c_i(u_i^n - u_i^{n-1}, u_i^n) + a_i(u_i^n, -u_i^n) + \frac{1}{\varepsilon}X_i^n \leq (f_i^n, u_i^n) \tag{9.114}$$

where

$$X_i^n = \sum_j \langle r_{ji}u_i^n - r_{ij}u_j^{n-1}, r_{ji}u_i^n \rangle_{H_{ij}} \tag{9.115}$$

We observe that

$$c_i\left(\frac{u_i^n - u_i^{n-1}}{\Delta t}\right) = \frac{1}{2\Delta t}c_i(u_i^n - u_i^{n-1}) + \frac{1}{2\Delta t}(c_i(u_i^n) - c_i(u_i^{n-1})) \tag{9.116}$$

so that

$$\sum_{n=1}^m c_i\left(\frac{u_i^n - u_i^{n-1}}{\Delta t}, u_i^n\right) = \frac{1}{2\Delta t}c_i(u_i^m) + \frac{1}{2\Delta t}\xi_{im} \tag{9.117}$$

where

$$\xi_{im} = \sum_{n=1}^m c_i(u_i^n - u_i^{n-1}) \tag{9.118}$$

We observe next that

$$X_i^n = \frac{1}{2}\sum \|r_{ji}u_i^n - r_{ij}u_j^{n-1}\|_{H_{ij}}^2 + \frac{1}{2}\sum_j \|r_{ji}u_i^n\|_{H_{ij}}$$
$$-\frac{1}{2}\sum_j \|r_{ij}u_j^{n-1}\|_{H_{ij}}^2\| \tag{9.119}$$

If we define

$$Y^n = \sum_{i,j} \|r_{ji}u_i^n\|_{H_{ij}} \tag{9.120}$$

then

$$\sum_i X_i^n = Z^n + Y^n - Y^{n-1} \tag{9.121}$$

where

$$Z^n = \frac{1}{2} \sum \|r_{ji} u_i^n - r_{ij} u_j^{n-1}\|_{H_{ij}}^2 \qquad (9.122)$$

Consequently, by summing (9.113) in i and in n, we obtain

$$\frac{1}{2\Delta t} \sum_i c_i(u_i^m) + \frac{1}{2\Delta t} \sum_i \xi_{im} + \sum_{n=1}^{m} \sum_i a_i(u_i^n)$$

$$+\frac{1}{\varepsilon} Y^m + \frac{1}{\varepsilon} \sum_{n=1}^{m} Z^n \leq \sum_{n=1}^{m} \sum_i (f_i^n, u_i^n) \qquad (9.123)$$

where notations $c_i(u_i, u_i) = c_i(u_i)$ and $a_i(u_i, u_i) = a_i(u_i)$ are used. But

$$(f_i^n, u_i^n) \leq \frac{1}{2} a_i(u_i^n) + \frac{c}{2} \|f_i^n\|_{V_i'}^2$$

so that (9.122), after multiplying by $2\Delta t$, gives

$$\sum_i c_i(u_i^m) + \sum_i \xi_{im} + \Delta t \sum_{n=1}^{m} \sum_i a_i(u_i^n)$$

$$+\frac{2\Delta t}{\varepsilon} Y^m + \frac{2\Delta t}{\varepsilon} \sum_{n=1}^{m} Z^n \leq \lambda \Delta t \sum_{n=1}^{m} \|f_i^n\|_{V_i'}^2 \leq \mu \qquad (9.124)$$

hence stability follows. Here, λ and μ are constants.

Remark 9.5 (i) Other time-discretization schemes could be used in (9.111).

(ii) An open problem. The previous methods do not apply (at least without new ideas) for nonlocal constraints; for example, for variational inequalities of the type

$$\left\langle \frac{\partial u}{\partial t}, \hat{u} - u \right\rangle_H + a(u, \hat{u} - u) + j(\hat{u}) - j(u) \geq (f, \hat{u} - u)$$

$$\forall \hat{u} \in V, u(t) \in V, \ u(0) = 0 \qquad (9.125)$$

where $V = H_0^1(\Omega)$, $H = L_2(\Omega)$, and where, for instance

$$j(\hat{u}) = \left(\int_\Omega |\nabla \hat{u}|^2 dx \right)^{1/2} \qquad (9.126)$$

(iii) One can extend the present method to some quasi-variational inequalities.

(iv) The examples of decomposition given here correspond to decomposition of domains. Other possibilities can be envisioned, e.g., multi-Galerkin methods, using replica equations (in our case, replica variational inequalities).

9.6 Obstacle Problem

In this section, we discuss obstacle problem in one and two dimensions. We have seen in Sect. 9.1, how a one-dimensional obstacle problem expressed as a boundary value problem can be written in the form of a variational inequality. We briefly mention how a two-dimensional obstacle problem can be modeled in the form of a variational inequality.

9.6.1 Obstacle Problem

Let us consider a body $A \subset R^2$, which we shall call the obstacle, and two points P_1 and P_2 not belonging to A (see Fig. 9.3); let us connect P_1 and P_2 by a weightless elastic string whose points can not penetrate A. Finding the shape assumed by the string is known as the one-dimensional obstacle problem.

We consider a system of Cartesian axes OXY with respect to which P_1 and P_2 have coordinates $(0, 0)$ and $(0, \ell)$, and the lower part of the boundary of the obstacle A is a Cartesian curve of equation $y = \Psi(x)$. By experience, we know that if $y = u(x)$ is the shape assumed by the string, then

$$u(0) = u(\ell) = 0 \tag{9.127}$$

Since the string connects P_1 and P_2

$$u(x) \leq \Psi(x) \tag{9.128}$$

because the string does not penetrate the obstacle.

$$u''(x) \geq 0 \tag{9.129}$$

because the string being elastic and weightless must assume a convex shape,
and

Fig. 9.3 Obstacle problem

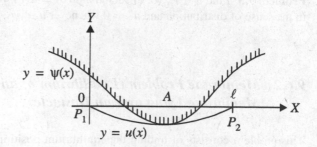

$$u(x) < \Psi(x) \Rightarrow u''(x) = 0 \qquad (9.130)$$

that is, the string takes a linear shape where it does not touch the obstacle.

Since the string tends to the shape with the minimum possible length (in particular, in the absence of an obstacle this would be ℓ), (9.126)–(9.129) are equivalent to (9.126)–(9.128) and

$$[u(x) - \Psi(x)]u''(x) = 0 \qquad (9.131)$$

A physicist may prefer to formulate the obstacle problem by saying that the configuration that the string takes such as to minimize the energy of the system. In the system made up of the string and the obstacle, given that the only energy involved is that due to elastic deformation (the string being weightless) and so by the principle of minimum energy, the configuration u taken by the string is that which minimizes the energy of the elastic deformation.

$$E(v) = \frac{1}{2} \int_0^\ell v'(x)^2 dx \qquad (9.132)$$

under given conditions. Let $K = \{v \in H_0^1(0, \ell)/v(x) \le \Psi(x) \text{ for all } x \in (0, \ell)\}$, and assume that $\Psi \in H^1(0, \ell)$ satisfies the conditions.

$$\max_\Omega, \psi > 0 \qquad (9.133)$$

$$\psi(0) < 0, \ \psi(\ell) < 0 \qquad (9.134)$$

The following formulations of the obstacle problem are equivalent:

Problem 9.1 Find $u \in K$ such that $E(u) \le E(v) \ \forall v \in K$.

Problem 9.2 Model 1 where $a(u, v) = \int_0^\ell u'v'dx$ and $F = 0$, $H = H_0^1(0, \ell)$, K as above.

Problem 9.3 Find $u \in H_0^1(0, \ell)$ such that $u(0) = u(\ell) = 0$, $u(x) \le \psi(x)$, $u'' \ge 0$ in the sense of distribution and $u'' = 0$ whenever $u < \psi$.

9.6.2   Membrane Problem (Equilibrium of an Elastic Membrane Lying over an Obstacle)

This problem consists of finding the equilibrium position of an elastic membrane, with tension τ, which

- passes through a curve Γ; that is, the boundary of an open set Ω of the horizontal plane of the coordinates (x_1, x_2),
- is subjected to the action of a vertical force of density $G = f$,
- must lie over an obstacle which is represented by a function $\chi : \overline{\Omega} \to R$

(see Fig. 9.4). This physical situation is represented by the following boundary value problem:

$$-\Delta u \geq f \text{ in } \Omega$$

$$u \geq \chi \text{ in } \Omega$$

$$(-\Delta u - f)(u - \chi) \text{ in } \Omega$$

$$u = 0 \text{ on } \Gamma$$

$$u = \chi \text{ on } \Gamma^*$$

$$\frac{\partial u}{\partial n} = \frac{\partial \chi}{\partial n} \text{ on } \Gamma^*$$

where Γ^* is the interface between the sets $\{x \in \Omega | u(x) = \chi(x)\}$ and
$\{x \in \Omega | u(x) > \chi(x)\}$ and $\partial/\partial n$ is the normal derivative operator along Γ^*. Γ^* is an unknown of the problem. Finding a solution to this boundary value problem is equivalent to solving Model 1 where $H = H_0^1(\Omega)$,
$K = \{v \in H_0^1(0, \ell)/v \geq \chi, a.e. \in \Omega\}$

$$a(u, v) = \sum_{i=1}^{2} \int_{\Omega} \frac{\partial u}{\partial x_i} \frac{\partial v}{\partial x_i} dx$$

$$\langle F, v \rangle = \int_{\Omega} fv \, dx$$

As an illustration, we verify that the following boundary value problem:

Fig. 9.4 Equilibrium of elastic string

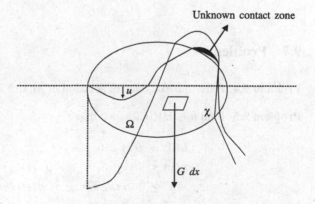

$$-\Delta u + u = -\sum_{i=1}^{2} \frac{\partial^2 u}{\partial x_i^2} + u = f \text{ on } \Omega \subset R^2$$

$$(\Omega \text{ is } boundary, \ subset, \ of, \ R^2)$$

$$u \geq 0, \ \frac{\partial u}{\partial n} \geq 0 \ on, \ \partial\Omega$$

$$u\frac{\partial u}{\partial n} = 0 \ on, \ \partial\Omega \tag{9.135}$$

is equivalent to solving Model 1, where $H = H_0^1(\Omega)$,

$K = \{v \in H^1(\Omega)/(restriction \ u \ on \ \partial\Omega)u/\partial\Omega \geq 0\}$, $u/\partial\Omega$ is understood in the sense of traces

$$a(u, v) = \int_{\Omega} \left(\sum_{i=1}^{2} \frac{\partial u}{\partial x_i} \frac{\partial v}{\partial x_i} + uv \right) dx$$

$$\langle F, v \rangle = \int_{\Omega} f(x)v(x) \, dx$$

Since K is a cone, the argument of Sect. 9.1 could be applied. By Green's theorem (in place of integration by parts), we have

$$\int_{\Omega} \left(-\sum_{i=1}^{2} \frac{\partial u}{\partial x_i} \right) + \int_{\Omega} \frac{\partial u}{\partial n} v ds \geq \int_{\Omega} fv dx$$

Following the arguments of Sect. 9.1, it can be shown that $\int_{\Omega} \frac{\partial u}{\partial n} v ds \geq 0$.

However, $v/\partial\Omega \geq 0 \ for \ all \ v \in K$ which implies that $\frac{\partial u}{\partial n} \geq 0$ on $\partial\Omega$. Thus, we have (BVP) (9.134). If u is a solution of (9.134), then on the lines of the Sect. 7.3, we can show that u is a solution of Model 1.

9.7 Problems

Problem 9.4 Prove the existence of solution of Model 3.

Problem 9.5 Find $u \in H_0^1(0, 2)$ such that

$$u(0) = u(2) = 0$$
$$u(x) \leq \Psi(x)$$
$$u'' \geq 0 \ in \ the \ sense \ of \ distribution$$
$$u'' = 0 \ whenever \ u < \Psi$$

Problem 9.6 Give a proof for Theorem 9.3.

Problem 9.7 Show that the variational inequality

$$a(u, v - u) \geq F(v - u) \; for \; all \; v \in K$$

has a unique solution where

$$a(u, v) = \int_{\Omega} \sum_{k=1}^{2} \frac{\partial u}{\partial x_k} \frac{\partial v}{\partial x_k} dx$$
$$K = \{v \in H_0^1(\Omega) / v \geq 0 \, a.e. \, in \, \Omega\}$$
$$H = H_0^1(\Omega)$$

Problem 9.8 Discuss the existence of solution of Model 4.

Problem 9.9 Obtain the Lax–Milgram Lemma (Theorem 3.39) from Theorem 9.2.

Problem 9.10 Sketch the proof of the existence of solution of Model 9.

Problem 9.11 Sketch the proof of the solution of Model 6.

Problem 9.12 Discuss the properties of the mapping $J : F \to u$ where u is a solution of Model 9 corresponding to $F \in H^*$.

Problem 9.13 Discuss a parallel algorithm for the following boundary value problem on the lines of Sect. 9.4.2

$$V = H_0^1(\Omega) \subset H = L_2(\Omega)$$
$$K = \{v / v \geq 0 \, in \, \Omega, \, v \in V\}$$
$$\langle u, v \rangle = \int_{\Omega} uv dx$$
$$a(u, v) = \int_{\Omega} \Delta u . \Delta v dx$$

Problem 9.14 Discuss the existence of solution of the variational inequality problem: Find $u \in L_2(0, T; H) \cap L_\infty(0, T; H)$, $u(t) \in K$ a.e. such that

A. $a(u(t), v - u'(t)) \geq \langle F, v - u'(t) \rangle$

B. $a(u'(t), v - u'(t)) \geq \langle F, v - u'(t) \rangle$

Problem 9.15 Let $\Omega = (0, 1)$ and

$$F(v) = c \int_0^1 v \, dx, \ c > 0, \tag{9.136}$$

$$K = \{v \in H_0^1(\Omega) / |v'| \le 1 \ a.e. \ on \ \Omega\}. \tag{9.137}$$

Write down the explicit form of the variational inequality.

$$a(u, v - u) \ge F(v - u) \ \forall \, v \in K \ u \in K$$

Chapter 10
Spectral Theory with Applications

Abstract This chapter deals with the introduction of the resolvent and the spectrum set of a bounded linear operator as well as introduction of inverse problems and their regularization.

Keywords Resolvent of a closed linear operator · Spectrum of a closed linear operator compact operator · Spectrum of compact operator · Hilbert–Schmidt theorem · Inverse problems · Singular value decomposition · Regularization · Morozov's discrepancy principle

10.1 The Spectrum of Linear Operators

We have studied properties of linear operators on Banach and Hilbert spaces in Chaps. 2 and 3. Concepts of compact operators, resolvent and spectrum of linear operators were introduced through problems 2.30–2.35. In this chapter, we recall the definition of the resolvent and the spectrum set of a bounded linear operator and present a few illustrative examples.

Extensions of classical results of spectral theory of linear operator to nonlinear operator are discussed in [4]. References [7, 63, 97, 100, 117] deal with basic results of the spectral theory.

Definition 10.1 Let T be a linear operator on a normed space X into itself. The resolvent set $\rho(T)$ is the set of complex numbers for which $(\lambda I - T)^{-1}$ is a bounded operator with domain which is dense in X. Such points of C are called regular points. The spectrum, $\sigma(T)$, of T is the complement of $\rho(T)$. We say that λ belongs to the continuous spectrum of T if the range of $(\lambda I - T)$ is dense in X, $(\lambda I - T)^{-1}$ exists, but it is unbounded. We say that λ belongs to the residual spectrum of T if $(\lambda I - T)^{-1}$ exists, but its domain is not dense in X. λ is called an eigenvalue if there is a nonzero vector satisfying $Tx = \lambda x$, and such a vector x is called an eigenvector of T.

© Springer Nature Singapore Pte Ltd. 2018

A. H. Siddiqi, *Functional Analysis and Applications*, Industrial and Applied Mathematics,
https://doi.org/10.1007/978-981-10-3725-2_10

Remark 10.1 In continuum mechanics, we often encounter operator equations of the form

$$x - T(\mu)x = f \qquad (10.1)$$

in a Banach space X, where $T(\mu)$ is a linear operator depending on a real or complex parameter μ. An important example is the equation governing the steady vibration of an elastic body with frequency $w = \lambda^{1/2}$, namely

$$\lambda x - Tx = f \qquad (10.2)$$

In particular, the natural vibration of a string is governed by the boundary value problem

$$\lambda x + x'' = 0, \quad x(0) = 0 = x \qquad (10.3)$$

Example 10.1 Consider a matrix operator defined on R^n (Example 2.25). This operator has only a point spectrum consisting of no more than n eigenvalues. All other points of the complex plane are regular points.

Example 10.2 Consider the differential operation

$$\frac{d}{dt}(\cdot)$$

on $C[a, b]$. Any point in the complex plane belongs to the point spectrum, since for any λ the equation

$$\frac{df}{dt} - \lambda f = 0 \qquad (10.4)$$

has a solution $f = e^{\lambda t}$. Thus, the operator has no regular points.

Example 10.3 Consider the following operator T defined on $C[a, b]$ by

$$Tu(t) = tu(t) \qquad (10.5)$$

It is clear that T has no eigenvalues. If $\lambda \notin [a, b]$, then the equation

$$\lambda u(t) - Tu(t) = f(t),$$

that is,

$$(\lambda - t)u(t) = f(t)$$

has the unique solution

$$u(t) = \frac{f(t)}{(\lambda - t)} \tag{10.6}$$

in $C[a, b]$. Thus, λ belongs to the resolvent set. If $\lambda \in [a, b]$, then the inverse (10.6) is defined for, that is, has the domain of, functions $f(t)$ of the form $f(t) = (t)z(t)$, where $z(t) \in C[a, b]$. This domain is not dense in $C[a, b]$, which means that points $\lambda \in [a, b]$ belong to the residual spectrum.

Example 10.4 Consider the boundary value problem

$$\frac{\partial^2 y}{\partial x^2} + \frac{\partial^2 y}{\partial y^2} = f(x, y), \; x, y \in \Omega$$
$$u = 0, \; x, y \in \partial\Omega \tag{10.7}$$

where Ω is the square $\Omega = (0, \pi) \times (0, \pi)$. Let $f(x, y) \in L_2(\Omega)$, then, given $\varepsilon > 0$, we can find N such that

$$\|f(x, y) - f_N(x, y)\| \le \varepsilon$$

where

$$f_n(x, y) = \sum_{m, x = 1} \sin mx \sin ny$$

Thus, the set S of all such $f_N(x, y)$ is dense in $L_2(\Omega)$, and

$$\|f\|_2^2 = \frac{\pi^2}{4} \sum_{m, n = 1}^{\infty} |f_{mn}|^2 < \infty$$

Consider Eq. (10.7) for $f_N(x, y)$. Suppose $\lambda \in C$, but λ is not on the negative real axis. The unique solution of (10.7) is

$$u_N(x, y) = - \sum_{m, n = 1}^{N} \frac{f_{mn}}{m^2 + \lambda n^2} \sin mx \sin ny$$

To show that $u \in L_2(\Omega)$, we need to show that $|m^2 + \lambda n^2|$ is bounded away from zero; that is, there is a $\delta > 0$ such that

$$|m^2 + \lambda n^2| \ge \delta > 0 \; for \; all \; m, n \tag{10.8}$$

If $\lambda = \lambda_1 + i\lambda_2$, then

$$|m^2 + \lambda n^2|^2 = (m^2 + \lambda_1 n^2)^2 + (\lambda_2 n^2)^2$$
$$\ge (m^2 + \lambda_1 n^2)^2 + \lambda_2^2$$

Thus if $\lambda_2 \neq 0$, then we may take $\delta = |\lambda_2|$. If $\lambda_2 = 0$, and $\lambda_1 \geq 0$, then $|m^2 + \lambda n^2|^2 \geq m^4 \geq 1$. Thus, (10.8) holds with

$$\delta = |\lambda_2| \; if \; \lambda_2 \neq 0$$
$$= 1 \; if \; \lambda_2 = 0$$

Thus

$$\|u_N\|_2^2 = \frac{\pi^2}{4} \sum_{m,n=1}^{N} \frac{|f_{mn}|^2}{|m^2 + \lambda n^2|} \leq \frac{\pi^2}{4\delta^2} \sum_{m,n=1}^{N} |f_{mn}|^2$$

$$\leq \frac{1}{\delta^2} \|f_N\|_2^2 \leq \frac{1}{\delta^2} \|f\|_2^2$$

so that

$$\|u\|_2^2 \leq \frac{1}{\delta^2} \|f\|_2^2.$$

This means that if Eq. (10.7) is written as the operator equation

$$T(\lambda)u = f$$

the inverse operation $(T(\lambda))^{-1}$ is a bounded linear operation on $L_2(\Omega)$; that is, λ belongs to the resolvent set $\rho(T)$.

It can be verified that if λ is on the negative real axis, that is

$$\lambda = -\frac{p^2}{q^2}$$

where p, q are integers; then, $\sin px \sin qy$ is a solution of

$$\frac{\partial^2 y}{\partial x^2} + \frac{\partial^2 y}{\partial y^2} = 0, \; x, y \in \Omega$$
$$u = 0, \; on \; \partial\Omega$$

so that λ is an eigenvalue of T.

It can be also checked that for $\lambda = -b^2/q^2$ where p and q are integers, there is no solution of (10.7) for $f(x, y) = \sin px \sin qy$.

Remark 10.2 1. An eigenvector is called an *eigenfunction* if the normed space X is taken as a function space say $L_2[a, b]$, $C(a, b)$ and $BV[a, b]$.
2. The set of all eigenvectors corresponding to one particular eigenvalue of an operator is a vector space.

3. Every eigenvector corresponds to exactly one eigenvalue, but there are always infinitely many eigenvectors corresponding to an eigenvalue. In fact, every scalar multiple of an eigenvector of a bounded linear operator corresponding to an eigenvalue is also an eigenvector.

10.2 Resolvent Set of a Closed Linear Operator

The concept of a closed linear operator is given in Definition 4.4. Here, we introduce the concept of the resolvent operator and study its properties.

Definition 10.2 Let T be a closed linear operator on a Banach space X into itself. For any $\lambda \in \rho(T)$, the resolvent operator denoted by $R(\lambda_0, T)$ is defined as

$$R(\lambda_0, T) = (\lambda_0 I - T)^{-1} \tag{10.9}$$

Definition 10.3 Let G be a nonempty open subset of C. The function $f(\lambda)$ is said to be holomorphic in G if at every point $\lambda_0 \in G$ it has a power series expansion

$$f(\lambda) = \sum_{n=0}^{\infty} c_n (\lambda - \lambda_0)^n \tag{10.10}$$

with nonzero radius of convergence.

Theorem 10.1 *The resolvent operator is a bounded linear operator on X into itself.*

Proof Let D, R denote the domain and range of $\lambda_0 I - T$. Thus, $D = D(\lambda_0 I - T)$, $R = R(\lambda_0 I - T) = D((\lambda_0 I - T)^{-1}) = D(R(\lambda_0, T))$. By the definition of the resolvent set, (λ_0, T) is bounded on S. Thus, there is $C > 0$ such that

$$\|(\lambda_0 I - T)^{-1} y\| \le C\|y\|, \ y \in R \tag{10.11}$$

our goal is to show that $D = R = X$. If $x \in D$, there is an $y \in R$ such that $y = (\lambda_0 I - T)x$, so that (10.11) gives

$$\|x\| \le C\|(\lambda I - T)x\|, \ x \in D \tag{10.12}$$

Let y be an arbitrary element in X: Since S is dense in X, we can find $\{y_n\} \in R$ such that

$$\lim_{n \to \infty} y_n = y$$

Since R = range of $(\lambda_0 I - T)$, we can find $\{x_n\} \in D$ such that

$$(\lambda_0 I - T)x_n = y_n$$

and thus

$$(\lambda_0 I - T)(x_n - x_m) = y_n - y_m$$

Applying the inequality (10.12) to $x_n - x_m$, we find

$$||x_n - x_m|| \le C||y_n - y_m|| \tag{10.13}$$

Since $y_n \to y$, $\{y_m\}$ is a Cauchy sequence and so $\{x_n\}$ is a Cauchy sequence by (10.13). Since X is a Banach space, there is an element $x \in X$ such that

$$\lim_{n \to \infty} x_n = y$$

Since $\lambda_0 I - T$ is a closed operator, we have

$$(\lambda_0 I - T)x = y$$

But y was an arbitrary element in X; thus, the range of $\lambda_0 I - T$ is R and the domain of $R(\lambda_0, T)$ is X: Thus, the inequality (10.11) holds on X, so that $R(\lambda_0, T)$ is continuous on X.

Theorem 10.2 *Let T be a closed linear operator on a Banach space X into itself. The resolvent set $\rho(T)$ is an open set of C, and the resolvent operator $R(\lambda, T)$ is holomorphic with respect to λ in $\rho(T)$.*

Proof Suppose $\lambda_0 \in \rho(T)$: By Theorem 10.1, $R(\lambda, T)$ is a continuous linear operator on X. Thus, the series

$$R(\lambda_0, T) + \sum_{n=1}^{m} (\lambda - \lambda_0)^n R^{n+1}(\lambda_0, T) \tag{10.14}$$

is convergent in the circle $|\lambda_0 - \lambda| \cdot ||R(\lambda_0, T)|| < 1$ of C and thus is a holomorphic function of in this circle. Multiplying the series by $\lambda I - T = (\lambda - \lambda_0)I + (\lambda_0 I - T)$, we obtain I; that is, (10.14) is $(\lambda I - T)^{-1} = R(\lambda, T)$.

10.3 Compact Operators

In this section, we prove some of the results concerning compact operators on Banach as well as on Hilbert spaces given in Chaps. 2 and 3 as remarks and problems.

In the first place, we prove the following theorem containing results announced in Remark 3.19.

Theorem 10.3 *Let T be a bounded linear operator on a Hilbert space X into another Hilbert space Y. Then*

(a) $R(T) = N(T^*)^\perp$.
(b) $R(T^*) = N(T)^\perp$.

Proof Suppose $y \in \overline{R(T)}$. This means that there is a sequence $\{y_n\} \in R(T)$ such that $y_n = Tx_n$ and $y_n \to y$. If $z \in N(T^*)$, then

$$T^*z = 0 \ and \ \langle y_n, z \rangle = \langle Tx_n, z \rangle = 0$$

Thus

$|\langle y, z \rangle| = |\langle y - y_n + y_n, z \rangle| = |\langle y - y_n, z \rangle| \leq ||y - y_n|| \ ||z|| \to 0$ so that $\langle y, z \rangle = 0$; that is, y is orthogonal to every $z \in N(T^*)$. $y \in N(T^*)^\perp$ which implies $R(T) \subset N(T^*)^\perp$.

Now suppose $y \notin \overline{R(T)}$. We will show that $y \notin N(T^*)^\perp$. $R(T)$ is a closed subspace of the Hilbert space Y. By Lemmas 3.2 and 3.3, there is a unique element z in $R(T)$ such that

$$||y - z|| = \min_{v \in \overline{R(T)}} ||y - v|| \neq 0$$

and

$$\langle y - z, v \rangle = 0 \text{ for all } v \in \overline{R(T)}.$$

Put $y - z = u$, then $\langle u, v \rangle = 0$ for all $v \in R(T)$ so that, in particular, $\langle u, z \rangle = 0$ and thus

$$\langle y, u \rangle = \langle u + z, u \rangle = \langle u, u \rangle + \langle z, u \rangle = ||u||^2 \neq 0$$

or

$$\langle y, u \rangle = ||u||^2 \neq 0. \tag{10.15}$$

But since $\langle z, v \rangle = 0$ for all $v \in R(T)$, it is zero for all $v \in R(T)$, that is

$$\langle z, Tx \rangle = \langle T^*z, x \rangle = 0 \text{ for all } x \in X.$$

Hence $Tz^* = 0$, so that $z \in N(T^*)$. Thus, Eq. (10.15) states that y is not orthogonal to a nonzero element $u \in N(T^*)$; that is, y is not in $N(T^*)^\perp$. Therefore, $N(T^*)^\perp \subset R(T)$ and so $N(T^*)^\perp = R(T)$.

(b) follows immediately by changing T to T^* in (a).

A useful characterization of compact operators (see Problems 2.30–2.34 for basic properties of compact operators) can be formulated in the following form.

Theorem 10.4 *A linear operator T on a normed space X into a normed space Y is compact if and only if it maps every bounded sequence $\{x_n\}$ in X onto a sequence $\{Tx_n\}$ in Y which has a convergent subsequence.*

A few authors consider characterization of Theorem 10.4 as the definition of the compact operator.

Example 10.5 (Example of Compact Operators) Let T be an operator on $L_2(a, b)$ defined by

$$(Tx)(s) = \int_a^b K(s, t)x(t)dt$$

where a and b are finite and $K(\cdot, \cdot)$ is continuous on $[a, b] \times [a, b]$. T is a compact operator that can be checked as follows.

Let $x_n \in L_2(a, b)$ and $||x_n|| \leq M$ for $n = 1, 2, 3, \ldots$, and some $M > 0$. Then

$$|(Tx_n(s))| \leq \int_a^b |K(s, t)||x_n(t)dt \leq M \max |K(s, t)|\sqrt{b - a}$$

and thus the sequence $\{Tx_n\}$ is uniformly bounded. Moreover, for every s_1, $s_2 \in [a, b]$, we have

$$|(Tx_n)(s_1) - (Tx_n)(s_2)| \leq \int_a^b |K(s_1, t) - K(s_2, t)| \, |x_n(t)|dt$$

$$\leq \sqrt{\int_a^b |K(s_1, t) - K(s_2, t)|} \sqrt{\int_a^b |x_n(t)|^2 \, dt}$$

$$\leq M\sqrt{b - a} \max_{t \in (a,b)} |K(s_1, t) - K(s_2, t)|.$$

Since K is uniformly continuous, the last inequality implies that the sequence $\{Tx_n\}$ is equicontinuous. Therefore, by Ascoli's Theorem (Theorem A.8), the set $\{Tx_n\}$ is compact.

We prove now two theorems giving properties of compact operators on a Hilbert space.

Theorem 10.5 *Let T_1, T_2, T_3, $\ldots$, be compact operators on a Hilbert space X into itself and let $||T_n - T|| \to 0$ as $n \to \infty$ for some operator T on H into itself, then T is compact.*

Proof Let $\{x_n\}$ be a bounded sequence in X. Since T_1 is compact, by Theorem 10.4, these exists a subsequence $\{x_1, n\}$ of $\{x_n\}$ such that $\{T_1x_1, n\}$ is convergent. Similarly, the sequence $\{T_2x_1, n\}$ contains a convergent subsequence $\{T_2x_2, n\}$. In general, for $k \geq 2$, let $\{T_kx_k, n\}$ be a subsequence of $\{x_{k1}, n\}$ such that $\{T_kx_k, n\}$ is convergent. Consider the sequence $\{x_k, n\}$. Since it is a subsequence of $\{x_n\}$, we can put $x_{p_n} = x_{n,n}$, where p_n is an increasing sequence of positive integers. Obviously, the sequence $\{T_kx_{p_n}\}$ converges for every $k \in N$. We will show that the sequence $\{Tx_{pn}\}$ also converges. Let $\varepsilon > 0$. Since $\|T_n - T\| \to 0$, there exists $k \in N$ such that $\|T_k - T\| < \frac{\varepsilon}{1/3M}$, where M is a constant such that $\|x_n\| \leq M$ for all $n \in N$: Next, let $k_1 \in \mathbb{N}$ be such that

$$\|T_kx_{p_n} - T_kx_{p_m}\| < \frac{\varepsilon}{3}$$

for all $n, m > k_1$. Then

$$\|Tx_{p_n} - Tx_{p_m}\| \leq \|Tx_{p_n} - T_kx_{p_n}\| + \|T_kx_{p_n} - T_kx_{p_m}\|$$
$$+ \|T_kx_{p_m} + Tx_{p_m}\|$$
$$< \frac{\varepsilon}{3} + \frac{\varepsilon}{3} + \frac{\varepsilon}{3} = \varepsilon$$

for sufficiently large n and m. Thus, $\{Tx_{p_n}\}$ is a Cauchy sequence in the Hilbert space and so it is convergent.

Theorem 10.6 *The adjoint of a compact operator is compact.*

Proof Let T be a compact operator on a Hilbert space X. Let $\{x_n\}$ be a bounded sequence in X, that is, $\|x_n\| \leq M$ for some M for all $n \in \mathbb{N}$. Define $y_n = T^*x_n$, $n = 1, 2, \ldots$. Since T^* is bounded, the sequence $\{y_m\}$ is bounded. It thus contains a subsequence $\{y_{k_n}\}$ such that the sequence $\{Ty_{k_n}\}$ converges in X. Now, for any $m, n \in \mathbb{N}$, we have

$$\|y_{k_m} - y_{k_n}\|^2 = \|Tx_k - Tx_{k_n}\|^2$$
$$= \langle T^*(x_{k_m} - x_{k_n}), T^*(x_{k_m} - x_{k_n}) \rangle$$
$$\leq \|TT^*(x_{k_m} - x_{k_n})\| \, \|x_{k_m} - x_{k_n}\|$$
$$\leq 2M\|Ty_{k_m} - Ty_{k_n}\| \to 0$$

as $m, n \to \infty$. Therefore, $\{y_k\}$ is a Cauchy sequence in X, which implies that $\{y_{k_n}\}$ converges.

This proves that T^* is compact.

Remark 10.3 It can be proved that an operator T defined on a Hilbert space X is compact if and only if $x_n \rightharpoonup x$ implies $Tx_n \to Tx$.

10.4   The Spectrum of a Compact Linear Operator

In this section, we prove theorems related to the spectrum of a compact operator $T(\mu)$ defined on a Hilbert space X.

Definition 10.4 The null space and range of $I - T(\mu)$ denoted by $N(\mu)$ and $R(\mu)$, respectively, are defined as $N(\mu) = \{x \in X : x - T(\mu)x = 0\}$ and

$$R(\mu) = \{y \in X : y = x - T(\mu)x; x \in X\}.$$

Similarly null space and range of $I - T^\star(\mu)$ are denoted by $N^\star(\mu)$ and $R^\star(\mu)$ and are defined as

$$N^\star(\mu) = \{x \in X : x - T^\star(\mu)x = 0\}$$
$$R^\star(\mu) = \{y \in X : y = x - T^\star(\mu)x; x \in X\}$$

Remark 10.4 (a) $N(\mu)$ and $N^\star(\mu)$ are finite dimensional.
(b) $N(\mu)^\perp$, $N^\star(\mu)^\perp$ denote the orthogonal complements of $N(\mu), N^\star(\mu)$ on X, respectively (for orthogonal complements, see Sect. 3.2).

Theorem 10.7 $N(\mu)^\perp = R^\star(\mu)$ and $N^\star(\mu)^\perp = R(\mu)$. *In the proof of this theorem, we require the following lemma.*

Lemma 10.1 *There are constants $M_1, M_2 > 0$ such that*

$$M_1\|x\| \leq \|x - T(\mu)x\| \leq M_2\|x\| \tag{10.16}$$

for all $x \in N(\mu)^\perp$.

Proof (Proof of Lemma 10.1) The right-hand inequality holds because $T(\mu)$, being compact, is bounded. Let us prove the left-hand inequality. Suppose that there is no such $m_1 > 0$ for all $x \in N(\mu)^\perp$. This means that there is a sequence $x_n \subset N(\mu)^\perp$ such that $\|x_n\| = 1$ and $\|x_n - T(\mu)x_n\| \to 0$ as $n \to \infty$. Because $T(\mu)$ is compact, the sequence $\{T(\mu)x_n\}$ contains a Cauchy subsequence. But this means that xn must also contain a Cauchy subsequence because

$$x_n = T(\mu)x_n + (x_n - T(\mu)x_n) \tag{10.17}$$

Let us rename this Cauchy subsequence $\{x_n\}$. Since $N(\mu)^\perp$ is complete, $\{x_n\}$ will converge to $x \in N(\mu)^\perp$; since $\|x_n\| = 1$, we have $\|x\| = 1$. On the other hand, $x_n \to x$ and T is continuous so that $T(\mu)x_n \to T(\mu)x$. But $\|x_n - T(\mu)x_n\| \to 0$ so that $x - T(\mu)x = 0$ that is, $x \in N(\mu)$. But $\|x\| = 1$; so that x is not zero, but it belongs to both of the mutually orthogonal sets $N(\mu)^\perp$, $N(\mu)$. This is impossible. Therefore, the left-hand inequality holds.

Proof (Proof of Theorem 10.7) To prove the first result, we must show that the equation

$$x - T^*(\mu)x = y \tag{10.18}$$

has a solution if and only if $y \in N(\mu)^\perp$.

Suppose that, for some y, Eq. (10.18) has a solution x, i.e. $y \in R^*(\mu)$. Let $z \in N(\mu)$, then, by using definition of adjoint we find

$$\langle z, y \rangle = \langle z, x - T^*(\mu)x \rangle - \langle z - T(\mu)z, x \rangle = 0$$

Now suppose that $y \in N(\mu)^\perp$. The functional $\langle x, y \rangle$ is linear and continuous on X and therefore on $N(\mu)^\perp$. The space $N(\mu)^\perp$ is a Hilbert space. Therefore, by Rieszs representation theorem, there is $y^* \in N(\mu)^\perp$ such that

$$\langle x, y \rangle = \langle x, y^* \rangle = \langle x - T(\mu)x, y^* - T(\mu)y^* \rangle, \quad x \in N(\mu)^\perp$$

This equality holds for $x \in N(\mu)^\perp$, but it also holds for $x \in X$. For if $x \in X$, we may write $x = x_1 + x_2$, where $x_1 \in N(\mu)^\perp$ and $x_2 \in N(\mu)$, and use

$$\langle x, y \rangle = \langle x_1 + x_2, y \rangle = \langle x_1, y \rangle$$
$$x - T(\mu)x = x_1 - T(\mu)x_1 + x_2 - T(\mu)x_2 = x_1 - T(\mu)x_1$$
$$\langle x, y \rangle = \langle x - T(\mu)x, z \rangle = \langle x, z - T^*(\mu)z \rangle, \quad x \in X$$

so that

$$z - T^*(\mu)z = y$$

In other words, if $y \in N(\mu)^\perp$, then Eq. (10.18) has a solution, i.e. $y \in R^*(\mu)$; thus, $N(\mu)^\perp \subset R^*(\mu)$ and $N(\mu)^\perp = R^*(\mu)$. The second part follows similarly from (10.1) applied to $T^*(\mu)$.

Remark 10.5 (a) $R(\mu) = X$ if and only if $N(\mu) = 0$.

(b) If $R(\mu) = X$, then $(I - T(\mu)^{-1}$ is continuous.

(c) A compact linear operator $T(\mu)$ in a Hilbert space X has only a point spectrum.

(d) The spaces $N(\mu)$, $N^*(\mu)$ have the same dimension.

(e) The set of eigenvalues of a compact linear operator $T(\mu) = \mu T$ has no finite limit point in C.

10.5 The Resolvent of a Compact Linear Operator

It can be verified that if the domain of a linear operator T on a Hilbert space is finite dimensional, then T is compact. We have seen (Theorem 10.2) that the resolvent $(I - T(\mu)^{-1}$ of a closed operator is a holomorphic operator function of μ in the resolvent set. We will examine here its behavior near the spectrum when T is a finite-dimensional operator, a special type of compact operator, namely T_n having the general form

$$T_n x = \sum_{k=1}^{n} \langle x, \alpha_k \rangle x_k, \quad \alpha_k, x_k \in X \tag{10.19}$$

where $x_1, x_2, x_3, \ldots, x_n$ are linearly independent.

For a general compact operator, we refer to Lebedev, Vorovich, and Gladwell [120, Theorem 7.4].

The equation

$$(I - \mu T_n)x = f$$

is

$$x - \mu \sum_{k=1}^{n} \langle x, \alpha_k \rangle x_k = f \tag{10.20}$$

Its solution has the form

$$x = f + \sum_{k=1}^{n} c_k x_k$$

and, on substituting this into (10.19), we find

$$f + \sum_{k=1}^{n} c_k x_k - \mu \sum_{k=1}^{n} \left\langle f + \sum_{k=1}^{n} c_j x_j, a_k \right\rangle x_k = f.$$

Since the x_k are linearly independent, we have

$$c_k - \mu \sum_{j=1}^{n} \langle x_j, a_k \rangle = \mu \langle f, a_k \rangle, \quad k = 1, 2, \ldots, n.$$

This system may be solved by Cramer's rule to give

$$c_k = \frac{D_k(\mu; f)}{D(\mu)}, \quad k = 1, 2, \ldots, n$$

and thus

$$x = \frac{D(\mu)f + \sum_{k=1}^{n} D_k(\mu; f)x_k}{D(\mu)}. \tag{10.21}$$

The solution is a ratio of two polynomials in μ of degree not more than n. All μ which are not eigenvalues of T are points where the resolvent is holomorphic; thus, they cannot be roots of D(). If μ_0 is an eigenvalue of T_n, then $D(\mu_0) = 0$. If this were not true, then (10.21) would be solution of (10.20) which means μ_0 is not an eigenvalue $R(\mu) = X$ implies $N(\mu) = 0$ by Remark (10.5). Thus, the set of all roots of $D(\mu)$ coincides with the set of eigenvalues of T_n, and so each eigenvalue of T_n is a pole of finite multiplicity of the resolvent $(I - \mu T_n)^{-1}$.

A general result can be formulated as follows:

Theorem 10.8 *Every eigenvalue of a compact linear operator T in a separable Hilbert space is a pole of finite multiplicity of the resolvent $(I - \mu A_n)^{-1}$.*

10.6   Spectral Theorem for Self-adjoint Compact Operators

We have examined certain properties of compact operators related to eigenvalues in Sects. 10.3 and 10.4, but results for existence of eigenvalues were missing. In this section, we present existence results for self-adjoint compact operators.

It may be recalled that a linear operator T on a Hilbert space X is self-adjoint if and only if $\langle Tx, x \rangle$ is real for all $x \in X$ (Theorem 3.25). Properties of eigenvalues (proper values) and eigenvectors (proper vectors) are given by Theorem 3.33.

Theorem 10.9 (Existence Theorem) *If T is a nonzero, compact, self-adjoint operator on a Hilbert X, then it has an eigenvalue λ equal to either $\|T\|$ or $-\|T\|$.*

Proof Let $\{u_n\}$ be a sequence of elements of X such that $\|u_n\| = 1$, for all $n \in \mathbb{N}$, and

$$\|Tu_n\| \to \|T\| \ as \ n \to X \tag{10.22}$$

then

$$
\begin{aligned}
\|T^2 u_n - \|Tu_n\|2u_n\|^2 &= \langle T^2 u_n - \|Tu_n\|^2 u_n, T^2 u_n - \|Tu_n\|u_n \rangle \\
&= \|T^2 u_n\|^2 - 2\|Tu_n\|^2 \langle T^2 u_n, u_n \rangle + \|Tu_n\|^4 \|u_n\|^2 \\
&= \|T^2 u_n\|^2 - 2\|Tu_n\|^2 \langle Tu_n, Tu_n \rangle + \|Tu_n\|^4 \|u_n\|^2 \\
&= \|T^2 u_n\|^2 - \|Tu_n\|^4 \\
&\leq \|T\|^2 \|Tu_n\|^2 - \|Tu_n\|^4 \\
&= \|Tu_n\|^2 (\|T\|^2 - \|Tu_n\|^2)
\end{aligned}
$$

Since $\|Tu_n\|$ converges to $\|T\|$, we obtain

$$\|T^2 u_n - \|T\| u_n^2\| \to 0 \ as \ n \to \infty \tag{10.23}$$

The operator T^2, being the product of two compact operators, is also compact. Hence, there exists a subsequence (u_{p_n}) of (u_n) such that $(T^2 u_{p_n})$ converges. Since $||T|| \neq 0$, the limit can be written in the form $||T||^2 v$, $v \neq 0$. Then, for every $n \in \mathbb{N}$, we have

$$\left\| ||T||^2 v - ||T||^2 u_{p_n} \right\| \leq \left\| ||T||^2 v - T^2 u_{p_n} \right\| + \left\| T^2 u_{p_n} - ||Tu_{p_n}||^2 u_{p_n} \right\|$$
$$+ \left\| ||Tu_{p_n}||^2 u_{p_n} - ||T||^2 u_{p_n} \right\|$$

Thus, by (10.22) and (10.23), we have

$$\left\| ||T^2|| v - ||T||^2 u_{p_n} \right\| \to 0 \ as \ n \to \infty$$

or

$$\left\| ||T^2|| (v - u_{p_n}) \right\| \to 0 \ as \ n \to \infty$$

This means that the sequence (u_{p_n}) converges to v, and therefore

$$T^2 v = ||T||^2 v$$

The above equation can be written as

$$(T - ||T|| i)(T + ||T|| i) v = 0.$$

If $w = (T + ||T|| i) v \neq 0$, then $(T - ||T|| i) w = 0$, and thus, $||T||$ is an eigenvalue of T. On the other hand, if $w = 0$, then $-||T||$ is an eigenvalue of T.

Corollary 10.1 *If T is a nonzero compact self-adjoint operator on a Hilbert space X, then there is a vector w such that $||w|| = 1$ and*

$$|\langle Tw, w \rangle| = \sup_{||x|| \leq 1} |\langle Tx, x \rangle|$$

Proof Let w, $||w|| = 1$, be an eigenvector corresponding to an eigenvalue such that $|\lambda| = ||T||$. Then

$$|\langle Tw, w \rangle| = |\langle \lambda w \rangle| = |\lambda| \ ||w||^2 = |\lambda| = ||T|| = \sup_{||x|| \leq 1} |\langle Tx, x \rangle|.$$

Corollary 10.2 *Theorem 10.9 guarantees the existence of at least one nonzero eigenvalue, but no more in general. The corollary gives a useful method for finding that eigenvalue by maximizing certain quadratic expression.*

Theorem 10.10 *The set of distinct nonzero eigenvalues (λ_n) of a self-adjoint compact operator is either finite or $\lim_{n \to \infty} \lambda_n = 0$.*

Proof Suppose T is a self-adjoint compact operator that has infinitely many distinct eigenvalues λ_n, $n \in \mathbb{N}$. Let un be an eigenvector corresponding to λ_n such that $\|u_n\| = 1$. By Theorem 3.33, $\{u_n\}$ is an orthonormal sequence. Since orthonormal sequences are weakly convergent to 0, Remark 10.3 implies

$$0 = \lim_{n \to \infty} \|Tu_n\|^2 = \lim_{n \to \infty} \langle Tu_n, Tu_n \rangle$$
$$= \lim_{n \to \infty} \langle \lambda_n u_n, \lambda_n u_n \rangle = \lim_{n \to \infty} \lambda_n^2 \|u_n\|^2 = \lim_{n \to \infty} \lambda_n^2.$$

Example 10.6 We will find the eigenvalues and eigenfunctions of the operator T on $L_2([0, 2\pi])$ defined by

$$(Tu)(x) = \int_0^{2\pi} K(x - t)u(t)dt$$

where K is a periodic function with period 2π, square integrable on $[0, 2\pi]$. As a trial solution, we take

$$u_n(x) = e^{inx}$$

and note that

$$(Tu_n) = \int_0^{2\pi} K(x - t)e^{int}dt = e^{inx} \int_{x-2\pi}^{x} K(s)e^{ins}ds$$

Thus

$$Tu_n = \lambda_n u_n, n \in Z$$

where

$$\lambda_n = \int_0^{2\pi} K(s)e^{ins}ds.$$

The set of functions $\{u_n\}$, $n \in Z$, is a complete orthogonal system in $L_2([0, 2\pi])$.

Note that T is self-adjoint if $K(x) = K(-x)$ for all x, but the sequence of eigenfunctions is complete even if T is not self-adjoint

Theorem 10.11 (Hilbert–Schmidt Theorem) *For every self-adjoint, compact operator T on an infinite-dimensional Hilbert space X, there exists an orthonormal system of eigenvectors (u_n) corresponding to nonzero eigenvalues (λ_n) such that every element $x \in X$ has a unique representation in the form*

$$x = \sum_{n=1}^{\chi} \alpha_n u_n + v \tag{10.24}$$

where $\alpha_n \in C$ *and* v *satisfies the equation* $Tv = 0$. *If* T *has infinitely many distinct eigenvalues* $\lambda_1, \lambda_2, \ldots$, *then* $\lambda_n \to 0$ *as* $n \to \infty$.

Proof By Theorem 10.9 and Corollary 10.1, there exists an eigenvalue λ_1 of T such that

$$|\lambda_1| = \sup_{\|x\| \le 1} |\langle Tx, x \rangle|$$

Let u_1 be a normalized eigenvector corresponding to λ_1. We set

$$Q_1 = \{x \in X : x \perp u_1\}$$

i.e., Q_1 is the orthogonal complement of the set $\{u_1\}$. Thus, Q_1 is a closed linear subspace of X. If $x \in Q_1$, then

$$\langle Tx, u_1 \rangle = \langle x, Tu_1 \rangle = \lambda_1 \langle x, u_1 \rangle = 0$$

which means that $x \in Q_1$ implies $Tx \in Q_1$. Therefore, T maps the Hilbert space Q_1 into itself. We can again apply Theorem 10.9 and Corollary 10.1 with Q_1 in place of X. This gives an eigenvalue λ_2 such that

$$|\lambda_2| = \sup_{\|x\| \le 1} \{ |\langle \lambda Tx, x \rangle| : x \in Q_n \}$$

Let u_2 be a normalized eigenvector of λ_2. Clearly, $u_1 \perp u_2$. Next, we set

$$Q_2 = \{x \in Q_1 : x \perp u_2\}$$

and repeat the above argument. Having eigenvalues $\lambda_1, \lambda_2, \ldots, \lambda_n$ and the corresponding normalized eigenvectors $u_1, \ldots, u_n$, we define

$$Q_n = \{x \in Q_{n-1} : x \perp u_n\}$$

and choose an eigenvalue λ_{n+1} such that

$$|\lambda_{n+1}| = \sup_{\|x\| \le 1} \{ |\langle Tx, x \rangle| : x \in Q_n \} \tag{10.25}$$

For u_{n+1}, we choose a normalized vector corresponding to λ_{n+1}.

This procedure can terminate after a finite number of steps. Indeed, it can happen that there is a positive integer k such that $\langle Tx, x \rangle = 0$ for every $x \in Q_k$.

Then, every element x of X has a unique representation

$$x = \alpha_1 u_1 + \cdots + \alpha_k u_k + v$$

where $Tv = 0$, and

$$Tx = \lambda_1 \alpha_1 u_1 + \cdots + \lambda_k \alpha_k u_k$$

which proves the theorem in this case.

Now suppose that the described procedure yields an infinite sequence of eigenvalues $\{\lambda_n\}$ and eigenvectors $\{u_n\}$. Then, $\{u_n\}$ is an orthonormal sequence, which converges weakly to 0. Consequently, by Remark 10.3, the sequence (Tu_n) converges strongly to 0. Hence

$$|\lambda_n| = ||\lambda_n u_n|| = ||Tu_n|| \to 0 \; as \; n \to \infty.$$

Denote by M the space spanned by the vectors $u_1, u_2, \ldots$. By the projection theorem (Theorem 3.10), every $x \in X$ has a unique decomposition $x = u + v$, or

$$x = \sum_{n=1}^{\infty} \alpha_n u_n + v$$

where $v \in M^{\perp}$. It remains to prove that $Tv = 0$ for all $v \in M^{\perp}$.

Let $v \in S^{\perp}$, $v \neq 0$: Define $w = v/||v||$. Then

$$\langle Tv, v \rangle = ||v||^2 \langle Tw, w \rangle$$

Since $w \in M^{\perp} \subset Q_n$ for every $n \in \mathbb{N}$, we have

$$|\langle Tv, v \rangle| = ||v||^2 \sup_{||x|| \leq 1} \{|\langle \lambda Tx, x \rangle| : x \in Q_n\} = ||v||^2 |\lambda_{n+1}| \to 0$$

This implies $\langle Tv, v \rangle = 0$ for every $v \in M^{\perp}$. Therefore by Corollary 3.3, the norm of T restricted to $M^{\perp}$ is 0; thus, $Tv = 0$ for all $v \in M^{\perp}$.

Theorem 10.12 (Spectral Theorem for Self-adjoint Compact Operators) *Let T be a self-adjoint, compact operator on an infinite-dimensional Hilbert space X. Then, there exists in X a complete orthonormal system (an orthonormal basis) $\{v_1, v_2, \ldots\}$ consisting of eigenvectors of T. Moreover, for every $x \in X$*

$$Tx = \sum_{n=1}^{\infty} \lambda_n \langle x, v_n \rangle v_n \tag{10.26}$$

where λ_n is the eigenvalue corresponding to v_n.

Proof Most of this theorem is already contained in Theorem 10.11. To obtain a complete orthonormal system $\{v_1, v_2, ...\}$, we need to add an arbitrary orthonormal basis of $M^\perp$ to the system $\{u_1, u_2, ...\}$ (defined in the proof of Theorem 10.11). The eigenvalues corresponding to those vectors from $M^\perp$ are all equal to zero. Equality (10.26) follows from the continuity of T.

10.7   Inverse Problems and Self-adjoint Compact Operators

10.7.1   Introduction to Inverse Problems

Problems possessing the following properties are known as the well-posed problems.

 (i) Existence, that is, the problem always has a solution.
 (ii) Uniqueness, namely the problem cannot have more than one solution.
(iii) Stability, that is, small change in the cause (input) will make only small in the effect (output).

These three characteristics of problems were identified at the turn of this century by the French mathematician Jacques Salomon Hadamard who lived during 1865–1963. A major part of research in theoretical mechanics and physics during the twentieth century has been devoted to showing that under appropriate conditions the classical problems in these fields are well posed. Only in the last four decades, it has been observed that there are important real-world problems which do not have some or all properties termed above as the well-posed problems. The problems, which are not well posed, that is, they fail to have all or some of three properties, are called ill-posed problems.

Examples of well-posed problems are of the type:

Find the effect of a cause, for example (i) find the effect on the shape of a structure (deformation in a structure) when a force is applied to it (ii) how heat diffuses through a body when a heat source is applied to a boundary. Well-posed problems are also known as the direct problems. Finding cause, when an effect is known, is often ill-posed, and such problems are called *inverse problems*, for example finding the force applied when the deformation is known. Identification of parameters in a differential equation when a solution is given is an example of an inverse problem.

Tikhonov and Arsenin [185] is an invaluable guide to the early literature on inverse problems. It uses methods of functional analysis and contains several instructive examples from the theory of Fredholm integral equations. Groetsch [91] introduces inverse problems with the help of examples taken from various areas of mathematics, physics, and engineering. It treats inverse problems in the form of Fredholm integral equations of the first kind, or more generally in the form of the operator equation $Tx = y$. Applications of the results from the spectral theory to inverse problems are very well treated in Refs. [77, 103, 120, 152].

It can be noted that many ill-posed and/or inverse problems can be reduced, maybe after some linearization, to the operator equation

$$Tx = y \tag{10.27}$$

where x, y belong to normed spaces X, Y, and T in an operator from X into Y. Properties of equation (10.27) are discussed in Sects. 1.1, 1.2, 2.3.3, 2.5, 3.7, 3.8, 4.6, 7.1, 7.2, 7.4, 7.5, 7.6, 8.1, 8.3, 9.2, 9.4. We close this section by introducing Moore–Penrose generalized inverse of an operator T.

Let T be a bounded linear operator on a Hilbert space X into a Hilbert space Y. The closure $\overline{R(T)}$ of the range of T is a closed subspace of Y. By Theorem 3.10, Y can be written as $Y = \overline{R(T)} + \overline{R(T)}^{\perp} = \overline{R(T)} + R(T)^{\perp}$ because orthogonal complement is always closed. Thus, the closure of $R(T) + R(T)^{\perp}$ in Y or, in other words, the subspace $R(T) + R(T)^{\perp}$ of Y is dense in Y. We show now how one can extend the inverse operator from $R(T)$ to $R(T) + R(T)^{\perp}$. Suppose $y \in R(T) + R(T)^{\perp}$, then its projection Py onto $R(T)$ is actually in $R(T)$. This implies that there is an $x \in X$ such that

$$Tx = Py \tag{10.28}$$

This Tx, being the projection of y onto $R(T)$, is the element of $R(T)$ which is closest to y, that is, by Lemma 3.2 and Remark 3.10

$$\|Tx - y\| = \inf_{u \in X} \|Tu - y\| \tag{10.29}$$

By the projection theorem (Theorem 3.10), any $y \in Y$ may be written as

$$y = m + n, \quad m \in \overline{R(T)}, \quad n \in R(T)^{\perp}$$

Here $m = Py$. By saying that $m \in R(T)$, we state that there is an $x \in X$ such that

$$y - Tx = n \in R(T)^{\perp} \tag{10.30}$$

Such an x is called a *least squares solution* of the equation because it minimizes the norms $\|Tu - y\|$. Suppose T does have a null space, so that solution of (10.28) is not unique. There will then be a subset M of solutions x satisfying

$$\|Tx\| \geq C\|x\| \tag{10.31}$$

It can be verified that M is closed and convex. By Lemma 3.1, there is a unique $x \in M$ which minimizes $\|x\|$ on M. x is called the generalized solution of Eq. (10.27); it gives a unique solution for $y \in R(T) + R(T)^{\perp}$ which is a dense subspace of Y. This solution is called the *least squares solution of minimum norm*.

The mapping T^T from $D(T^\dagger) = R(T) + R(T)^\perp$ into $D(T)$ which associates y to the unique least squares solution of minimum norm, $T^\dagger y$, is called the Moore–Penrose generalized inverse of T.

Remark 10.6 Let T be a continuous linear operator on Hilbert space X into Hilbert space Y. Suppose $y \in D(T^\dagger)$. Then $T^\dagger(y)$ is the unique least squares solution in $N(T)^\perp$.

Theorem 10.13 *Let X, Y be Hilbert spaces and T be continuous linear operator from X into Y. The generalized inverse $T^\dagger$ on Y is a closed operator. It is continuous if and only if $R(T)$ is closed.*

Proof We recall Definition 4.4 and reword it for our case. $T^\dagger$ is closed if and only if the three statements

$$\{y_n\} \subset D(T^\dagger), \ y_n \to y, \ T^\dagger y_n \to x$$

together imply

$$y \in D(T^\dagger) \ and \ x = T^\dagger y$$

Note Remark 10.6 states that $x_n = T^\dagger y_n$ is the unique solution of $T^\star T x_n = T^\star y_n$ in $N(T)^\perp$. Thus, $\{x_n\} \subset N(T)^\perp$, $x_n \to x$, and $N(T)^\perp$ is closed, imply $x \in N(T)^\dagger$. Also $T^\star T x_n \to T^\star T x$ and $T^\star T x_n = T^\star y_n \to T^\star y$ imply $T^\star T x = T^\star y$. But $T^\star T x = T^\star y$ implies $T^\star(T x - y) = 0$, so that $T x - y \in N(T) = R(T)^\perp$ and $y \in R(T) + R(T)^\perp = D(T^\dagger)$ and x is a solution of $A^\star T x = T^\star y$ in $N(T)^\perp$. Again Remark 10.6 states that $x = T^\dagger y$. Thus $T^\dagger$ is a closed operator.

Now suppose that $T^\dagger$ is continuous. Let $\{y_n\}$ be a convergent sequence in $R(T)$, converging to y. Let $x_n = T^\dagger y_n$, then $x_n \in N(T)^\perp$. Since $T^\dagger$ is continuous and $N(T)^\perp$ is closed, $x_n \to x = T^\dagger y \in N(T)^\dagger$. On the other hand, $T x_n = y_n$ so that $T x_n \to x$ and $y_n \to T x \in R(T)$. Thus, $y = T x$ and $y \in R(T)$. Therefore, $R(T)$ is closed.

Now suppose $D(T)$ is closed, then $D(T) = R(T) + R(T)^\perp = X$, so that $T^\dagger$ is a closed linear operator on X into Y, so that, by Theorems 4.22 and 4.25, it is continuous.

10.7.2 *Singular Value Decomposition*

Let T be a compact linear operator on a Hilbert space X into another Hilbert space Y. $T^\star T$ and $T T^\star$ are compact self-adjoint linear operators on X and Y, respectively, keeping in mind Theorem 10.6 and the fact that product of two compact operators is a compact operator. It can be also verified that $T^\star T$ and $T T^\star$ are positive operators and both have the same positive eigenvalue say $\lambda > 0$, that is, $(T^\star T)x = \lambda x$, so that

$Tx \neq 0$. Then $T(T^{\star}Tx) = (TT^{\star})(Tx) = \lambda(Tx)$, so Tx is an eigenvector of $TT^{\star}$ corresponding to λ. The converse can be checked similarly.

By Theorem 10.12, existence of positive eigenvalues $\lambda_1 \geq \lambda_2 \geq \lambda_3 \geq - - -$ and finite or infinite set of orthonormal eigenvectors corresponding to eigenvalues $\lambda_1, \lambda_2, \ldots \lambda_n \ldots$ can be checked.

Theorem 10.14 *Let $\mu_j = \sqrt{\lambda_j}$ and $u_j = \mu_j^{-1} T u_j$, then*

$$T^{\star}u_j = \mu_j^{-1} T^{\star}T u_j = \mu_j^{-1}\lambda_j u_j = \mu_j v_j \tag{10.32}$$

so that

$$TT^{\star}j_j = \mu_j T v_j = \mu_j^2 u_j = \lambda_j u_j \tag{10.33}$$

Thus $\{u_j\}$ form an orthonormal set of eigenvectors for $TT^{\star}$, and Theorem 10.13 states that they are complete in the closure $R(TT^{\star}) = N(TT^{\star})^{\perp} = N(T^{\star})^{\perp}$.

Definition 10.5 The system $\{v_i, u_j, \mu_j\}$ is called a singular system for the operator T, and the members μ_j are called singular values of T.

$$Tx = \sum_{j=1}^{\infty} \alpha_j T v_j = \sum_{j=1}^{\infty} \alpha_j \mu_j u_j \tag{10.34}$$

where $\alpha_j = \langle x, v_j \rangle$, is called the *singular value decomposition* (SVD) of the operator T.

Remark 10.7 (a) Let $y \in R(T)$ then

$$x = \sum_{j=1}^{\infty} \frac{\langle y, u_j \rangle}{\mu_j} v_j + v \tag{10.35}$$

where $v \in N(T)$, is a solution of the operator equation $Tx = y$ provided

$$\sum_{j=1}^{\infty} \lambda_j^{-1} |\langle y, u_j \rangle|^2 < \infty \tag{10.36}$$

(b) The operator equation $Gx = y$ has a solution if and only if $y \in R(T)$ and (10.36) are satisfied.

(c) Condition (10.36) is often called Picard's existence criterion.

(d) It may be observed that in order for the operator equation $Tx = y$ to have a solution equivalently Picard's existence criterion to hold, $|\langle y, u_j \rangle|$ must tend to zero faster than μ_j.

A Regular Scheme for Approximating $Ty^{\dagger}$ We have seen earlier that for $y \in R(T) + R(T)^{\perp}$, $T^{\dagger}y$ gives the unique element in $N(T)^{\perp}$ which satisfies (10.29).

Equation (10.36) shows that when T is compact this solution is the one obtained by taking $v = 0$. Thus

$$T^\dagger y = \sum_{j=1}^{\infty} \frac{\langle y, u_j \rangle}{\mu_j} v_j \qquad \qquad (10.37)$$

If we denote this by x, then

$$Tx = \sum_{j=1}^{\infty} \langle y, u_j \rangle v_j = Py$$

in the notation of (10.28).

Equation (10.37) shows that if there are an infinity of singular values, then $T^\dagger$ is unbounded because, for example, $||u_k|| = 1$, while

$$||T^\dagger u_k|| = \frac{1}{\mu_k} \to 0 \ as \ k \to \infty$$

In order to obtain an approximation to $T^\dagger y$, we may truncate the expansion (10.37) and take the nth approximation as

$$x_n = \sum_{j=1}^{n} \frac{\langle y, u_j \rangle}{\mu_j} v_j \qquad \qquad (10.38)$$

then

$$||x_n - T^\dagger y|| \to 0 \ as \ n \to \infty$$

Suppose that, instead of evaluating (10.38) for y, we actually evaluate it for some near by y^δ such that $||y - y^\delta|| \le \delta$. We will obtain a bound for the difference between the x_n formed from y^δ, which we will call x_n^δ, and the true x_n formed from y; that is, we estimate $||x_n - x_n^\delta||$.

$$||x_n - x_n^\delta||^2 \sum_{j=1}^{n} \frac{\langle y - y^\delta, u_j \rangle}{\mu_j} ||v_j||^2 = \sum_{j=1}^{n} \frac{|\langle y - y^\delta, u_j \rangle|^2}{\mu_j^2}$$

$$\le \frac{1}{\mu_n^2} \sum_{j=1}^{n} |\langle y - y^\delta, u_j \rangle|^2 \le \frac{\delta_2}{\mu_n^2} \qquad (10.39)$$

A. This bound on the solution error exhibits the characteristic properties of a solution to an ill-posed problem, namely for fixed n, the error decreases with δ, but for a given δ the error tends to infinity as $n \to \infty$. The inequality (10.39) implies that in choosing an n, say $n(\delta)$, corresponding to a given data error δ, we must do so in such a way that

$$\delta_{\mu_n}^{-1} \to 0 \ as \ \delta \to 0$$

Thus, there are two conflicting requirements on n; namely, it must be large enough to mark $||x_n - T^\dagger y||$ small, but not so large as to make $\delta_{\mu_n}^{-1}$

B. A choice of $n(\delta)$ such that

$$x_n^\delta \to T^\dagger y \ as \ \delta \to 0$$

is called a regular scheme for approximating $T^\dagger y$.

10.7.3 Regularization

Let α be a positive parameter. Let us consider the problem of finding $x \in X$(X an arbitrary Hilbert space) which minimizes

$$F(u) = ||Tu - y||_Y^2 + \alpha ||u||_X \tag{10.40}$$

where Y is another Hilbert space.

It can be verified that $H = X \times Y$ is a Hilbert space with respect to the following inner product.

$$\langle z_1, z_2 \rangle_H = \langle y_1, y_2 \rangle_Y + \alpha \langle x_1, x_2 \rangle_X \tag{10.41}$$

and

$$||z||_H^2 = ||y||_Y^2 + ||x||_X^2 \tag{10.42}$$

Remark 10.8 If $(x_n, T x_n) \in R(T_H)$ converges to (x, y) in the norm of (10.42), then $Tx = y$; that is, $R(T_H)$ is closed (range of an operator defined on H is closed).

Remark 10.9 Using Lemma 3.2 (see also Remark 3.2), Theorems 3.10, 10.3, and 3.19, it can be shown that

$$x = x\alpha = (T^\star T + \alpha I)^{-1} T^\star y \tag{10.43}$$

is the unique solution of

$$(T^\star T + \alpha I)x = T^\star y \tag{10.44}$$

Theorem 10.15 x_α *given by (10.43) converges to* $T^\dagger y$ *as* $\alpha \to 0$ *when* y *satisfies Picard's condition, that is, if* $T^\dagger y$ *exists.*

Proof We note that

$$\alpha x = T^\star y - T^\star y - T^\star T x \in R(T^\star) \tag{10.45}$$

But $\overline{R(T)} = N(T)^\perp$, so that $x \in N(T)^\intercal$. But we showed that the v_j span $N(T)^\perp$ so that we many write

$$x = x_\alpha = \sum_{j=1}^\infty c_j v_j, \quad c_j = \langle x, v_j \rangle$$

Substituting this into (10.44), we find

$$\sum_{j=1}^\infty (\lambda_j + \alpha) c_j v_j = T^\star y$$

so that

$$(\lambda_j + \alpha) c_j = \langle T^\star y, v_j \rangle = \langle y, T v_j \rangle = \mu_j \langle y, u_j \rangle$$

and hence

$$x_\alpha = \frac{\mu_j \langle y, u_j \rangle v_j}{\lambda_j + \alpha} \tag{10.46}$$

To show that $x_\alpha \to T^\dagger y$, we proceed in two steps; first, we show that this operator which gives x_α in terms of y is bounded. We note that since the λ_j are positive and tend to zero, we can find $N \geq 0$ such that

$$\lambda_j \geq 1 \, for \, j \leq N, \, 0 < \lambda_j < 1 \, for \, j > N$$

Thus when $j \leq N$

$$\frac{\mu_j}{\lambda_j + \alpha} < \frac{1}{\mu_j} \leq 1$$

while if $j > N$

$$\frac{\mu_j}{\mu_j + \alpha} < \frac{\mu_j}{\alpha} < \frac{1}{\alpha}$$

so that if $\beta = \max(1, 1/\alpha)$, we may write

$$\|x_\alpha\|^2 \le \beta^2 \sum_{j=1}^{\infty} |\langle y, u_j \rangle|^2 \le \beta^2 \|y\|^2 \tag{10.47}$$

Now we show the convergence of x_α to $T^\dagger y$, for those y for which Picard's existence criterion holds. We note that

$$T^\dagger - x_\alpha = \sum_{j=1}^{\infty} \left(\frac{1}{\mu_j} - \frac{\mu_j}{\lambda_j + \alpha} \right) \langle \lambda y, y_j \rangle v_j$$

so that

$$\|T^\dagger - x_\alpha\|^2 = \sum_{j=1}^{\infty} \left(\frac{\alpha}{\mu_j(\lambda_j + \alpha)} \right)^2 |\langle y, u_j \rangle|^2$$

$$\le \sum_{j=1}^{\infty} \frac{|\langle y, u_j \rangle|^2}{\lambda_j} < \infty \tag{10.48}$$

Choose $\varepsilon > 0$. Since the series in (10.48) converges, the sum from $N+1$ to ∞ must tend to zero as N tends to infinity. Therefore, we can find N such that sum is less than $\varepsilon/2$ and

$$\|T^\dagger - x_\alpha\|^2 < \sum_{j=1}^{\infty} \left(\frac{\alpha}{\mu_j(\lambda_j + \alpha)} \right)^2 |\langle y, u_j \rangle|^2 + \varepsilon/2$$

But now the sum on the right is a finite sum, and we can write

$$\|T^\dagger - x_\alpha\|^2 \le \alpha^2 \sum_{j=1}^{N} \varepsilon \frac{|\langle y, u_j \rangle|^2}{\lambda_j^3} + \frac{\varepsilon}{2} = S_N \alpha^2 + \frac{\varepsilon}{2} \tag{10.49}$$

Finally, we choose α so that $\alpha < \sqrt{\varepsilon/(2S_N)}$, then $\|T^\dagger y - x_0\|^2 < \varepsilon$, so that

$$\lim_{0 \to 0} \|T^\dagger - x_\alpha\| = 0$$

We have proved that, for any α, for any α, x_α is a continuous operator and that, for those y for which $T^\dagger y$ exists, x converges to $T^\dagger y$ as $\alpha \to 0$. Let the data y is subject to error. This means that instead of solving (10.44) for y, we actually find below a bound for the difference between the x_α formed from y^δ, which we call x_α^δ, and the actual x_α formed from y. Thus, we wish to estimate $\|x_\alpha - x_\alpha^\delta\|$. We have

$$x_\alpha - x_\alpha^\delta = \sum_{j=1}^{\infty} \frac{\mu_j}{\lambda_j + \alpha} \langle y - y^\delta, u_j \rangle v_j$$

so that, by proceeding as in (10.46), (10.47), we have

$$||x_\alpha - x_\alpha^\delta||^2 = \sum_{j=1}^{\infty} \frac{\lambda_j}{(\lambda_j + \alpha)^2} |\langle y - y^\delta, u_j \rangle|^2 \le \beta^2 ||y - y^\delta||^2$$

where $\beta = max(1, 1/\alpha)$. Since the series convergence, we may, for any given $\varepsilon > 0$, find N such that the sum from $N + 1$ to ∞ is less than ε. Now

$$\frac{\lambda_j}{(\lambda_j + \alpha)^2} = \frac{\lambda_j}{(\lambda_j + \alpha)} \cdot \frac{\lambda_j}{(\lambda_j + \alpha)} < 1 \cdot \frac{1}{\alpha}$$

so that

$$||x_\alpha - x_\alpha^\delta||^2 < \frac{1}{\alpha} \sum_{j=1}^{\infty} |\langle y - y^\delta, u_j \rangle|^2 + \varepsilon < \frac{||y - y^\delta||^2}{\alpha} + \varepsilon$$

and hence, since this is true for all $\varepsilon > 0$, we must have

$$||x_\alpha - x_\alpha^\delta|| \le \frac{||y - y^\delta||}{\sqrt{\alpha}} \le \frac{\delta}{\sqrt{\alpha}} \tag{10.50}$$

Again, this bound on the solution error illustrates the characteristic properties of a solution to an ill-posed problem: for fixed α, the error decreases with δ, but for a given δ, the error tends to infinity as $\alpha \to 0$. The inequality (10.50) implies that in choosing an α, say $\alpha(\delta)$, corresponding to a given data error δ, we must do so in such a way that

$$\frac{\delta}{\sqrt{\alpha}} \to 0 \ as \ \delta \to 0 \tag{10.51}$$

when we choose $\alpha(\delta)$ so that (10.51) holds, the difference between $x_{\alpha(\delta)}^\delta$ and $T^\dagger y$ satisfies the inequality

$$||x_{\alpha(\delta)}^\delta - T^\dagger y|| \le ||x_{\alpha(\delta)}^\delta - x_{\alpha(\delta)}|| + ||x_{\alpha(\delta)} - T^\dagger y||$$

$$\le \frac{\delta}{\sqrt{\alpha}} + ||x_{\alpha(\delta)} - T^\dagger y|| \tag{10.52}$$

and we have already shown that the second term tends to zero with α. A choice of $\alpha(\delta)$ such that

$$x_{\alpha(\delta)}^\delta \Rightarrow T^\dagger y \ as \ \delta \to 0$$

is called a regular scheme for approximating $T^\dagger y$.

Remark 10.10 The inequality (10.52) gives a bound for the error in $T^\dagger y$. The error has two parts: the first is that due to the error in the data, while the second is that due to using α rather than the limit as $\alpha \to 0$. It is theoretically attractive to ask whether we can choose the way in which α depends on delta, i.e. $\alpha(\delta)$, so that both error terms are of the same order.

To bound the second term, we return to the inequality (10.49). This holds for arbitrary ε. If we take $\varepsilon = 2S_N\alpha^2$, we find

$$\|T^\dagger y - x_\alpha\|^2 \le 2S_N\alpha^2 \ as \ \alpha \to 0$$

so that

$$\|T^\dagger y - x_\alpha\| = O(\alpha)$$

This means that if we use the simple choice $\alpha(\delta) = k\delta$, then the first term in (10.52) will be of order δ, while the second will be of order δ. On the other hand, if we take $\alpha(\delta) = C\delta^{2/3}$, then $\delta/\alpha(\delta)$ and α will both be of order $\delta^{2/3}$, so that

$$\|x_{\alpha(\delta)}^\delta - T^\dagger y\| = O(\delta^{2/3})$$

Remark 10.11 A. The solution of the problem

$$\|Tx - y\|^2 + \alpha\|x\|^2 = \inf_{u \in X}\{\|Tu - y\|^2 + \alpha\|u\|^2\}$$

converges to $T^\dagger y$.
B. The effect of error (10.50)

$$\|x_\alpha - x_\alpha^\delta\| \le \frac{\delta}{\sqrt{\delta}} \to 0 \ as \ \delta \to 0$$

$$\le \frac{\delta}{\sqrt{\delta}} \to \infty \ as \ \alpha \to 0$$

10.8 Morozov's Discrepancy Principle

In 1984, Morozov {Mo 84} put forward a discrepancy principle in which the choice of α is made so that the error in the prediction of y^δ, that is, $\|Tx_\alpha^\delta - y^\delta\|$ is equal to

$$\|Tx_\alpha^\delta - y^\delta\| = \|y - y^\delta\| = \delta \tag{10.53}$$

Theorem 10.16 *For any $\delta > 0$, there is a unique value of α satisfying (10.53).*

Proof It can be verified that

$$y^\delta = \sum_{j=1}^{\infty} \langle y^\delta, u_j \rangle u_j + P y^\delta$$

where $P y^\delta$ is the projection of y^δ on $R(T)^\perp$.

Applying T to x_α^δ given by (10.45) to get

$$T x_\alpha^\delta = \sum_{j=1}^{\infty} \lambda_j \cdot \frac{\langle \lambda y^\delta, u_j \rangle u_j}{\lambda_j + \alpha}$$

$$because \ \ ||T x_\alpha^\delta||^2 = \sum_{j=1}^{\infty} \left(\frac{\lambda_j}{(\lambda_j + \alpha)} \right) / |\langle y^\delta, y_j \rangle|^2 \le ||y^\delta||^2$$

Thus

$$y^\delta - T x_\alpha^\delta = \sum_{j=1}^{\infty} \frac{\alpha \langle y^\delta, u_j \rangle u_j}{\lambda_j + \alpha} + P y^\delta$$

and

$$||T x_\alpha^\delta - y^\delta||^2 = \sum_{j=1}^{\infty} \left(\frac{\alpha}{\lambda_j + \alpha} \right)^2 |\langle y^\delta, u_j \rangle|^2 + ||P y^\delta||^2.$$

This equation shows that $f(\alpha) = ||T x_\alpha^\delta - y^\delta||$ is a monotonically increasing function of α for $\alpha > 0$. In order to show the existence of a unique value of α such that $f(\alpha) = \delta$, we must show that

$$\lim_{\alpha \to \infty} f(\alpha) \le \delta \ and \ \lim_{\alpha \to \infty} f(\alpha) > \delta$$

Since

$$y \in R(T), \ \ Py = 0 \ and \ thus$$
$$\lim_{\alpha \to 0} f(\alpha) = ||P y^\delta|| = ||P(y - y^\delta)|| \le ||y - y^\delta|| \le \delta$$

on the other hand, by Theorem 3.18

$$\lim_{\alpha \to \infty} f(\alpha) = ||y^\delta|| > \delta$$

Theorem 10.17 *Choosing $\alpha(\delta)$ according to the discrepancy principle does provide a regular scheme for approximating $T^\dagger y$, that is*

$$x_{\alpha(\delta)}^\delta \to T^\dagger y \ as \ \delta \to 0 \tag{10.54}$$

or equivalently Morozov's discrepancy principle provides a regular scheme for solving $Tx = y$.

Proof Without loss of generality, we may take $y \in R(T)$ so that there is a unique $u \in N(T)^{\perp}$, which we call $x = T^{\dagger}y$, such that $y = Tx$. Since we have shown that α is uniquely determined, we may write $x^0_{\alpha(\delta)}$ as $x(\delta)$.

First, we show that the $x(\delta)$ are bounded. We find $x(\delta)$ as the minimum of

$$F(u) = ||Tu - y^{\delta}||^2 + \alpha||u||^2$$

for all $u \in X$. Thus if $u \in X_1$, then

$$F(x(\delta)) \leq F(u)$$

so that in particular

$$F(x(\delta)) \leq F(x)$$

But we choose $x(\delta)$ so that $||Tx(\delta) - y^{\delta}|| = \delta$, so that

$$F(x(\delta)) = \delta^2 + \alpha||x(\delta)||^2$$

while

$$F(x) = ||Tx - y^{\delta}||^2 + \alpha||x||^2$$
$$= ||y - y^{\delta}||^2 + \alpha||x||^2 = \delta^2 + \alpha||x||^2$$

from which we conclude that

$$||x(\delta)|| \leq ||x||$$

i.e., the $x(\delta)$ are bounded. Now suppose that $\{y_n\}$ is a sequence converging to y and that $||y_n - y|| = \delta_n$. Each such pair y_n, δ_n will determine an $\alpha_{(\delta_n)}$ and a corresponding $x^{\alpha}(\delta_n)$ which we will call x_n.

It can be proved that there is subsequence of (δ_n) for which the x_n converge to $x = T^{\top}y$ provided T is compact linear operator on X into Y.

Thus, we have shown that Morozov's discrepancy principle provides a regular scheme for solving the operator equation $Tx = y$, where T is a compact linear operator on X into Y.

10.9   Problems

Problem 10.1 (a) Let T be a compact operator on a Hilbert space X and let S be a bounded operator on X, then show that TS and ST are bounded.

(b) Show that $TT^\star$ and $T^\star T$ are self-adjoint compact operator whenever T is a compact operator on a Hilbert space X.

Problem 10.2 Prove that a bounded set A in a finite-dimensional normed space is precompact; that is, its closure $\bar{A}$ is compact

Problem 10.3 Show that a linear operator T on a normed space X into another normed space Y is compact if and only if it maps bounded sets of X onto precompact sets of Y.

Problem 10.4 Let T be a linear operator on normed space X into normed space Y: Show that

(a) T is compact whenever it is bounded and dim $R(T)$ is finite.
(b) T is compact provided dim of the domain of T is finite.

Problem 10.5 Prove Remark 10.3.

Problem 10.6 Prove the last assertion; namely, there is a subsequence of δ_n for which the $x_n \to x = T^\dagger y$, in the proof of Theorem 10.17.

Chapter 11
Frame and Basis Theory in Hilbert Spaces

Abstract Duffin and Schaeffer introduced in 1951 a tool relaxing conditions on the basis and named it frame. Every element of an inner product space can be expressed as a linear combination of elements in a given frame where linear independence is not required. In this chapter, frame and basis theory are presented.

Keywords Orthonormal basis · Bessel sequence · Riesz bases · Biorthogonal system · Frames in Hilbert spaces

11.1 Frame in Finite-Dimensional Hilbert Spaces

In the study of vector spaces, a basis is among the most important concepts, allowing every element in the vector space to be written as a linear combination of the elements in the basis. It may be observed that conditions to be a basis are quite restrictive such as no linear dependence between the elements is possible, and sometimes we even want the orthogonality of elements with respect to an inner product. Due to these limitations, one might look for a more flexible concept. In 1951, Duffin and Schaeffer [66] introduced a tool, relaxing conditions on the basis, called a frame. A frame for an inner product space also allows each element in the space to be written as a linear combination of the elements in the frame, but linear independence between the frame elements is not required. In this chapter, we present basic properties of frames in finite-dimensional Hilbert spaces.

Definition 11.1 Let H be a finite-dimensional inner product space. A sequence $\{f_k\}$ in H is called a basis if the following two conditions are satisfied

(a) $H = span\{f_k\}$, that is, every element f of H is a linear combination of f_k;
(b) $\{f_k\}_{k=1}^{m}$ is linearly independent, that is, if $\sum_{k=1}^{m} \alpha_k f_k = 0$ for some scalar coefficients α_k, $k = 1, 2, 3, \ldots m$, then $\alpha_k = 0$ for all $k = 1, 2, \ldots m$.

Definition 11.2 A basis $\{f_k\}_{k=1}^{m}$ is called an orthonormal basis if

$$\langle f_k, f_j \rangle = \delta_{k,j} = 1 \; if \; k = j$$
$$= 0 \; if \; k \neq j$$

© Springer Nature Singapore Pte Ltd. 2018
A. H. Siddiqi, *Functional Analysis and Applications*, Industrial and Applied Mathematics,
https://doi.org/10.1007/978-981-10-3725-2_11

Theorem 11.1 *If $\{f_k\}$ is an orthonormal basis, then any element $f \in H$ can be written as*

$$f = \sum_{k=1}^{m} \langle f, f_k \rangle f_k$$

Proof

$$\langle f, f_j \rangle = \left\langle \sum_{k=1}^{m} \alpha_k f_k, f_j \right\rangle$$

$$= \sum_{k=1}^{m} \alpha_k \langle f_k, f_j \rangle$$

$$= \alpha_j$$

so

$$f = \sum_{k=1}^{m} \alpha_j f_j = \sum_{k=1}^{m} \langle f, f_j \rangle f_j$$

Definition 11.3 A sequence of elements $\{f_k\}_{k \in I}$ in a finite-dimensional inner product space H is called a frame if there exist constants $A, B > 0$ such that

$$A\|f\|^2 \le \sum_{k=1}^{\infty} |\langle f, f_k \rangle|^2 \le B\|f\|^2, \; for \; all \; f \in H \tag{11.1}$$

The constants A and B are called frame bounds.

Remark 11.1 A. Constants A and B are not unique.
B. In a finite-dimensional space, it is somehow artificial to consider sequences having infinitely many elements. Therefore, we focus here only finite families $\{f_k\}_{k=1}^{m}$, $m \in N$. With this restriction in finite-dimensional inner product spaces, the upper frame condition is automatically satisfied.
C. In order that the lower condition in (11.1) is satisfied, it is necessary that span $\{f_k\}_{k=1}^{m} = H$. This condition turns out to be sufficient; in fact, every finite sequence is a frame for its span.

Theorem 11.2 *Let $\{f_k\}_{k=1}^{m}$ be a sequence in a finite inner product space H. Then $\{f_k\}_{k=1}^{m}$ is a frame for span $\{f_k\}_{k=1}^{m}$.*

Proof We can assume that not all f_k are zero. As we have seen, the upper frame condition is satisfied with $B = \sum_{k=1}^{m} \|f_k\|^2$. Now let

$$M = span\{f_k\}_{k=1}^{m}$$

and consider the continuous mapping

$$\phi : M \to R, \phi(f) = \sum_{k=1}^{m} |\langle f, f_k \rangle|^2$$

The unit ball in M is compact, so we can find $g \in M$ with $\|g\| = 1$ such that

$$A := \sum_{k=1}^{m} |\langle g, f_k \rangle|^2 = \inf \left\{ \sum_{k=1}^{m} |\langle f, f_k \rangle|^2 : f \in M, \|f\| = 1 \right\}$$

It is clear that $A > 0$. Now given $f \in M$, $f \neq 0$, we have

$$\sum_{k=1}^{m} |\langle f, f_k \rangle|^2 = \sum_{k=1}^{m} \left\| \left\langle \frac{f}{\|f\|}, f_k \right\rangle \right\|^2 \|f\|^2 \geq \|f\|^2$$

Remark 11.2 Remark 11.1 shows that a frame might contain more elements than needed to be a basis. In particular, if $\{f_k\}_{k=1}^{m}$ is a frame for H and $\{g_k\}_{k=1}^{n}$ is an arbitrary finite collection of vectors in H, then $\{f_k\}_{k=1}^{m} \cup \{g_k\}_{k=1}^{n}$ is also a frame for H. A frame which is not a basis is said to be over complete or redundant.

Remark 11.3 Consider now a vector space H equipped with a frame $\{f_k\}_{k=1}^{m}$ and define

(i) a linear mapping

$$T : C^m \to H, \quad T\{c_k\}_{k=1}^{m} = \sum_{k=1}^{m} c_k f_k \tag{11.2}$$

T is usually called the *preframe operator* or the *synthesis operator*. The adjoint operator is given by

$$T^\star : H \to C^m, \quad T^\star = \{\langle f, f_k \rangle\}_{k=1}^{m} \tag{11.3}$$

(ii)

$$S : H \to H, \quad Sf = TT^\star f = \sum_{k=1}^{m} \langle f, f_k \rangle f_k \tag{11.4}$$

is called the *frame operator*.

$$\langle Sf, f \rangle = \sum_{k=1}^{m} |\langle f, f_k \rangle|^2, \quad f \in V \tag{11.5}$$

the lower frame condition can thus be considered as some kind of "lower bound" on the frame operator.

(iii) A frame $\{f_k\}_{k=1}^m$ is tight if we can choose $A = B$ in the definition, i.e., if

$$\sum_{k=1}^m |\langle f, f_k\rangle|^2 = A\|f\|^2, \ \forall f \in V \tag{11.6}$$

For a tight frame, the exact value A in (11.6) is simply called the *frame bound*.

Theorem 11.3 *Let* $\{f_k\}_{k=1}^m$ *be a frame for H with frame operator S. Then*

(i) *S is invertible and self-adjoint.*
(ii) *Every $f \in H$ can be represented as*

$$f = \sum_{k=1}^m \langle f, S^{-1} j_k\rangle f_k = \sum_{k=1}^m \langle f, f_k\rangle S^{-1} j_k \tag{11.7}$$

(iii) *If $f \in H$ also has the representation $f = \sum_{k=1}^m c_k f_k$ for some scalar coefficients $\{c_k\}_{k=1}^m$, then*

$$\sum_{k=1}^m |c_k|^2 = \sum_{k=1}^m |\langle f, S^{-1} j_k\rangle|^2 + \sum_{k=1}^m |c_k - \langle f, S^{-1} j_k\rangle|^2.$$

Proof Since $S = TT^\star$, it is clear that S is self-adjoint. We now prove (ii). Let $f \in H$ and assume that $Sf = 0$. Then

$$0 = \langle Sf, f\rangle = \sum_{k=1}^m |\langle f, f_k\rangle|^2$$

implying by the same condition that $f = 0$: That S is injective actually implies that S surjective, but let us give a direct proof. The frame condition implies by Remark 11.1 that $\mathrm{span}\{f_k\}_{k=1}^m = H$, so the preframe operator T is surjective. Given $f \in H$, we can therefore find $g \in H$ such that $Tg = f$; we can choose $g \in N_T^\perp = R_{T^\star}$, so it follows that $R_S = R_{TT^\star} = H$. Thus, S is surjective, as claimed. Each $f \in H$ has the representation

$$f = SS^{-1}f$$
$$= TT^\star S^{-1}f$$
$$= \sum_{k=1}^m \langle S^{-1}f, f_k\rangle f_k,$$

using that S is self-adjoint, we arrive at

$$f = \sum_{k=1}^{m} \langle f, S^{-1} f_k, \rangle f_k.$$

The second representation in (11.7) is obtained in the same way, using that $f = S^{-1} S f$. For the proof of (iii), suppose that $f = \sum_{k=1}^{m} c_k f_k$. We can write

$$\{c_k\}_{k=1}^{m} = \{c_k\}_{k=1}^{m} - \{\langle f, S^{-1} f_k, \rangle\}_{k=1}^{m} + \{\langle f, S^{-1} f_k \rangle\}_{k=1}^{m}.$$

By the choice of $\{c_k\}_{k=1}^{m}$, we have

$$\sum_{k=1}^{m} (c_k - \langle f, S^{-1} f_k \rangle) f_k = 0$$

i.e., $\{c_k\}_{k=1}^{m} - \{\langle f, S^{-1} f_k \rangle\}_{k=1}^{m} \in N_T = R_T^{\perp}$; since

$$\{\langle f, S^{-1} f_k \rangle\}_{k=1}^{m} = \{\langle S^{-1} f, f_k \rangle\}_{k=1}^{m} \in R^{T\star}$$

we obtain (iii).

Theorem 11.4 *Let $\{f_k\}_{k=1}^{m}$ be a frame for a subspace M of the vector space H. Then the orthogonal projection of H onto M is given by*

$$Pf = \sum_{k=1}^{m} \langle f, S^{-1} f_k \rangle f_k \tag{11.8}$$

Proof It is enough to prove that if we define Pf by (11.8), then

$$Pf = f \text{ for } f \in M \text{ and } Pf = 0 \text{ for } f \in M^{\perp}$$

The first equation follows by Theorem 11.3, and the second by the fact that the range of S^{-1} equals M because S is a bijection on M.

Example 11.1 Let $\{f_k\}_{k=1}^{2}$ be an orthonormal basis for a two-dimensional vector space H with inner product. Let

$$g_1 = f_1, \ g_2 = f_1 - f_2, \ g_3 = f_1 + f_2.$$

Then $\{g_k\}_{k=1}^{3}$ is a frame for H. Using the definition of the frame operator

$$Sf = \sum_{k=1}^{3} \langle f, g_k \rangle g_k$$

we obtain that

$$Sf_1 = f_1 + f_1 - f_2 + f_1 + f_2 = 3f_1$$
$$Sf_2 = (f_1 - f_2) + f_2 = 2f_2$$

Thus,

$$S^{-1}f_1 = \frac{1}{3}f_1, \ S^{-1}f_2 = \frac{1}{2}f_2.$$

Therefore, the canonical dual frame is

$$\{S^{-1}g_k\}_{k=1}^3 = \{\frac{1}{3}f_1, \frac{1}{3}f_1 - \frac{1}{2}f_2, \frac{1}{3}f_1 + \frac{1}{2}f_2\}$$

By Theorem 11.3, the representation of $f \in V$ in terms of the frame is given by

$$f = \sum_{k=1}^3 \langle f, S^{-1}g_k \rangle g_k$$
$$= \frac{1}{3}\langle f, f_1 \rangle f_1 + \langle f, \frac{1}{3}f_1 - \frac{1}{2}f_2 \rangle (f_1 - f_2) + \langle f, \frac{1}{3}f_1 + \frac{1}{2}f_2 \rangle (f_1 + f_2)$$

11.2 Bases in Hilbert Spaces

11.2.1 Bases

The concept of orthonormal basis in a Hilbert space is discussed in Sect. 3.5. We introduce here bases in Banach spaces in general and Hilbert spaces in particular. For classical work on bases theory, we refer to [129, 175].

Definition 11.4 Let X be a Banach space. A sequence of elements $\{f_k\}_{k=1}^\infty$ of X is called a basis (very often Schauder basis) for X if, for each $f \in X$, there exists unique scalar coefficients $\{\alpha_k\}_{k=1}^\infty$ such that

$$f = \sum_{k=1}^\infty \alpha_k f_k \tag{11.9}$$

Remark 11.4 A. $\alpha_k's$ depend on f.
  B. Sometimes one refers to (11.9) as the expansion of f in the basis $\{f_k\}_{k=1}^\infty$ (11.9) means that the series on the right side converges f in the norm of X with respect to the chosen order of the elements.

C. Besides the existence of an expansion of each $f \in X$. Definition 11.4 demands for uniqueness. This can be obtained by imposing the condition of linear independence on $\{f_k\}_{k=1}^{\infty}$. In infinite-dimensional Banach spaces, different concepts of independence exist [40, p. 46]; however, we discuss here only classical linear independence.

Definition 11.5 Let $\{f_k\}_{k=1}^{\infty}$ be a sequence in a Banach space X. $\{f_k\}_{k=1}^{\infty}$ is called linearly independent if every finite subset of $\{f_k\}_{k=1}^{\infty}$ is linearly independent. $\{f_k\}_{k=1}^{\infty}$ is called a basis in X if (11.9) is satisfied and $\{f_k\}_{k=1}^{\infty}$ is linear independent

Definition 11.6 Sequence of vector $\{f_k\}_{k=1}^{\infty}$ in a Banach space X is called complete if

$$\overline{span}\{f_k\}_{k=1}^{\infty} = X \tag{11.10}$$

Theorem 11.5 *A complete sequence of nonzero vectors $\{f_k\}_{k=1}^{\infty}$ in a Banach space X is a basis if and only if there exists a constant K such that for all $m, n \in N$ with $m \leq n$*

$$\left\| \sum_{k=1}^{\infty} \alpha_k f_k \right\| \leq K \left\| \sum_{k=1}^{\infty} \alpha_k f_k \right\| \tag{11.11}$$

for all scalar sequences $\{\alpha_k\}_{k=1}^{\infty}$.

Corollary 11.1 *The coefficient functionals $\{\alpha_k\}$ associated to a basis $\{f_k\}_{k=1}^{\infty}$ for X are continuous and can be as elements in the dual X^*. If there exists a constant $C > 0$ such that $\|f_k\| \geq C$ for all $k \in \mathcal{N}$, then the norms of $\{\alpha_k\}_{k=1}^{\infty}$ are uniformly bounded.*

The proof of Theorem 11.5 and Corollary 11.1 we refer to [40, pp. 47–50].

Definition 11.7 *(Bessel Sequences)* A sequence $\{f_k\}$ in a Hilbert space H is called a *Bessel sequence* if there exists a constant $B > 0$ such that

$$\sum_{k=1}^{\infty} |\langle f, f_k \rangle|^2 \leq B \|f\|^2, \; for all \; f \in H \tag{11.12}$$

contact B occurring in (11.12) is called a Bessel bound for $\{f_k\}_{k=1}^{\infty}$.

Theorem 11.6 *A sequence $\{f_k\}_{k=1}^{\infty}$ in a Hilbert space H is a Bessel sequence with Bessel bound B if and only if*

$$T : \{\alpha_k\}_{k=1}^{\infty} \to \sum_{k=1}^{\infty} \alpha_k f_k$$

is a well-defined bound operator from l_2 into H and $\|T\| \leq B$.

We require the following lemma in the proof.

Lemma 11.1 *Let* $\{f_k\}_{k=1}^\infty$ *be a sequence in H, and suppose that convergent for all* $\{f_k\}_{k=1}^\infty \in l_2$. *Then*

$$T : l_2 \to \mathcal{H}, \quad T\{\alpha_k\}_{k=1}^\infty := \sum_{k=1}^\infty \alpha_k f_k \tag{11.13}$$

defines a bounded linear operator. The adjoint operator is given by

$$T^* : \mathcal{H} \to l_2, \quad T^* f = \{\langle f, f_k \rangle\}_{k=1}^\infty \tag{11.14}$$

Furthermore,

$$\sum_{k=1}^\infty |\langle f, f_k \rangle|^2 \le \|T\|^2 \|f\|^2, \quad \forall f \in \mathcal{H} \tag{11.15}$$

Proof Consider the sequence of bounded linear operators

$$T_n : l_2 \mathbb{N} \to \mathcal{H}, \quad T_n\{\alpha_k\}_{k=1}^\infty := \sum_{k=1}^\infty \alpha_k f_k$$

Clearly $T_n \to T$ pointwise as $n \to \infty$, so T is bounded by Theorem 4.21. In order to find the expression for T^*, let $f \in H$, $\{\alpha_k\} \in l_2$. Then

$$\langle f, T\{\alpha_k\}_{k=1}^\infty \rangle_{\mathcal{H}} = \left\langle f, \sum_{k=1}^\infty \alpha_k f_k \right\rangle_{\mathcal{H}} = \sum_{k=1}^\infty \langle f, f_k \rangle \overline{\alpha}_k \tag{11.16}$$

We mention two ways to find $T^* f$ from here:

1. The convergence of the series $\sum_{k=1}^\infty \langle f, f_k \rangle \overline{\alpha}_k$ for all $\{\alpha_k\}_{k=1}^\infty \in l^2$ implies that $\{\langle f, f_k \rangle\}_{k=1}^\infty \in l^2$. Thus, we can write

$$\langle f, T\{\alpha_k\}_{k=1}^\infty \rangle_{\mathcal{H}} = \langle \{\langle f, f_k \rangle\}, \{\alpha_k\} \rangle$$

   and conclude that

$$T^* f = \{\langle f, f_k \rangle\}_{k=1}^\infty$$

2. Alternatively, when $T : l_2 \to H$ is bounded we already known that T^* is a bounded operator from $\mathcal{H}$ to l_2. Therefore, the kth coordinate function is bounded from $\mathcal{H}$ to $\mathbb{C}$; by Riesz' representation theorem, T^* therefore has the form

$$T^* f = \{\langle f, g_k \rangle\}_{k=1}^\infty$$

for some $\{g_k\}_{k=1}^{\infty}$ in $\mathscr{H}$. By definition of $T^{\star}$, (11.16) now shows that

$$\sum_{k=1}^{\infty}\langle f, g_k\rangle\overline{\alpha}_k = \sum_{k=1}^{\infty}\langle f, f_k\rangle\overline{\alpha}_k, \ \forall\{\alpha_k\}_{k=1}^{\infty}\in l^2(\mathbb{N}), \ f\in\mathscr{H}$$

It follows from here that $g_k = f_k$.

First assume that $\{f_k\}_{k=1}^{\infty}$ is a Bessel sequence with Bessel bound B. Let $\{\alpha_k\}_{k=1}^{\infty}\in l_2$. First we want to show that $T\{\alpha_k\}_{k=1}^{\infty}$ is well defined, i.e., that $\alpha_k\}_{k=1}^{\infty}$ is convergent. Consider $n, m\in N$, $n > m$. Then

$$\left\|\sum_{k=1}^{n}\alpha_k f_k - \sum_{k=1}^{m}\alpha_k f_k\right\| = \left\|\sum_{k=m+1}^{n}\alpha_k f_k\right\|$$

$$= \sup_{\|g\|=1}\left|\langle\sum_{k=m+1}^{n}\alpha_k f_k, g\rangle\right|$$

$$\geq \sup_{\|g\|=1}\sum_{k=m+1}^{n}|\alpha_k\langle f_k, g\rangle|$$

$$\geq \left(\sum_{k=m+1}^{n}|\alpha_k|^2\right)^{1/2}\sup_{\|g\|=1}\left(\sum_{k=m+1}^{n}|\langle f_k, g\rangle|^2\right)^{1/2}$$

$$\leq \sqrt{B}\left(\sum_{k=m+1}^{n}|\alpha_k|^2\right)^{1/2}$$

Since $\{\alpha_k\}_{k=1}^{\infty}\in l_2$, we know that $\left\{\sum_{k=m+1}^{n}|\alpha_k|^2\right\}_{n=1}^{\infty}$ is a Cauchy sequence in $\mathbb{C}$. The above calculation now shows that $\left\{\sum_{k=1}^{n}\alpha_k f_k\right\}_{n=1}^{\infty}$ is a Cauchy sequence in $\mathscr{H}$, and therefore convergent. Thus, $T\{\alpha_k\}_{k=1}^{\infty}$ is well defined. Clearly, T is linear; since $\|T\{\alpha_k\}_{k=1}^{\infty}\| = \sup_{\|g\|=1}|\langle T\{\alpha_k\}_{k=1}^{\infty}, g\rangle|$, a calculation as above shows that T is bounded and that $\|T\|\leq\sqrt{B}$. For the opposite implication, suppose that T is well defined and that $\|T\|\leq\sqrt{B}$. Then (11.5) shows that $\{f_k\}_{k=1}^{\infty}$ is a Bessel sequence with Bessel Bound B.

11.3   Riesz Bases

Definition 11.8 A Riesz basis for a Hilbert space H is a family of the form $\{Uf_k\}_{k=1}^{\infty}$, where $\{f_k\}$ is an orthonormal basis for H and $U : H \to H$ is a bounded bijective operator.

The following theorem gives an alternative definition of a Riesz basis which is convenient to use (see Definition 12.5).

Theorem 11.7 [40] *For a sequence $\{f_k\}_{k=1}^{\infty}$ in a Hilbert space H, the following conditions are equivalent:*

(i) $\{f_k\}_{k=1}^{\infty}$ *is a Riesz basis for H*

(ii) $\{f_k\}_{k=1}^{\infty}$ *is complete in H, and there exist constants $A, B > 0$ such that for every finitely nonzero sequences $\{\alpha_k\} \in l_2$ one has*

$$A \sum_{k=1}^{\infty} |\alpha_k|^2 \leq \left\| \sum_{k=1}^{\infty} \alpha_k f_k \right\|^2 \leq B \sum_{k=1}^{\infty} |\alpha_k|^2 \qquad (11.17)$$

(iii) $\{f_k\}$ *is complete, and its Gram matrix $\{\langle f_k, f_j \rangle\}_{j,k=1}^{\infty}$ defines a bounded, invertible operator on l_2.*

(iv) $\{f_k\}_{k=1}^{\infty}$ *is a complete Bessel sequence, and it has a complete biorthogonal sequence $\{g_k\}_{k=1}^{\infty}$ which is also a Bessel sequence.*

Remark 11.5 1. Any sequence $\{f_k\}_{k=1}^{\infty}$ satisfying (11.17) for all finite sequence $\{\alpha_k\}$ is called a Riesz sequence. By Theorem 11.10, a Riesz sequence $\{f_a\}_{k=1}^{\infty}$ is a Riesz basis for span $\{f_k\}_{k=1}^{\infty}$, which might just be a subsequence of H. It is clear that if (11.17) is satisfied by a sequence $\{f_k\}_{k=1}^{\infty}$, then it will be satisfied by its every subsequence. This gives us the result: "Every subsequence of a Riesz basis is itself a Riesz basis."

2. Let $\overline{span}\{f_k\}_{k=1}^{\infty} = H, B = 1$ and let equality hold on the right-hand side of (11.17) for all finite scalar sequences $\{\alpha_k\}$. Then $\{f_k\}_{k=1}^{\infty}$ is an orthonormal basis for H.

Definition 11.9 (*Biorthogonal Systems*) A sequence $\{g_k\}_{k=1}^{\infty}$ in H is called biorthogonal to a sequence $\{f_k\}$ in H if

$$\langle g_k, f_j \rangle = \delta_{k,j}, \text{ where}$$
$$\delta_{nj} = 0, if \ k \neq j$$
$$= 1, \ if \ k = j$$

Very often we say that $\{g_k\}_{k=1}^{\infty}$ is the biorthogonal system associated with $\{f_k\}_{k=1}^{\infty}$.

The following theorem guarantees the existence of a biorthogonal system.

Theorem 11.8 *Let $\{f_k\}$ be a basis for the Hilbert space H. Then there exists a unique family $\{g_k\}_{k=1}^{\infty}$ in H for which*

$$f = \sum_{k=1}^{\infty} \langle f, g_k \rangle f_k, \ for \ all \ f \in H \qquad (11.18)$$

$\{g_k\}_{k=1}^{\infty}$ *is a basis for H, and $\{f_k\}$ and $\{g_k\}$ are biorthogonal.*

Definition 11.10 The basis $\{g_k\}$ satisfying (11.18) is called the dual basis, or the biorthogonal basis, associated to $\{f_k\}$.

11.4 Frames in Infinite-Dimensional Hilbert Spaces

The concept of frame in a finite-dimensional Hilbert space discussed in Sect. 11.1 can be extended to infinite-dimensional case.

Definition 11.11 Let $\{f_k\}$ be sequence of elements in an infinite-dimensional Hilbert space.

A. $\{f_k\}$ is called a frame if there exist constants $A, B > 0$ such that

$$A\|f\|^2 \leq \sum_{k=1}^{\infty} |\langle f, f_k \rangle|^2 \leq B\|f\|^2, \ for \ all \ f \in H \quad (11.19)$$

B. If $A = B$, then $\{f_k\}$ is called a tight frame.
C. If a frame ceases to be a frame when an arbitrary element is removed, it is called an exact frame.
D. $\{f_k\}_{k=1}^{\infty}$ is called a frame sequence if it is a frame for $\overline{span}\{f_k\}_{k=1}^{\infty}$.

Example 11.2 Let $\{f_k\}_{k=1}^{\infty}$ be an orthonormal basis for a Hilbert space $\mathscr{H}$.

A. By repeating each element in $\{f_k\}_{k=1}^{\infty}$ twice, we obtain

$$\{f_k\}_{k=1}^{\infty} = \{f_1, f_1, f_2, f_2, \ldots\}$$

which is a tight frame with frame bound $A = 2$. If only f_1 is repeated, we obtain

$$\{f_k\}_{k=1}^{\infty} = \{f_1, f_1, f_2, f_3, \ldots\}$$

which is a frame with bounds $A = 1, B = 2$.
B. Let

$$\{f_k\}_{k=1}^{\infty} = \left\{ f_1, \frac{1}{\sqrt{2}} f_2, \frac{1}{\sqrt{2}} f_2, \frac{1}{\sqrt{3}} f_3, \frac{1}{\sqrt{3}} f_3, \frac{1}{\sqrt{3}} f_3, \ldots \right\}$$

$$\sum_{k=1}^{\infty} |\langle f, g_k \rangle|^2 = \sum_{k=1}^{\infty} k \left\| \left\langle f, f \frac{1}{\sqrt{k}} f_k \right\rangle \right\|^2$$

so $\{g_k\}_{k=1}^{\infty}$ is a tight frame for $\mathscr{H}$ with frame bound $A = 1$.
C. If $I \subset \mathbb{N}$ is a pure subset, then $\{f_k\}_{k \in I}$ is not complete in $\mathscr{H}$ and cannot be a frame for $\mathscr{H}$. However, $\{f_k\}_{k \in I}$ is a frame for $\overline{span}\{f_k\}_{k \in I}$, i.e., it is a frame sequence.
D. $\{f_k\}_{k=1}^{\infty}$ is called a frame sequence if it is a frame for $\overline{span}\{f_k\}_{k \in I}$. Since a frame $\{f_k\}_{k=1}^{\infty}$ is a Bessel sequence, this operator

$$T : l_2 \to H, T\{\alpha_k\}_{k=1}^{\infty} = \sum_{k=1}^{\infty} \alpha_k f_k \quad (11.20)$$

is a bounded by Theorem 11.6. T is called the preframe operator or the synthesis operator. The adjoint is given by

$$T^* : H \to l_2, \ T^* = \{f\langle f, f_k\rangle\}_{k=1}^{\infty} \tag{11.21}$$

in view of Lemma 11.1.

T^* is called the analysis operator. By composing T and T^*, we obtain the frame operator

$$S : H \to H, Sf = TT^* f = \sum_{k=1}^{\infty} \langle f, f_k\rangle f_k \tag{11.22}$$

Theorem 11.9 *Let $\{f_k\}$ be a frame with frame operator S. Then*

$$f = \sum_{k=1}^{\infty} \langle f, S^{-1} f_k\rangle f_k, \ for \ all \ f \in H \tag{11.23}$$

Remark 11.6 It may observed that (11.23) means that $\left\| f - \sum_{k=1}^{\infty} \langle f, S^{-1} f_k\rangle f_k \right\| \to 0$ as $n \to \infty$. This is possible if f_k is replaced by $f_{\sigma(k)}$ where σ is any permutation of the natural numbers.

We require the following for the proof.

Lemma 11.2 *Let $\{f_k\}_{k=1}^{\infty}$ be a frame with frame operator S and frame bounds A, B. Then the following holds:*

A. *S is bounded, invertible, self-adjoint, and positive.*
B. *$\{S^{-1} f_k\}_{k=1}^{\infty}$ is a frame with bounds B^{-1}, A^{-1}, if A, B are the optimal bounds for $\{f_k\}_{k=1}^{\infty}$, then the bounds B^{-1}, A^{-1} are optimal for $\{S^{-1} f_k\}_{k=1}^{\infty}$. The frame operator for $\{S^{-1} f_k\}_{k=1}^{\infty}$ is S^{-1}.*

Proof A. S is bounded a composition of two bounded operators. By Theorem 11.6

$$\|S\| = TT^*\| = \|T\| \, \|T^*\| = \|T\|^2 \leq B$$

Since $S^* = (TT^*) = TT^* = S$, the operator S is self-adjoint. The inequality (11.19) means that $A\|f\|^2 \leq \langle Sf, f\rangle \leq B\|f\|^2$ for all $f \in H$, or, $AI \leq S \leq BI$; thus, S is positive. Furthermore, $0 \leq I - B^{-1}S \leq \frac{B-A}{B} < I$, and consequently

$$\|I - B^{-1}S\| = \sum_{\|f\|=1} |\langle(I - B^{-1}S)f, f\rangle| \leq \frac{B-A}{B} < I$$

which shows that S is invertible.

B. Note that for $f \in \mathcal{H}$

$$\sum_{k=1}^{\infty} |\langle f, S^{-1} f_k \rangle|^2 = \sum_{k=1}^{\infty} |\langle S^{-1} f, f_k \rangle|^2$$
$$\leq B \| S^{-1} f \|^2$$

for the remaining part see [40, 11.91].

Proof (Proof of Theorem 11.9) Let $f \in H$. By Lemma 11.2

$$f = S S^{-1} f = \sum_{k=1}^{\infty} |\langle S^{-1} f, f_k \rangle| f_k = \sum_{k=1}^{\infty} \langle f, S^{-1} f_k \rangle f_k$$

$\{f_k\}_{k=1}^{\infty}$ is a Bessel sequence and $\{\langle f, S^{-1} f_a \rangle\}_{k=1}^{\infty} \in l_2$.

Remark 11.7 It can checked that a Riesz basis in H is a frame, and the Riesz basis bounds coincide with the frame bounds.

The following interesting characterization of a frame without involving frame bounds has been proved by Christensen [40].

Theorem 11.10 *A sequence* $\{f_k\}_{k=1}^{\infty}$ *in a Hilbert space* $\mathcal{H}$ *is a frame if and only*

$$T : \{\alpha_k\}_{k=1}^{\infty} \to \sum_{k=1}^{\infty} \alpha_k f_k$$

is a well-defined mapping of l_2 *onto* $\mathcal{H}$.

Proof First, suppose $\{f_k\}_{k=1}^{\infty}$ is a frame. By Theorem 11.6, T is a well-defined bounded operator from l2 into H, and by Lemma 11.2, the frame operator $S = TT^*$ is surjective. Thus, T is surjective. For the opposite implication, suppose that T is a well-defined operator from $l_2(\mathbb{N})$ onto $\mathcal{H}$. Then Lemma 11.1 shows that T is bounded and that $\{f_k\}_{k=1}^{\infty}$ is a Bessel sequence. Let $T^{\dagger} : H \to l_2$ denote the pseudo-inverse of T (see [40, A7] for Definition or Chap. 13). For $f \in H$, we have

$$f = TT^{\dagger} f = \sum_{k=1}^{\infty} (T^{\dagger} f)_k f_k$$

where $(T^{\dagger} f)_k$ denotes the kth coordinate of $(T^{\dagger} f)$. Thus,

$$\|f\|^4 = |\langle f, f \rangle|^2$$

$$= \left\| \left\langle \sum_{k=1}^{\infty} (T^\dagger f)_k f_k, f \right\rangle \right\|^2$$

$$\leq \sum_{k=1}^{\infty} |(T^\dagger f)_k|^2 \sum_{k=1}^{\infty} |\langle f, f_k \rangle|^2$$

$$\leq \|T^\dagger\|^2 \|f\|^2 \sum_{k=1}^{\infty} |\langle f, f_k \rangle|^2$$

we conclude that

$$\sum_{k=1}^{\infty} |\langle f, f_k \rangle|^2 \geq \frac{1}{\|T^\dagger\|^2} \|f\|^2$$

11.5 Problems

Problem 11.1 Prove Theorem 11.5.

Solution 11.1 We write $s = |\alpha_1| + \cdots + |\alpha_n|$. If $s = 0$, all α_j are zero, so that (11.11) holds for any K. Let $s > 0$. Then (11.11) is equivalent to the inequality which we obtain from (11.11) inequality of Theorem 11.5 ($f_k = x_k$) by dividing by s and writing $\beta_j = \alpha_j / s$, that is

$$\|\beta_1 x_1 + \beta_2 x_2 + \cdots + \beta_n x_n\| \geq \left(\sum_{j=1}^{n} |\beta_j x_j| \right) \tag{11.24}$$

Hence, it suffices to prove the existence of a $K > 0$ such that (11.24) holds for every n-tuple of scalars $\beta_1, \ldots \beta_n$ with $\sum |\beta_j| = 1$.

Suppose that this is false. Then there exists a sequence (y_m) of vectors

$$y_m = \beta_1^m x_1 + \beta_2^m x_2 + \cdots + \beta_n^m x_n \quad \left(\sum_{j=1}^{n} |\beta_1^m| = 1 \right) \tag{11.25}$$

such that

$$\|y_m\| \to 0 \ as \ m \to \infty$$

Now we reason as follows. Since $\sum |\beta_j^{(m)}| = 1$, we have $|\beta_j^m| \leq 1$. Hence for each fixed j, the sequence

$$(\beta_j^{(m)}) = (\beta_j^{(1)}, \beta_j^{(2)}, \ldots)$$

is bounded. Consequently, by the Bolzano–Weierstrass theorem, $(\beta_1^{(m)})$ has a convergent subsequence. Let β_1 denote the limit of that subsequence, and let $(y_{1,m})$ denote the corresponding subsequence of (y_m). By the same argument, $(y_{1,m})$ has a subsequence $(y_{2,m})$ for which the corresponding subsequence of scalars $\beta_2^{(m)}$ converges; let β_2 denote the limit. Continuing in this way, after n steps we obtain a subsequence $(y_{n,m}) = (y_{n,1}, y_{n,2}, \ldots)$ of (y_m) whose terms are of the form

$$y_{n,m} = \sum_{j=1}^{n} \gamma_j^{(m)} x_j \left(\sum_{j=1}^{n} |\gamma_j^{(m)}| = 1 \right)$$

with scalars $\gamma_j^{(m)}$ satisfying $\gamma_j^{(m)} \to \beta_j$ as $m \to \infty$. Hence, as $m \to \infty$

$$y_{n,m} \to y = \sum_{j=1}^{n} \beta_j x_j$$

where $\sum |\beta_j| = 1$, so that not all β_j can be zero. Since $\{x_1, \ldots, x_n\}$ is a linearly independent set, we thus have $y \neq 0$. On the other hand, $y_{n,m} \to y$ implies $||y_{n,m}|| \to ||y||$, by the continuity of the norm. Since $||y_m|| \to 0$ by the assumption and $(y_{n,m})$ is a subsequence of (y_m), we must have $||y_{n,m}|| \to 0$. Hence, $||y|| = 0$, so that $y = 0$ by the second axiom of the norm. This contradicts $y \neq 0$, and the desired result is proved.

Problem 11.2 Let $\{g_m(x)\}$ be a sequence of functions in $L_2[a, b]$ and suppose that there is a sequence $\{f_n(x)\}$ in $L_2[a, b]$ biorthogonal to $\{g_n(x)\}$. Then show that $\{g_n(x)\}$ is linearly independent.

Solution 11.2 Let $\{\alpha_k\}_{k=1}^{\infty} \in l_2$, and satisfy

$$\sum_{n=1}^{\infty} \alpha_n g_n(x) = 0$$

in $L_2[a, b]$. Then for each $m \in \mathbb{N}$.

$$0 = \langle 0, f_m(x) \rangle = \left\langle \sum_{n=1}^{\infty} \alpha_n g_n(x), f_m(x) \right\rangle$$

$$= \sum_{n=1}^{\infty} \alpha_n \langle g_n(x), f_m(x) \rangle = \alpha_n$$

by virtue of biorthogonality. Therefore, $\{g_n(x)\}$ is linearly independent.

Problem 11.3 Show that the vectors $\{f_k\}_{k=1}^\infty$ defined by

$$f_k(j) = \frac{1}{\sqrt{n}} e^{2\pi i (j-1)\frac{k-1}{n}}, \ j = 1, 2, 3, \ldots n$$

that is

$$f_k = \frac{1}{\sqrt{n}} \begin{pmatrix} 1 \\ e^{2\pi i \frac{k-1}{n}} \\ e^{4\pi i \frac{k-1}{n}} \\ . \\ . \\ . \\ e^{2\pi i (n-1)\frac{k-1}{n}} \end{pmatrix}, k = 1, 2, \ldots, n$$

Solution 11.3 Since $\{f_k\}_{k=1}^\infty$ are n vectors in an n-dimensional vector space, it is enough to prove that they constitute an orthonormal system. It is clear that $\|f_k\| = 1$ for all k. Now, given $k \ne l$

$$\langle f_k, f_l \rangle = \frac{1}{n} \sum_{j=1}^n e^{2\pi i \frac{k-1}{n}} e^{-2\pi i (j-1)\frac{l-1}{n}} = \frac{1}{n} \sum_{j=0}^{n-1} e^{2\pi i j \frac{k-l}{n}}$$

Using the formula $(1-x)(1+x+\cdots+x^{n-1})$ with $x = e^{2\pi i \frac{k-1}{n}}$, we get

$$\langle f_k, f_l \rangle = \frac{1}{n} \frac{1 - (e^{2\pi i \frac{k-l}{n}})^n}{1 - e^2 \pi i \frac{k-l}{n}} = 0$$

Problem 11.4 Let $\{f_k\}_{k=1}^\infty$ be a frame for a finite-dimensional inner product space H with bounds A, B and let P denote the m orthogonal projection of H onto a subspace M. Prove that $\{Pf_k\}_{k=1}^m$ is a frame for M with frame bounds A, B.

Problem 11.5 Let $m > n$ and define the vectors $\{f_k\}_{k=1}^\infty$ in C^m by

$$f_k = \frac{1}{\sqrt{m}} \begin{pmatrix} 1 \\ 2\pi i \frac{k-1}{n} \\ e \\ . \\ . \\ . \\ e^{2\pi i (n-1)\frac{k-1}{m}} \end{pmatrix}, k = 1, 2, \ldots, m$$

Then prove that $\{f_k\}_{k=1}^m$ is a tight frame for C^m with frame bound equal to one, and $\|f_k\| = \sqrt{\frac{n}{m}}$ for all k.

Problem 11.6 Prove that for a sequence $\{f_k\}_{k=1}^{\infty}$ in a finite-dimensional inner product space H, the following are equivalent:

A. $\{f_k\}$ is a Bassel sequence with bounds B.
B. The Gram matrix associated to $\{f_k\}_{k=1}^{\infty}$, namely $\{\langle f_k, f_j \rangle\}_{j,k=1}^{\infty}$ defines a bounded operator on l_2, with norm at most B.

Problem 11.7 Let $\{f_k\}_{k=1}^{\infty}$ be an orthonormal basis and consider the sequence $\{g_k\}_{k=1}^{\infty} = \{e_1 + \frac{1}{k}e_k, e_k\}_{k=2}^{\infty}$.

A. Prove that $\{g_k\}$ is not a Bessel sequence.
B. Find all possible representation of f_1 as linear combinations of $\{g_k\}_{k=1}^{\infty}$.

Chapter 12
Wavelet Theory

Abstract This chapter deals with wavelet theory developed in the eighties that is the refinement of Fourier analysis. This has been developed by the serious interdisciplinary efforts of mathematicians, physicists, and engineers. A leading worker of this filed, Prof. Y. Meyer, has been awarded Abel Prize of 2017. It has interesting and significant applications in diverse fields of science and technology such as oil exploration and production, brain studies, meteorology, earthquake data analysis, astrophysics, remote sensing, tomography, biometric.

Keywords Continuous wavelet transform · Haar wavelets · Gabor wavelet
Mexican hat wavelet · Parsevals formula for wavelet transform
Calderon–Grossman wavelet theorem · Discrete wavelet transform
Wavelet coefficients · Wavelet series · Multiresolution analysis

12.1 Introduction

The study of wavelets was initiated by Jean Morlet, a French geophysicist in 1982. The present wavelet theory is the result of joint efforts of mathematicians, scientists, and engineers. This kind of work created a flow of ideas that goes well beyond the construction of new bases and transforms. Joint efforts of Morlet and Grossman yielded useful properties of continuous wavelet transforms, of course, without realization that similar results were obtained by Caldéron, Littlewood, Paley, and Franklin more than 20 years earlier. Rediscovery of the old concepts provided a new method for decomposing signals (functions). In many applications, especially in the time–frequency analysis of a signal, the Fourier transform analysis is inadequate because Fourier transform of the signal does not contain any local information. It is a serious drawback of the Fourier transform as it neglects the idea of frequencies changing with time or, equivalently, the notion of finding the frequency spectrum of a signal locally in time. As a realization of this flaw, as far back as 1946, Dennis Gabor first introduced the windowed Fourier transform (short-time Fourier transform), commonly known as the Gabor transform, using a Gaussian distribution function as the window function. The Gabor transform faced some algorithmic problems which were

© Springer Nature Singapore Pte Ltd. 2018
A. H. Siddiqi, *Functional Analysis and Applications*, Industrial and Applied Mathematics,
https://doi.org/10.1007/978-981-10-3725-2_12

resolved by Henric Malvar in 1987. Malvar introduced a new wavelet called Malvar wavelet, which is more effective and superior to Morlet–Grossmann and Gabor wavelets.

In 1985, Yves Meyer accidentally found a dissertation by Stephane Mallat on wavelets. Being the expert of the Caldrón–Zygmund operators and the Littlewood–Paley theory, he succeeded to lay the mathematical foundation of wavelet theory. Daubechies, Grossman, and Meyer received a major success in 1986 for constructing a painless nonorthogonal expansion. A collaboration of Meyer and Lemarié yielded the construction of smooth orthonormal wavelet during 1985–86 in the setting of finite-dimensional Euclidean spaces. As anticipated by Meyer and Mallat that the orthogonal wavelet bases could be constructed systematically from a general formalism, they succeeded in inventing the multiresolution analysis in 1989. Wavelets were constructed applying multiresolution analysis. Decomposition and reconstruction algorithms were developed by Mallat. Under the inspiration of Meyer and Coifmann, Daubechies made a remarkable contribution to wavelet theory by constructing families of compactly supported orthonormal wavelets with some degree of smoothness. Her work found a lot of applications in digital image processing.

Coifman and Meyer constructed a big library of wavelets of various duration, oscillation, and other behavior. With an algorithm developed by Coifman and Victor Wickerhauser, it became possible to do very rapidly computerized searches through an enormous range of signal representations in order to quickly find the most economical transcription of measured data [20, 51, 76, 110, 113, 116, 197]. This development allowed, for example, the FBI to compress a fingerprint database of 200 terabytes into a less than 20 terabytes, saving millions of dollars in transmission and storage costs. As explained by the author of this book during the course of a seminar in the Summer of 1989 at the Kaiserslautern University, Germany, Coifman, Meyer, and Wickerhauser introduced the concept of "wavelet packets."

The wavelet method is a refinement of Fourier method which enables to simplify the description of a cumbersome function in terms of a small number of coefficients. In wavelet method, there are fewer coefficients compared to the Fourier method.

The remaining sections of this chapter are devoted to continuous and discrete wavelets (Sect. 12.2), Multiresolution Analysis and Wavelets Decomposition and Reconstruction (Sect. 12.3), Wavelets and Smoothness of Functions (Sect. 12.4), Compactly supported Wavelets (Sect. 12.5), Wavelet Packets (Sect. 12.6), Problems (Sect. 12.7), and References.

12.2 Continuous and Discrete Wavelet Transforms

12.2.1 Continuous Wavelet Transforms

Definition 12.1 Let $\psi \in L_2(R)$. It is called a *wavelet* if it has zero average on $(-\infty, \infty)$, namely

$$\int_{-\infty}^{\infty} \Psi(t)dt = 0 \tag{12.1}$$

Remark 12.1 If $\psi \in L_2(R) \cap L_1(R)$ satisfying

$$c_\psi = 2\pi \int_R \frac{\hat{\psi}(w)}{w} dw < \infty \tag{12.2}$$

where $\hat{\psi}(w)$ is the Fourier transform of ψ, and then, (12.1) is satisfied. The condition (12.2) is known as the *wavelet admissibility condition.*

Verification By the Riemann–Lebesgue theorem (Theorem A.27), $\lim w \to \infty \hat{\psi}$ $(w) = 0$ and the Fourier transform is continuous which implies that $0 = \hat{\psi}(w)(0) = \int_{-\infty}^{\infty} \psi(t)dt$.

The following lemma gives a method to construct a variety of wavelets:

Lemma 12.1 *Let φ be a nonzero n-times ($n \geq 1$) differentiable function such that $\varphi^n(x) \in L_2(R)$. Then*

$$\psi(x) = \varphi^{(n)}(x) \tag{12.3}$$

is a wavelet.

Proof From the property of the Fourier transform (Corollary A.1), $|\hat{\psi}(w)| = |w|^k |\hat{\varphi}(w)|$. Then

$$\begin{aligned}
c_\psi &= \int_R \frac{\hat{\psi}(w)}{w} dw \\
&= \int_R \frac{|w|^{2k} |\hat{\varphi}(w)|^2}{|w|} dw \\
&= \int_{-1}^{1} |w|^{2k-1} \hat{\varphi}(w)|^2 dw + \int_{|w|>1} \frac{|w|^{2k} |\hat{\varphi}(w)|^2}{|w|} dw \\
&\leq 2\pi (\|\varphi\|_{L_2}^2 \|\varphi\|_{L_2}^k) < \infty
\end{aligned}$$

Hence, (12.2) holds which proves that ψ is a wavelet by Remark 12.1.

Lemma 12.2 *Let $0 \neq \psi \in L_1(R) \cap L_2(R)$ with $\int_R \psi(t) = 0$ and $\int_R |x|^\beta \psi(x)dx < \infty$ for a $\beta > 1/2$. Then ψ is a wavelet.*

Proof Without loss of generality, we may assume that $1/2 < \beta \leq 1$. From this, it follows that $1 + |x|^\beta \geq (1 + |x|)^\beta$ and $\int_R (1 + |x|)^\beta |\psi(x)|dx < \infty$.

We know that the function $\varphi(x) = \int_{-\infty}^{x} \psi(t)dt$ is differentiable almost everywhere and $\varphi(x) = \psi(x)$. For $x \leq 0$, we find (Fig. 12.1)

Fig. 12.1 Haar wavelet

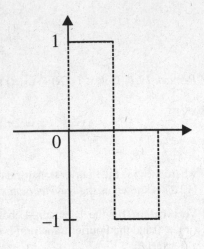

$$|\varphi(x)| \leq \int_{-\infty}^{x} (1 + |t|)^{-\beta}(1 + |t|)^{\beta}\psi(t)dt$$

$$|\varphi(x)| \leq \frac{1}{(1 + |x|)^{\beta}} \int_{R}(1 + |t|)^{\beta}\psi(t)dt \tag{12.4}$$

If $x > 0$, the zero mean value of ψ implies that $\varphi(x) = -\int_{x}^{\infty}\psi(t)dt$. This shows that (12.4) holds for all $x \in R$, and therefore, $\varphi \in L_2(R)$. Since $\varphi = \psi \in L_2(R)$, by Lemma 12.1, ψ is a wavelet.

Corollary 12.1 *Let $\psi \neq 0$, $\psi \in L_2(R)$ with compact support, the following statements are equivalent:*

A. *The function ψ is a wavelet.*
B. *Relation (12.2) is satisfied.*

Proof $(b) \Rightarrow (a)$ (Remark 12.1). $(a) \Rightarrow (b)$: Let ψ be a wavelet, that is, $\psi \in L_2(R)$ and $\int_R \psi(t)dt = 0$. Since $\psi \in L_2(R)$, and ψ has compact support, $|\psi(x)| \in L_1(R)$; that is

$$\int_{R} |x|^{\beta}|\psi(x)|dt < \infty \text{ for all} \beta \geq 0$$

In particular

$$\int_{R} |x|^{\beta}|\psi(x)|dt < \infty \text{ for all} \beta > \frac{1}{2}$$

As seen in the proof of Lemmas 12.1 and 12.2, we have the desired result.

Examples of Wavelet

Example 12.1 (Haar Wavelet) Let

$$\psi(x) = 1, \quad 0 \le x < \frac{1}{2}$$
$$= -1, \quad \frac{1}{2} \le x < 1$$
$$= 0, \quad \textit{otherwise}$$

It is clear that

$$\int_{-\infty}^{\infty} \psi(x)dx = 0$$

and $\psi(x)$ has the compact support $[0, 1]$ (Fig. 12.2). It can be verified that

$$\hat{\psi} = \frac{1}{\sqrt{2\pi}} \frac{(\sin\frac{w}{4})^2}{\frac{w}{4}} e^{-i(w-\pi)/2}$$

and

$$\int_{-\infty}^{\infty} \frac{|\hat{\psi}(w)|^2}{|w|} dw = \frac{8}{\pi} \int_{-\infty}^{\infty} \frac{|\sin\frac{w}{4}|^4}{|w|^3}.$$

$\psi(x)$ defined above is called the Haar wavelet in honor of the Hungarian mathematician Alfred Haar who studied it in 1909. The Haar wavelet is discontinuous at $x = 0, \frac{1}{2}, 1$.

Example 12.2 (Mexican Hat Wavelet) The function defined by the equation

$$\psi(x) = (1 - x^2)e^{-x^2/2}$$

is called the *Mexican hat wavelet*. $\psi(x)$ satisfies Eq. (12.3) of Lemma 12.1, namely

Fig. 12.2 Mexican hat wavelet

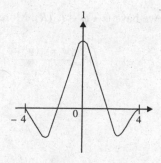

$$\psi(x) = -\frac{d^2}{dx^2}e^{-x^2/2} = (1 - x^2)e^{-x^2/2}$$

We can observe that the Mexican hat wavelet has no discontinuity. $\psi(x)$ is a wavelet by Lemma 12.1.

Example 12.3 (*Gabor Wavelet*) A Gabor wavelet, with width parameter w and frequency parameter v, is defined as

$$\psi(x) = w^{-1/2}e^{-\pi(x/w)^2}e^{i2\pi vx/w}$$

It is a complex-valued function. Its real part $\psi_R(x)$ and imaginary part ψ_I are given by

$$\psi_R(x) = w^{-1/2}e^{-\pi(x/w)^2}\cos(2\pi vx/w)$$
$$\psi_I(x) = w^{-1/2}e^{-\pi(x/w)^2}\sin(2\pi vx/w)$$

With the help of the following theorem, one can obtain new wavelets:

Theorem 12.1 *Let ψ be a wavelet and φ a bounded integrable function, then the convolution function $\psi \star \varphi$ is a wavelet.*

Proof Since

$$\int_{-\infty}^{\infty}|\psi \star \varphi(x)|^2 dx = \int_{-\infty}^{\infty}\left|\int_{-\infty}^{\infty}\psi(x-t)\varphi(t)dt\right|^2 dx$$

$$\leq \int_{-\infty}^{\infty}\left(\int_{-\infty}^{\infty}|\psi(x-t)||\varphi(t)|dt\right)^2 dx$$

$$= \int_{-\infty}^{\infty}\left(\int_{-\infty}^{\infty}|\psi(x-t)|^2|\varphi(t)|dt\int_{-\infty}^{\infty}|\varphi(t)|dt\right)dx$$

$$\leq \int_{-\infty}^{\infty}|\varphi(t)|dt\int_{-\infty}^{\infty}\int_{-\infty}^{\infty}|\psi(x-t)|^2|\varphi(t)|dxdt$$

$$= \left(\int_{-\infty}^{\infty}|\varphi(t)|dt\right)^2\int_{-\infty}^{\infty}|\psi(x)|^2dx < \infty$$

we have $\psi \star \varphi \in L_2(R)$. Moreover

$$\int_{-\infty}^{\infty}\frac{|\widehat{\psi \star \varphi}(w)|^2}{|w|}dw = \int_{-\infty}^{\infty}\frac{\hat{\psi}(w)\hat{\varphi}(w)}{|w|}dw \text{ by } Theorem\ F13$$

$$= \int_{-\infty}^{\infty}\frac{|\hat{\psi}(w)|}{|w|}|\hat{\varphi}(w)|^2dw$$

$$\leq \int_{-\infty}^{\infty}\left(\frac{|\hat{\psi}(w)|^2}{|w|}dw\right)\sup|\hat{\varphi}(w)|^2 < \infty$$

Fig. 12.3 Convolution of
Haar wavelet and
$\varphi(x) = e^{-x^2}$

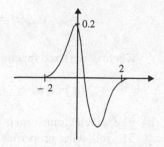

Therefore, $\psi \star \varphi$ is a wavelet.

Example 12.4 By convolving the Haar wavelet with $\varphi(x) = e^{-x^2}$, we get the function
represented in Fig. 12.3.

It is interesting to observe that the set of nonzero wavelets with compact support
is a dense subset of $L_2(R)$; that is, we have the following theorem:

Theorem 12.2 *Let*

$$A = \left\{ \psi \in L_2(R)/\psi \neq 0, \ \psi \ has \ compact \ support \ and \ \int_R \psi(t)dt = 0 \right\}$$

then A is a dense subset of $L_2(R)$.

Proof Let $h \in L_2(R)$, then $\hat{h} \in L_2(R)$. Let h_ε be defined by

$$\hat{h}_\varepsilon(w) = \hat{h}(w) \cdot \ |w| \geq \varepsilon$$
$$= 0 \ \ |w| > \varepsilon$$

Then, for every ε, h satisfies (12.2) and so by Corollary 12.1, h_ε is a wavelet. Since
by Theorem A.32, $||h||_{L_2} = ||\hat{h}||_{L_2}$, it follows that $||h_\varepsilon - h||_{L_2}^2 = \int_{-\varepsilon}^{\varepsilon} |\hat{h}(w)|^2 dw \to 0$.
This means that every function h in L_2 can be considered as a limit of a sequence of
wavelets and hence A is dense in $L_2(R)$.

Definition 12.2 (*Continuous Wavelet Transform*) The continuous wavelet transform
T_ψ of a function $f \in L_2(R)$ with respect to the wavelet ψ is defined as

$$f(a, b) = T_\psi f(a, b) = |a|^{-1/2} \int_R f(t) \overline{\psi} \left(\frac{t - b}{a} \right) dt \qquad (12.5)$$

where $a \in R/0$, $b \in R$, and $\overline{\psi}$ denotes complex conjugate.

We note that for real ψ, $\overline{\psi} = \psi$. We consider here real wavelet.

Remark 12.2   (i) If we consider $\psi_{a,b}(t)$ as a family of functions given by

$$\psi_{a,b}(t) = |a|^{-1/2} \psi \left(\frac{t-b}{a} \right), \ a > 0, \ b \in R \tag{12.6}$$

where ψ is a fixed function, often called *mother wavelet*, then (12.5) takes form:

$$T_\psi f(a, b) = \langle f, \psi_{a,b} \rangle = \text{the inner product of } f \text{ with } \psi_{a,b} \tag{12.7}$$

(ii) The wavelet transform is linear as it can be written in terms of the inner product. The following properties hold in view of properties of the inner product given in Sect. 3.1.

Let ψ and ϕ be wavelets and let $f, g \in L_2(R)$. Then, the following relations hold:

(a) $T_\psi(\alpha f + \beta g)(a, b) = \alpha T_\psi f(a, b) + \beta T_\psi g(a, b)$ for any $\alpha, \beta \in R$.
(b) $T_\psi(S_c f)(a, b) = T_\psi f(a, b - c)$, where S_c is a translation operator defined by $S_c f(t) = f(t - c)$.
(c) $T_\psi(D_c f)(a, b) = (1/c) T_\psi f(a/c, b/c)$, where c is a positive number and D_c is the dilation operator defined by $D_c f(t) = (1/c) f(t/c)$.
(d) $T_\psi \phi(a, b) = T_\phi \psi(1/a, -b/a), \ a \neq 0$.
(e) $T_{\psi + \phi}(a, b) = \bar{\alpha} T_\psi f(a, b) + \bar{\beta} T_\phi f(a, b)$, for any scalar α, β.
(f) $T_{A\psi} A f(a, b) = T_\psi f(a, -b)$ where A is defined by $A\psi(t) = \psi(-t)$.
(g) $(T_{S_c} \psi(f)(a, b)) = T_\psi f(a, b + ca)$.
(h) $(T_{D_c} \psi f)(a, b) = 1/c (T_\psi f)(ac, b), \ c > 0$.

Remark 12.3 We note that the wavelet transform $T_\psi f(a, b)$ is a function of the scale or frequency a and the spatial position or time b. The plane defined by the variables (a, b) is called the *scale-space* or *time–frequency plane*. $T_\psi f(a, b)$ gives the variation of f in a neighborhood of b. If ψ is a compactly supported wavelet, then wavelet transform $T_\psi f(a, b)$ depends upon the value of f in a neighborhood of b of size proportional to the scale a.

For a small scale, $T_\psi f(a, b)$ gives localized information such as localized regularity of $f(x)$: The local regularity of a function (or signal) is often measured with Lipschitz exponents. The global and local Lipschitz regularity can be characterized by the asymptotic decay of wavelet transformation at small scales. For example, if f is differentiable at b, $T f(a, b)$ has the order $a^{3/2}$ as $a \to 0$. For more details, see [59] and Sect. 12.4.

We prove now analogue of certain well-known results of Fourier analysis such as Parseval' s formula, Isometry formula, and Inversion formula.

Theorem 12.3 (Parseval' s Formula for Wavelet Transforms) *Suppose ψ belongs to $L_2(R)$ satisfying (12.2), that is, ψ is a wavelet. Then, for all $f, g \in L_2(R)$, the following formula holds:*

$$\langle f, g \rangle = \frac{1}{c_\psi} \int\limits_{-\infty}^{\infty} \int_{-\infty}^{\infty} (T_\psi f)(a, b) \overline{T_\psi g}(a, b) \frac{db\, da}{a^2} \tag{12.8}$$

Proof By Parseval's formula for the Fourier transforms (Theorem A.32), we have

$$
\begin{aligned}
(T_\psi f)(a, b) &= \langle f, \psi_{a,b} \rangle \\
&= \langle \hat{f}, \hat{\psi}_{a,b} \rangle \\
&= \int_{-\infty}^{\infty} \hat{f}(x) |a|^{1/2} e^{ibx} \overline{\hat{\psi}(ax)} dx \\
&= (2\pi)^{1/2} \mathscr{F}\{|a|^{1/2} \hat{f}(x) \overline{\hat{\psi}(ax)}\}(-b)
\end{aligned}
$$

and

$$
\begin{aligned}
\overline{(T_\psi f)(a, b)} &= \int_{-\infty}^{\infty} \overline{g(t)} |a|^{-1/2} \psi \left(\frac{t-b}{a} \right) dt \\
&= \int_{-\infty}^{\infty} \overline{\hat{g}(y)} |a|^{1/2} e^{-iby} \hat{\psi}(ay) dy \\
&= (2\pi)^{1/2} \overline{\mathscr{F}\{|a|^{1/2} \hat{g}(x) \overline{\hat{\psi}(ax)}\}(-b)}
\end{aligned}
$$

Then

$$
\begin{aligned}
\int_{-\infty}^{\infty} \int_{-\infty}^{\infty} & (T_\psi f)(a, b) \overline{T_\psi g}(a, b) \frac{db\, da}{a^2} \\
&= 2\pi \int_{-\infty}^{\infty} \int_{-\infty}^{\infty} \mathscr{F}\{\hat{f}(x) \overline{\hat{\psi}(ax)}\}(-b) \overline{\mathscr{F}\{\hat{g}(x) \overline{\hat{\psi}(ax)}\}}(-b) \frac{db\, da}{a} \\
&= 2\pi \int_{-\infty}^{\infty} \int_{-\infty}^{\infty} \hat{f}(x) \overline{\hat{g}(x)} |\hat{\psi}(ax)|^2 dx \frac{da}{a}
\end{aligned}
$$

by Parseval's formula for Fourier transform (Theorem A.32)

$$
= 2\pi \int_{-\infty}^{\infty} |\hat{\psi}(ax)|^2 \frac{da}{a} \int_{-\infty}^{\infty} \hat{f}(x) \overline{\hat{g}(x)} dx
$$

by Fubini's theorem (Theorem A.17)

$$
\begin{aligned}
&= 2\pi \int_{-\infty}^{\infty} \frac{|\hat{\psi}(\omega x)|^2}{|w|} dw \langle \hat{f}, \hat{g} \rangle \\
&= c_\psi \langle f, g \rangle,
\end{aligned}
$$

by Parseval's formula for Fourier transform (Theorem A.32)

Theorem 12.4 (Calderon, Grossman, Morlet) *Let* $\psi \in L_2(R)$ *satisfying* (12.2). *Then, for any* $f \in L_2(R)$, *the following relations hold:*
Inversion formula

$$
f(t) = \frac{1}{c_\psi} \int_{-\infty}^{\infty} \int_{-\infty}^{\infty} (T_\psi f)(a, b) |a|^{-1/2} \psi \left(\frac{t-b}{a} \right) db \frac{da}{a^2} \tag{12.9}
$$

Isometry

$$\int_{-\infty}^{\infty} |f(t)|^2 dt = \frac{1}{c_\psi} \int_{-\infty}^{\infty} \int_{-\infty}^{\infty} |(T_\psi f)(a,b)|^2 db \frac{da}{a^2} \tag{12.10}$$

Equation (12.9) can be written as

$$\cdot \qquad \|f\|_{L_2} = \|T_\psi f(a,b)\|_{L_2}\left(R^2, \frac{dadb}{a^2}\right) \tag{12.11}$$

Proof For any $g \in L_2(R)$, we have

$$c_\psi \langle f, g \rangle = \langle T_\psi f, T_\psi g \rangle$$

$$= \int_{-\infty}^{\infty} \int_{-\infty}^{\infty} T_\psi f(a,b) \overline{T_\psi g(a,b)} \frac{dadb}{a^2}$$

by Theorem 10.3

$$= \int_{-\infty}^{\infty} \int_{-\infty}^{\infty} T_\psi f(a,b) \overline{\int_{-\infty}^{\infty} g(t)\overline{\psi_{a,b}(t)}dt} \frac{dadb}{a^2}$$

$$= \int_{-\infty}^{\infty} \int_{-\infty}^{\infty} \int_{-\infty}^{\infty} T_\psi f(a,b)\psi_{a,b}(t) \frac{dadb}{a^2} \overline{g(t)}dt$$

by Fubini's theorem

$$= \langle \int_{-\infty}^{\infty} \int_{-\infty}^{\infty} T_\psi f(a,b)\psi_{a,b}(t) \frac{dadb}{a^2}, g \rangle$$

or

$$\langle c_\psi f - \int_{-\infty}^{\infty} \int_{-\infty}^{\infty} T_\psi f(a,b)\psi_{a,b}(t) \frac{dadb}{a^2}, g \rangle = 0 \; for \; all \; g \in L_2(R)$$

By Remark 3.1(7), we get

$$c_\psi f - \int_{-\infty}^{\infty} \int_{-\infty}^{\infty} T_\psi f(a,b)\psi_{a,b}(t) \frac{dadb}{a^2} = 0$$

and thus we have (12.8).

We now prove (12.9). Since Fourier transform in b of $(T_\psi f)(\xi, b)$ is $\hat{f}(w + \xi)\hat{\psi}(w)$, Theorem A.32 concerning isometry of the Fourier transform applied to the right-hand side of (12.9) gives

$$\frac{1}{c_\psi} \int_{-\infty}^{\infty} \int_{-\infty}^{\infty} |(T_\psi f)(\xi, b)|^2 db \frac{d\xi}{\xi} = \frac{1}{c_\psi} \int_{-\infty}^{\infty} \frac{1}{c_\psi} \int_{-\infty}^{\infty} |\hat{f}(w + \xi)\hat{\psi}(w)|^2 dw \, d\xi$$

By the Fubini theorem (Theorem A.17) and Theorem A.32, we get

$$\frac{1}{c_\psi} \int_{-\infty}^{\infty} |\hat{f}(w + \xi)|^2 d\xi = ||f||^2$$

This proves (12.9).

It may be remarked that Theorem 12.4 was first proved by Caldéron in 1964 in Mathematica Studia, a journal started by the proponent of Functional Analysis Stefan Banach, without the knowledge of the concept of wavelets in explicit form. Grossman and Morlet rediscovered it in the context of wavelet in their paper of 1984 without the knowledge of Caldéron's result

12.2.2 Discrete Wavelet Transform and Wavelet Series

In practical applications, especially those involving fast algorithms, the continuous wavelet transform can only be computed on a discrete grid of point (a_n, b_n), $n \in Z$. The important issue is the choice of this sampling so that it contains all the information on the function f. For a wavelet ψ, we can define

$$\psi_{m,n}(t) = a_0^{n/2} \psi(a_0^n t - b_0 m), \quad m, n \in Z$$

where $a_0 > 1$ and $b_0 > 0$ are fixed parameters. For such a family, two important questions can be asked?

A. Does the sequence $\{\langle f, \psi_{m,n}\rangle\}_{m,n \in Z}$ completely characterize the function f?
B. Is it possible to obtain f from this sequence in a stable manner?

These questions are closely related to the concept of frames which we introduce below.

Definition 12.3 (*Frames*) A sequence $\{\varphi_n\}$ in a Hilbert space H is called a frame if there exist positive constants α and β such that

$$\alpha||f||^2 \leq \sum_{n=1}^{\infty} |\langle f, \varphi_n\rangle|^2 \leq \beta||f||^2 \ \forall \ f \in H \tag{12.12}$$

The constants α and β are called *frame bounds*.

If $\alpha = \beta$, then equality holds in (12.12). In this case, the frame is often called the **tight frame**. It may be observed that the frame is an orthonormal basis if and only if $\alpha = \beta = 1$.

Definition 12.4 (*Frame Operator*) Let $\{\varphi_n\}$ be a frame in a Hilbert space H: Then the operator F from H into ℓ_2 defined as

$$F(f) = \langle f, \varphi_n \rangle, \; n \in Z$$

F is called a **frame operator**.

It may be checked that the frame operator F is linear, invertible, and bounded. Consider the adjoint operator $F^\star$ of a frame operator F associated with the frame φ_n. For arbitrary $\{\alpha_n\} \in \ell_2$, we have

$$\langle F^\star\{\alpha_n\}, f \rangle = \langle \{\alpha_n\}, Ff \rangle = \sum_{n=1}^{\infty} \alpha_n \langle \varphi_n, f \rangle$$

$$= \langle \alpha_n \varphi_n, f \rangle$$

Therefore, the adjoint operator $F^\star$ of a frame operator F has the form

$$F^\star(\{\alpha_n\}) = \sum_{n=1}^{\infty} \alpha_n \varphi_n \tag{12.13}$$

Since $\sum_{n=1}^{\infty} |\langle f, \varphi_n \rangle|^2 = ||Ff||^2 = \langle F^\star F f, f \rangle$, (12.12) can be expressed as

$$AI \le F^\star F \le BI \tag{12.14}$$

where I is the identity operator and $\le$ is an ordering introduced in Definition 3.9.

It can be checked that $F^\star F$ has a bounded inverse.

Theorem 12.5 *Let $\{\varphi_n\}$ be a frame with frame bounds A and B and F the associated frame operator. Let $\tilde{\varphi}_n = (F^\star F)^{-1}\varphi_n$. Then $\{\tilde{\varphi}_n\}$ is a frame with bounds $1/\beta$ and $1/\alpha$.*

The sequence $\{\tilde{\varphi}_n\}$ is called the dual frame of the $\{\varphi_n\}$.

Proof We have

$$(F^\star F)^{-1} = ((F^\star F)^{-1})^\star$$

by Theorem 3.32(8). Hence

$$\langle f, \tilde{\varphi}_n \rangle = \langle f, (F^\star F)^{-1}\varphi_n \rangle = \langle (F^\star F)^{-1}f, \varphi_n \rangle$$

We can check that

$$\sum_{n=1}^{\infty} |\langle f, \varphi_n \rangle|^2 = \sum_{n=1}^{\infty} |\langle (F^\star F)^{-1}f, \varphi_n \rangle|^2$$

$$= ||F(F^\star F)^{-1}f||^2$$

$$= \langle F(F^\star F)^{-1}f, F(F^\star F)^{-1}f \rangle$$

$$= \langle (F^\star F)^{-1}f, f \rangle$$

By virtue of (12.14), it can be checked that

$$\frac{1}{B}I \le (F^*F)^{-1} \le \frac{1}{A}I \tag{12.15}$$

This implies that

$$\frac{1}{B}\|f\|^2 \sum_{n=1}^{\infty} |\langle f, \tilde{\varphi}_n \rangle|^2 \le \frac{1}{A}\|f\|^2$$

Thus, $\tilde{\varphi}_n$ is the dual frame of φ_n with bounds $\frac{1}{A}$ and $\frac{1}{B}$.
The sequence $\tilde{\varphi}_n$ is called the *dual frame*.
Verification of (12.5) It follows from the following results.

A. Inverse T^{-1} of T is positive, if T is an invertible positive operator.
B. If T is a positive operator on a Hilbert space H such that $AI \le T \le BT$ for some $0 < A < B$, then

$$\frac{1}{B}I \le T^{-1} \le \frac{1}{A}I.$$

Lemma 12.3 *Let F be a frame operator associated with the frame $\{\varphi_n\}$ and $\tilde{F}$ the frame operator associated with its dual frame $\tilde{\varphi}_n$. Then*

$$\tilde{F}^*F = I = F^*F$$

Proof By

$$F(F^*F)^{-1}f = (\langle (F^*F)^{-1}f, \varphi_n \rangle) = \{\langle f, \tilde{\varphi}_n \rangle\}$$
$$= \tilde{F}f$$

we obtain

$$\tilde{F}^*F = (F(F^*F)^{-1})^*F = (F^*F) - F^*F = I$$

and

$$F^*\tilde{F} = F^*F(F^*F)^{-1} = I.$$

Theorem 12.6 *Let $\{\varphi_n\}$ be a frame in a Hilbert space H and $\tilde{\varphi}_n$ the dual frame. Then*

$$f = \sum_{n=1}^{\infty} \langle f, \varphi_n \rangle \tilde{\varphi}_n \tag{12.16}$$

$$f = \sum_{n=1}^{\infty} \langle f, \tilde{\varphi}_n \rangle \varphi_n \tag{12.17}$$

for any $f \in H$.

Proof Let F be the frame operator associated with frame $\{\varphi_n\}$ and $\tilde{F}$ the frame operator associated with the dual frame $\{\tilde{\varphi}_n\}$. Since $I = \tilde{F}^*F$ by Lemma 12.3, for any $f \in H$, we have

$$f = \tilde{F}^*Ff = \tilde{F}^*(\langle f, \varphi_n \rangle) = \sum_{n=1}^{\infty} \langle f, \varphi_n \rangle \tilde{\varphi}_n \; by \; (12.13)$$

Therefore, we get (12.16). Equation (12.17) can be proved similarly.

Remark 12.4 For tight frame, then $\tilde{\varphi}_n = A^{-1}\varphi_n$; so (12.17) becomes

$$f = \frac{1}{A} \sum_{n=1}^{\infty} \langle f, \varphi_n \rangle \varphi_n$$

For orthonormal basis instead of frame, we have

$$f = \sum_{n=1}^{\infty} \langle f, \varphi_n \rangle \varphi_n.$$

Definition 12.5 (*Riesz Basis*) A sequence of vectors $\{\varphi_n\}$ in a Hilbert space H is called a Riesz basis if the following conditions are satisfied:

(a)   There exist constants α and β, $0 < \alpha \le \beta$ such that

$$A||\alpha|| \le \left\| \sum_{n \in N} \alpha_n \varphi_n \right\| \le B||\alpha||$$

where $||\alpha|| = \left(\sum_{n \in N} |x_n|^2 \right)^{1/2}$, $x = (x_1, x_2, \ldots, x_n, \ldots)$
(b) $[\{\varphi_n\}] = H$; that is, H is spanned by $\{\varphi_n\}$.

A sequence $\{\varphi_n\}$ in H satisfying (a) is called *Riesz sequence*.

It may be observed that a Riesz basis is a special case of frames and an orthonormal basis is a particular case of a Riesz basis, where $\alpha = \beta = 1$. Such cases can be obtained for the particular choice $a_0 = 2$ and $b_0 = 1$, $m = j$ and $n = k$; that is, $\psi_{j,k}(t) = 2^{j/2}\psi(2^j t - k)$.

Definition 12.6 A function $\psi \in L_2(R)$ is a wavelet if the family of functions $\psi_{j,k}(t)$ defined by

$$\psi_{j,k}(t) = 2^{j/2}\psi(2^j t - k), \tag{12.18}$$

where j and k are arbitrary integers, is an orthonormal basis in the Hilbert space $L_2(R)$.

It may be observed that the admissibility condition (12.2) is a necessary condition under which $\psi_{j,k}(t)$; that is, $\psi_{j,k}(t) = 2^{j/2}\psi(2^j t - k)$ is a frame, in general, and an orthonormal basis, in particular. For more discussion, one can see [59, 65].

Definition 12.7 (*Wavelet Coefficients*) Wavelet coefficients of a function $f \in L_2(R)$, denoted by $d_{j,k}$, are defined as the inner product of f with $\psi_{j,k}(t)$; that is

$$d_{j,k} = \langle f, \psi_{j,k}(t) \rangle = \int_R f(t)\psi_{j,k}(t)dt \qquad (12.19)$$

The series

$$\sum_{j \in Z} \sum_{k \in Z} \langle f, \psi_{j,k}(t) \rangle \psi_{j,k}(t) \qquad (12.20)$$

is called the *wavelet series* of f. The expression

$$f = \sum_{j \in Z} \sum_{k \in Z} \langle f, \psi_{j,k}(t) \rangle \psi_{j,k}(t)$$

is called the wavelet representation of f.

Remark 12.5 A. $\psi_{j,k}(t)$ is more suited for representing finer details of a signal as it oscillates rapidly. The wavelet coefficients $d_{j,k}$ measure the amount of fluctuation about the point $t = 2^{-j}k$ with a frequency measured by the dilation index j.

B. It is interesting to observe that $d_{j,k} = T_\psi f(2^{-j}, k2^{-j})$. Wavelet transform of f with wavelet ψ at the point $(2^{-j}, k2^{-j})$.

We conclude this section by a characterization of Lipschitz α class, $0 < \alpha < 1$, in terms of the wavelet coefficients.

Theorem 12.7 $f \in Lip\alpha$; *that is,* $|f(x) - f(y)| \le K|x - y|^\alpha$, $0 < \alpha < 1$ *if and only if*

$$|d_{j,k}| =\le K2^{(\frac{1}{2}+\alpha)j}$$

where K is a positive constant and ψ is smooth and well localized.

Proof Let $f \in C^\alpha$. Then

$$|d_{j,k}| = \left\| \int_R f(x)\psi_{j,k}(x)dx \right\|$$

$$= \left\| \int_R (f(x) - f(k2^{-j}))\psi_{j,k}(x)dx \right\|$$

$$\le K \int_R |(x - k2^{-j}|^\alpha \frac{2^{j/2}}{(1 + |(x - k2^{-j}|^2)}dx$$

$$\le K2^{(\frac{1}{2}+\alpha)j}$$

as $\int_R f(k2^{-j})\psi_j(x)dx$ vanishes and $|\psi(x)| \le \frac{1}{(1+|x|)^2}$ in view of the assumption.

Conversely, if $|d_{j,k}| \le K2^{-(\frac{1}{2}+\alpha)j}$, then

$$|f(x) - f(y)| \le K \sum_j \sum_k 2^{-(\frac{1}{2}+\alpha)j}|\psi_{j,k}(x) - \psi_{j,k}(y)|$$

Let J be such that $2^{-J} \le |x - y| < 2^{-J+1}$. Using the Mean Value Theorem

$$\sum_{j \le J} \sum_k 2^{-(\frac{1}{2}+\alpha)j}|\psi_{j,k}(x) - \psi_{j,k}(y)|$$

$$\le \sum_{j \le J} \sum_k K2^{-\alpha j}2^j|x - y| \sup\left(\frac{1}{(1 + |(x - k2^{-j}|^2)}, \frac{1}{(1 + |(y - k2^{-j}|^2)}\right)$$

The sum in k is bounded by a constant independent of j, and the sum in j is bounded by $K2^{(1-\alpha)J}|x - y| \le K|x - y|^\alpha$. Furthermore

$$\sum_{j > J} \sum_k 2^{-(\frac{1}{2}+\alpha)j}|\psi_{j,k}(x) - \psi_{j,k}(y)|$$

$$\le \sum_{j > J} \sum_k 2^{-(\frac{1}{2}+\alpha)j}(|\psi_{j,k}(x)| + |\psi_{j,k}(y)|)$$

$$\le 2K \sum_{j > J} \sum_k \frac{2^{-\alpha J}}{(1 + |(y - k2^{-j}|^2)} \le K2^{-\alpha j} \le K|x - y|^\alpha$$

by the given property of ψ.

Theorem 12.7 is about global smoothness. Using localization of the wavelets, a similar result for pointwise smoothness of f will be proved in Sect. 12.4.

12.3 Multiresolution Analysis, and Wavelets Decomposition and Reconstruction

12.3.1 Multiresolution Analysis (MRA)

The concept of multiresolution analysis (MRA) is called as the heart of wavelet theory.

Definition 12.8 (*Multiresolution Analysis, Mallat, 1989*) A multiresolution analysis is a sequence $\{V_j\}$ of subspaces of $L_2(R)$ such that

(a) $\cdots \subset V_{-1} \subset V_0 \subset V_1 \subset \cdots$,
(b) $span \bigcup\limits_{j \in Z} V_j = L_2(R)$,
(c) $\bigcap\limits_{j \in Z} V_j = \{0\}$,
(d) $f(x) \in V_j$ if and only if $f(2^j x) \in V_0$,
(e) $f(x) \in V_0$ if and only if $f(x - m) \in V_0$ for all $m \in Z$, and
(f) There exists a function $\varphi \in V_0$ called *scaling function*, such that the system $\{\varphi(t - m)\}_{m \in Z}$ is an orthonormal basis in V_0.

For the sake of convenience, we shall consider real functions unless explicitly stated.

Some authors choose closed subspace V_j in Definition 12.8.

Definition 12.9 (1) *Translation Operator*: A translation operator T_h acting on functions of R is defined by

$$T_h(f)(x) = f(x - h)$$

for every real number h.
(2) *Dilation Operator*: A dyadic dilation operator J_j acting on functions defined on R by the formula

$$J_j(f)(x) = f(2^j x)$$

for an integer j.
It may be noted that (i) we can define dilation by any real number, not only by 2^j, (ii) T_h and J_j are invertible and $T_h^{-1} = T_{-h}$ and $J_j^{-1} = J_{-j}$, and (iii) T_h and $2^{j/2} J_j$ are isometries on $L_2(R)$.

Remark 12.6 (a) Conditions (*i*) to (*iii*) of Definition 12.8 signify that every function in $L_2(R)$ can be approximated by elements of the subspaces V_j, and precision increases as j approaches ∞.
(b) Conditions (*iv*) and (*v*) express the invariance of the system of subspaces $\{V_j\}$ with respect to the dilation and translation operators. These conditions can be

expressed in terms of T_h and J_j as (iv)' $V_j = J_j(V_0)$ for all $j \in Z$ (v)' $V_0 = T_n(V_0)$ for all $n \in Z$. (vi)' Since $2^s J_s$ and T_n are isometries, (vi) can be rephrased as: For each $j \in Z$ the system $\{2^{j/2}\varphi(2^j x - k)\}_{k \in Z}$ is an orthonormal basis in V_j.

(c) Condition (vi) implies Condition (vi).
(d) If we define the dilation operator D_a, $a > 0$, as $D_a f(x) = a^{1/2} f(ax)$, translation operator as above for $b \in R$, modulation operator E_c as

$$E_c f(x) = e^{2\pi i c x} f(x), \ f(x) \in L_1 \text{ or } L_2 \text{ on } R, \ c \in R$$

then the following results can be easily verified

(1) $D_a T_b f(x) = a^{1/2} f(ax - b)$
(2) $D_a T_b f(x) = T_{a^{-1}b} D_a f(x)$
(3) $\langle f, T_a g \rangle = \langle D_{a^{-1}} f, g \rangle$
(4) $\langle f, T_b g \rangle = \langle T_{-b} f, g \rangle$
(5) $\langle D_a f, D_a g \rangle = \langle f, g \rangle$
(6) $\langle T_b f, T_b g \rangle = \langle f, g \rangle$
(7) $T_b E_c f(x) = e^{2\pi i b c} E_c T_b f(x)$
(8) $\langle f, E_c g \rangle = \langle E_c f, g \rangle$

The following theorems describe the basic properties of an MRA whose proofs can be found in any of the standard references on wavelet theory (see, e.g., [199]).

Theorem 12.8 *Let $\varphi \in L_2(R)$ satisfy*

(i) $\{\varphi(t - m)\}_{m \in Z}$ *is a Riesz sequence in $L_2(R)$*
(ii) $\varphi(x/2) = \sum_{k \in Z} a_k \varphi(x - k)$ *converges on $L_2(R)$*
(iii) $\hat{\varphi}(\xi)$ *is continuous at 0 and $\hat{\varphi}(0) \neq 0$.*

Then the spaces $V_j = span\{\varphi(2^j x - k)\}_{k \in Z}$ with $j \in Z$ form an MRA.

Theorem 12.9 *Let $\{V_j\}$ be an MRA with a scaling function $\varphi \in V_0$. The function $\psi \in W_0 = V_1 \ominus V_0$ ($W_0 \oplus V_0 = V_1$) is a wavelet if and only if*

$$\hat{\psi}(\xi) = e^{i\xi/2} v(\xi) \overline{m_\phi(\xi/2 + \pi)} \hat{\varphi}(\xi/2) \tag{12.21}$$

for some 2π-periodic function $v(\xi)$ such that $|v(\xi)| = 1$ a.e., where $m_\varphi(\xi) = \frac{1}{2} \sum_{n \in Z} a_n e^{-n\xi}$. Each such wavele ψ has the property that spans $\{\psi_j\}_{k \in Z}, j < s = V_s$ for every $s \in Z$.

Remark 12.7 (i) For a given MRA $\{V_j\}$ in $L_2(R)$ with the scaling function φ, a wavelet is obtained in the manner described below; it is called the wavelet associated with the MRA $\{V_j\}$. Let the subspace W_j of $L_2(R)$ be defined by the condition

$$V_j \oplus W_j = V_{j+1}, \ V_j \perp W_j \ \forall j$$

Since $2^{j/2}J_j$ is an isometry, from Definition 12.8, $J_j(V_1) = V_{j+1}$. Thus

$$V_{j+1} = J_j(V_0 \oplus W_0) = J_j(V_0) \oplus J_j(W_0) = V_j \oplus J_j(W_0)$$

This gives

$$W_j = J_j(W_0) \; for \; all \; j \in Z \tag{12.22}$$

From conditions (i)–(iii) of Definition 12.8, we obtain an orthogonal decomposition

$$L_2(R) = \oplus \sum_{j\in Z} W_j = W_1 \oplus W_2 \cdots \oplus W_n \oplus \cdots$$

$$= \oplus \sum_{j\in Z} W_j \tag{12.23}$$

We need to find a function $\psi \in W_0$ such that $\{\psi(t-m)\}_{m\in Z}$ is an orthonormal basis in W_0. Any such function is a wavelet; it follows from (12.22) and (12.23).
(ii) Let φ be a scaling function of the MRA $\{V_j\}$, then

$$\psi(x) = \sum_{n\in Z} a_n(-1)^n\varphi(2x+n+1) \tag{12.24}$$

where

$$a_n = \sum_{-\infty}^{\infty} \varphi(x/2)\varphi(x-n) \tag{12.25}$$

is a wavelet. In fact, assuming that integer translates of generate an orthonormal basis for V_0, wavelet ψ can be constructed as follows: By conditions (i) and (ii) of MRA, there exists c_n such that (see Sect. 12.4.3)

$$\varphi(x) = \sum_{n\in Z} c_n\varphi(2x-n)$$

Then, $\psi(x)$ is given by

$$\psi(x) = \sum_{n\in Z}(-1)^n c_{n+1}\varphi(2x+n)$$

(iii) It should be observed that the convention of increasing subspaces $\{V_j\}$ used here is not universal, as many experts like Daubechies and Mallat use exactly the opposite convention by choosing decreasing sequence $\{V_j\}$, where (iv) is replaced by $f(x) \in V_j$ if and only if $f(2^j x) \in V_0$ and $V_j \oplus W_j = V_{j+1}\}, V_j \perp W_j$

is replaced by $V_j \oplus W_j = V_{j1}, V_j \perp W_j$. Broadly speaking, in the convention of increasing subspaces adapted by Meyer, the functions in V_j scale like 2^{-j}, whereas in the convention of decreasing subspaces followed by Daubechies and Mallat, they scale like 2^j.

(vi) With a smooth wavelet ψ, we can associate an MRA. More precisely, let be an $L_2(R)$ function such that $\{2^{j/2}\psi(2^j x - k), j \in Z, k \in Z\}$ is an ONB of $L_2(R)$. Is ψ the mother wavelet of an MRA? For any ψ, the answer is no. However, under mild regularity conditions, the answer is yes (for details, see references in [94], p. 45).

12.3.2   Decomposition and Reconstruction Algorithms

Decomposition Algorithm Let $c_{j,k}$ and $d_{j,k}$ denote, respectively, the scaling and wavelet coefficients for j and $k \in Z$ defined by

$$c_{j,k} = \int_R f(x)\varphi_{j,k}(x)dx \tag{12.26}$$

and

$$d_{j,k} = \int_R f(x)\psi_{j,k}(x)dx \tag{12.27}$$

where

$$\varphi_{j,k}(x) = 2^{j/2}\varphi(2^j x - k)$$
$$\psi_{j,k}(x) = 2^{j/2}\psi(2^j x - k)$$

$\varphi(x)$ and $\psi(x)$ are, respectively, the scaling function (often called the father wavelet) and the wavelet (mother wavelet).

Since $\varphi_{j,k}(x) = 2^{j/2}\varphi(2^j x - k)$, there exists h_ℓ such that

$$\begin{aligned}
\varphi_{j,k}(x) &= \sum_{\ell \in Z} h_\ell 2^{j/2}\varphi_{1,\ell}(2^j x - k) \\
&= \sum_{\ell \in Z} h_\ell 2^{(j+1)/2}\varphi(2^{j+1} x - 2k - \ell) \\
&= \sum_{\ell \in Z} h_\ell \varphi_{j+1,\ell+2k}(x) \\
&= \sum_{\ell \in Z} h_{\ell-2k}\varphi_{j+1,\ell}(x) \tag{12.28}
\end{aligned}$$

Substituting this value into (12.26), we get

$$c_{j,k} = \int_R f(x) \sum_{\ell \in Z} h_{\ell-2k}\varphi_{j+1,\ell}(x)$$

$$= \sum_{\ell \in Z} h_{\ell-2k} \int_R f(x)\varphi_{j+1,\ell}(x)dx$$

$$= \sum_{\ell \in Z} h_{\ell-2k}c_{j+1,\ell}$$

or

$$c_{j,k} = \sum_{\ell \in Z} h_{\ell-2k}c_{j+1,\ell} \tag{12.29}$$

Since $V_0 \subset V_1$, every $\varphi \in V_0$ also satisfies $\varphi \in V_1$. Since $\{\varphi_{1,k}, k \in Z\}$ is an orthonormal basis for V_1, there exists a sequence $\{h_k\}$ such that

$$\varphi(x) = \sum_{k \in Z} h_k\varphi_{1,\ell}(x) \tag{12.30}$$

and that the sequence element may be written in the form

$$h_k = \langle \varphi, \varphi_{1,k} \rangle \ and \ \{h_k\} \in \ell_2$$

The two-scale relationship (12.29), relating functions with differing scaling factors, is also known as the dilation equation or the refinement equation. It can be seen that for the Haar basis

$$h_k = \frac{1}{\sqrt{2}}, \quad k = 0, 1$$

$$= 0, \quad otherwise \tag{12.31}$$

Scaling function φ (father wavelet) and ψ (mother wavelet) are related with the following relation:

$$\psi(x) = \sum_{k \in Z} (-1)^k h_{-k+1}\varphi_{1,k}(x) \tag{12.32}$$

Substituting the value of $\psi(x)$ from (12.31) in (12.27), we obtain

$$d_{j,k} = \sum_{\ell \in Z} (-1)^l h_{-\ell+2k+1}c_{j+1,\ell} \tag{12.33}$$

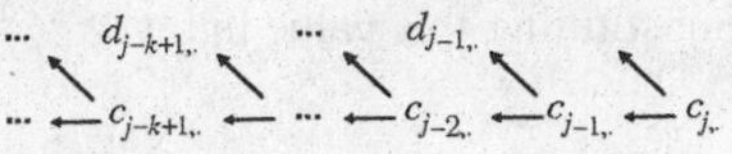

Fig. 12.4 Schematic
representation of the
decomposition algorithm

$$\cdots \quad \overset{d_{j-k+1,\cdot}}{\diagdown} \quad \cdots \quad \overset{d_{j-1,\cdot}}{\diagdown} \quad \diagdown$$
$$\cdots \leftarrow c_{j-k+1,\cdot} \leftarrow \cdots \leftarrow c_{j-2,\cdot} \leftarrow c_{j-1,\cdot} \leftarrow c_{j,\cdot}$$

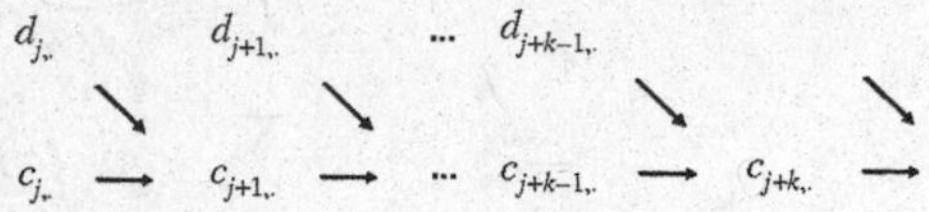

Fig. 12.5 Schematic
representation of the
reconstruction algorithm

$$d_{j,\cdot} \qquad d_{j+1,\cdot} \qquad \cdots \quad d_{j+k-1,\cdot}$$
$$\diagdown \qquad \diagdown \qquad \qquad \diagdown \qquad \diagdown$$
$$c_{j,\cdot} \longrightarrow c_{j+1,\cdot} \longrightarrow \cdots \quad c_{j+k-1,\cdot} \longrightarrow c_{j+k,\cdot} \longrightarrow$$

All lower-level scaling coefficients ($j > i$) can be computed from scaling function coefficients applying (12.33). Given scaling coefficients at any level $j > i$ can be computed recursively applying (12.29).

Figure 12.4 is schematic representation of the decomposition algorithm for scaling and wavelet coefficients $d_{j,\cdot}$ and $c_{j,\cdot}$, respectively, at level j. Equations (12.29) and (12.33) share an interesting feature; that if, any one of them, the dilation index k is increased by one, then indices of the $\{h_\ell\}$ are all offset by two. It may be observed that computation by the decomposition algorithm yields fewer coefficient at each level. Mallat named it *pyramid algorithm*, while Daubechies called it the *cascade algorithm*.

Reconstruction Algorithm

Let $\{\varphi_{j,k}\}_{k\in Z}$ and $\{\psi_{j,k}\}_{k\in Z}$ be generated, respectively, by the father wavelet φ and the mother wavelet ψ; that is, $\varphi_{j,k}(x) = 2^{j/2}\varphi(2^j x - k)$ and $\psi_{j,k}(x) = 2^{j/2}\psi(2^j x - k)$ form the orthonormal basis, respectively, of V_j and W_j of a given MRA for each k (Fig. 12.5). Further, let

$$a_{2k} = \langle \varphi_{1,0}, \varphi_{0,k} \rangle a_{2k-1} = \langle \varphi_{1,0}, \varphi_{0,k} \rangle$$
$$b_{2k} = \langle \varphi_{1,0}, \varphi_{0,k} \rangle b_{2k-1} = \langle \varphi_{1,0}, \varphi_{0,k} \rangle$$

where $a_k = h_{-k}$ and $b_k = (-1)^k h_{k+1}$. Then

$$c_{j,k} = \sum_{\ell\in Z} a_{2\ell-k} c_{j-1,\ell} + b_{2\ell-k} d_{j-1,\ell} \tag{12.34}$$

or

$$c_{j,k} = \sum_{\ell\in Z} h_{k-2\ell} c_{j-1,\ell} + (-1)^k h_{2\ell-k+1} d_{j-1,\ell}. \tag{12.35}$$

Verification of Equation (12.35)

$$\varphi_{1,0}(x) = \sum_{k\in Z} (a_{2k}\varphi_{0,k}(x) + b_{2k}\psi_{0,k}(x)) \tag{12.36}$$

and

$$\varphi_{1,1}(x) = \sum_{k \in Z}(a_{2k-1}\varphi_{0,k}(x) + b_{2k-1}\psi_{0,k}(x)) \tag{12.37}$$

By (12.36) and (12.37), we can write a similar expression for any $\varphi_{1,k}$. First, we derive the formula for even k

$$\varphi_{1,1}(x) = \varphi_{1,0}\left(x - \frac{k}{2}\right)$$

$$= \sum_{\ell \in Z} a_{2\ell}\varphi_{0,\ell}\left(x - \frac{k}{2}\right) + b_{2\ell}\psi_{0,\ell}$$

$$= \sum_{\ell \in Z}\left(a_{2\ell}\varphi_{0,\frac{k}{2}+\ell} + b_{2\ell}\psi_{0,\frac{k}{2}+\ell}\right)$$

$$= \sum_{\ell \in Z}(a_{2\ell-k}\varphi_{0,\ell}(x) + b_{2\ell-k}\psi_{0,\ell}(x)) \tag{12.38}$$

A similar result holds for odd k. For even (odd) k, only the even-indexed (oddindexed) elements of the sequences $\{a_\ell\}$ and $\{b_\ell\}$ are accessed. On similar lines, an expression relating each scaling function $\varphi_{j,k}$ to scaling functions and wavelets at level $j - 1$ can be derived; that is, we have

$$\varphi_{j,k}(x) = \sum_{\ell \in Z}(a_{2\ell-k}\varphi_{j-1,\ell}(x) + b_{2\ell-k}\psi_{j-1,\ell}(x)) \tag{12.39}$$

Equation (12.35) is obtained from (12.26). We obtain (12.36) by substituting the values of $a_{2\ell-k}$ and $b_{2\ell-k}$ in terms of h_k.

Remark 12.8   (i) The scaling coefficients at any level can be computed from only one set of low-level scaling coefficients and all the intermediate wavelet coefficients by applying (12.36) recursively.
(ii) Each wavelet basis is completely characterized by two-scale sequence $\{h_k\}$.
(iii) h_k will consist of finite number of elements for wavelets having compact support.

12.3.3 Wavelets and Signal Processing

Signals are nothing but functions which represent real-world problems arising in different fields. Signal processing deals with denoising, compression, economic storage, and communication synthesis, and the signal processing has been used in brain studies, global warming, and prediction by calamities. For a lucid introduction and updated literature, we refer to [43, 44, 50, 51, 55, 58, 59, 94, 98, 133, 145, 147, 149, 179, 188, 191, 192, 197, 199]. Let $f(t)$, $t \in R$ denote a signal. A signal $f(t)$ is said to have a finite energy, if $\int_{-\infty}^{\infty}|f(t)|^2 dt < \infty$, equivalently, $f \in L_2(R)$.

$E(f) = \left(\int\limits_{-\infty}^{\infty} |f(t)|^2 dt \right)^2$ is called the *energy of signal* f. The main idea is to store
or transmit certain values of $f(t)$ instead of entire values. It has been shown that
wavelet orthonormal system yields very economical results. Let $\{\varphi_n\}$ be an ONB in
$L_2(R)$, then we can write

$$f = \sum_{n \in N} \langle f, \varphi_n \rangle \varphi_n$$

Thus, instead of transmitting the function f, it suffices to transmit the sequence of
coefficients $\{\langle f, \varphi_n \rangle\}$ and let the recipient sum the series himself. If the ONB $\{\varphi_n\}$ is
given by a compactly supported wavelet $\{\psi_{i,j}\}$ (Definition 12.6 or wavelet associated
with an MRA), then, in view of Remark 12.8, we achieve our objective.

Now, we express decomposition and reconstruction algorithms in terms of a concept of the signal processing called the filtering processing. A filter can be considered
as an operator on ℓ_2 into itself. Thus, applying a filter (operator or sequence) to a
signal of ℓ_2 (discrete form of $L_2(R)$) results in another signal. It may be observed
that the term filter is taken from its real-life uses. We know that a filter is used in a
laboratory either to purify a liquid from solid impurities (if the liquid is of interest)
or to remove a solid from a suspension in a liquid (if the solid is of interest), and our
filter applied to a pure signal contaminated with noise might attempt either to isolate
the pure signal or to extract the noise, depending on our primary interest, signal or
noise. Applying the decomposition algorithm involves a down-sampling operation.
The operation of filtering, called subband filtering, which we consider here, operates
exactly as the decomposition algorithm; namely, applying a subband filter to a signal
yields a signal with length half that of the original signal. In practice, we deal only
with signals having a finite number of nonzero terms; usually, we consider an even
number of nonzero terms.

In general, a filter A is defined by

$$A f_k = \sum_{\ell \in Z} a_{2\ell - k} f_\ell \tag{12.40}$$

where $f = (f_1, f_2, \ldots, f_n, \ldots)$. $\{A f_k\} \in \ell_2$; that is, the filtering process consists of
a discrete convolution of the filter sequence with the signal. Let H be a filter defined
by the relation

$$H c_{j,.} = c_{j-1,.} \tag{12.41}$$

which corresponds to (12.29), where j is replaced by $j - 1$; that is

$$c_{j-1,k} = \sum_{\ell \in Z} h_{2\ell - k} c_{j,\ell}$$

Applying the filter H m-times, we get

$$H^m c_{j,.} = c_{j-m,.} \tag{12.42}$$

Let G be another filter defined by

$$Gc_{j,.} = d_{j-1,.} \tag{12.43}$$

which corresponds to (12.33) where j is replaced by $j - 1$, that is

$$d_{j-1,k} = \sum_{\ell \in Z} (-1)^\ell h_{-\ell+2k+1} c_{j,\ell} \tag{12.44}$$

By (12.41) and (12.43), we get

$$d_{j-m,.} = GH^{m-1} c_{j,.} \tag{12.45}$$

This means that the wavelet coefficients at any lower level can be computed from scaling function coefficients at level j. Thus, the decomposition algorithm has been written in terms of filter H and G. In engineering literature, H is called the *low-pass filter* while G is called *high-pass filter*. Both are examples of quadrature mirror filters of signal processing. Broadly speaking, low-pass filters correspond to the averaging operations which yield trend, while high-pass filters correspond to differencing and signify difference or fluctuation.

Example 12.5 Let us consider the Haar wavelet in which $h_0 = h_1 = 1/\sqrt{2}$. Let $f = (f_1, f_2, f_3, \ldots, f_k, \ldots) = \{f_k\} \in \ell_2$. Then

$$Hf = \tilde{f}_k$$

where $\tilde{f}_k = \frac{1}{\sqrt{2}}(f_{2k} + f_{2k-1})$, and

$$Gf = f_k^\star$$

where $f_k^\star = \frac{1}{\sqrt{2}}(f_{2k} + f_{2k-1})$. Choose $f = (4, 6, 10, 12, 8, 6, 5, 5)$. Then

$$Hf = (5\sqrt{2}, 11\sqrt{2}, 7\sqrt{2}, 5\sqrt{2})$$
$$Gf = (-\sqrt{2}, -\sqrt{2}, -\sqrt{2}, 0).$$

12.3.4 The Fast Wavelet Transform Algorithm

Cooley and Tukey algorithm developed in 1965 known as fast Fourier transform is treated as a revolution in computational techniques. It reduces computational cost to O (n log n), a very significant development, see [145]. The pertinent point is to reduce the number of computations applying recursively computing the discrete

Fourier transform of subsets of given data. This is obtained by reordering the given data to find advantage of some redundancies in the usual discrete Fourier transform algorithm. Proceeding on the lines of fast Fourier transform, we could choose appropriate wavelet transforms by replacing the function f in the definition of wavelet coefficients by an estimate such as

$$d_{j,k} = \int\limits_{-\infty}^{\infty} g(x)\psi_{j,k}(x)dx$$

where

$$g(t) = X_k, \ k - 1 < t \le k$$
$$= 0, \ otherwise$$

Similar technique could be used to evaluate the following scaling function coefficients as well:

$$\tilde{c}_{j,k} = \int\limits_{-\infty}^{\infty} g(x)\varphi_{j,k}(x)dx$$
$$\simeq X_\ell \varphi_{j,k}(x)$$

Let us begin with a set of high-level scaling function coefficients, which are finite. This assumption is appropriate, as any signal $f \in L_2(R)$ must have rapid decay in both directions; so $d_{j,k}$ can be neglected for large $|k|$. Rescale the original function, if necessary, so that the scaling function coefficients at level m are given by $c_{m,0}, \ldots, c_{m,n-1}$. Computing the scaling function and wavelet coefficients at level $m - 1$ is accomplished via Eqs. (12.29) and (12.33). As noticed earlier, the $\{h_k\}$ sequence used in these calculations has only finite number of nonzero values if the wavelets are compactly supported; otherwise, h_k values decay exponentially; so it can be approximated by finite number of terms. In either case, let K denote the number of nonzero terms used in the sequence, possibly truncated. Computing a single coefficient at level $m - 1$ according to either (12.29) or (12.33) would take at most K operations. If scaling function coefficients $c_{m,k}$ for $k \ne \{0, \ldots, n-1\}$ are set to zero, they give exactly n nonzero coefficients at level m and then the number of nonzero scaling function coefficients at level $m - 1$ would be at most

$$\left[\frac{n}{2}\right] + \left[\frac{K}{2}\right] + 2$$

where $[x]$ denotes the greatest integer less than or equal to x. The total number of nonzero scaling function coefficients at level $m - 1$ is approximately $\frac{n}{2}$, and the total number of operations required to compute the one-level-down wavelet and scaling function coefficients is approximately $2K\frac{n}{2}$. Let n_1 be the number of nonzero scaling function coefficients at level $m - 1$. Applying the decomposition again requires no more than

$$\left[\frac{n_1}{2}\right] + \left[\frac{K}{2}\right] + 2$$

operations to compute the scaling function coefficients and no more than the same number of operations to compute the wavelet coefficients. There will be approximately $n_1/2$ or $n/4$ nonzero scaling function coefficients at level $m - 2$, and the computation will require approximately $2K.n_1/2$ or $2K.n/4$ operations.

Continuing in this manner, we find that the total number of operations required to do all decompositions is approximately

$$2K\left(\frac{n}{2} + \frac{n}{4} + \frac{n}{8} + \cdots\right) = O(n)$$

Thus, the fast wavelet transform algorithm described above is faster than the fast Fourier transform algorithm.

12.4 Wavelets and Smoothness of Functions

In this section, we will discuss a relationship between smoothness of functions and properties of wavelets.

12.4.1 Lipschitz Class and Wavelets

Definition 12.10 (*Vanishing Moments*) A wavelet ψ is said to have n vanishing moments if

$$\int_{-\infty}^{\infty} t^k \psi(t)dt = 0 \ for \ 0 \le k < n \tag{12.46}$$

A wavelet ψ has n vanishing moments and is C_n with derivatives that have a fast decay means that for any $0 \le k \le n$ and $m \in N$, there exists M_m such that

$$\forall t \in R, |\psi^{(k)}(t)| \le \frac{M_m}{1 + |t|^m} \tag{12.47}$$

Definition 12.11 (*Pointwise Lipschitz Regularity*) A function f is pointwise Lipschitz $\alpha, \alpha \ge 0$, at v if there exists $K > 0$ and a polynomial p_v of degree $m = [\alpha]$

such that

$$|f(t) - p_v(t)| \leq K|t - v|^\alpha, \ \forall t \in R \tag{12.48}$$

Theorem 12.10 (Jaffard, 1991) *If* $f \in L_2(R)$ *is Lipschitz* α, $\alpha \leq n$ *at* v, *then there exists a positive constant* M *such that*

$$|T_\psi f(a, b)| \leq Ma^{\alpha+1/2} \left(1 + \left|\frac{b-v}{a}\right|^\alpha\right),$$

$$\forall(a, b) \in R^+ \times R \tag{12.49}$$

Conversely, if $\alpha < n$ *is not an integer and there exist* M *and* $\alpha' < \alpha$ *such that*

$$|T_\psi f(a, b)| \leq Ma^{\alpha+1/2} \left(1 + \left|\frac{b-v}{a}\right|^{\alpha'}\right),$$

$$\forall(a, b) \in R^+ \times R \tag{12.50}$$

then f *is Lipschitz* α *at* v.

Proof Since f is Lipschitz α at v, there exists a polynomial p_v of degree $[\alpha] < n$ and a positive constant K such that

$$|f(t) - p_v(t)| \leq K|t - v|^\alpha$$

A wavelet with n vanishing moments is orthogonal to polynomials of degree $n-1$ (clear from Definition 12.10). Since $\alpha < n$, the polynomial pv has degree at most $n-1$. With the change of variable $t' = (t-u)/s$, we check that

$$|T_\psi p_v(a, b)| = \int_{-\infty}^{\infty} p_v(t) \frac{1}{\sqrt{a}} \psi \frac{t-b}{a} dt = 0 \tag{12.51}$$

In view of (12.51), we get

$$|T_\psi f(a, b)| = \left|\int_{-\infty}^{\infty} (f(t) - p_v(t)) \frac{1}{\sqrt{a}} \psi \frac{t-b}{a} dt\right|$$

$$\leq K|t - v|^\alpha \frac{1}{\sqrt{a}} \left|\psi \frac{t-b}{a}\right| dt$$

The change of variable $x = \frac{t-b}{a}$ yields

$$|T_\psi f(a,b)| \le \sqrt{a} \int_{-\infty}^{\infty} K|ax + b - v|^\alpha |\psi(x)| dx$$

Since $|a + b|^\alpha \le 2^\alpha(|a|^\alpha + |b|^\alpha)$,

$$|T_\psi f(a,b)| \le K2^\alpha \sqrt{a} \left(a^\alpha \int_{-\infty}^{\infty} |x|^\alpha |\psi(x)| dx + |b - v|^\alpha \int_{-\infty}^{\infty} |\psi(x)| dx \right)$$

$$\le Ma^{\alpha+1/2} \left(1 + \left| \frac{b-v}{a} \right|^\alpha \right)$$

where M is a constant and that is the desired result (12.49).

Converse: f can be decomposed by (12.8) in the form

$$f(t) = \sum_{j=-\infty}^{\infty} \Delta_j(t) \tag{12.52}$$

with

$$\Delta_j(t) = \frac{1}{c_\psi} \int_{-\infty}^{\infty} \int_{2^j}^{2^{j+1}} T_\psi f(a,b) \frac{1}{\sqrt{a}} \psi\left(\frac{t-b}{a}\right) \frac{da}{a^2} db \tag{12.53}$$

Let $\Delta_j^{(k)}$ be its kth order derivative. In order to prove that f is Lipschitz α at v, we shall approximate f with a polynomial that generalizes the Taylor polynomial

$$p_v(t) = \sum_{k=0}^{[\alpha]} \left(\sum_{j=-\infty}^{\infty} \Delta_j^{(k)}(v) \right) \frac{(t-v)^k}{k!} \tag{12.54}$$

If f is n times differentiable at v, then p_v corresponds to the Taylor polynomial, but this may not be true. We shall first prove that $\sum_{j=-\infty}^{\infty} \Delta_j^{(k)}(v)$ is finite by getting upper bounds on $|\Delta_j^{(k)}(t)|$.

To simplify the notation, let K be a generic constant which may change value from one line to the next but that does not depend on j and t. The hypothesis (12.50) and the asymptotic decay condition (12.47) imply that

$$|\Delta_j(t)| \leq \frac{1}{c_\psi} \int\limits_{-\infty}^{\infty} \int\limits_{2^j}^{2^{j+1}} K a^\alpha \left(1 + \left|\frac{b-v}{a}\right|^{\alpha'}\right) \frac{M_m}{1 + |(t-b)/a|^m} \frac{da}{a^2} db$$

$$\leq K \int\limits_{-\infty}^{\infty} 2^{\alpha_j} \left(1 + \left|\frac{b-v}{a}\right|^{\alpha'}\right) \frac{1}{1 + |(t-b)/2^{\alpha_j}|^m} \frac{1}{2^j} db \qquad (12.55)$$

Since $|b-v|^{\alpha'} \leq 2^{\alpha'}(|b-t|^{\alpha'} + |t-v|^{\alpha'})$, the change of variable $b' = 2^{-j}(b-t)$ gives

$$|\Delta_j(t)| \leq K 2^{\alpha_j} \int\limits_{-\infty}^{\infty} \frac{1 + |b'|^{\alpha'} + |v - t/2^j|^{\alpha'}}{1 + |b'|^m} db'$$

Choosing $m = \alpha' + 2$ yields

$$|\Delta_j(t)| \leq K 2^{\alpha_j} \left(1 + \left|\frac{v.-t}{2^j}\right|^{\alpha'}\right)$$

The same derivations applied to the derivatives of $\Delta_j(t)$ give

$$|\Delta_j^{(k)}(t)| \leq K 2^{(\alpha-k)_j} \left(1 + \left|\frac{v-t}{2^j}\right|^{\alpha'}\right) \quad \forall k \leq [\alpha] + 1$$

At $t = v$, it follows that

$$|\Delta_j^{(k)}(v)| \leq K 2^{(\alpha-k)_j} \quad \forall k \leq [\alpha] \qquad (12.56)$$

This guarantees a fast decay of $|\Delta_j^{(k)}(v)|$ when 2^j goes to zero, because α is not an integer so $\alpha > [\alpha]$. At large scales 2^j, since $|T_\psi f(a,b)| \leq \|f\| \|\psi\|$ with the change of variable, $b' = \frac{t-b}{a}$ in (12.28), we have

$$|\Delta_j^{(k)}(v)| \leq \frac{\|f\| \|\psi\|}{c_\psi} \int\limits_{-\infty}^{\infty} |\psi^k(b')db'| \int\limits_{2^j}^{2^{j+1}} \frac{da}{a^{3/2+k}} \qquad (12.57)$$

and hence $|\Delta_j^k(v)| \leq K 2^{-(k+1/2)j}$, which together with (12.58) proves that the polynomial p defined in (12.54) has finite coefficients. With (12.52), we compute

$$|f(t) - p_v(t)| = \left|\sum_{j=-\infty}^{\infty} \left(\Delta_j(t) - \sum_{k=0}^{[\alpha]} \Delta_j^{(k)}(v) \frac{(t-v)^k}{k!}\right)\right| \qquad (12.58)$$

The sum over scales is divided into two at $2J$ such that $2^J \geq |tv|2^{J1}$. For $j \geq J$, we can use the classical Taylor theorem to bound the Taylor expansion of Δ_j as follows:

$$I = \sum_{j=J}^{\infty} \left| \Delta_j(t) - \sum_{k=0}^{[\alpha]} \Delta_j^{(k)}(v) \frac{(t-v)^k}{k!} \right|$$

$$\leq \sum_{j=J}^{\infty} \frac{(t-v)^{[\alpha]+1}}{([\alpha]+1)!} \sup_{h \in [t,v]} |\Delta_j^{[\alpha]+1}(h)|$$

$$I \leq K|t-v|^{[\alpha]+1} \sum_{j=J}^{\infty} 2^{-j([\alpha]+1-\alpha)} \left| \frac{v-t}{2^j} \right|^{\alpha'}$$

and since $2^J \geq |t-v| \geq 2^{J-1}$, we get $I \leq K|v-t|^\alpha$.

Let us now consider the case $j < J$.

$$II = \sum_{j=-\infty}^{J-1} \left| \Delta_j(t) - \sum_{k=0}^{[\alpha]} \Delta_j^{(k)}(v) \frac{(t-v)^k}{k!} \right|$$

$$\leq K \sum_{j=-\infty}^{J-1} \left(2^{\alpha j} \left(1 + \left| \frac{v-t}{2^j} \right|^{\alpha'} + \sum_{k=0}^{[\alpha]} \frac{(t-v)^k}{k!} \right) 2^{(\alpha-k)j} \right)$$

$$\leq K \left(2^{\alpha J} + 2^{(\alpha-\alpha')J} |t-v|^{\alpha'} + \sum_{k=0}^{[\alpha]} \frac{(t-v)^k}{k!} 2^{(\alpha-k)j} \right)$$

and since $2^J \geq |t-v| \geq 2^{J-1}$, we get $II \leq K|v-t|^\alpha$. This proves that $|f(t) - p_v(t)| \geq K|v-t|^\alpha$; hence, f is Lipschitz α at v.

12.4.2 Approximation and Detail Operators

First of all, we discuss approximation and detail operators in the context of the Haar wavelet.

Definition 12.12 (a) For each pair of integers, j, k, let

$$I_{j,k} = [2^{-j}k, 2^{-j}(k+1)]$$

The collection of all such intervals is known as the *family of dyadic intervals*.

(b) A *dyadic step function* is a step function $f(x)$ with the property that for some integer j, $f(x)$ is constant on all dyadic intervals $I_{j,k}$, for any integer k. For any interval I, a dyadic step function on I is a dyadic step function that is supported on I.

(c) Given a dyadic interval at scale j, $I_{j,k}$, we write $I_{j,k} = I_{j,k}^{\ell} \cup I_{j,k}^{r}$, where $I_{j,k}^{\ell}$ and $I_{j,k}^{r}$ are dyadic intervals at scale $j + 1$, to denote the left half and right half of the interval $I_{j,k}$. In fact, $I_{j,k}^{\ell} = I_{j+1,2k}$ and $I_{j,k}^{r} = I_{j+1,2k+1}$.

Definition 12.13 (a) Let $\varphi(x) = \chi[0, 1)$, and for each $j, k \in Z$ define $\varphi_{j,k}(x) = 2^{j/2}\varphi(2^j x - k) = D_{2^j}T_k\varphi(x)$. The collection $\varphi_{j+k}(x)_{j,k\in Z}$ is called the *system of Haar scaling functions* at scale j.

(b) Let $\psi(x) = \chi_{[0,1/2)}(x) - \chi_{[1/2,1]}(x)$, and for each $j, k \in Z$ define $\psi_{j,k}(x) = 2^{j/2}\psi(2^j x - k) = D_{2^j}T_k\psi(x)$. The collection $\psi_{j,k}(x)_{j,k\in Z}$ is called the *Haar wavelet system*.

(c) For each $j \in Z$, we define the *approximation operator* P_j for functions $f(x) \in L_2$ and for Haar scaling system by

$$P_j f(x) = \sum_k \langle f, \varphi_{j,k}\rangle \varphi_{j,k} \tag{12.59}$$

(d) For each $j \in Z$, we define the approximation space V_j by

$$V_j = \overline{span}\{\varphi_{j,k}(x)\}_{k\in Z}$$

Remark 12.9 (i) By Example 3.13, $P_j f(x)$ is the function in V_j best approximating $f(x)$ in L_2.

(ii) Since $\varphi_{j,k}(x) = 2^{j/2}\xi_{I_{j,k}}(x)$

$$\langle f, \varphi_{j,k}\rangle \varphi_{j,k}(x) = \left(2^j \int_{I_{j,k}} f(t)dt\right) \chi_{I_{j,k}}(x)$$

Thus, $P_j f(x)$ is the average value of $f(x)$ on $I_{j,k}$. Due to this reason, one may consider the function $P_j f(x)$ as containing the features of $f(x)$ at resolution or scale 2^{-j}.

Lemma 12.4 (i) *For $j \in Z$, P_j is a linear operator on L_2.*

(ii) *P_j is idempotent, that is, $P_j^2 = P_j$.*

(iii) *Given integers j, j' with $j \leq j'$, and $g(x) \in V_j$, $P_{j'}g(x) = g(x)$.*

(vi) *Given $j \in Z$ and $f(x) \in L_2$*

$$\|P_j f\|_{L_2} \leq \|f\|_{L_2}$$

Proof (a) and (b) follow from Example 3.13 and (c) from (b); we prove here (d). Since $\{\varphi_{j,k}(x)\}_{k\in Z}$ is an orthonormal system (Lemma 12.9), by Theorem 3.18(c)

$$\|P_j f\|_{L_2}^2 = \sum_k |\langle f, \varphi_{j,k}\rangle|^2$$

$$= \sum_k |2^{j/2} \int_{I_{j,k}} f(t)dt|^2$$

By the Cauchy–Schwartz inequality (Hölders inequality for $p = 2$)

$$\left| 2^{j/2} \int_{I_{j,k}} f(t)dt \right|^2 \leq \left(\int_{I_{j,k}} 2^j dt \right) \left(\int_{I_{j,k}} |f(t)|^2 dt \right) = \int_{I_{j,k}} |f(t)|^2 dt$$

Therefore

$$\|P_j f\|_{L_2}^2 \leq \sum_k \int_{I_{j,k}} |f(t)|^2 dt = \int_R |f(t)|^2 dt = \|f\|_{L_2}^2$$

Lemma 12.5 *Let f be a continuous function on R with compact support. Then*

(i) $\lim\limits_{j \to \infty} \|P_j f - f\|_{L_2} = 0$

(ii) $\lim\limits_{j \to \infty} \|P_j f\|_{L_2} = 0$

Proof (i) Let $f(x)$ have compact support $[-2^N, 2^N]$ for some integer N. By Problem 12.13, there exists an integer p and a function $g(x) \in V_J$ such that

$$\|f - g\|_\infty = \sup_{x \in R} |f(x) - g(x)| < \frac{\varepsilon}{\sqrt{2^{N+3}}}$$

If $j \geq p$, then by Lemma 12.4(c), $P_j g(x) = g(x)$ and by Minkowski's inequality and Lemma 12.4(d)

$$\begin{aligned}
\|P_j f - f\|_{L_2} &\leq \|P_j f - P_j g\|_{L_2} + \|P_j g - g\|_{L_2} + \|g - f\|_{L_2} \\
&= \|P_j(f - g)\|_{L_2} + \|g - f\|_{L_2} \\
&\leq 2\|g - f\|_{L_2}
\end{aligned} \tag{12.60}$$

Since

$$\|g - f\|_{L_2}^2 = \int_{-2^N}^{2^N} |g(x) - f(x)|^2 dx \leq \int_{-2^N}^{2^N} \frac{\varepsilon^2}{2^{N+3}} dx = \frac{\varepsilon^2}{4}$$

$$\|g - f\|_{L_2} < \frac{\varepsilon}{2} \tag{12.61}$$

The desired result follows by (i) and (ii) of MRA.

Part (b) follows from the fact that f has compact support on R and Minkowski's inequality.

Definition 12.14 For each $j \in Z$, we define the *detail operator* Q_j on $L_2(R)$ into R, by

$$Q_j f(x) = P_{j+1} f(x) - P_j f(x) \tag{12.62}$$

The proof of the following lemma is on the lines of the previous lemma:

Lemma 12.6 A. *The detail operator Q_j on $L_2(R)$ is linear*
B. *Q_j is idempotent.*
C. *If $g(x) \in W_j$ and if j' is an integer with $j' \neq j$ then*

$$Q_{j'}g(x) = 0$$

d. *Given $j \in Z$, and $f(x) \in L_2(R)$, $\|Q_j f\|_{L_2} \leq \|f\|_{L_2}$.*

Lemma 12.7 *Given $j \in Z$ and a continuous function $f(x)$ with compact support on R:*

$$Q_j f(x) = \sum_k \langle f, \psi_{j,k}\rangle \psi_{j,k} \tag{12.63}$$

where the sum is finite.

Proof (Proof of Lemma 12.7) Let $j \in Z$ be given and let $f(x)$ have compact support on R. Consider $Q_j f(x)$ for $x \in I_{j,k}$. We observe that

$$P_{j+1} f(x) = 2^{j+1} \int\limits_{I_{j,k}^{\ell}} f(t)dt \ \ if \ x \in I_{j,k}^{\ell}$$

$$= 2^{j+1} \int\limits_{I_{j,k}^{r}} f(t)dt \ \ if \ x \in I_{j,k}^{r}$$

and that

$$P_j f(x) = 2^j \int\limits_{I_{j,k}} f(t)dt \ if \ x \in I_{j,k}$$

For $x \in I_{j,k}^{\ell}$

$$Q_j f(x) = P_{j+1} f(x) - P_j f(x)$$

$$= 2^j \left(2\int\limits_{I_{j,k}^{\ell}} f(t)dt - \int\limits_{I_{j,k}^{\ell}} f(t)dt - \int\limits_{I_{j,k}^{r}} f(t)dt \right)$$

$$= 2^j \left(\int\limits_{I_{j,k}^{\ell}} f(t)dt - \int\limits_{I_{j,k}^{r}} f(t)dt \right)$$

$$= 2^{j/2} \langle f, \psi_{j,k}\rangle$$

by the definition of the Haar wavelet, and on $I_{j,k}^r$

$$Q_j f(x) = 2^j \left[-\int_{I_{j,k}^\ell} f(t)dt + \int_{I_{j,k}^r} f(t)dt \right] = -2^{j/2}\langle f, \psi_{j,k}\rangle$$

Since

$$\psi_{j,k}(x) = 2^{j/2}, \; if \; x \in I_{j,k}^\ell$$
$$= 2^{j/2}, \; if \; x \in I_{j,k}^r$$
$$= 0, \quad otherwise$$
$$Q_j f(x) = \langle f, \psi_{j,k}\rangle \psi_{j,k}$$

Remark 12.10 (i) As we have seen that in passing from $f(x)$ to $P_j f(x)$, the behavior of $f(x)$ on the interval $I_{j,k}$ is reduced to a single number, the average value of $f(x)$ on $I_{j,k}$. In this sense, $P_j f(x)$ can be thought of as a blurred version of $f(x)$ at scale 2^{-j}; that is, the details in $f(x)$ of size smaller than 2^{-j} are invisible in the approximation of $P_j f(x)$, but features of size larger than 2^{-j} are still discernible in $P_j f(x)$.

(ii) The wavelet space W_j for every $j \in Z$ is given by

$$W_j = \overline{span}\{\psi_{j,k}(x)\}_{k \in Z}$$

Since $\{\psi_{j,k}(x)\}_{k \in Z}$ is an orthonormal system on R, in light of Lemma 12.7 and Example 3.13, $Q_j f(x)$ is the function in Wj best approximating $f(x)$ in the L_2 sense.

As discussed $P_j f(x)$, the blurred version of $f(x)$ at scale 2^{-j}, we can interpret $Q_j f(x)$ as containing those features of $f(x)$ that are of size smaller than 2^{-j} but larger than 2^{-j-1}. This means that, $Q_j f(x)$ has those details invisible to the approximation $P_j f(x)$ but visible to the approximation $P_{j+1} f(x)$.

Approximation and Detail Operators for an Arbitrary Scaling Function $\varphi(x)$
and $\psi(x)$ be, respectively, general scaling and wavelet functions of an MRA $\{V_j\}$.

Definition 12.15 For each $j, k \in Z$, let

$$\varphi_{j,k} = 2^{j/2}\varphi(2^j x - k) = D_{2^j}T_k\varphi(x)$$
$$P_j f(x) = \sum_k \langle f, \varphi_{j,k}\rangle \varphi_{j,k} \tag{12.64}$$
$$Q_j f(x) = P_{j+1} f(x) - P_j f(x) \tag{12.65}$$

$\{\varphi_{j,k}(x)\}_{j,k \in Z}$ is an orthonormal basis for V_j [Lemma 12.9].

Lemma 12.8 *For all continuous $f(x)$ having compact support on R*

$$\lim_{j \to \infty} ||P_j f - f||_{L_2} = 0 \qquad (12.66)$$

$$\lim_{j \to \infty} ||P_j f||_{L_2} = 0 \qquad (12.67)$$

Proof (a) Let $\varepsilon > 0$. By Definition 12.8(ii), there exists $p \in Z$ and $g(x) \in V_p$ such that $||f - g||_{L_2} < \varepsilon/2$. By Definition 12.8(i), $g(x) \in V_j$ and $P_j g(x) = g(x)$ for all $j \geq p$. Thus

$$
\begin{aligned}
||f - P_j f||_{L_2} &= ||f - g - P_j g - P_j f||_{L_2} \\
&\leq ||f - g||_{L_2} + ||P(f - g)||_{L_2} \\
&\leq 2||f - g||_{L_2} < \varepsilon
\end{aligned}
$$

by Minkowski' s and Bessel' s inequality.

Since this inequality holds for all $j \geq p$, we get

$\lim_{j \to \infty} ||P_j f - f||_{L_2} = 0$.

(b) Let $f(x)$ be supported on $[-M, M]$ and let $\varepsilon > 0$. By the orthonormality of $\{\varphi_{j,k}(x)\}$, and applying the Cauchy–Schwarz (Hölder' s inequality for $p = 2$) and Minkowski inequalities

$$
\begin{aligned}
||P_j f||_{L_2}^2 &= \left\| \sum_k \langle f, \varphi_{j,k}\rangle \varphi_{j,k} \right\|_{L_2}^2 \\
&= \sum_k |\langle f, \varphi_{j,k}\rangle|^2 \\
&= \sum_k \left| \int_{-M}^{M} f(x) 2^{j/2} \varphi(2^j x - k) dx \right|^2 \\
&\leq \sum_k \left(\int_{-M}^{M} |f(x)|^2 dx \right) 2^j \left(\int_{-M}^{M} |\varphi(2^j x - k)|^2 dx \right) \\
&= ||f||_{L_2}^2 \sum_k \int_{-2^j M - k}^{2^j M - k} |\varphi(x)|^2 dx
\end{aligned}
$$

We need to show that

$$\lim_{j \to -\infty} \sum_k \int_{-2^j M - k}^{2^j M - k} |\varphi(x)|^2 dx = 0$$

For this, let $\varepsilon > 0$ and choose K so large that

$$\sum_{|k| \geq K} \int_{-1/2-k}^{1/2-k} |\varphi(x)|^2 dx = \int_{|x| > J} \tag{12.68}$$

Therefore, if $2^j M < 1/2$, then

$$\sum_k \int_{-2^j M-k}^{2^j M-k} |\varphi(x)|^2 dx \leq \sum_{|k| \geq K} \int_{-1/2-k}^{1/2-k} |\varphi(x)|^2 dx < \varepsilon$$

Since for each $k \in Z$, $\displaystyle\lim_{j \to -\infty} \int_{-2^j M-k}^{2^j M-k} |\varphi(x)|^2 dx = 0$

$$\lim_{j \to -\infty} \|Pf\|_{L_2}^2 \leq \|f\|_{L_2}^2 \lim_{j \to -\infty} \int_{-2^j M-k}^{2^j M-k} |\varphi(x)|^2 dx$$

$$= \|f\|_{L_2}^2 \lim_{j \to -\infty} \left(\sum_{|k| \geq K} \int_{-2^j M-k}^{2^j M-k} |\varphi(x)|^2 dx + \sum_{|k| > K} \int_{-2^j M-k}^{2^j M-k} |\varphi(x)|^2 dx \right)$$

$$\leq \|f\|_{L_2}^2 \lim_{j \to -\infty} \left(\varepsilon + \sum_{|k| > K} \int_{-2^j M-k}^{2^j M-k} |\varphi(x)|^2 dx \right)$$

$$= \|f\|_{L_2}^2 \varepsilon$$

Since $\varepsilon > 0$ was arbitrary, the result follows.

12.4.3   Scaling and Wavelet Filters

In this subsection, we prove existence of scaling filters, present a construction of wavelet with a given scaling filter, and prove certain properties of scaling and wavelet filters.

Theorem 12.11 *Let φ be a scaling function of the MRA, $\{V_j\}$. Then there exists a sequence $h_k \in \ell_2$ such that*

$$\varphi(x) = \sum_k h_k 2^{1/2} \varphi(2x - k)$$

is a function in $L_2(R)$. Moreover, we may write

$$\hat{\varphi}(t) = m_\varphi(t/2)\hat{\varphi}(t/2)$$

where

$$m_\varphi(t) = \frac{1}{\sqrt{2}} \sum_k h_k e^{-2\pi i k t}$$

We require the following lemma in the proof:

Lemma 12.9 *For each $j \in Z$, $\{\varphi_{j,k}(x)\}_{j,k \in Z}$ given in Sect. 10.3 is an orthonormal basis.*

Proof (Proof of Lemma 12.9) Since $\varphi_{0,k} \in V_0$ for all k, Definition 12.8(iv) implies that $D_{2^j}\varphi_{0,k}(x) \in V_j$ for all k. Also, since $\{\varphi_{0,k}(x)\}$ is an orthonormal sequence of translates, Remark 12.6(d) (v) implies that

$$\langle \varphi_{0,k}, \varphi_{0,m} \rangle = \langle D_{2^j}\varphi_{0,k}, D_{2^j}\varphi_{0,m} \rangle = \langle \varphi_{j,k}, \varphi_{j,m} \rangle = \delta_{k-m}$$

Hence, $\{\varphi_{j,k}(x)\}$ is an orthonormal sequence.

Given $f(x) \in V_j$, $D_{2^{-j}}f(x) \in V_0$ so that by Definition 12.8(v) and Remark 12.6(d)

$$D_{2^{-j}}f(x) = \sum_k \langle D_{2^{-j}}f(x), \varphi_{0,k}(x) \rangle \varphi_{0,k}(x)$$

$$= \sum_k \langle f, D_{2^j}\varphi_{0,k} \rangle \varphi_{0,k}(x)$$

Applying D_{2^j} to both sides of this equation, we obtain

$$f(x) = D_{2^j} D_{2^{-j}}f(x)$$

$$= D_{2^j} \sum_k \langle f, D_{2^j}\varphi_{0,k} \rangle D_{2^j}\varphi_{0,k}(x)$$

$$= \sum_k \langle f, \varphi_{j,k} \rangle \varphi_{j,k}(x)$$

This proves the theorem since $\{\varphi_{j,k}(x)\}$ is an orthonormal sequence and every element of V_j is its linear combination.

Proof (Proof of Theorem 12.11) Since $\varphi \in V_0 \subset V_1$, and since by Lemma 12.9, $\{\varphi_{1,k}(x)\}_{k \in Z}$ is an orthonormal basis for V_1

$$\varphi(x) = \sum_k \langle \varphi, \varphi_{1,k} \rangle 2^{1/2} \varphi(2x - k)$$

Thus, (12.68) holds with $h_k = \langle \varphi, \varphi_{1,k} rangle$, which is in ℓ_2 by Theorem 3.15.

By taking the Fourier transform of both sides of (12.68), we get (12.69).

Definition 12.16 The sequence $\{h_k\}$ in Theorem 12.11 is called the *scaling filter* associated with the scaling function φ of MRA V_j. The function $m_\varphi(t)$ defined by (12.69) is called the auxiliary function associated with $\varphi(x)$. Equation (12.68) which is nothing but Eq. (12.29) is called the *refinement or two-scale difference (dilation) equation*. $g_k = (-1)^k h_{1-k}$ is called the wavelet filter.

Theorem 12.12 *Let $\{V_j\}$ be an MRA with scaling function $\varphi(x)$, and let $\{h_k\}$ and $\{g_k\}$ be, respectively, the scaling and wavelet filter. Let*

$$\psi(x) = \sum_k g_k 2^{1/2} \varphi(2x - k)$$

Then $\{\psi_{j,k}(x)\}_{j,k \in Z}$ is a wavelet orthonormal basis on R. We refer to $\{W a \, 00\}$ for the proof.

Theorem 12.13 *Let $\{V_j\}$ be an MRA with scaling filter h_k and wavelet filter g_k, then*

(i) $\sum_n h_n = \sqrt{2}$

(ii) $\sum_n g_n = 0$

(iii) $\sum_k h_k h_{k-2n} = \sum_k g_k g_{k-2n} = \delta_n$

(vi) $\sum_k g_k h_{k-2n} = 0$ *for all $n \in Z$.*

(v) $\sum_k h_{m-2k} h_{n-2k} + \sum_k g_{m-2k} g_{n-2k} = \delta_{n-m}$

We require the following lemma in the proof.

Lemma 12.10 *Let $\varphi(x)$ be a scaling function of an MRA, $\{V_j\}$ belonging to $L_1(R) \cap L_2(R)$; and let ψ be the associated wavelet defined by (12.70) and ψ also belong to $L_1(R)$. Then*

(i) $\left\| \sum_R \varphi(x) dx \right\| = 1$

(ii) $\left\| \sum_R \psi(x) dx \right\| = 0$

(iii) $\hat{\varphi}(n) = 0$ *for all integers $n \neq 0$.*

(vi) $\sum_n \varphi(x + n) = 1$

Proof (i) Let $f(x)$ be given such that $\|f\|_{L_2} = 1, \hat{f}(t)$ is continuous and supported in $[-\alpha, \alpha], \alpha > 0$. It can be verified that

$$\hat{\varphi}_{j,k} = 2^{j/2} \hat{\varphi}(2^{-j}t) e^{-2\pi i k 2^j t}$$

(by using properties of the Fourier transform)

By Parseval' s formula (Theorem A.32)

$$||P_j f||_{L_2}^2 = \sum_k |\langle f, \varphi_{j,k} \rangle|^2 = \sum_k |\langle \hat{f}, \hat{\varphi}_{j,k} \rangle|^2$$

$$= \sum_k \left| \int_R \hat{f}(t) 2^{-j/2} \hat{\varphi}(2^{-j}t) e^{-2\pi i k 2^j t} \right|^2$$

Since $\{2^{-j/2} e^{-2\pi i k 2^j t}\}_{k \in Z}$ is a complete orthonormal system on the interval $[-2^{j-1}, 2^{j-1}]$, therefore as long as $2^{-j} > \alpha$, the above sum is the sum of the squares of the Fourier coefficients of the period 2^j extension of the function $\hat{f}(t) \hat{\varphi}(2^{-j}t)$.

By the Plancherel Theorem (Theorem F18), we have

$$||P_j f||_{L_2}^2 = \int_{-\alpha}^{\alpha} |\hat{f}(t)|^2 |\hat{\varphi}(2^{-j}t)|^2 dt$$

Since $\varphi \in L_1(R)$, $\hat{\varphi}(t)$ is continuous on $R^{\cdot}$ by the Riemann–Lebesgue theorem (Theorem A.27). It follows that

$$\lim_{j \to \infty} \hat{\varphi}(2^{-j}t) = \hat{\varphi}(0) \ uniformly \ on \ [-\alpha, \alpha]$$

By taking the limit under the integral sign, we conclude that

$$\lim_{j \to \infty} \int_{-\alpha}^{\alpha} |\hat{f}(t)|^2 |\hat{\varphi}(2^{-j}t)|^2 dt = |\hat{\varphi}(0)|^2 \int_{-\alpha}^{\alpha} |\hat{f}(t)|^2 dt$$

Since $\lim_{j \to \infty} ||P_j f||_{L_2} = ||f||_{L_2}$

$$||f||_{L_2}^2 = \lim_{j \to \infty} ||P_j f||_{L_2}^2$$

$$= \lim_{j \to \infty} \int_{-\alpha}^{\alpha} |\hat{f}(t)|^2 |\hat{\varphi}(2^{-j}t)|^2 dt$$

$$= |\hat{\varphi}(0)|^2 \int_{-\alpha}^{\alpha} |\hat{f}(t)|^2 dt$$

$$= |\hat{\varphi}(0)|^2 ||f||_{L_2}^2$$

Hence, $|\hat{\varphi}(0)|^2 = 1$, and since $\varphi(x) \in L_1$

Since $\hat{\varphi}(t) = m_\varphi(t/2)\hat{\varphi}(t/2)$ and $\hat{\varphi}(0) \neq 0$ as we have seen above, $m_\varphi(0) = 1$. By taking the Fourier transform of both sides in Eq. (12.70), we get

$$\hat{\psi}(t) = m_1(t/2)\hat{\varphi}(t/2) \tag{12.69}$$

where

$$m_1(t) = \frac{1}{\sqrt{2}} \sum g_n e^{-2\pi i t n}$$
$$= e^{-2\pi i t + 1/2} \overline{m_\varphi(t + 1/2)} \tag{12.70}$$

(m_φ is given by (12.69).

Thus, we can write
$\hat{\psi}(t) e^{-2\pi i (t/2 + 1/2)} \overline{m_\varphi(t/2 + 1/2)} \hat{\varphi}(t/2)$ and since by the orthonormality of $\{T_k \varphi(x)\}$

$$|m_\varphi(t)|^2 + |m_\varphi(t + 1/2)|^2 = 1 (see\ for\ example[Wo97])$$

$m_\varphi(1/2) = 0$, and hence $\hat{\psi}(0) = 0$. Therefore, we have the desired result as $\psi(x) \in L_1(R)$.

Step 1: First, we prove that the sequence $\{T_n g(x)\}$, where $g(x) \in L_2(R)$ and T_n is as in Definition 12.9(a) for $h = n$, is orthonormal if and only if for all $t \in R$

$$\sum_n |\hat{g}(t + n)|^2 = 1 \tag{12.71}$$

We observe that

$$\langle T_k g, T_\ell g \rangle = \langle g, T_{\ell-k} g \rangle = \delta_{k-\ell}$$

if and only if $\langle g, T_k g \rangle = \delta_k$. By Parsevals formula

$$\int_R g(t)\overline{g(t-k)}dt = \int_R \hat{g}(t)\overline{\hat{g}(t)}e^{-2\pi i k t}dt$$

$$= \int_R |\hat{g}(t)|^2 e^{-2\pi i k t} dt$$

$$= \sum_n \int_n^{n+1} |\hat{g}(t)|^2 e^{-2\pi i k t} dt$$

$$= \int_0^1 \sum_n |\hat{g}(t + n)|^2 e^{-2\pi i k t} dt$$

By the uniqueness of Fourier series

$$\int_0^1 \sum_n |\hat{g}(t+n)|^2 e^{-2\pi i k t}\,dt = \delta_k \ for\ all\ k \in Z$$

if and only if

$$\sum_n |\hat{g}(t+n)|^2 = 1 \ for\ all\ t \in R$$

Step 2: In view of Step 1

$$\sum_n |\hat{\varphi}(t+n)|^2 = 1 \ for\ all\ t \in R$$

In particular, $\sum_n |\hat{\varphi}(n)|^2 = 1$ by choosing $t = 0$. By part (i), $\hat{\varphi}(0) = 0$ which implies

$$\sum_{n \neq 0} |\hat{\varphi}(n)|^2 = 0$$

Hence, $\hat{\varphi}(n) = 0$ for $n \neq 0$.

We observe that $\sum_n \varphi(x+n) \in L_1[0, 1)$ and has period 1. By parts (i) and (iii), we have

$$\hat{\varphi}(0) = 1 \ and \ \hat{\varphi}(k) = 0 \ for\ all\ integers\ k \neq 0$$

Therefore, for each $k \in Z$

$$\int_0^1 e^{-2\pi i k t} \sum_n \varphi(x+n)\,dx = \sum_n \int_0^1 e^{-2\pi i k t} \varphi(x+n)\,dx$$

$$= \sum_n \int_n^{n+1} e^{-2\pi i k t} \varphi(x)\,dx$$

$$= \hat{\varphi}(k) = \delta_k$$

The only function with period 1 and Fourier coefficients equal to δ_k is the function that is identically 1 on $[0, 1)$. Therefore, we get Eq. (12.76).

Proof (*Proof of Theorem* 12.13)

(i) By Lemma 12.10(i), $\int_R \varphi(x)dx \neq 0$ so that

$$\int_R \varphi(x)dx = \int_R \sum_n h_n 2^{1/2}\varphi(2x-n)dt$$

$$= \sum_n h_n \int_R 2^{1/2}\varphi(2x-n)dx$$

$$= \sum_n h_n 2^{-1/2} \int_R \varphi(x)dx$$

Canceling the nonzero factor $\int_R \varphi(x)dx$ from both sides, we get

$$\int_n h_n = \sqrt{2}$$

(ii) By Lemma 12.10(ii), $\int_R \psi(x)dx = 0$ so that

$$0 = \int_R \psi(x)dx$$

$$= \int_R \sum_n g_n 2^{1/2}\varphi(2x-n)dx$$

$$= \sum_n g_n \int_R 2^{1/2}\varphi(2x-n)dx$$

$$= \sum_n g_n 2^{-1/2} \int_R \varphi(x)dx$$

$$= \frac{1}{\sqrt{2}}g_n \text{ as } \int_R \varphi(x)dx = 1$$

Thus, $\sum_n g_n = 0$.

(iii) Since $\{\varphi_{0,n}(x)\}$ and $\{\varphi_{1,n}(x)\}$ are orthonormal systems on R, we have

$$\delta_n = \int_R \varphi(x)\varphi(x-n)$$

$$= \int_R \sum h_k \varphi_{1,k}(x) \sum h_m \varphi_{1,m}(x-n)dx$$

$$= \sum_n \sum_m h_k h_{m-2n} \int_R \varphi_{1,k}(x)\varphi_{1,m}(x-n)dx$$

$$= \sum h_k h_{k-2n}$$

Therefore, $\sum_k h_k h_{k-2n} = \delta_n$.

The above argument gives us that

$$\sum g_k g_{k-2n} = \delta_n$$

is an orthonormal system of R.

(vi) Since $\langle \varphi_{0,n}(x), \varphi_{0,m}(x) \rangle = 0$ for all $n, m \in Z$, the above argument yields

$$\sum g_k h_{k-2n} = 0$$

Proof Proof of (v) Since for any signal (sequence) $c_{0,n}$

$$c_{0,n} = \sum_k c_{1,k} h_{n-2k} + \sum_k d_{1,k} g_{n-2k}$$

where

$$c_{0,n} = \sum_k c_{0,m} h_{m-2k}$$

$$d_{1,k} = \sum_k c_{0,m} g_{m-2k} \ (see \ also \ Sect. 10.3.2)$$

it follows that

$$c_{0,n} = \sum_k \sum_m c_{0,m} h_{m-2k} h_{n-2k} + \sum_k \sum_m \sum_m c_{0,m} g_{m-2k} g_{n-2k}$$

$$= \sum_m c_{0,m} \left(\sum_k h_{m-2k} h_{n-2k} + \sum_k g_{m-2k} g_{n-2k} \right)$$

Hence, we must have

$$\sum_k h_{m-2k} h_{n-2k} + \sum_k g_{m-2k} g_{n-2k} = \delta_{n-m}.$$

12.4.4 Approximation by MRA-Associated Projections

The main goal of this section is to discuss relationship between smoothness of functions measured by modulus of continuity and properties of wavelet expansions

Definition 12.17 Let f be a real-valued function defined on R. For $1 \le p \le \infty$ and $\delta > 0$, let

$$w_p(f; \delta) = \sup_{0 < |h| \le \delta} \|f(x) - f(x - h)\|_p \qquad (12.72)$$

where $w_p(f; \delta)$ is called the *p-modulus of continuity of f*, and we say that f has a *p*-modulus of continuity if $w_p(f; \delta)$ is finite for some $\delta > 0$ (equivalently, for all δ). The set of all functions having a p-modulus of continuity is denoted by $W_p(R)$.

Remark 12.11 (a) For each f and p the function $w_p(f; \delta)$ is an increasing function of δ.
(b) $w_p(f; \delta) \to 0$ as $\delta \to 0$ if either $f \in L_p$, $1 \le p < \infty$ or f is continuous and has compact support. For any positive integer m, $w_p(f; \delta) \le m w_p(f; \delta)$.
(c) For each $\delta > 0$

$$w_p(f; \delta) \le 2\|f\|_{L_p}$$

(d) If $\lim_{s \to 0} s^{-1} w_p(f; \delta)$, then

$$w_p(f; \delta) = 0 \qquad (12.73)$$

Conversely, if $w_p(f; \delta) = 0$, then f is a constant function.
(e) For translation and dilation operators (Definition 12.9(a) and (b)).

$$\text{(a)} \quad w(T_h f; \delta) = w_p(f; \delta) \qquad (12.74)$$

$$\text{(b)} \quad w_p(J_a f; \delta) = 2^{-a/p} w_p(f; 2^a \delta) \qquad (12.75)$$

(f) f satisfies Hölder' s condition with exponent α, $0 \le \alpha \le 1$ if $w_\infty(f; \delta) \le c\delta^\alpha$, $c > 0$ constant.
(g) For each $\delta > 0$, $w_p(f; \delta)$ is a seminorm on $W_p(R)$, that is

$$w_p(\alpha f + \beta g; \delta) \le |\alpha| w_p(f; \delta) + |\beta| w_p(g; \delta) \qquad (12.76)$$

and

$$w_p(f; \delta) = 0 \ if \ and \ only \ if \ f = constant \qquad (12.77)$$

With each MRA, $\{V_j\}$ we can associate projections P_j defined by the following equation:

$$P_j f(x) = \int_{-\infty}^{\infty} f(t) 2^j \varphi(2^j t, 2^j x) dt \qquad (12.78)$$

where

$$\varphi(t, x) = \sum_{k \in Z} \varphi(t - k) \varphi(x - k) \qquad (12.79)$$

and $\varphi(x)$ is a real scaling function satisfying the conditions

$$|\varphi(x)| \leq C(1 + |x|)^{-\beta}, \beta > 3$$
$$|\varphi'(x)| \leq C(1 + |x|)^{-\beta}, \beta > 3$$
$$\int_{-\infty}^{\infty} \varphi(x) dx = 1 \qquad (12.80)$$

It may be observed that (12.91) is automatically satisfied in view of (12.89) and Lemma 12.10(i).
It follows from (12.89) that

$$|\varphi(t, x)| \leq C \sum_{k \in Z} \frac{1}{(1 + |t - k|^\beta)} \frac{1}{(1 + |x - k|^\beta)} \qquad (12.81)$$

or

$$|\varphi(t, x)| \leq C \frac{1}{(1 + |t - x|^{\beta-1})} \qquad (12.82)$$

From (12.91) and (12.76), it follows that

$$\int_{-\infty}^{\infty} \varphi(x, t) dt = 1 \qquad (12.83)$$

Theorem 12.14 (Jackson' s Inequality) *There exists a constant C such that for any $f \in W_p(R)$*

$$\|f - P_j f\| \leq C w_p(f, 2^{-j}) \ for \ all \ j \in Z \qquad (12.84)$$

It may be observed that in view of Remark 12.11(iii) for $f \in L_p$, $1 \leq p \leq \infty$ (12.97) takes the form

$$\|f - P_j f\|_{L_p} \leq C \|f\|_{L_p} \qquad (12.85)$$

Proof Let (12.97) hold for $j = 0$ and some constant C. Then in view of the relations

$$||J_s f||_p = 2^{-s/p} ||f||_{L_p}$$

$P_j J_r = J_r P_{j-r}$, where J_s is the dilation operator (Definition 10.9(b)) and (12.84), we obtain

$$\begin{aligned}
||f - P_j f||_{L_p} &= 2^{-j/p} ||J_{-j} f - P_0 J_{-j} f||_{L_p} \\
&\leq C 2^{-j/p} w_p(J_{-j} f; 1) = C w_p(f, 2^{-j})
\end{aligned}$$

so it suffices to consider the case $j = 0$.

From (12.87) and (12.94), we can write

$$f(x) - P_0 f(x) = \int\limits_{-\infty}^{\infty} [f(x) - f(t)] \varphi(t, x) dt$$

From (12.93), we get

$$\begin{aligned}
||f - P_0 f||_{L_p}^p &= \int\limits_{-\infty}^{\infty} \left| \int\limits_{-\infty}^{\infty} [f(x) - f(t)] \varphi(t, x) dt \right|^p dx \\
&\leq C \int\limits_{-\infty}^{\infty} \left(\int\limits_{-\infty}^{\infty} \frac{|f(x) - f(t)| dt}{1 + |t - x|^{\beta-1}} \right)^p dx \\
&= C \int\limits_{-\infty}^{\infty} \left(\int\limits_{-\infty}^{\infty} \frac{|f(x) - f(x+u)| du}{(1 + |u|)^{\beta-1}} \right)^p dx
\end{aligned}$$

Writing $\beta - 1 = a + b$ with $a, b \geq 0$, $ap > p + 1$ and $bq > 1$ where $\frac{1}{p} + \frac{1}{q} = 1$ and applying Hölder's inequality to the inside integral, we get

$$\begin{aligned}
||f - P_0 f||_p &\leq C \int\limits_{-\infty}^{\infty} \int\limits_{-\infty}^{\infty} \frac{|f(x) - f(x+u)|^p}{(1 + |u|^{ap})} du \left(\int\limits_{-\infty}^{\infty} \frac{du}{1 + |u|^{bp}} \right)^{p/q} dx \\
&\leq C \int\limits_{-\infty}^{\infty} \frac{1}{(1 + |u|^{ap})} \int\limits_{-\infty}^{\infty} |f(x) - f(x+u)|^p dx du \\
&\leq C \int\limits_{-\infty}^{\infty} \frac{1}{(1 + |u|^{ap})} w_p(f; |u|^p) du
\end{aligned}$$

By splitting the last integral into two parts and estimating each part separately as follows, we get the desired result:

$$\int_{-1}^{1} w_p(f, |u|)^p \frac{du}{1 + |u|^{ap}} \leq C w_p(f; 1)^p$$

and

$$\int_{-\infty}^{-1} + \int_{1}^{\infty} w_p(f, |u|)^p \frac{du}{1 + |u|^{ap}}$$

$$\leq 2 \int_{1}^{\infty} w_p(f; u)^p \frac{du}{(1 + u)^{ap}}$$

$$\leq C \int_{1}^{\infty} u^p w_p(f; 1)^p \frac{du}{(1 + u)^{ap}} (Remark\ 12.11(ii))$$

$$\leq C w_p(f; 1)^p \int_{1}^{\infty} \frac{u^p du}{(1 + u)^{ap}}$$

$$\leq C w_p(f; 1)^p$$

The last inequality holds from choice of a. The case $p = \infty$ requires the standard modifications.

12.5   Compactly Supported Wavelets

12.5.1   Daubechies Wavelets

Daubechies (see for details [58, 59] and Pollen in [44]) has constructed, for an arbitrary integer N, an orthonormal basis for $L_2(R)$ of the form

$$2^{j/2} \psi(2^j x - k), \ j, k \in Z$$

having the following properties: The support of ψ_N is contained in $[-N + 1, N]$. To emphasize this point, ψ is often denoted by ψ_N.

$$\int\limits_{-\infty}^{\infty} \psi_N(x)dx = \int\limits_{-\infty}^{\infty} x\psi_N(x)dx = \cdots = \int\limits_{-\infty}^{\infty} x^N\psi_N(x)dx = 0 \qquad (12.86)$$

$$\psi_N(x) \text{ has } \gamma N \text{ continuous derivatives, where the}$$
$$\text{positive constant } \gamma \text{ is about } 1/5 \qquad (12.87)$$

In fact, we have the following theorem:

Theorem 12.15 (Daubechies) *There exists a constant K such that for each $N = 2, 3, \ldots,$ there exists an MRA with the scaling function φ and an associated wavelet ψ such that*

A. *$\varphi(x)$ and $\psi(x)$ belong to C^N.*
B. *$\varphi(x)$ and $\psi(x)$ are compactly supported and both suppφ and suppψ(x) are contained in $[-KN, KN]$.*
C. $\int\limits_{-\infty}^{\infty} \psi_N(x)dx = \int\limits_{-\infty}^{\infty} x\psi_N(x)dx = \cdots = \int\limits_{-\infty}^{\infty} x^N\psi_N(x)dx = 0$

We refer to {Da 92} for a proof of the theorem. Here, we present a construction of the Daubechies scaling function and wavelet on $[0, 3]$ due to Pollen (for details, see [199]).

Theorem 12.16 *Let $D = \bigcup\limits_{j \in Z} D_j$ where $D_j = \{k2^{-j}/k \in Z\}$ (D is a ring, that is, sums, difference and product of elements of D are also in D. It is a dense subset of R). Then there exists a unique function $\varphi : D \to R$ having the following properties:*

$$\varphi(x) = \{a\varphi(2x) + (1 - b)\varphi(2x - 1)$$
$$+ (1 - a)\varphi(2x - 2) + b\varphi(2x - 3)\} \qquad (12.88)$$
$$\sum_{k \in Z} \varphi(k) = 1 \qquad (12.89)$$
$$\varphi(d) = 0 \text{ if } d < 0 \text{ or } d > 3 \qquad (12.90)$$

where $a = \frac{1+\sqrt{3}}{4}, b = \frac{1-\sqrt{3}}{4}$ (It is clear that $\frac{1}{2} < a < 1$ and $\frac{-1}{4} < b < 0$).

Theorem 12.17 *The function φ defined in Theorem 12.16 extends to a continuous function on R which we also denote by φ: This continuous function φ has the following properties:*

$$\int\limits_{-\infty}^{\infty} \varphi(x)dx = 1$$

and

$$\int_{-\infty}^{\infty} \varphi(x)\varphi(x-k)dx = 1 \ if \ k = 0$$

$$= 0 \ if \ k \neq 0 \qquad (12.91)$$

In other words, φ is a scaling function.

Theorem 12.18 *The function ψ defined as*

$$\psi(x) = -b\varphi(2x) + (1-a)\varphi(2x-1) - (1-b)\varphi(2x-2) + a\varphi(2x-3) \qquad (12.92)$$

satisfies the following conditions:

$$supp\psi(x) \subset [0,3]$$

$$\int_{-\infty}^{\infty} \psi(x)\psi(x-k)dx = 1 \ if \ k = 0$$

$$= 0 \ if \ k \neq 0 \qquad (12.93)$$

$$\int_{-\infty}^{\infty} \varphi(x)dx = 1 \qquad (12.94)$$

Thus, $\{2^{j/2}\psi(2^j t - k)\}_{j\in Z, k\in Z}$ is an orthonormal basis in $L_2(R)$.

To prove Theorem 12.18, we need the following lemmas (D is as in Theorem 12.17).

Lemma 12.11 *For every $x \in D$, we have*

$$\sum_{k\in Z} \varphi(x-k) = 1$$

and

$$\sum_{k\in Z} \left(\frac{3-\sqrt{3}}{2} + k \right) \varphi(x-k) = x$$

Lemma 12.12 *If $x \in D$ and $0 \leq x \leq 1$, then*

$$2\varphi(x) + \varphi(x+1) = x + \frac{1+\sqrt{3}}{2} \qquad (12.95)$$

$$2\varphi(x+2) + \varphi(x+1) = -x + \frac{3-\sqrt{3}}{2} \qquad (12.96)$$

$$\varphi(x) + \varphi(x+2) = x + \frac{-1+\sqrt{3}}{2} \qquad (12.97)$$

Lemma 12.13 *For $0 \leq x \leq 1$ and $x \in D$, the following relations hold:*

$$\varphi\left(\frac{0+x}{2}\right) = a\varphi(x)$$

$$\varphi\left(\frac{1+x}{2}\right) = b\varphi(x) + ax + \frac{2+\sqrt{3}}{4}$$

$$\varphi\left(\frac{2+x}{2}\right) = a\varphi(1+x) + bx\frac{\sqrt{3}}{4}$$

$$\varphi\left(\frac{3+x}{2}\right) = a\varphi(x) - ax + \frac{1}{4}$$

$$\varphi\left(\frac{4+x}{2}\right) = a\varphi(2+x) - bx + \frac{3-2\sqrt{3}}{4}$$

$$\varphi\left(\frac{5+x}{2}\right) = b\varphi(2+x).$$

Lemma 12.14 *Suppose that $m(\xi) = \sum_{k=M}^{N} a_k e^{-ik\xi}$ is a trigonometric polynomial such that*

$$|m(\xi)|^2 + |m(\xi + \pi)|^2 = 1 \ for \ all \ \xi \in R \qquad (12.98)$$

$$m(0) = 1 \qquad (12.99)$$

$$m(\xi) \neq 0 \ for \ \xi \in \left[\frac{-\pi}{2}, \frac{\pi}{2}\right] \qquad (12.100)$$

Then the infinite product

$$\theta(\xi) = \prod_{j=1}^{\infty} m(2^{-j}\xi) \qquad (12.101)$$

converges almost uniformly. The function $\theta(\xi)$ is thus continuous. Moreover, it belongs to $L_2(R)$. The function φ given by $\hat{\varphi} = \frac{1}{\sqrt{2\pi}}\theta(\xi)$ has the support contained in $[M, N]$ and is a scaling function of an MRA. In particular, $\{\varphi(x - k)\}$ is an ONB in $L_2(R)$. The function $\psi(x)$ defined by

$$\psi(x) = 2\sum_{k=M}^{N}(-1)^k \bar{a}_k \varphi(2x + k + 1) \qquad (12.102)$$

is a compactly supported wavelet with $supp\psi \subset \left[\frac{M-N-1}{2}, \frac{N-M-1}{2}\right]$.

We give an outline of the proofs of Theorems 12.17 and 12.18 and refer to [199] for more details and proofs of Lemmas.

Proof (Proof of Theorem 12.17) Let K be a nonlinear operator acting on the space of functions on R. Let us define $K(f)$ for $x \in [0, 1]$ by the following set of conditions:

$$K(f)\left(\frac{0+x}{2}\right) = af(x)$$

$$K(f)\left(\frac{1+x}{2}\right) = bf(x) + ax + \frac{2+\sqrt{3}}{4}$$

$$K(f)\left(\frac{2+x}{2}\right) = af(1+x) + bx\frac{\sqrt{3}}{4}$$

$$K(f)\left(\frac{3+x}{2}\right) = af(x) - ax + \frac{1}{4}$$

$$K(f)\left(\frac{4+x}{2}\right) = af(2+x) - bx + \frac{3-2\sqrt{3}}{4}$$

$$K(f)\left(\frac{5+x}{2}\right) = bf(2+x)$$

This definition gives two values of $K(f)$ at the points $0, \frac{1}{2}, 1, \frac{3}{2}, 2, \frac{5}{2}, 3$. Let us denote by $\varphi_j, j = 0, 1, 2, \ldots$, the continuous, piecewise linear function on R which on D_j equals φ. The function $K(\varphi_j)$ is well defined at each point, and in fact, $K(\varphi_j) = \varphi_{j+1}$. Let $x \in [0, 3]$ and $j > 0$, then we immediately get from the definition of $K(f)$

$$\varphi_{j+1}(x) - \varphi_j(x) = K(\varphi_j)(x) - K(\varphi_{j-1})(x)$$
$$= \eta(\varphi_j(y) - \varphi_{j-1}(y))$$

where $\eta = a$ or $\eta = b$ and $y \in R$ is a point depending on x. Since $K(f)(x) = 0$ for $x \notin [0, 3]$ and $max(|a|, |b|) = a$, from (12.116), we get

$$||\varphi_{j+1} - \varphi_j||_\infty \leq a||\varphi_j - \varphi_{j-1}||_\infty$$

so by induction, we get

$$||\varphi_{j+1} - \varphi_j||_\infty \leq a^j||\varphi_1 - \varphi_0||_\infty$$

Since $||\varphi_1 - \varphi_0||_\infty$ is finite, the sequence $\{\varphi_j\}$ converges uniformly to a continuous function which is denoted by φ. This proves the first part of the theorem.

We know that (12.106) holds for all $x \in R$. Since $supp\varphi \subset [0, 3]$ for each $x \in R$, there are at most three nonzero terms in the $\sum_{k \in Z} \varphi(x - k)$. For a positive integer M, let

$$F_M(x) = \sum_{k=-M}^{M} \varphi(x - k)$$

From (12.101), we conclude that

$$|F_M(x)| \leq C \; for \; some \; constant \; C$$

and

$$F_M(x) = 1 \; if \; |x| \leq M - 3$$
$$= 0 \; if \; |x| \geq M + 3$$

Thus, for every integer M

$$2(M - 3) - 12C \leq \int_{-\infty}^{\infty} F_M(x)dx \leq 2(M - 3) + 12C \qquad (12.103)$$

From the definition of F_M, we also conclude that

$$\int_{-\infty}^{\infty} F_M(x)dx = (2M + 1) \int_{-\infty}^{\infty} \varphi(x)dx \qquad (12.104)$$

Since (12.118) and (12.119) hold for every positive integer M, making $M \to \infty$, we obtain (12.101). To prove (12.102), let

$$L_k = \int_{-\infty}^{\infty} \varphi(x)\varphi(x - k)dx \qquad (12.105)$$

Since $supp\varphi \subset [0, 3]$, we find that

$$L_k = 0 \; for \; |k| \geq 3 \qquad (12.106)$$

It is clear that

$$L_k = L_k \qquad (12.107)$$

By a change of variable, we see that, for any $\ell, m, n \in Z$

$$\int_{-\infty}^{\infty} \varphi(2x - m)\varphi(2x - 2\ell - n)dx = \frac{1}{2}L_{2\ell+n-m} \qquad (12.108)$$

Substituting value of $\varphi(x)$ given by (12.119) into (12.119) for $\ell = 0, 1, 2$ and applying (12.120) and (12.121), we obtain the following equations

$$(a(1-a)+b(1-b))L_0 = (1-ab)L_1 + (a(1-a)+b(1-b))L_2$$
$$2L_1 = (a(1-a)+b(1-b))L_0 + (1-b)L_1$$
$$+ ((1-b)^2 + (1-a)^2 + b^2)L_2$$
$$2L_2 = abL_1 + (a(1-a)+b(1-b))L_2$$

From given values of a and b (see Theorem 12.16), we have

$$a(1-a)+b(1-b) = 0 \qquad\qquad (12.109)$$

Thus, the above system of equations becomes

$$0 = (1-ab)L_1$$
$$2L_1 = (1-ab)L_1 + ((1-b)^2 + (1-a)^2 + b^2)L_2$$
$$2L_2 = abL_1$$

This implies that $L_1 = L_2 = 0$. Thus, $L_k = 0$ for $k \neq 0$. By (12.101), (12.100), and (12.102), we can compute

$$1 = \int_{-\infty}^{\infty} \varphi(x)dx \int_{-\infty}^{\infty} \varphi(x) \sum_{k \in Z} \varphi(x-k)dx$$
$$= \sum_{k \in Z} L_k = L_0$$

This proves (12.102).

Proof (*Proof of Theorem* 12.18) Supp $\psi(x) \subset [0,3]$ follows immediately from (12.103) and the fact that $supp\varphi \subset [0,3]$. To obtain (12.104), substitute $\psi(x)$ given by (12.103) into the left-hand side of (12.104). We find (12.104) using (12.122), (12.102), (12.123) and values of a and b. To obtain (12.105), we proceed similarly, but we substitute both $\psi(x)$ given by (12.103) and $\varphi(x)$ given by (12.98) into the left-hand side of (12.105). It follows directly from (12.104) and (12.105) that $\{2^{j/2}\psi(2^j t - k)\}_{j \in Z, k \in Z}$ is orthonormal.

12.5.2 *Approximation by Families of Daubechies Wavelets*

Let P_n denote the orthogonal projection of $L_2(R)$ onto V_n, and let Q_n denote the orthogonal projection of $L_2(R)$ onto W_n (recall $L_2(R) = V_n \oplus \sum_{j \geq n} \oplus W_j = \oplus_k W_k$, for any $n \in Z$).

For every integer $N \geq 1$, Daubechies constructed a pair of functions φ and ψ that are, respectively, $\chi[0, 1]$ and the Haar function for $N = 1$ and that generalize these functions for $N > 1$. The construction is on the following lines. *Step 1* Construct a finite sequence $h_0, h_1, \ldots, h_{2N-1}$ satisfying the conditions

$$\sum_k h_k h_{k+2m} = \delta_m \ for \ every \ integer \ m \tag{12.110}$$

$$\sum_k h_k = \sqrt{2} \tag{12.111}$$

$$\sum_k g_k k^m = 0 \tag{12.112}$$

whenever

$$0 \leq m \leq N - 1, where \ g_k = (-1)^k h_{1-k}$$

It can be observed that (12.123) and (12.124) imply (12.125) for $m = 0$.
Step 2 Construct the trigonometric polynomial $m_0(y)$ by

$$m_0(y) = \sqrt{2} \sum_k h_k e^{iky} \tag{12.113}$$

Step 3 Construct the scaling function φ so that its Fourier transform satisfies

$$\varphi(x) = (1/\sqrt{2\pi}) \prod_{k \geq 1} m_0(2^{-k} y) \tag{12.114}$$

Step 4 Construct the wavelet function ψ by

$$\psi(x) = \sum_k g_k \varphi(2x - k) \tag{12.115}$$

Definition 12.18 (*Coifman Wavelets or Coiflets*) For $N > 1$, $\varphi(x)$ and $\psi(x)$ define sets and subspaces of MRA V_n and W_n of $L_2(R)$ having the following additional properties:

$$\varphi_{n,k} = \sum_j h_{j-2k} \varphi_{n+1,j} \ and$$

$$\psi_{n,k} = \sum_j g_{j-2k} \varphi_{n+1,j}$$

$$for \ any \ integer \ n \tag{12.116}$$

$$supp \varphi_{n,k} = [2^{-n}k, 2^{-n}(k + 2N - 1)]$$

Fig. 12.6 Scaling functions
(–) and wavelet functions
(...) (top: $N = 2$, middle
$N = 3$, bottom $N = 4$)

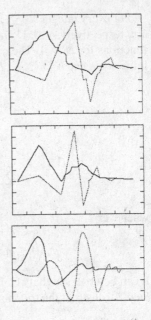

$$supp\,\psi_{n,k} = [2^{-n}(k + 2N - 1), 2^{-n}(k + N)] \qquad (12.117)$$

$$\sum \psi_{(j,k)}(x)x^m dx = 0 \text{ for all integers j and k}$$

$$and \; any \; integer \; 0 \le m < N - 1 \qquad (12.118)$$

$$\varphi_{j,k} \; and \; \psi_{(j,k)} \in C^{\lambda(n)} = Lip\lambda(n) \qquad (12.119)$$

with exponent $\lambda(n)$, where $\lambda(2) = 2 - \log_2(1 + \sqrt{3}) \simeq .5500$, $\lambda(3) \simeq 1.087833$,
$\lambda(4) \simeq 1.617926$, and $\lambda(N) \simeq .3485N$ for large N.

Graphs of the Daubechies scaling functions and wavelet functions for $2 \le N \le 4$
are given in Fig. 12.6.

For $N = 2$

$$h_0 = \frac{1 + \sqrt{3}}{4\sqrt{2}}, \; h_1 = \frac{3 + \sqrt{3}}{4\sqrt{2}}$$

$$h_2 = \frac{3 - \sqrt{3}}{4\sqrt{2}}, \; h_3 = \frac{1 - \sqrt{3}}{4\sqrt{2}}$$

$$m_0(y) = \frac{1}{\sqrt{2}} \sum_{k=0}^{3} h_k e^{-iky}$$

Theorem 12.19 $f \in C_0^{\infty}(R)$, φ and $\psi \in H^m(R)$, then there exists a $K > 0$ such
that

$$||f - P_n(f)||_{H^m(R)} \le K2^{-n(N-m)} \qquad (12.120)$$

where $P_n(f)$ denotes the orthogonal projection of $L_2(R)$ on V_n and N is the order of Daubechies wavelet ψ. For proof, see Problem 13.1

Coifman Wavelets [Coiflets]

Coifman wavelets are similar to Daubechies wavelets in that they have maximal number of vanishing moments; however in Coifman wavelets, the vanishing moments are equally distributed between the scaling function and the wavelet function. These are very useful for numerical solutions of partial differential equations as they have very good order of approximation (see $\{ReWa\ 98\}$ for a comprehensive account).

An orthonormal wavelet system with compact support is called a Coifman wavelet system of degree N if the moments of associated scaling function φ and wavelet ψ satisfy the conditions

$$Mom_\ell(\varphi) = \sum x^\ell \varphi(x)dx = 1 \ if \ \ell = 0 \tag{12.121}$$

$$Mom_\ell(\varphi) = \sum x^\ell \varphi(x)dx = 0 \ if \ \ell = 1, 2, 3, \ldots, N \tag{12.122}$$

$$Mom_\ell(\psi) = \sum x^\ell \psi(x)dx = 0 \ if \ \ell = 0, 1, 2, \ldots, N \tag{12.123}$$

It may be observed that (12.134), (12.135), and (12.136) are equivalent to the following conditions:

$$\sum_k (2k)^\ell h_{2k} = \sum_k (2k+1)^\ell h_{2k+1} = 0, \ for \ \ell = 1, 2, \ldots, N \tag{12.124}$$

$$\sum_k h_{2k} = \sum_k h_{2k+1} = 1 \tag{12.125}$$

where h_k is the scaling filter of $\varphi(x)$.

Lemma 12.15 *Let $\varphi(x)$ be a continuous Coifman scaling function of degree N, then*

$$\sum_k (x-k)^\ell \varphi(x-k) = 0, \ for \ \ell = 0, 1, 2, \ldots, N \tag{12.126}$$

Proof We prove it by the principle of finite induction. Let (12.139) hold for $\ell \leq n$, where $0 \leq n \leq N - 1$, and define

$$f(x) = \sum_k (x-k)^{n+1} \varphi(x-k)$$

Then, $f(x)$ is well defined and continuous; moreover

$$2^{n+1} f(x) = 2^{n+1} \sum_k ((x-k)^{n+1} \varphi(x-k))$$

$$= 2^{n+1} \sum_k \sum_m (h_m(x-k)^{n+1} \varphi(2x-2k-m))$$

$$= 2^{n+1} \sum_k \sum_i (h_{i-2k}(x-k)^{n+1} \varphi(2x-i))$$

$$= \sum_i \sum_k (h_{i-2k}(2x-i+i-2k)^{n+1} \varphi(2x-i))$$

$$= \sum_i \sum_k \sum_p \left(h_{i-2k} \binom{n+1}{p} (2x-i)^p (i-2k)^{n+1-p} \times \varphi(2x-i) \right)$$

$$= \sum_i \sum_p \left(\binom{n+1}{p} \left(\sum_k (i-2k)^{n+1-p} h_{i-2k} \right) \times (2x-i)^p \varphi(2x-i) \right)$$

Applying (12.137) and (12.138), $\sum_k (i-2k)^{n+1-p} h_{i-2k} = 0^{n+1-p}$. Thus

$$2^{n+1} f(x) = \sum_i (2x-i)^{n+1} \varphi(2x-i) = f(2x)$$

Since $f(x)$ is continuous and $\int_0^1 f(x)dx = 0$

$$f(x) = \sum_k (x-k)^{n+1} \varphi(x-k) = 0$$

It is clear that the result is true for $n = \ell = 1$.

Theorem 12.20 (Tian and Wells Jr., 1997) *For an orthogonal Coifman wavelet system of degree N with scaling function $\varphi(x)$, let $\{h_k\}$ be finite. For $f(x) \in C^N(R)$ having compact support, define*

$$S^{(j)} f(x) = 2^{-j/2} \sum_{k \in Z} f\left(\frac{k}{2^j}\right) \varphi_{j,k}(x) \; \forall j \in Z \tag{12.127}$$

then

$$\|f(x) - S^{(j)} f(x)\|_{L_2} \le C 2^{-jN} \tag{12.128}$$

where C depends only on $f(x)$ and $\varphi(x)$.

Proof By the Taylor expansion of f at the point x, we have

$$f\left(\frac{k}{2^j}\right) = \sum_{q=0}^{N-1}\left(\frac{1}{q!}f^{(q)}(x)\left(\frac{k}{2^j}-x\right)^q\right) + \frac{1}{N!}f^{(N)}(\alpha_k)\left(\frac{k}{2^j}-x\right)^N$$

for some α_k on the line segment joining x and $\frac{k}{2^j}$. By Lemma 12.15, for $1 \leq q \leq N$, we get

$$\sum_{k\in Z}\left(\frac{k}{2^j}-x\right)^q \varphi_{j,k}(x) = 2^{j/2-jq}\sum_{k\in Z}(k-2^j x)^q\varphi(2^j x - k) = 0 \quad (12.129)$$

and

$$\sum_{k\in Z}\varphi_{j,k}(x) = 2^{j/2}\sum_{k\in Z}\varphi(2^j x - k) = 2^{j/2}$$

Assume $supp(f) \subset [-K, K]$, and $supp(\varphi) \subset [-K, K]$ for some positive number K, and then

$$\sum_{|k|\leq(2^j+1)K}\left(\frac{k}{2^j}-x\right)^q f^{(q)}(x)\varphi_{j,k}(x) = 0, \quad for \ 1 \leq q \leq N$$

$$\sum_{|k|\leq(2^j+1)K} f(x)\varphi_{j,k}(x) = 2^{j/2}f(x)$$

and

$$S^{(j)}f(x) = 2^{-j/2}\sum_{k\in Z}f\left(\frac{k}{2^j}\right)\varphi_{j,k}(x)$$

$$= 2^{-j/2}\sum_{|k|\leq(2^j+1)K}f\left(\frac{k}{2^j}\right)\varphi_{j,k}(x)$$

Putting the Taylor expansion of f at x, we get

$$S^{(j)}f(x) = 2^{-j/2}\sum_{|k|\leq(2^j+1)K}\left(\sum_{q=0}^{N-1}\frac{1}{q!}f^{(q)}(x)\left(\frac{k}{2^j}-x\right)^q\right)$$

$$+ \frac{1}{N!}f^{(N)}(\alpha_k)\left(\frac{k}{2^j}-x\right)^N \varphi_{j,k}(x)$$

$$= 2^{-j/2}\left(\sum_{q=0}^{N-1}\sum_{|k|\leq(2^j+1)K}\frac{1}{q!}f^{(q)}(x)\left(\frac{k}{2^j}-x\right)^q\varphi_{j,k}(x)\right)$$

$$+ 2^{-j/2} \sum_{|k| \le (2^j+1)K} \left(\frac{1}{N!} f^{(N)}(\alpha_k) \left(\frac{k}{2^j} - x \right)^N \varphi_{j,k}(x) \right).$$

$$= f(x) + 2^{-j/2} \sum_{|k| \le (2^j+1)K} \left(\frac{1}{N!} f^{(N)}(\alpha_k) \left(\frac{k}{2^j} - x \right)^N \varphi_{j,k}(x) \right)$$

Thus

$$S^{(j)} f(x) - f(x) = 2^{-j/2} \sum_{|k| \le (2^j+1)K} \left(\frac{1}{N!} f^{(N)}(\alpha_k) \left(\frac{k}{2^j} - x \right)^N \varphi_{j,k}(x) \right)$$

and

$$\|f - S^{(j)} f(x)\|_{L_2} = \left\| 2^{-j/2} \sum_{|k| \le (2^j+1)K} \left(\frac{1}{N!} f^{(N)}(\alpha_k) \left(\frac{k}{2^j} - x \right)^N \varphi_{j,k}(x) \right) \right\|_{L_2}$$

$$= \frac{2^{-j(1/2+N)}}{N!} \left\| \sum_{|k| \le (2^j+1)K} (f^{(N)}(\alpha_k)(k-y)^N \varphi(y-k)) \right\|_{L_2}$$

where we make the substitution $y = 2^j x$. Let

$$g_k(y) = f(N)(\alpha_k)(k-y)^N \varphi(y-k)$$

then $g_k(y)$ has compact support $[-K+k, K+k]$, and $\|g_k(y)\|_{L_2}$ is uniformly bounded, $\|g_k(y)\|_{L_2} \le C$, where C depends on $f(x)$ and $\varphi(x)$. This gives

$$\|f(x) - S^{(j)} f(x)\|_{L_2} = \frac{2^{-j(1/2+N)}}{N!} \left(\int_R \left(\sum_{|k_1|,|k_2| \le (2^j+1)K} g_{k_1}(y) g_{k_2}(y) \right) dy \right)^{1/2}$$

$$= \frac{2^{-j(1/2+N)}}{N!} \left(\int_R \left(\sum_{|k_1|,|k_2| \le (2^j+1)K, |k_1-k_2| \le 2K} g_{k_1}(y) g_{k_2}(y) \right) dy \right)^{1/2}$$

$$= \frac{2^{-j(1/2+N)}}{N!} \left(2(2^j+1)K \cdot 4K \left(\max_{k_1,k_2 \in Z} \left\{ \left| \int_R g_{k_1}(y) g_{k_2}(y) dy \right| \right\} \right) \right)^{1/2}$$

$$= \frac{2^{-j(1/2+N)}}{N!} (8(2^j+1)K^2 \cdot C^2)^{1/2} \le C 2^{-jN}$$

Definition 12.19 $S_j(f)(x)$ defined by Eq. (12.140) is called the wavelet *sampling* approximation of the function $f(x)$ at the level j.

Corollary 12.2 *Let*

$$P_j(t) = \sum_{k \in Z} \left(\int_R f(x)\varphi_{j,k}(x)dx \right) \varphi_{j,k}(x)$$

Under the hypotheses of Theorem 12.20

$$\|f(x) - P_j(f)(x)\|_{L_2} \leq \lambda 2^{-jN} \tag{12.130}$$

where λ is a positive constant which depends only on f and the scaling factor h_k.

Proof By Theorem 3.8, we have

$$\|f - S^{(j)}(f)\|_{L_2}^2 = \|f - P_j(f)\|_{L_2}^2 + \|P_j(f) - S^j(f)\|_{L_2}^2 \tag{12.131}$$

Remark 12.12 This theorem holds for nonorthogonal wavelets. More precisely in the following theorem:

Theorem 12.21 [{ResWa 98}] *Suppose $\{h_k\}$ is a finite filter. Let $\{h_k\}$ satisfy the following conditions:*

$$\sum_{k \in Z} (2k)^q h_{2k} = \sum_{k \in Z} (2k + 1)^q h_{2k+1} = 0$$

$$for \; q = 1, 2, 3, \ldots, N \sum_{k \in Z} h_{2k} = \sum_{k \in Z} h_{2k+1} = 1 \tag{12.132}$$

Define

$$F_0(y) = \frac{1}{2} \sum_{k \in Z} h_k e^{iky} \tag{12.133}$$

and

$$\hat{\varphi}(y) = \prod_{j=1}^{\infty} F_0(2^{-j}y) \tag{12.134}$$

If $\varphi(x) \in L_2(R)$, then for any function $f(x) \in C^N(R)$ with compact support

$$\|f(x) - S^{(j)}(f)\|_{L_2} \leq \lambda 2^{-jN} \tag{12.134}$$

where λ depends on f and the sequence $\{h_k\}$, and $S_j(f)(x)$ is the wavelet sampling approximation.

Error estimation can be obtained in terms of the order of smoothness of the scaling function {Res 98}.

Theorem 12.22 *For $\varphi(x) \in C^n(R)$, $0 \leq n \leq N$, under the conditions of Theorem 12.21, the inequality*

$$||f(x) - S^{(j)}(f)||_{H^n} \leq \lambda 2^{-j(N-n)} \tag{12.135}$$

holds, where λ depends on f and $\{h_k\}$.

Remark 12.13 Existence of Coifman wavelets of degree N, say N = 9, can be proved by using a fundamental result of Kantorovich (see, e.g., Res 98, pp. 216–218). It is clear from definition that the degree 0 orthogonal Coifman wavelet system is exactly the same as the Haar wavelet system $\{h_0 = 1, h_1 = 1\}$. The Coifman wavelet system of degree $N = 1, 2, 3$, etc., can be computed. Interested readers may find details in {Res 98}.

12.6   Wavelet Packets

Wavelet packets are generalizations of wavelets. They are organized into collections, and each collection is an orthonormal basis for $L_2(R)$. This makes possible comparison of the advantages and disadvantages of the various possible decompositions of a signal in these orthonormal bases and selection of the optimal collection of wavelet packets for representing the given signal. The Walsh system is the simplest example of wavelet packets.

Definition 12.20 (*Wavelet Packets*) Let $\{h_k\}$ and $\{g_k\}$ be two sequences of ℓ_2 such that

$$\sum_{n \in Z} h_{n-2k} h_{n-2\ell} = \delta_{k-\ell} \tag{12.136}$$

$$\sum_{n \in Z} h_n = \sqrt{2} \tag{12.137}$$

$$g_k = (-1)^k h_{\ell-k} \tag{12.138}$$

Furthermore, let $\varphi(x)$ be a continuous and compactly supported real-valued function on R that solves the equation

$$\varphi(0)\varphi(x) = 2^{1/2} \sum_k h_k \varphi(2x - k) \tag{12.139}$$

with $\varphi(0) = 1$.

Let $\psi(x)$ be an associated function defined by

$$\psi(x) = 2^{1/2} \sum_k g_k \varphi(2x - k) \tag{12.140}$$

A family of functions $\omega_n \in L_2(R)$, $n = 0, 1, 2, \ldots$, defined recursively from φ and ψ as follows, is called the wavelet packet

$$\omega_0 k = \varphi_x, \ \omega_1(x) = \psi(x)$$
$$\omega_{2n}(x) = 2^{1/2} \sum_k h_k \omega_n(2x - k)$$

$$\omega_{2n+1}(x) = 2^{1/2} \sum_k g_k \omega_n(2x - k) \tag{12.141}$$

As in the case of wavelets, $\varphi(x)$ and $\psi(x)$ are often called, respectively, father and mother wavelets.

It has been proved that $\{\omega_n(x - k)\}$, $k \in Z$, is an orthonormal basis of $L_2(R)$ for all $n \geq 0$ where

$$\omega_n(x - k) = \frac{1}{\sqrt{2}} \sum_i h_{k-2i} \omega_{2n} \left(\frac{x}{2} - i \right)$$

$$+ \frac{1}{\sqrt{2}} \sum_i g_{k-2i} \omega_{2n+1} \left(\frac{x}{2} - i \right) \tag{12.142}$$

For $f \in L_2(R)$

$$f(x) = \sum_{n=-\infty}^{\infty} \sum_{k=-\infty}^{\infty} c_{n,k} \omega_n(x - k) \tag{12.143}$$

where

$$c_{n,k} = \langle f, \omega_n(x - k) \rangle \tag{12.144}$$

is called the wavelet packet series and $c_{n,k}$ are called wavelet packet coefficients of f.

The Walsh system $\{W_n\}_{n=0}^{\infty}$ is defined recursively on $[0, 1]$ by $W_0(x) = \chi[0, 1)(x)$ and $W_{2n+\alpha}(x) = W_n(2x) + (-1)^\alpha W_n(2x - 1)$, $\alpha = 0, 1; n = 0, 1, \ldots$.

This is an emerging area; for a comprehensive account, we refer to Wickerhauser $\{Wi\ 30\}$ and Meyer $\{Me\ 93\}$.

12.7 Problems

Problem 12.1 Let ψ_1 and ψ_2 be two wavelets, then show that $\psi_1 + \psi_2$ is a wavelet.

Problem 12.2 Verify that the Haar function is a wavelet.

Problem 12.3 Check that the characteristic function $\chi_{[0,1]}(x)$ of the closed interval [0, 1] is the scaling function of the Haar function.

Problem 12.4 Compute the wavelet transform of $f(x) = sinx$ for the Haar and the Mexican hat wavelet.

Problem 12.5 Analyze a meteorological data using Haar wavelet.

Problem 12.6 Describe the procedure for constructing mother wavelets from a father wavelet.

Problem 12.7 Examine whether B-spline is a wavelet?

Problem 12.8 Explain the concept of the multiresolution with the help of the signal given in Example 12.5.

Problem 12.9 Let $T_h f(x) = f(x-h), D_a f(x) = a^{1/2} f(ax), E_c f(x) = e^{2\pi icx} f(x)$, $a, b, c \in R, a > 0$. Then show that for $f, g \in L_2$:

(1) $D_a T_b f(x) = a^{1/2} f(ax - b)$
(2) $D_a T_b f(x) = T_{a^{-1}b} D_a f(x)$
(3) $\langle f, D_a g \rangle = \langle D_{a^{-1}} f, g \rangle$
(4) $\langle f, T_b g \rangle = \langle T_{-b} f, g \rangle$
(5) $\langle f, D_a T_b g \rangle = \langle T_{-b} D_{a^{-1}} f, g \rangle$
(6) $\langle D_a f, D_a g \rangle = \langle f, g \rangle$
(7) $\langle T_b f, T_b g \rangle = \langle f, g \rangle$
(8) $TbEc f(x) = e^{-2\pi ibc} E_c T_b f(x)$
(9) $\langle f, E_c g \rangle = \langle E_{-c} f, g \rangle$
(10) $\langle f, T_b E_c g \rangle = \langle T_{-b} E_{-c} f, g \rangle e^{2\pi ibc}$

Problem 12.10 Given $j_0, k_0, j_1, k_1 \in Z$, with either $j_0 \neq j_1$ or $k_0 \neq k_1$, then show either

(1) $I_{j_1,k_1} \cap I_{j_0,k_0} = \phi$
(2) $I_{j_1,k_1} \subseteq I_{j_0,k_0}$ or
(3) $I_{j_0,k_0} \subseteq I_{j_1,k_1}$

Problem 12.11 Verify that the Haar system on R is an orthonormal system.

Problem 12.12 Show that for each integer $N \geq 0$, the scale N Haar system on [0, 1] is a complete orthonormal system on [0, 1].

Problem 12.13 Given $f(x)$ continuous on [0, 1], and $\varepsilon > 0$, there is $N \in Z$, and a scale N dyadic function $g(x)$ supported in [0, 1] such that $|f(x) - g(x)| < \varepsilon$, for all $x \in [0, 1]$; that is, $\sup_{x} |f(x) - g(x)| < \varepsilon$.

Problem 12.14 Let the scaling function $\varphi(x)$ be given by $\varphi(x) = \frac{sin\pi x}{\pi x}$. Then find the corresponding wavelet ψ.

Problem 12.15 Prove that if $\{\ell_n\}$ is a tight frame in a Hilbert space H with frame bound A, then

$$A\langle f, g \rangle = \sum_{n=1}^{\infty} \langle f, \ell_n \rangle \langle \ell_n, g \rangle$$

for all $f, g \in H$.

Problem 12.16 Let $\varphi(x)$ be defined by

$$\begin{aligned} \varphi(x) &= 0, \quad x < 0 \\ &= 1, \quad 0 \le x \le 1 \\ &= 0, \quad x \ge 1 \end{aligned}$$

Draw the graph of the wavelet obtained by taking the convolution of the Haar wavelet with $\varphi(x)$.

Problem 12.17 Prove that $\{w_n(\cdot, -k)\}_{k \in \mathbb{Z}}, 0 \le n < 2^j$, where $\{w_n(\cdot, \cdot)$ denotes a family of wavelet packets, is an orthonormal basis of $L_2(R)$.

Chapter 13
Wavelet Method for Partial Differential Equations and Image Processing

Abstract In this chapter, applications of wavelet theory to partial differential equations and image processing are discussed.

Keywords Wavelet-based Galerkin method · Parabolic problems · Viscous Burger equations · Korteweg–de Vries equation · Hilbert transform and wavelets · Error estimation using wavelet basis · Representation of signals by frames · Iterative reconstruction · Frame algorithm · Noise removal from signals · Threshold operator · Model and algorithm · Wavelet method for image compression · Linear compression · Nonlinear compression

13.1 Introduction

There has been a lot of research papers on applications of wavelet methods; see, for example, [2, 14–16, 20, 30, 33, 34, 52, 55, 56, 60, 65, 70, 80, 84, 99, 128, 132, 134, 145]. Wavelet analysis and methods have been applied to diverse fields like signal and image processing, remote sensing, meteorology, computer vision, turbulence, biomedical engineering, prediction of natural calamities, stock market analysis, and numerical solution of partial differential equations. Applications of wavelet methods to partial differential equations (PDEs) and signal processing will be discussed in this chapter.

We require trial spaces of very large dimension for numerical treatment of PDEs by Galerkin methods which means we have to solve large systems of equations. Wavelets provide remedy for removing obstructions in applying Galerkin methods. It has been observed that the stiffness matrix relative to wavelet bases is quite close to sparse matrix. Therefore, efficient sparse solvers can be used without loss of accuracy. These are obtained from the following results:

1. Weighted sequence norm of wavelet expansion coefficients is equivalent to Sobolev norms in a certain range, depending on the regularity of the wavelets.
2. For a large class of operators, in the wavelet basis are nearly diagonal.
3. Smooth part of a function is removed by vanishing moments of wavelets.

© Springer Nature Singapore Pte Ltd. 2018
A. H. Siddiqi, *Functional Analysis and Applications*, Industrial and Applied Mathematics,
https://doi.org/10.1007/978-981-10-3725-2_13

As discussed in Sect. 12.3.4, a signal is a function of one variable belonging to $L_2(R)$ and wavelet analysis is very useful for signal processing. An image is treated as a function f defined on the unit square $Q = [0, 1] \times [0, 1]$. We shall see here that image processing is closely linked with wavelet analysis. The concept of wavelets in dimension 2 is relevant for this discussion. Let φ be a scaling function and $\psi(x)$ be the corresponding mother wavelet, then the three functions

$$\psi_1(x, y) = \psi(x)\psi(y)$$
$$\psi_2(x, y) = \psi(x)\varphi(y)$$
$$\psi_3(x, y) = \varphi(x)\psi(y)$$

form, by translation and dilation, an orthonormal basis for $L_2(R^2)$; that is

$$\{2^{j/2}\psi_m(2^j x - k_1, 2^j y - k_2)\}, \ j \in Z, \ k = (k_1, k_2) \in Z^2$$

$m = 1, 2, 3$ is an orthonormal basis for $L_2(R^2)$. Therefore, each $f \in L_2(R^2)$ can be represented as

$$f = \sum_{j,k \in Z} d_{j,k}\psi_{j,k} \tag{13.1}$$

where ψ is any one of the three $\psi_m(x, y)$ and

$$d_{j,k} = \langle f, \psi_{j,k} \rangle.$$

13.2　Wavelet Methods in Partial Differential and Integral Equations

13.2.1　Introduction

It has been shown that wavelet methods provide efficient algorithms to solve partial differential equations in the following sense:

(a) exact solution of given PDE must be very near to approximate solution. In mathematical language

$$\inf_{u \in \mathscr{A}} d(u, v) \ll 1 \tag{13.2}$$

($\ll$ means very small) and $\mathscr{A}$, set of approximate solutions needs to be small enough to allow the computation of the numerical solution.

(b) The algorithm should be fast, that is, takes less time-consuming. It may be observed that in order to reduce time of computation, the algorithm needs to select the minimal set of approximations at each step so that the computed solution remains close to the exact solution. This point is called adaptivity, which means no unnecessary quantity is computed. Some properties of wavelets are quite appropriate for adaptive algorithms. For example, if the solution of the partial differential equation we wish to compute is smooth in some regions, only a few wavelet coefficients will be needed to get a good approximation of the solution in those regions. Practically, only the wavelet coefficients of low frequencies wavelets whose supports are in these regions will be required. On the other hand, the greatest coefficients (in absolute value) will be localized near the singularities; this allows us to define and implement easily criteria of adaptivity through time evaluation.

13.2.2   General Procedure

General Procedure 1

Suppose U is a finite-dimensional subspace of a Sobolev space H in which we search for a weak solution of a partial differential equation (PDE). Then as seen in Chaps. 7 and 8, the differential equation reduces to a matrix equation in U. A solution of the matrix equation is an approximate solution to the partial differential equation and we shall try to find that solution of the matrix equation and in fact, that matrix for which the distance between the solutions of the partial differential equation and the associated matrix equation is minimum. More precisely, let Ω be a bounded open set in R^2 with Lipschitz boundary; that is, the boundary $\partial\Omega$ is a Lipschitz function. Let

$$U_j = \{f \in X_j / supp f \cap \overline{\Omega} \neq \phi\}$$

Here, $X_j = V_j \oplus V_j$ which is spanned by the products $\varphi_{j,k}(x)\varphi_{j,k}(y)$, for $k, \ell \in Z$ whereas usual $\varphi_{j,k}(x) = 2^j/2\varphi(2^j x - k)$, $\varphi_{j,k}(y) = 2^j/2\varphi(2^j y - k)$, where $\varphi(x)$ is a scaling function.

Since Ω is bounded, U_j is a finite-dimensional subspace of $V_j \oplus V_j$; U_j is the Galerkin approximation space in this context. If a Sobolev space $H_1(\Omega)$ is considered, then $U_j \subset H_1(\Omega)$ and hence, we can use elements of U_j to represent approximations to a solution of a differential equation which are in $H_1(\Omega)$. It may be noted that if we use the wavelet ψ associated with scaling function φ, we obtain a system

$$A_J U_J = F_J$$

where U_J is the coordinate vector of U_J in the basis $\{\psi_j, k(x)\}$, $F_j = (\langle f, \psi_\lambda rangle)$ $|\lambda| < J$ and $A_J = (\langle A\psi_\lambda, \psi_\mu rangle)|\lambda|, |\mu| < J$ is the associated stiffness matrix with the operator A (see Sect. 8.1). One can choose $H = H_0^1(\Omega)$, $A = -\Delta$ or $A = I - \Delta$ or $A = \Delta(\Delta) = \Delta^2$.

Example 13.1 (*Wavelet-based Galerkin Method*) We consider here for solution of example to illustrate the application of wavelets in solving differential equations. It is clear from the MRA that any function of $L_2(R)$ can be approximated arbitrarily well by the piece-wise constant functions from V_j provided j is large enough. V_j is the space of piece-wise constant functions $L_2(R)$ with breaking at the dyadic integers $k \cdot 2^{-j}$, $j, k \in Z$. Let us consider the following Dirichlet boundary value problem:

$$-u''(x) + Cu(x) = f(x), \; x \in \Omega = (0, 1)$$
$$u(0) = u(1) = 0 \tag{13.3}$$

with $C > 0$, a constant, $f \in H^1(\Omega)$ and find its solution. We apply the variational method of approximation (Galerkin method) for solving Eq. (13.3). As the existence of the solution of variational problem is guaranteed only in a complete space, we consider the Sobolev space $H^1(\Omega)$. The objective is to solve variational equation on a finite-dimensional subspace of $H^1(\Omega)$. In variational form, the solution $u \in H^1(\Omega)$ of the above equation satisfies

$$\int_\Omega (u'v' + uv) = \int_\Omega fv \tag{13.4}$$

for all $v \in H^1(\Omega)$ and $C = 1$.

To approximate u by the Galerkin's method, we choose a finite-dimensional subspace of $H^1(\Omega)$ which is a space spanned by wavelets defined on the interval $[0, 1]$. We have already discussed the wavelet bases $\phi_{j,k}$, $\psi_{j,k}$ and spaces V_j, W_j generated by them respectively in Chap. 12. For getting numerical solution of (13.3), we choose a positive value m and approximate u by an element $u_m \in V_m$ that satisfies

$$\int_\Omega (u_m'v' + u_mv) = \int_\Omega f(x)vdx, \; v \in V_m \tag{13.5}$$

where u_m can be written as

$$u_m = P_0u_m + \sum_{k=0}^{m-1} Q_ku_m \tag{13.6}$$

P_k is the projection from $H^1(\Omega)$ onto V_k and $Q_{k1} = P_k - P_{k-1}$. Therefore

$$u_m = C_0\phi_{0,0} + \sum_{j=0}^{m-1} \sum_{k=0}^{2^j-1} d_{j,k}\psi_{j,k} \tag{13.7}$$

Here, $\phi_{0,0}$ is identically 'one' on Ω, and C_0 is the average of u_m on Ω. We have used a "multiresolution" approach; i.e., we have written a successive coarser and coarser approximation to u_m.

Therefore, for all practical purposes, u_m can be approximated to an arbitrary precision by a linear combination of wavelets. We need now to determine $d_{j,k}$. Equation (13.7) together with (13.5) gives the following:

$$\sum_{j=0}^{m-1}\sum_{k=0}^{2^j-1} d_{j,k}\int_{\Omega}(\psi_{j,k}\psi_{j',k'} + \psi_{j,k}\psi_{j',k'})dx = \int_{\Omega} f\psi_{j',k'}dx \qquad (13.8)$$

for $k' = 0, \ldots, 2^{j'} - 1$ and $j' = 0, \ldots, m - 1$ or, more precisely

$$AU = F \qquad (13.9)$$

where A is the stiffness matrix relative to the wavelet basis functions $\psi_{j,k}$ and U, F are corresponding vectors $U = (d_{j,k})$, $f_k = \langle f, \psi_{j,k}\rangle$.

Example 13.2   Consider the following Neumann problem:

$$-\alpha\frac{d^2u}{dx^2} + \beta u + \gamma\frac{du}{dx} = f \ in \ (0, 1), \ f \in H^{-1}(0, 1)$$

$$\alpha\left(\frac{du}{dx}\right)_{x=0} = c, \ \alpha\left(\frac{du}{dx}\right)_{x=0} = d \qquad (13.10)$$

Proceeding on the lines of Sect. 7.3, it can be seen that (13.10) can be written in the following form. Find $u \in U$ such that

$$a(u, v) = L(v) \ for \ all \ v \in U, \ where \ U = H^1(0, 1)$$

$$a(u, v) = \int_0^1 \alpha(x)\frac{du}{dx}\frac{dv}{dx} + \int_0^1 \beta(x) \, uv \, dx \qquad (13.11)$$

$$+ \int_0^1 \gamma(x)\frac{du}{dx}v dx \ for \ all \ u, v \in U$$

$$L(v) = \int_0^1 fv dx + dv(1) + cv(0) \ for \ all \ v \in U$$

By the Lax–Milgram Lemma (3.39), Eq. (13.11) has a unique solution if

$$0 < \alpha_0 \le \alpha(x)\alpha_M \ a.e. \ on \ (0, 1) \qquad (13.12)$$
$$0 < \beta_0 \le \beta(x)\beta_M \ a.e. \ on \ (0, 1) \qquad (13.13)$$
$$\gamma \in L_2(0, 1), \ ||\gamma||_{L_2(0,1)} < min(\alpha_0, \beta_0) \qquad (13.14)$$

Let $\varphi(x)$ be a scaling function and let us consider the Daubechies wavelet for $N = 3$ (see Sect. 12.1). Let n be any integer and V_n be as in Sect. 10.5.2. Let $V_n(0, 1)$ denote the set of all elements of V_n restricted to $(0, 1)$. In Theorem 12.19, every function in $H^1(0, 1)$ can be approximated very closely by an element in $V_n(0, 1)$ for sufficiently large n; hence

$$\overline{\cup_n V_n(0, 1)} = H^1(0, 1) \tag{13.15}$$

Therefore, the family of subspaces $V_n(0, 1)$ of $H^1(0, 1)$ is very appropriate for Galerkin solution of the Neumann problems (13.9) and (13.10) formulated as given in (13.11)–(13.12). The Galerkin formulation of this problem: For any integer n, find $u_n \in V_n(0, 1)$ such that

$$a(u_n, v) = L(v) \; for \; all \; v \in V_n(0, 1) \tag{13.16}$$

Since $(0, 1)$ is bounded, $V_n(0, 1)$ is finite-dimensional and is a closed subspace of $H^1(0, 1)$. This implies that (13.15) has a unique solution. In view of Corollary 8.1 and (13.14)

$$\lim_{n \to \infty} ||u_n - u|| = 0 \; in \; H^1(0, 1) \tag{13.17}$$

For any integer n, the solution u_n of approximate problem (13.16) can be represented as

$$u_n = \sum_{k=1}^{p} u_{n,k} \varphi_{n,k-2N+1} \tag{13.18}$$

where $p = 2^n + 2N - 2$ and $u_1, u_2, \ldots, u_p \in R$, where functions are considered to be restricted to $(0, 1)$.

This yields the following system of linear equations in p unknowns:

$$\sum_{k=1}^{p} a(\varphi_{n,k-2N+1}, \varphi_{n,j-2N+1}) u_{n,k} = L(\varphi_{n,j-2N+1}) \tag{13.19}$$

for every $j = 1, 2, \ldots, p$. This can be written in matrix form as

$$AU = F \tag{13.20}$$

where

$$A = a_{i,j}, a_{i,j} = a(\varphi_{n,j-2N+1}, \varphi_{n,i-2N+1}) \tag{13.21}$$
$$F = (f_i), \; f_i = L(\varphi_{n,i-2N+1})$$

and

$$U = (u_{n,i}) \tag{13.22}$$

In the case where the bilinear form a $(\cdot, \cdot)$ defined by (13.11) is elliptic on H1 $(0, 1)$, the above matrix A is positive, implying the existence of a unique solution of (13.19). If the bilinear form $a(\cdot, \cdot)$ is symmetric, then (13.19) can be solved by the conjugate gradient method.

Remark 13.1 The evaluation of $a_{i,j}$ and f_i given by (13.20) and (13.21), which are required to compute the solution of the matrix equation (13.19), can be performed using well-known numerical quadrature methods.

13.2.3 Miscellaneous Examples

Example 13.3 (*Parabolic Problems*) We discuss in this example problems of the type

$$\frac{\partial u}{\partial t} + Au + B(u) = 0 \tag{13.23}$$

$$u(\cdot, 0) = u_0(\cdot) \tag{13.24}$$

where A is an elliptic operator and B is a possibly nonlinear function of u or a first derivative of u.

Special cases are the heat equation, the viscous Burgers equation, and the Korteweg-de-Vries equation.

Heat Equation

$$\frac{\partial u}{\partial t} = \frac{\partial^2 u}{\partial x^2} + f \text{ for } t > 0 \text{ } and \text{ } x \in (0, 1) \tag{13.25}$$

$$u(0, t) = u(1, t), \frac{\partial u}{\partial x}(0, t) = \frac{\partial u}{\partial x}(1, t) \tag{13.26}$$

$$u(x, 0) = u_0(x) \tag{13.27}$$

Viscous Burgers Equation For $v > 0$

$$\frac{\partial u}{\partial t}(x,t) = v\frac{\partial^2 v}{\partial x^2} + u(x,t)\frac{\partial u}{\partial x}(x,t)$$

$$\text{for } t > 0 \text{ and } 0 < x < 1 \tag{13.28}$$

$$u(x,0) = u_0(x) \tag{13.29}$$

Reaction diffusion equation : $\dfrac{\partial u}{\partial t} = u\dfrac{\partial^2 v}{\partial x^2} + u^p$

$$p > 0, \ v > 0 \tag{13.30}$$

with appropriate boundary conditions.

Korteweg-de Vries Equation

$$\frac{\partial u}{\partial t} + \alpha u\frac{\partial u}{\partial x} + \beta\frac{\partial^3 u}{\partial x^3} = 0 \tag{13.31}$$

with appropriate boundary conditions.

We consider here the solution of (13.24)–(13.26). Let f (t) and u(t) denote, respectively, the functions $x \to f(x,t)$ and $x \to u(x,t)$. It is assumed that, for almost every $t > 0$, $f(t) \in V^*$ (dual of $V = H_1(R)$) and $u(t) \in V$. A variational formulation is obtained by multiplying Eq. (13.24) by $v \in H_0^1(0,1)$ and integrating by parts with respect to x. This yields

$$\left(\frac{\partial u}{\partial x}, v\right) + \int_0^1 \frac{\partial u}{\partial t}\frac{\partial v}{\partial x}dx = (f,v), \ \text{for every } v \in H_0^1(0,1) \tag{13.32}$$

Here, $(\cdot,\cdot)$ denotes the duality pairing between V^* and V which reduces to the inner product $\langle\cdot,\cdot\rangle$ of L_2 when both arguments are in L_2. It may be observed that outside [0, 1], the functions $v \in H_0^1(0,1)$ are defined by periodicity with period 1.

Let $N \geq 3$, $n \geq 1$, and $V_n = H_n(0,1)$ be the subspace of $V = H_1(0,1)$. Approximate problem of (13.24)–(13.26): Find $u_n(t)$ satisfying for almost every $t > 0$

$$\int_0^1 \frac{\partial u_n}{\partial t}(t)vdx + \int_0^1 \frac{\partial u_n}{\partial t}(t)\frac{\partial v}{\partial x}dx = \int_0^1 f_n(t)vdx$$

$$\text{for all } v \in V_n \tag{13.33}$$

$$u_n(0) = u_{0,n} \tag{13.34}$$

where $u_{0,n}$ is the L_2 projection $P_n(u_0)$ of the initial data u_0 on V_n and where $f_n(t) = P_n^*(f(t))$ is the projection of $f(t)$ on V_n^* identified with V_n in the sense of Definition 2.6; indeed $f_n(t)$ is the unique element of V_n satisfying

$$\int\limits_{0}^{1} f_n(t)vdt = \langle f(t), v \rangle \; for \; all \; v \in V_n \tag{13.35}$$

The problem described by (13.31) and (13.33) is equivalent to a system of first-order ordinary differential equations obtained by substituting v in (13.31) with elements of a basis in $H_n(0, 1)$. The system is equivalent to the following initial value problem in $H_n(0, 1)$.

$$\frac{\partial u_n}{\partial t} + A_n u_n = f_n \; for \; t > 0 \tag{13.36}$$

$$u_n(0) = u_{0,n} \tag{13.37}$$

where $A_n = P_n^* A P_n$ is the approximation of $A = -\frac{\partial^2}{\partial x^2}$.

The problem (13.34)–(13.35) has the following closed-form solution:

$$u_n(t) = e^{(-tA_n)u_{0,n}} + \int\limits_{0}^{1} e^{-A_n(t-s)} f_n(s)ds \tag{13.38}$$

Conventional numerical schemes are now obtained by expanding the evolution operator $e^{-A_n(t-s)}$. For example, the Taylor expansion for $u_n(t + \Delta t)$ by the following quadratic polynomial in Δt gives us

$$
\begin{aligned}
u_n(t + \Delta t) \simeq u_n(t) + \Delta[f_n(t) - A_n u_n(t)] \\
+ 0.5\Delta t^2[\partial f_n(t)/\partial t - A_n(f_n(t) - A_n u_n(t))]
\end{aligned} \tag{13.39}
$$

Any such approximation ε to $e^{-\Delta t A_n}$ provides

$$u(n\Delta t) \simeq u^n = \varepsilon^n u^o + \sum_{i=1}^{n} \varepsilon^{n-i} f \tag{13.40}$$

as a discrete counter part of (13.36).

$[u = u_k(t) \simeq u(k2^{-j}, t))_{k=0}^{2^{j-1}}, f = \{f_k(t)\}_{k=0}^{2^{j-1}}, f_k(t) = f(k2^{-j}, t)]$. The simplest examples are $\varepsilon = \left(I - \frac{\Delta t}{2} A_n\right)^{-1} \left(I + \frac{\Delta t}{2} A_n\right)$, and they correspond to the Crank-Nicholson scheme. Suppose we are interested in long-time solutions of the heat equation. This requires high powers of ε. In particular, the powers ε^{2^t} can be obtained by repeated squaring. Setting $S_m = \varepsilon^{2m}$, $C_m = \sum\limits_{i=1}^{2^m-1} \varepsilon^i f$, and noting that

$$\sum_{i=1}^{2^m-1} \varepsilon^i = I + \varepsilon + \varepsilon^2(I + \varepsilon) + \varepsilon^4(I + \varepsilon + \varepsilon^2 + \varepsilon^3) + \cdots$$

$$+ \varepsilon^{2m-1}(I + \varepsilon + \cdots + 2^{m-1} - 1) \qquad (13.41)$$

the following algorithm approximates the solution at time $t = 2^m \Delta t$ after m steps.

Algorithm 13.1 Set $S_0 = \varepsilon$, $C_0 = f$. For $i = 1, 2, \ldots, m$:

$$S_i = S_{i-1}^2$$
$$C_i = (I + S_{i-1})C_{i-1}$$

Then $u^{(2^m)} = S_m u^{(0)} + C_m$ is an approximate solution of (13.37) and (13.38) at times $2^m \Delta t$.

Transform Algorithm 13.1 in such a way that the S_i become sparse (within some tolerance). For this we exploit the fact that wavelet representations of Caldéron-Zygmund operator T defined below (and their powers) are nearly sparse:

$$\langle Tf, g \rangle = \int_R \int_R K(x, y)f(y)\overline{g(x)}dydx$$

where T is continuous on $L_2(R)$ and

$$|K(x, y)| \le \frac{C}{|x - y|}, \ x \ne y, \ C > 0$$

$$|K(x, y) - K(x', y)| + |K(y, x) - K(y, x')| \le C\frac{|x - x'|^\delta}{|x - y|^{1+\delta}}, \qquad (13.42)$$

$$\delta > 0$$

$|x - x'| \le |x - y|/2$ and f and g have disjoint compact support.

Example 13.4 (*Integral Operators and Wavelets*) Let T be defined by

$$Tf(x) = \int_0^1 (1 - x)yf(y)dy + \int_x^1 (1 - y)xf(y)dy$$

$$= \int_0^1 K(x, y)f(y)dy$$

where

$$K(x, y) = (1 - x)y \ \ if \ 0 < y \le x$$
$$= (1 - y)x \ \ if x < y \le 1$$

The wavelet coefficients

$$m_{j,k,j',k'} = \int\limits_0^1 \int\limits_0^1 K(x, y)\psi_{j,k}(x)\psi_{j',k'}(y)dxdy$$

are zero if $\psi_{j,k}$ and $\psi_{j',k'}$ have disjoint supports, provided that $\psi_{j,k}$ or $\psi_{j',k'}$ is orthogonal to polynomials of degree 1. For $j \geq j', k \geq k'$, and $\psi_{j,k}$ belonging to Lip α which is orthogonal to polynomials of degree $n + 2$ with $n \geq \alpha - 1$, we get

$$|m_{j,k,j',k'}| = \left| \int\limits_0^1 T\psi_{j,k}(x)\psi_{j',k'}(x)dx \right|$$

$$\leq \|T\psi_{j,k}(x)\|_{L_1} \inf_{g \in P_n} \|\psi_{j',k'} - g\|_{L_\infty(suppT\psi_{j,k})}$$

where P_n is the space of polynomials of degree less than or equal to n, or $|m_{j,k,j',k'}| \leq 2^{-(\alpha+5/2)|j-j'|}2^{-2(j+j')}$.

Example 13.5 (*Hilbert Transform and Wavelets*) The Hilbert transform of a function $f(x)$, denoted by $Hf(x)$, is defined as

$$Hf(x) = \lim_{\varepsilon \to 0} \frac{1}{\pi} \int\limits_{|x-t| \geq \varepsilon} \frac{f(t)}{x - t}dt = \lim_{\varepsilon \to 0} \frac{1}{\pi} \int\limits_{|t| \geq \varepsilon} \frac{f(x - t)}{t}dt$$

provided that the limit exists in some sense. It is often written as

$$Hf(x) = \frac{1}{\pi}p.v. \int\limits_R \frac{f(t)}{x - t}dt$$

where p.v. means principal value; that is,

$$p.v. \int\limits_R f(x)dx = \lim_{\varepsilon \to 0} \int\limits_{R\setminus(-\varepsilon,\varepsilon)} f(x)dx$$

Let $H_{(j,k),(\ell,m)} = \langle H\psi_{\ell,m}, \psi_{j,k}\rangle$; see Definition 12.6 for $\psi_{j,k}(x)$. We show by using the Haar wavelet that $H_{(j,k),(\ell,m)}$ exhibit a decay with increasing distance of scales.

Let ψ be Haar wavelet and

$$\langle \varphi_{0,0}, \psi_{j,k}\rangle = \int\limits_0^2 \psi_{j,k}(x)dx = 0, \ k = 0, \ldots, 2^j - 1, \ j \geq 0$$

Suppose now that $2^{-\ell}(m+1) < 2^{-j}k$, and $\ell > j$, i.e., the support of $\psi_{\ell,m}$ and $\psi_{j,k}$ is disjoint. Then, by the above relation and Taylor's expansion around $y = 2^{-\ell}m$, one obtains

$$
\pi|H_{(j,k),(\ell,m)}| = \left| \int_{2^{-j}k}^{2^{-j}(k+1)} \left\{ \int_{2^{-\ell}m}^{2^{-\ell}(k+1)} \left(\frac{1}{x-y} - \frac{1}{x-2^{\ell}m} \right) \psi_{\ell,m}(y)dy \right\} \psi_{(j,k)}(x)dx \right|
$$

$$
= \left| \int_{2^{-\ell}m}^{2^{-\ell}(m+1)} \left\{ \int_{2^{-j}k}^{2^{-j}(k+1)} \frac{y-2^{\ell}m}{(x-y_{\ell,m})^2} \psi_{(j,k)}dx \right\} \psi_{\ell,m}(y)dy \right|
$$

for some $y_{\ell,m}$ in the support $[2^{-\ell}m, 2^{-\ell}(m+1)]$ of $\psi_{\ell,m}$. Repeating the same argument, one can subtract a constant in x which yields

$$
\pi|H_{(j,k),(\ell,m)}| = \left| \int_{2^{-\ell}m}^{2^{-\ell}(m+1)} \left\{ \int_{2^{-j}k}^{2^{-j}(k+1)} \left(\frac{y-2^{\ell}m}{(x-y_{\ell,m})^2} - \frac{y-2^{\ell}m}{(2^{-j}k-y_{\ell,m})^2} \right) \psi_{j,k}(x)dx \right\} \psi_{\ell,m}(y)dy \right|
$$

On account of Taylor's expansion around $x = 2^{-j}k$, the factor in front of $\psi_{j,k}(x)$ can be written as $-2(y-2^{-\ell}m)(x-2^{-j}k)/(x_{j,k}-y_{\ell,m})^3$, where $x_{j,k}$ is some point in the support $[2^{-j}k, 2^{-j}(k+1)]$ of the wavelet $\psi_{j,k}$. Noting that

$$
\int_{\mathbb{R}} |\psi_{(j,k)}(x)|dx \leq 2^{-j/2}
$$

a straightforward estimate provides

$$
\pi|H_{(j,k),(\ell,m)}| \leq 2^{-(n+j)^{3/2}} |2^{-j} - 2^{-\ell}m|^{-3}
$$

$$
= \frac{2^{-|j-\ell|3/2}}{|k-2^{j-\ell}m|^3}. \tag{13.43}
$$

13.2.4 Error Estimation Using Wavelet Basis

In this section, we prove results concerning the elliptic equation with Neumann boundary condition which indicate that order of approximation is improved if Coifman or Daubechies wavelets are used as basis functions. Consider the elliptic equation

$$
-\Delta u + u = f, \; in \; \Omega \subset R^2 \tag{13.44}
$$

with the Neumann boundary condition

$$\frac{\partial u}{\partial n} = g, \; on \; \partial \Omega \tag{13.45}$$

where n is the unit outward normal vector of $\partial \Omega$. If u is a solution of (13.42) and (13.43) and if h is a test function in $H^1(\Omega)$, then multiplying (13.42) by h and integration by parts over $Omega$, one has from (13.43) that

$$\int_\Omega \nabla u \nabla h dx dy + \int_\Omega u h dx dy = \int_\Omega f h dx dy + \int_\Omega g h ds \tag{13.46}$$

Solving (13.42) and (13.43) is equivalent to finding $u \in H^1(\Omega)$ so that (13.46) is satisfied for all $h \in H^1(\Omega)$. Let u_j be a solution (13.44) where $u_j \in U_j \subset H^1(\Omega)$ and (13.46) is satisfied for all $h \in U_j$. Let $N \geq 2$ be the degree of the Coifman wavelet.

Theorem 13.1 *Let u be a solution to (13.42) and (13.43), then*

$$||u - u_j||_{H^1{}_0} \leq \lambda 2^{-j(N-1)}$$

where λ depends on the diameter of Ω, on N, and on the maximum modulus of the first and second derivatives of u.

Proof We first show that if $v \in U_j$, then

$$||u - u^j||_{H^1(\Omega)} \leq ||u - v||_{H^1(\Omega)} \tag{13.47}$$

To prove this, let $v \in U_j$, $w = u_j - v$. Then, since u and u_j both satisfy (13.44) for $h \in U_j$ and $w \in U_j$

$$\int_\Omega (\nabla u - \nabla u^j) \cdot \nabla w dx dy + \int_\Omega (u - u^j) w dx dy = 0 \tag{13.48}$$

Therefore, by (13.44) and the Cauchy–Schwartz–Bunyakowski inequality

$$
\begin{aligned}
||u - u^j||^2_{H^1(\Omega)} &= \int_\Omega |\nabla u - \nabla u^j|^2 dxdy + \int_\Omega (u - u^j)^2 dxdy \\
&= \int_\Omega (\nabla u - \nabla u^j) \cdot (\nabla u - \nabla v - \nabla w) dxdy \\
&\quad + \int_\Omega (u - u^j)(u - v - w) dxdy \\
&= \int_\Omega (\nabla u - \nabla u^j) \cdot (\nabla u - \nabla v) dxdy \\
&\quad + \int_\Omega (u - u^j)(u - v) dxdy
\end{aligned}
\tag{13.49}
$$

or

$$
||u - u^j||^2_{H^1(\Omega)} \leq ||u - u^j||_{H^1(\Omega)} ||u - v||_{H^1(\Omega)}
\tag{13.50}
$$

If $u = u^j$, (13.47) is evident; otherwise (13.48) implies (13.45). Suppose now that $v = S_j(u)$ is an approximation of u given by Theorem 12.20, then $S_j(u) \in U_j$. The proof follows from the following two-dimensional analogue of Theorem 12.20. "Let $f \in C^2(\overline{\Omega})$, where Ω is open and bounded in R^2. If

$$
S_j(f)(x) = \sum_{p,q \in \Omega} f(p, q) \varphi_{j,p}(x) \varphi_{j,p}(y), \quad (x, y) \in \Omega
$$

then

$$
||f - S_j(f)||_{L_2(\Omega)} \leq \lambda 2^{-jN}
$$

and, more generally

$$
||f - S_j(f)||_{H^1(\Omega)} \leq \lambda 2^{-j(N-1)}
$$

for some constant λ".

Corollary 13.1 *Let u be a solution to (13.42) and (13.43), and let u^j be the wavelet Galerkin solution to the following equation:*

$$
\int_\Omega \nabla u \nabla h dxdy + \int_\Omega u dxdy = \int_\Omega f^j h dxdy + \int_\Omega g^j h ds
\tag{13.51}
$$

where f_j and g_j are, respectively, wavelet approximation for f and g of order j. Then

$$||u - u^j||_{H^1(\Omega)} \leq \lambda 2^{-j(N-1)} \tag{13.52}$$

where λ depends on the diameter of Ω, on N, and on the maximum modulus of the first derivatives of u.

Proof From (13.44) and (13.45), we obtain

$$\int_{\Omega} (\nabla u - \nabla u^j) \cdot \nabla w dx dy + \int_{\Omega} (u - u^j) w dx dy$$

$$= \int_{\Omega} (f - f^j) w dx dy + \int_{\Omega} (g - g^j) w ds$$

Therefore, similar to the derivation of (13.48), we have

$$||u - u^j||^2_{H^1(\Omega)} \leq ||u - u^j||_{H^1(\Omega)} ||u - v||_{H^1(\Omega)}$$

$$+ \int_{\Omega} |(f - f^j)(u^j - v)| dx dy$$

$$+ \int_{\Omega} |(g - g^j)(u^j - v)| ds \tag{13.53}$$

Since

$$\int_{\Omega} |f| ds \leq \lambda ||f||_{H^1(\Omega)} \tag{13.54}$$

where $\partial \Omega$ is Lipschitz continuous and λ depends only on the dimension of the underlying space and Ω. From (13.50), (13.51), putting estimates of $f - f^j$ and $g - g^j$ from the two-dimensional analogue of Theorem 12.20 mentioned above, replacing v by $\tilde{u}^j$, and applying the Cauchy–Schwartz–Bunyakowski inequality, we get (13.51).

It may be observed that Theorem 13.1 holds true for many other wavelets including Daubechies wavelet.

13.3 Introduction to Signal and Image Processing

Signals are functions of one variable while functions of two variables represent images. These concepts are essential concepts in information technology. Tools of functional analysis and wavelet methods are quite helpful to prove interesting results in fields of signal analysis and image processing. There exists a vast literature but we present here applications of frame wavelets and orthonormal wavelets to denoising (removal of unwanted element from the signal). Section 13.5 is devoted to representation of signals by frames while we discuss denoising in Sect. 13.6. One treats the image compression problem as one of approximating f by another function $\tilde{f}$ ($\tilde{f}$ represents compressed image). The goal of a compression algorithm (procedure or method), an important ingredient of image processing, is to represent certain classes of pictures with less information than was used to represent original pictures. For a lossless algorithm, the original and compressed images will be the same, and the error between them will be zero. Very often algorithms, where original and compressed images are different, are studied. This kind of algorithms are of vital importance for minimizing storing space in a computer disk or diskettes or minimizing transmission time from one source to the other by sacrificing a bit of the quality of the image obtained after decompression of the compressed image. In mathematical language, it means that algorithms are related to:

A. In what metric or norm the error $||f - \tilde{f}||$ be measured?
B. How should one measure the efficiency of algorithms?
C. For which pictures does an algorithm give good results?
D. Is there an optimal level of compression that cannot be exceeded within a given class of compression algorithms and pictures?
E. What are near-optimal algorithms?

L_p norm, $0 < p < \infty$, and Besov norm have been used to study these problems. Usually $\tilde{f}$ is represented by a wavelet series (12.20) or its truncated version or series associated with scaling function φ, namely $\sum \sum c_{j,k} \varphi_{j,k}(x)$ [see Eq. (12.26) for $c_{j,k}$] or its partial sum of certain order.

Theorem 12.7, Decomposition and Reconstruction Algorithms of Mallat [Sect. 10.3], Theorems 12.10, 12.19, 12.20, Corollary 12.2, and Theorem 12.22 are important results in the area of image compression. In Sect. 12.5, Besov space and linear and nonlinear image compression are introduced. Readers interested in indepth study of this field are referred to [2, 20, 30, 34, 126, 128, 149, 197] and references therein.

13.4   Representation of Signals by Frames

13.4.1   Functional Analytic Formulation

We discuss here representation of signals on a Hilbert space H and $L_2(R)$ in particular in the context of frames (orthonormal wavelets are special type of frames).

Let $T : H \rightarrow \ell_2$ be an operator defined by

$$Tf = \langle f, w_n \rangle \tag{13.55}$$

where $\{w_n\}$ is a frame in H. T is called the *frame representation* or frame discretization operator. $T^\star : \ell_2 \rightarrow H$ defined by

$$T^*a = \sum a_n w_n \tag{13.56}$$

where $\{a_n\} \in \ell_2$ is the adjoint of T.

Let an operator S associated with the frame w_n be defined on H as

$$Sf = \sum_n \langle f, w_n \rangle w_n \tag{13.57}$$

It can be easily checked that

$$Sf = \sum_n \langle f, w_n \rangle w_n = T^*Tf$$

or

$$S = T^*T \tag{13.58}$$

S is called the *frame operator*. It may be called a generalized frame operator to avoid confusion with Definition 12.4.

Definition 13.1 Let $\{w_n\}$ be a frame for a Hilbert space H with frame representation operator T. The *frame correlation operator* is defined as $\mathbf{R} = TT^*$.

Remark 13.2 1. The frame representation operator T is injective (one-to-one).
2. $T(H)$ is closed.
3. T^* is surjective (onto).

Definition 13.2 Let $\mathbf{R}^{-1}$ be the inverse of $\mathbf{R}$, then

$$R^t = R^{-1} P_{T(H)} \tag{13.59}$$

where $P_{T(H)}$ denotes the orthogonal projection operator on to the range of T, namely $T(H)$, is called a pseudo-inverse of frame correlation $\mathbf{R}$.

Remark 13.3 (i) The frame correlation has matrix representation

$$\mathbf{R} = (\mathbf{R}_{m,n}) = (\langle w_m, w_n \rangle) \tag{13.60}$$

(ii) $\mathbf{R}$ maps $T(H)$ bijectively to itself. This implies that $\mathbf{R}^{-1}$ exists.
(iii) $\mathbf{R} = P_{T(H)}\mathbf{R} = \mathbf{R}P_{T(H)}$.
(iv) $\mathbf{R}$ is self-adjoint.
(v) $\mathbf{R} \geq 0$.
(vi) If A and B are frame bounds (Definition 12.3), then

    (a) $A = ||\mathbf{R}^t||^{-1}$.
    (b) $B = ||\mathbf{R}||$.

Remark 13.4 It can be verified that

$$\mathbf{R}^t\mathbf{R} = \mathbf{R}\mathbf{R}^t = P_{T(H)} \tag{13.61}$$

13.4.2   *Iterative Reconstruction*

Let f be an arbitrary element of a Hilbert space H and let $\{w_n\}$ be a frame for H with frame bounds A and B, frame representation T, and frame correlation R. We describe here an iterative method for the recovery of a signal f from its frame representation Tf. The iterative method generates a sequence $\{c_n\}$ in $T(H)$ that converges to a $c \in T(H)$ such that $f = T^*c$. Moreover, the sequence converges at an exponential rate.

Algorithm 13.2   (*Frame Algorithm*)
Let $c_0 = Tf$. Set $d_0 = 0$. For $\lambda = 2/(A + B)$ define d_n and g_n as

$$d_{n+1} = d_n + (I - \lambda R)^n c_0$$
$$g_n = \lambda T^* c_n$$

Then

A. $\lim g_n = f$, and
B. $\|g_n - f\|/\|f\| < B/A \cdot \alpha_n$, where $\alpha = \|I - \lambda R\|_{T(H)} < 1$.

The following lemma is required in the proof of (*a*) and (*b*).

Lemma 13.1   *For $\lambda = 2/(A + B)$, the following relations hold*

A. $f = \sum\limits_{j=0}^{\infty} (I - \lambda S)^j (\lambda S) f$

B. $\|I - \lambda R\|_{T(H)} = \sup\limits_{c \in T(H)} \dfrac{|\langle (I - \lambda R)c, c \rangle|}{\langle c, c \rangle} \le max\{|1 - \lambda A|, |1 - \lambda A|, |1 - \lambda B|\} < 1$

C. $f = \lambda \sum\limits_{j=0}^{\infty} T^*(I - \lambda R)^j Tf$, *where* $T^*c = \sum\limits_{k} c_n w_n$ *for* $c = \{c_n\}$.

Proof   A.   Since $\left\| I - \frac{2}{A+B}S \right\| \le \frac{B-A}{A+B} < 1$, so that by the Neumann expansion

$$S^{-1} = \frac{2}{A + B} \sum_{j=0}^{\infty} \left(I - \frac{2}{A + B} S \right)^j, (See\ Theorem\ 2.11) \tag{13.62}$$

where I is the identity operator. By applying this to Sf, we get the desired result.
B. It can be verified that

$$1 - \lambda B \le \frac{\langle (I - \lambda R)c, c \rangle}{\langle c, c \rangle} \le 1 - \lambda A$$

Since $I - \lambda R$ is self-adjoint and $|1 - \lambda A| = |1 - \lambda B| = \frac{B-A}{A+B} < 1$, we get the desired result keeping in mind Theorem 2.6.

C. Since $\langle Tg, c \rangle = \langle g, T^*c \rangle$ and

$$\langle Tg, c \rangle = \sum \bar{c} \langle g, w_n \rangle$$
$$= \langle g, \sum c_n w_n \rangle$$

Because of (i) and relation (13.58), it is sufficient to prove

$$\lambda \sum_{j=0}^{\infty} T^*(I - \lambda \mathbf{R})^j T f = \sum_{j=0}^{\infty} (I - \lambda T^*T)^j (\lambda T^*T) f \qquad (13.63)$$

This is proved by the principle of induction. It is true for $j = 0$ as terms on both sides are the same.

Let it be true for $j = m$; i.e.,

$$\lambda T^*(I - \lambda R)^m T f = (I - \lambda T^*T)^m (\lambda T^*T) f \qquad (13.64)$$

Then, applying (13.63), we obtain

$$\lambda T^*(I - \lambda \mathbf{R})^{m+1} T f = \lambda T^*(I - \lambda \mathbf{R}) m T f - \lambda T^*(I - \lambda \mathbf{R})^m \lambda R T f$$
$$= \lambda (I - \lambda T^*T)^j (I - \lambda T^*T) T^*T f$$
$$= \lambda (I - \lambda T^*T)^{m+1} T^*T f$$

Thus, (13.63) is true for $m + 1$. Hence, (13.62) is true.

Proof (*Proof of Algorithm* 13.1)

A. By induction argument, we have

$$g_{n+1} = \lambda T^* \left(\sum_{j=0}^{n} (I - \lambda \mathbf{R})^j \right) c_0, \ for \ all \ n$$

Lemma 13.1 (iii) implies that

$$\lim_{n \to \infty} g_n = f$$

B. We can write

$$\|g_n - f\| = \|(g_{n+1} - g_n) + (g_{n+2} - g_{n+1}) + (g_{n+3} - g_{n+2}) + \ldots\|$$

$$\leq \sum_{k \geq n} \|g_{k+1} - g_k\|$$

$$= \sum_{k \geq n} \|\lambda T^*(I - \lambda R)^k T f\|$$

$$\leq \sum_{k \geq n} \lambda \|T^*\| \, \|(I - \lambda R)^k\|_{T(H)} \|T\| \, \|f\|$$

$$\leq \lambda B \left(\sum_{k \geq n} \alpha^k \right) \|f\|$$

$$= \left(\frac{\alpha^n}{1 - \alpha} \right) \lambda B \|f\|$$

$$\leq \frac{B}{A} \alpha^n \|f\|, \ by \ Lemma \ 13.1(ii) \ and \ using \tag{13.65}$$

relationships between frame bounds, $\|T\|$, $\|T^*\|$, and $\|f\|$.

13.5 Noise Removal from Signals

13.5.1 *Introduction*

An unwanted element of a signal is known as noise, for example, cock-pit noise in an aircraft or thermal noise in a weak radio transmission. The noise of a signal is noncoherent part. In audio signals, noise can be readily identified with incoherent hissing sound. The main goal of this section is to discuss functional analytic methods for the suppression of noise. For comprehensive account, we refer to Teolis [184].

With respect to frame or wavelet representations of analog signals, there are certain domains in which noise may perturb a signal, namely the (analog) signal domain and the (discrete) coefficient domain. Let these two domains be specified as a Hilbert space $H \subseteq L_2(R)$, and the signal domain's image under T; that is, $T(H) \subseteq \ell_2$, where T is the frame representation operator for the frame w_n.

A signal is said to be coherent with respect to a set of functions if its inner product representation with respect to that set is succinctive in the sense that relatively few coefficients in the representation domain have large magnitudes (see Definition 13.4) and Remark 13.5 (iii). In the contrary case, the signal will be called incoherent.

Let $f^* \in H$ be coherent with respect to the frame $\{w_n\}$ and let f be a noise corrupted version of f^*; that is,

$$f = f^* + \sigma \cdot u \tag{13.66}$$

where u is noncoherent with respect to w_n and σ is known as *noise level*.

Definition 13.3 (*Threshold Operator*) Let $\{w_n\}$ be a frame for H and let $a^* \in \ell_2$ and δ, called threshold, > 0. A threshold operator $F_\delta = F_{\delta, a^*}$ is defined on $T(H)$ into ℓ_2 by

$$(F_\delta a)_n = a_n, \quad |a_n^*| \geq \delta$$
$$= 0, \quad otherwise \tag{13.67}$$

where T is the frame representation operator of $\{w_n\}$.

It may be observed that a threshold element of $T(H)$ is in general not in $T(H)$. It can be verified that F_δ is a linear continuous operator and $||F_\delta|| = 1$. Often, the threshold operator F_δ, a for $a \in T(H)$, is written as $F_{\delta, f}$, where a is given by the unique coefficient sequence satisfying $f = T^* a$.

Definition 13.4 Let $\{w_n\}$ be a frame for the Hilbert space H with the frame representation operator T. Let $f^* \in H$ be a fixed element. Coherence functional, denoted by Coh_δ, is a mapping which maps $f \in H$ to

$$\frac{||F_{\delta, f_*} Tf||^2}{||Tf||^2} \in [0, \infty)$$

The coherence distribution, denoted by $Coh_\delta f_*$ is a function of δ given by

$$Coh_\delta f_* = \frac{||F_{\delta, f_*} Tf||^2}{||Tf||^2}$$

Remark 13.5 A. As a function δ, the coherence distribution describes how the energy in a δ-truncated representation decays.

B. A reconstruction procedure that starts from the truncated sequence $F_{\delta, f_*} Tf^*$ gives

$$f_\delta = T^* R^t F_{\delta, f_*} Tf \tag{13.68}$$

The operator $T^* R^t F_{\delta, f_*}$ is used to recover coherent portion (pure signal) and suppress the incoherent portion (noise) (see Algorithm 13.2).

C. Thresholding of the discrete representation of $T_\psi f$, where T_ψ is the frame representation operator for wavelet frame may be identified with suppression of noise. Let

$$(Tf)_{n,m} = \langle f, \psi_{n,m} \rangle = \langle f^*, \psi_{n,m} \rangle + \sigma \cdot \langle u, \psi_{n,m} \rangle \tag{13.69}$$

where $\psi_{n,m}$ is associated with ψ as in Eq. (10.2.18).

By the property of coherence, $\langle f_*, \psi_{n,m} \rangle$ must be large and $\langle u, \psi_{n,m} \rangle$ must be small. Thus, for an appropriate choice of δ, the contribution of noise will be nullified while the pure signal will be preserved. The iterative Algorithm 13.2 is the technique for the removal or suppression of noise.

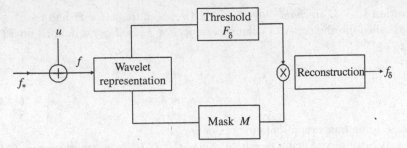

Fig. 13.1 Noise suppression processing model

13.5.2   *Model and Algorithm*

Figure 13.1 presents essential ingredients of the Noises Suppression Processing Model.

Step 1: (Pure signal f^*). A pure signal, $f^* \in H$ is input into the model.

Step 2: (Noise u). Noise u is mixed with f^* and represents an incoherent disturbance with respect to wavelet system $\{\psi_{n,m}\}$.

Step 3: (Wavelet Representation T_ψ). The observed polluted signal f is transformed to the wavelet domain to yield the wavelet coefficients $\langle f, \psi_{n,m} \rangle = (T_\psi f)_{m,n}$ or $d_{n,m}$.

Step 4: (Threshold Operator F_δ). Thresholding is performed so as to eliminate coefficients of small magnitude and to preserve coefficients having large magnitudes. In this case, we choose $a^* = a$ in Definition 13.3 to get

$$(F_\delta a)_{m,n} = a_{m,n}, \ \ |a_{m,n} > \delta|$$
$$= 0, \ otherwise$$

where $a \in T(H) \subseteq \ell_2$.

Step 5: (Masking Operator M). Mask operator M is defined by

$$(Ma)_{m,n} = a_{m,n}, \ \ if \ (m,n) \in Q$$
$$= 0, \ otherwise$$

where $Q \subseteq N \times N$ is fixed but arbitrary.

It is clear that the concept of mask operator is more general than that of threshold operator.

Step 6: (Reconstruction $T_\psi^* R_\psi^t$). Threshold coefficients are used to construct a noise removal version via an appropriate reconstruction algorithm (Algorithm 13.2).

Remark 13.6   1. For all $\delta \geq 0, 0 \leq Coh_\delta f \leq 1$ (boundedness)

2. $\delta_1 < \delta_2$ implies $Coh_{\delta_1} f \geq Coh_{\delta_2} f$ (Monotonicity decreasing)

3. (a) $\lim\limits_{\delta \to 0} Coh_\delta f = 1$ (Closed range)

    (b) $\lim\limits_{\delta \to \|Tf\|_\infty} Coh_\delta f = 0$

(c) $Coh_\delta(\alpha f) = Coh_{(\delta|\alpha|^{-1})} f$ (Scaling) for all scalar α (real or complex).

Theorem 13.2 *For a signal $f^\star \in H$, let $\{w_n\}$ be a frame for H with representation operator T, frame correlation R, and frame bounds A and B. Then for all $\delta \geq 0$*

$$\frac{||f_\star - f_\delta||^2}{||f_\star||^2} \leq \frac{B}{A}[1 - Coh_\delta f_\star]$$

where $f_\delta = T^\star R^t F_{\delta, f_\star} T f^\star$ and $Coh_\delta f^\star$ is the coherence distribution for the frame $\{w_n\}$.

Proof By (13.61)

$$T f_\delta = TT^\star R^t F_{\delta, f_\star} T f\star = P_{T(H)} F_{\delta, f_\star} T f_\star$$

Since T is a frame representation for all $g \in H$

$$A||g||^2 \leq ||Tg||^2 \leq B||g||^2$$

In particular, choosing $g = f_\delta - f_\delta$ yields

$$\begin{aligned}
A||f_\star - f_\delta||^2 &\leq ||T(f_\star - f_\delta)||^2 \\
&= ||Tf_\star - Tf_\delta||^2 \\
&= ||Tf_\star - P_{T(H)} F_{\delta, f_\star} T f_\star||^2 \\
&= ||P_{T(H)}(I - F_{\delta, f_\star}) T f_\star||^2 \\
&\leq ||P_{T(H)}|| \cdot ||(I - F_{\delta, f_\star}) T f_\star||^2 \\
&= ||(I - F_{\delta, f_\star}) T f_\star||^2 \\
&= [1 - Coh_\delta f_\star]||Tf_\star||^2 \\
&\leq [1 - Coh_\delta f_\star] B ||f_\star||^2
\end{aligned}$$

This gives the desired result.

It may be observed that while making above calculation, we have used linearity of T at $(f_\star - f_\delta)$, $Tf_\star = P_{T(H)} Tf_\star$ and $||P_{T(H)}|| = 1$ and properties of norm, frame bounds, and coherence distribution ($||(I - F_{\delta, f_\star}) T f_\star|| \leq [1 - Coh_\delta f_\star]||Tf_\star||$).

Corollary 13.2 *If the frame $\{w_n\}$ in Theorem 11.5.1 is orthonormal wavelet, then*

$$\frac{||f_\star - f_\delta||^2}{||f_\star||^2} = 1 - Coh_\delta f_\star$$

Proof In this case, $A = B$, $T(H) = \ell_2$, $||f_\star - f_\delta||^2 = ||T(f_\star - f_\delta)||^2$, and $P_{T(H)} = I$, and so we have the result.

Now, we modify Algorithm 13.1 so as to raise the problem of initialization with a coefficient sequence outside the range $T(H)$. We may recall that Algorithm 13.1

deals with the reconstruction of a signal from its frame representation. However, this does not converge for arbitrary initial data. We now present the modified algorithm.

Algorithm 13.3 Let $\{w_n\}$ be a frame for a Hilbert space H with frame representation T, correlation $\mathbf{R}$, and bounds A and B. Suppose $\tilde{c} \in \ell_2$ is the polluted frame representation of a signal $f \in H$. Set $c_0 = \mathbf{R}\tilde{c}$ and $d_0 = 0$. If $\lambda = \sqrt{2}/(A+B)$ and d_n and g_n are defined as

$$d_{n+1} = d_n + (I - (\lambda(\mathbf{R})^2)^n c_0$$
$$g_n = \lambda^2 T^\star d_n$$

then

A. $\lim_{n \to \infty} g_n = f_t = T^\star \mathbf{R}^t \tilde{c}$, and
B. $||f_n - g_n|| < M\alpha^n$, where $M < \infty$ and $\alpha = ||I - (\lambda\mathbf{R})^2|| < 1$. We need the following lemma in the proof of Algorithm 13.2.

Lemma 13.2 Let $\lambda = \sqrt{2}/(A+B)$, then $\mathbf{R}^t = \lim_{n \to \infty} \lambda^2 \sum_{k=0}^{\infty} (I - (\lambda(\mathbf{R})^2)^k R$.

Proof Choose $V = \lambda R$ and $H_1 = H_2 = T(H)$ in Solved Example 3.5. Then

$$||V|| = \lambda||R|| = \sqrt{2}/(A+B)B < \sqrt{2}$$

and

$$\inf_{-1} ||\lambda\mathbf{R}c|| > 0$$

since by Remark 13.3(ii), R is $1 - 1$ on $T(H)$ and $||c|| = 1$. By virtue of Solved Example 3.5, we get

$$||I - (\lambda R)^2||_{T(H)} < 1$$

which, in turn, implies that the series $\lambda^2 \sum_{k=0}^{\infty} (I - (\lambda(\mathbf{R})^2)^k$ converges to R^{-2} on $T(H)$. With $\tilde{c}$ an arbitrary element from ℓ_2 and $c_0 = \mathbf{R}\tilde{c}$

$$\lambda^2 \sum_{k=0}^{\infty} (I - (\lambda\mathbf{R})^2)^k c_0$$

converges to $(R^{-2})R\tilde{c}$ as $c_0 \in T(H)$. Thus

$$\lambda^2 \sum_{k=0}^{\infty} (I - (\lambda\mathbf{R})^2)^k R$$

is the pseudo-inverse R^t of R.

Proof (*Proof of Algorithm* 13.2) As in the proof of Algorithm 13.1, the principle of induction is used to show that for all n

$$g_{n+1} = \lambda^2 T^\star \sum_{j=0}^{\infty} \left((I - (\lambda \mathbf{R})^2)^j \right) c_0$$

This implies that

$$\lim_{n \to \infty} f_n = T^\star \mathbf{R}^t \tilde{c}$$

We have

$$
\begin{aligned}
\|g_{n+1} - g_n\| &= \|\lambda T^\star (I - (\lambda \mathbf{R})^2)^n \tilde{c}\| \\
&\leq \lambda^2 \|T^\star\| (\|(I - (\lambda \mathbf{R})^2)\|)^n \|R\tilde{c}\|) \\
&< M'\alpha^n
\end{aligned}
$$

where $M' = \lambda_2 B^{3/2} \|\tilde{c}\| < \infty$ since $\tilde{c} \in \ell_2$. Therefore

$$\|f_t - g_n\| \leq \sum_{k \geq n} M'\alpha^k = M' \frac{\alpha^n}{1 - \alpha} M\alpha^n.$$

13.6   Wavelet Methods for Image Processing

Function spaces, specially a generalization of Sobolev space known as Besov space, are quite appropriate for studying image processing; see [35].

13.6.1   *Besov Space*

We introduce here the notion of Besov space and equivalence of the Besov norm with a norm defined by wavelet coefficients. For any $h \in R^2$, we define

$$\Delta_h^0 f(x) = f(x)$$
$$\Delta_h^1 f(x) = \Delta(\Delta_h^0) f(x) = f(x - h) - f(x)$$
$$\Delta_h^2 f(x) = \Delta(\Delta_h^1) f(x) = f(x - 2h) - 2f(x + h) + f(x)$$

$$\cdots\cdots\cdots\cdots\cdots\cdots\cdots\cdots\cdots\cdots\cdots\cdots\cdots\cdots\cdots\cdots$$
$$\cdots\cdots\cdots\cdots\cdots\cdots\cdots\cdots\cdots\cdots\cdots\cdots\cdots\cdots\cdots\cdots$$

$$\Delta_h^{k+1} f(x) = \Delta_h^k f(x + h) - \Delta_h^k f(x), \quad k = 1, 2, 3 \ldots$$

Now, we define the rth modulus of continuity in L_p as

$$w_r(f,t)_p = \sup_{|h| \le t} \left(\int_{I_{rh}} |\Delta_h^r f(x)|^p dx \right)^{1/p} \tag{13.70}$$

where $I_{rh} = \{x \in I/x + rh \in I, \ I = [0,1] \times [0,1]\}$, $0 \le p \le \infty$; with usual change to an essential supremum when $p = \infty$.

Given $\alpha > 0, 0 < p \le \infty, 0 < q \le \infty$, choose $r \in Z$ with $q > \alpha \ge r - 1$. Then the space $B_q^{\alpha,r}(L_p(I))$, called Besov space, consists of those functions f for which the norm $\|f\|_{B_q^{\alpha,r}(L_p(I))}$ defined by

$$\|f\|_{B_q^{\alpha,r}(L_p(I))} = \|f\|_{L_p(I)} + \left(\int_0^\infty [t^{-\alpha} w_r(f,t)_p]^q \frac{dt}{t} \right)^{1/q} < \infty,$$

$$\text{when } q < \infty \tag{13.71}$$

and

$$\|f\|_{B_\infty^{\alpha,r}(L_p(I))} = \|f\|_{L_p(I)} + \sup_{t>0}[t^{-\alpha} w_r(f,t)_p] \text{ is finite when } q = \infty$$

Remark 13.7 A. If $0 < p < 1$ or $0 < q < 1$, then, $\| \cdot \|_{B_q^{\alpha,r}(L_p(I))}$ does not satisfy the triangle inequality. However, there exists a constant C such that for all $f, g \in B_q^{\alpha,r}(L_p(I))$

$$\|f + g\|_{B_q^{\alpha,r}(L_p(I))} \le C \left(\|f\|_{B_q^{\alpha,r}(L_p(I))} + \|g\|_{B_q^{\alpha,r}(L_p(I))} \right) \tag{13.72}$$

B. Since, for any $r > \alpha, r' > r, \|f\|_{B_q^{\alpha,r}(L_p(I))}$ and $\|f\|_{B_q^{\alpha,r'}(L_p(I))}$ are equivalent norms, we define the Besov space $B_q^\alpha(L_p(I))$ to be $B_q^{\alpha,r}(L_p(I))$ for any $r > \alpha$.

C. For $p = q = 2$, $B_2^\alpha(L_2(\Omega))$ is the Sobolev space $H\alpha(L_2(\Omega))$.

D. For $\alpha < 1, 1 \le p < \infty$, and $q = \infty$, $B_p^\alpha(\Omega)$ is Lip$(\alpha, L_p(\Omega)) = \{f \in L_p(I)/\|f(x+h) - f(x)\|Lp \le kh^\alpha, k > 0$ constant$\}$.

E. $\|f\|_{B_2^\alpha(L_2(I))}$ is equivalent to the norm

$$\left(\sum_k \sum_j 2^{\alpha k} |d_{j,k}|^q \right)^{1/q}$$

F. $\|f\|_{B_q^\alpha(L_p(I))}$ is equivalent to the norm

$$\left(\sum_k \sum_j |d_{j,k}|^q \right)^{1/q}$$

where $\frac{1}{q} = \frac{\alpha}{2} + \frac{1}{2}$.

13.6.2   *Linear and Nonlinear Image Compression*

We discuss here wavelet-based image compression of observed pixel values. In digitized image, the pixel values (observations) are samples which depend on the measuring device of an intensity field $F(x)$ for x in $I = [0, 1] \times [0, 1]$. In the simplest case, the pixel samples are modeled by averages of the intensity function F over small squares. In this case, one may choose a wavelet, say the Haar wavelet on the square. We assume that 2^{2m} pixel values p_j are indexed by $j = (j_1, j_2), 0 \le j_1, j_2 < 2^m$ in the usual arrangement of rows and columns, and that each measurement is the average value of F on the subsquare covered by that pixel. To fix notation, we note that the jth pixel covers the square $I_{j,m}$ with sidelength 2^{-m} and lower left corner at the point $j/2^m$. We denote the characteristic function of I by $\chi = \chi I$ and the $L_2(I)$ normalized characteristic function of $I_{j,m}$ by $\chi_{j,m} = 2^m \chi I_{j,m} = 2^m \chi(2^m x - j)$.

One can write each pixel value as

$$p_j = 2^{2m} \sum \chi(2^m x - j) F(x) dx$$
$$= 2^m \langle \chi_{j,m}, F \rangle$$

The normal practice in wavelet-based image processing is to use the observed pixel values p_j to construct the function

$$f_m = \sum_j p_j \chi(2^m x - j)$$
$$= \langle \chi_{j,m}, F \rangle \chi_{j,m}$$

which we call the observed image. Thus, if the wavelet expansion of the intensity field F is

$$F = \sum_{0 \le k} \sum_j d_{j,k} \psi_{j,k}$$

then the wavelet expansion of image f is

$$f_m = \sum_{0 \le k < m} \sum_j d_{j,k} \psi_{j,k}$$

The important point is that f_m is the $L_2(I)$ projection of F onto $span\{\chi_{x,j}\} = span\{\psi_{j,k}\}_{0 \le k < m, n}$. Moreover, if F is in any function space whose norm is determined by a sequence norm of Haar wavelet coefficients, then so is f_m, and the sequence space norm of f_m is less than the sequence space norm of F. In practice, we use smoother wavelets than the Haar wavelet.

Linear Compression

Let ψ be a wavelet and let F be an observed image mentioned above. Let $fN =$, that is, we include in the approximation all coefficients dj,k with frequency less than 2^N, $N \leq m$ (f_N is the projection onto V_m). f_N is called the wavelet approximation of F. We have

$$\|F - f_N\|^2_{L_2(I)} = \sum_{f \geq N} \sum_j |d_{j,k}|^2$$

$$\leq \sum_{f \geq N} \sum_j \frac{2^{2\alpha_k}}{2^{2\alpha_N}} |d_{j,k}|^2$$

$$\leq 2^{-\alpha_N} \sum_k \sum_j 2^{2\alpha_k} |d_{j,k}|^2$$

$$2^{-N} \|F\|^2_{H^\alpha(L_2(I))} \quad (By\ Remark\ 11.6.1(e))$$

Therefore

$$\|F - f_N\|_{(L_2(I))} \leq 2^{-\alpha_N} \|F\|_{H^\alpha(L_2(I))}$$

Nonlinear Compression

For nonlinear compression algorithm, we take

$$f_\lambda = \sum_{k < m, |d_{j,k}| \geq \lambda} d_{j,k} \psi_{j,k}$$

Thus, we consider all large coefficients without consideration of frequency but $k < m$. If we assume that $F \in B^\alpha_q(L_p(I))$, $\frac{1}{q} = \alpha/2 + 1/2$, then N, the number of coefficients greater than λ, satisfies

$$N\lambda^q \leq \sum_{j,k} |d_{j,k}|^q = \|F\|^q_{B^\alpha_q(L_p(I))}$$

So

$$N \leq \lambda^{-q} \|F\|^q_{B^\alpha_q(L_p(I))} \tag{13.73}$$

and

$$\|f_\lambda - f_m\|^2_{L_2(I)} \leq \sum_{|d_{j,k}|<\lambda} |d_{j,k}|^2$$

$$\leq \sum_{|d_{j,k}|<\lambda} \lambda^{2-q}|d_{j,k}|^q$$

$$= \lambda^{2\alpha/(\alpha+1)}\|F\|^q_{B^\alpha_q(L_p(I))} \qquad (13.74)$$

as $2 - q = 2\alpha/1 + \alpha$.

If N is nonzero, then (13.75) implies that

$$\lambda \leq N^{-1/q}\|F\|^q_{B^\alpha_q(L_p(I))} \qquad (13.75)$$

By (13.74) and (13.75), we get

$$\|f_\lambda - f_m\|^2_{L_2(I)} \leq N^{-\alpha/2}\|F\|^q_{B^\alpha_q(L_p(I))} \qquad (13.76)$$

It may be remarked that the above analysis can be applied to any compression scheme that satisfies

$$\tilde{f} = \sum \tilde{d}_{j,k}\psi_{j,k}, \ \tilde{d}_{j,k} = \langle \tilde{f}, \psi_{j,k} \rangle$$

with

$$|d_{j,k} - \tilde{d}_{j,k}| \leq \lambda \ and \ |d_{j,k}| < \lambda \ implying \ \tilde{d}_{j,k} = 0$$

It may also be noted that a given image, say f, will have greater smoothness in one of the nonlinear smoothness spaces $B^\alpha_q(L_p(I))$, $\frac{1}{p} = \alpha/2 + 1/2$ than in the Sobolev spaces H(L2(I)); that is, if a given image f is in $H^\alpha(L_2(I))$, then it is also in $B^\alpha_p(L_p(I))$, $\frac{1}{p} = \beta/2 + 1/2$, for $\beta \geq \alpha$.

13.7 Problems

Problem 13.1 Let Pn denote the orthogonal projection of $L_2(R)$ onto a scaling subspace V_n, then show that

$$\|f - P_nf\|_{H_m} \leq K2^{-n(N-m)}$$

where K is a positive constant, $f \in H_m(R)$, for Daubechies wavelet of order N

Problem 13.2 Let V_m be an r-regular MRA and let $f \in B^\alpha_q(L_p(I))$. Then prove that for all $n \geq 0$

$$\|f - P_nf\|_{B^\alpha_q(L_p(I))} \leq C\|f\|_{B^\alpha_q(L_p(I))}2^{-n\alpha}$$

Problem 13.3 Show that $f \in L_p(R)$ if and only if

$$\left[\sum_{j,k} |\langle f, \psi_{j,k}\rangle|^2 |\psi_{j,k}(x)|^2\right]^{1/2} \in L_p(R)$$

Problem 13.4 Write down the orthogonal wavelet decomposition of a time series, xt, for $t = 1, \ldots, n$ and apply it to analyze a real-world data

Problem 13.5 Apply the wavelet method for solving the boundary value problem

$$-(\alpha u')' + \beta u + \gamma u' = f$$
$$u(0) = c, \ u(1) = d, \ where \ f \in H^{-1}(0, 1)$$

Problem 13.6 Apply the wavelet method for solving the Regularized Burgers Equation defined with $\mu > 0$ by

$$\frac{\partial u}{\partial t}(x, t) - u(x, t)\frac{\partial u}{\partial x}(x, t) = \mu \frac{\partial^2 u}{\partial x^2}(x, t) \ for \ t > 0 \ and \ 0 < x < 1$$
$$u(x, 0) = u_0(x)$$
$$\frac{\partial u}{\partial t}(0, t) = 0, \ u(1, t) = 1$$

Problem 13.7 Show that for $f \in BV(R)$, the space of functions of bounded variation

$$\left\{\sum_j \sum_k |d_{j,k}|^p\right\}^{1/p} \leq \left(\frac{c}{p-1}\right) \|f\|_{BV}$$

$1 \leq p < \infty, c$ is a positive constant.

Problem 13.8 Show that the Besov space $B_2^\alpha(L_2(R))$ is a Hilbert space.

Problem 13.9 Prove that $f \in B_2^\alpha(R))$ if and only if

$$\int_{-\infty}^{\infty} |\hat{f}(\xi)|^2 (1 + \xi)^{2\alpha} d\xi < \infty$$

Problem 13.10 Let $\psi(x)$ be a wavelet such that

$$|\psi(x)| \leq C(1 + |x|)^{-K}, \ C > 0 \ and \ K > 0$$

are constants. Show that

$$\left\|\sum_{k\in Z} d_{j,k}\psi_{j,k}\right\|_{\infty} \leq 2^{j/p}\left\|\sum_{k\in Z} d_{j,k}\psi_{j,k}\right\|_{p}$$

Problem 13.11 Verify Remark 13.2.

Problem 13.12 Let $\{w_n\}$ be a frame for H with representation T. Show that

$$\|f\| = \|(\mathbf{R}^t)^{1/2}Tf\| \; for \; all \; f \in H$$

where $\mathbf{R}^t$ denote the pseudo-inverse of correlation R.

Problem 13.13 Verify Remark 13.3.

Problem 13.14 Verify Remark 13.6.

Problem 13.15 Let $\varphi(x)$ be defined by

$$\begin{aligned}
\varphi(x) &= 0 \; x < 0 \\
&= 1 \; 0 \leq x \leq 1 \\
&= 0 \; x \geq 1
\end{aligned}$$

Draw the graph of the wavelet obtained by taking the convolution of the Haar wavelet with $\varphi(x)$.

Problem 13.16 Prove that $\{w_n(\cdot, -k)\}_{k\in Z}$, $0 \leq n < 2^j$, where $\{w_n(\cdot, \cdot)\}$ denotes a family of wavelet packets, is an orthonormal basis of $L_2(R)$.

Chapter 14
Wavelet Frames

Abstract Wavelet frames are introduced in this chapter.

Keywords Wavelet frame · Dyadic wavelet frames · Frame multiresolution analysis

14.1 General Wavelet Frames

In this section, we shall discuss how we can choose a discrete subset and a function ψ such that

$$\psi_{(a,b)}(x) = (T_b D_b \psi)(x) \tag{14.1}$$

is a frame of $L_2(R)$.

For the sake of convenience, we consider the case where the points (a, b) are restricted to discrete sets of the type $\{(a_j, kba_j)\}_{j,k \in Z}$ where $a > 1, b > 0$, a is the dilation parameter or scaling parameter and b is the translation parameter.

Definition 14.1 Let $a > 1, b > 0$, and $\psi \in L_2(R)$. If the sequence $\{a^{j/2}\psi(a^j x - k^b)\}_{j,k \in Z}$ satisfies the condition of a frame (Definition 11.11), then it is called a general wavelet frame or a wavelet frame.

The main goal of this section is to present the following sufficient conditions in terms of $G_0(\gamma)$ and $G_1(\gamma)$ defined below for $\{a^{j/2}\psi(a^j x - k^b)\}_{j,k \in Z}$ to be a frame. Let

$$G_0(\gamma) = \sum_{j \in Z} |\hat{\psi}(a^j \gamma)|^2 \tag{14.2}$$

$$G_1(\gamma) = \sum_{k \neq 0} \sum_{j \in Z} \left| \hat{\psi}(a^j \gamma) \hat{\psi}\left(a^j \gamma + \frac{k}{b}\right) \right|, \ \gamma \in \mathbb{R} \tag{14.3}$$

A. H. Siddiqi, *Functional Analysis and Applications*, Industrial and Applied Mathematics, https://doi.org/10.1007/978-981-10-3725-2_14

Theorem 14.1 *Let $a > 1, b > 0$, and $\psi \in L_2(R)$ be given. Suppose that*

$$B := \frac{1}{b} \sup_{|\gamma| \in [1,\alpha]} \sum_{j,k \in \mathbb{Z}} \left| \hat{\psi}(a^j \gamma) \hat{\psi}\left(a^j \gamma + \frac{k}{b}\right) \right| < \infty \qquad (14.4)$$

Then $\{a^{j/2} \psi(a^j x - k^b)\}_{j,k \in Z}$ is a Bessel sequence with bound B, and for all functions $f \in L_2(R)$ for which $\hat{f} \in C_c(\mathbb{R})$

$$\sum_{j,k \in \mathbb{Z}} |\langle f, D_{\alpha^j} T_{kb} \psi \rangle|^2$$

$$= \frac{1}{b} \int_{-\infty}^{\infty} |\hat{f}(\gamma)|^2 \sum_{j \in \mathbb{Z}} |\hat{\psi}(a^j \gamma)|^2 d\gamma$$

$$+ \frac{1}{b} \sum_{k \neq 0} \sum_{j \in \mathbb{Z}} \int_{-\infty}^{\infty} \hat{f}(\gamma) \overline{\hat{f}(\gamma - a^j k/b)} \hat{\psi}(a^{-j} \gamma) \hat{\psi}(a^{-j} \gamma - k/b) d\gamma \qquad (14.5)$$

If furthermore

$$A := \frac{1}{b} \inf_{|\gamma| \in [1,\alpha]} \left(\sum_{j \in \mathbb{Z}} |\hat{\psi}(a^j \gamma)|^2 - \sum_{k \neq 0} \sum_{j \in \mathbb{Z}} |\hat{\psi}(a^j \gamma) \hat{\psi}(a^j \gamma + k/b)| \right) > 0$$

then $\{a^{j/2} \psi(a^j x - k^b)\}_{j,k \in Z}$ is a frame for $L_2(\mathbb{R})$ with bounds A, B.

The proof of the theorem is quite technical and we refer to Christensen [40] for its proof.

We have the following theorem providing sufficient conditions for small values b.

Theorem 14.2 *Let $\psi \in L_2(R)$ and $a > 1$ be given. Assume that*

A. $\inf\limits_{|\gamma| \in [1,a]} \sum\limits_{j \in \mathbb{Z}} |\hat{\psi}(a^j \gamma)|^2 > 0.$

B. *There exists a constant $C > 0$ such that*

$$|\hat{\psi}(\gamma)| \leq C \frac{|\gamma|}{(1 + |\gamma|^2)^{3/2}} \quad a.e. \qquad (14.6)$$

Then $\{a^{j/2} \psi(a^j x - k^b)\}_{j,k \in Z}$ is a frame for $L_2(\mathbb{R})$ for all sufficiently small translation parameters $b > 0$.

The following lemmas are required in the proof.

Lemma 14.1 *Let $x, y \in \mathbb{R}$. Then, for all $\delta \in [0, 1]$*

$$\frac{1}{1 + (x + y)^2} \leq 2 \left(\frac{1 + x^2}{1 + y^2} \right)^{\delta}$$

Proof Given $x, y \in R$, the function $\delta \to 2 \left(\frac{1+x^2}{1+y^2} \right)^\delta$ is monotone, so it is enough to prove the result for $\delta = 0$ and $\delta = 1$. The case $\delta = 0$ is clear; for $\delta = 1$, we use that $2ab \le a^2 + b^2$ for all $a, b \in R$ to obtain that

$$
\begin{aligned}
1 + y^2 &= 1 + ((y + x) - x)^2 \\
&= 1 + (y + x)^2 + x^2 - 2x(y + x) \\
&\le 1 + 2((y + x))^2 + x^2) \\
&\le 2(1 + (y + x)^2)(1 + x^2)
\end{aligned}
$$

Lemma 14.2 *Let $\psi \in L_2(\mathbb{R})$ and assume that there exists a constant $C > 0$ such that*

$$
|\hat{\psi}(\gamma)| \le C \frac{|\gamma|}{(1 + |\gamma|^2)^{3/2}} \ a.e.
$$

Then, for all $a > 1$ and $b > 0$

$$
\sum_{k \ne 0} \sum_{j \in \mathbb{Z}} |\hat{\psi}(a^j \gamma) \hat{\psi}(a^j \gamma + k/b)| > 0
$$

$$
\le 16 C^2 b^{4/3} \left(\frac{a^2}{a - 1} + \frac{a}{a^{2/3} - 1} \right) \tag{14.7}
$$

Proof The decay condition on ψ gives that

$$
\begin{aligned}
|\hat{\psi}(a^j \gamma) \hat{\psi}(a^j \gamma + k/b)| &\le C^2 \frac{|a^j \gamma|}{(1 + |a^j \gamma|^2)^{3/2}} \frac{|a^j \gamma + k/b|}{(1 + |a^j \gamma + k/b|^2)^{3/2}} \\
&\le C^2 \frac{|a^j \gamma|}{(1 + |a^j \gamma|^2)^{3/2}} \frac{(1 + |a^j \gamma + k/b|^2)^{1/2}}{(a + |a^j \gamma + k/b|^2)^{3/2}} \\
&= C^2 \frac{|a^j \gamma|}{(1 + |a^j \gamma|^2)^{3/2}} \frac{1}{(1 + |a^j \gamma + k/b|^2)^{3/2}}
\end{aligned}
$$

Applying Lemma 14.1 on $(1 + |a^j \gamma + k/b|^2)1$ with $\delta = \frac{2}{3}$ gives

$$
\begin{aligned}
|\hat{\psi}(a^j \gamma) \hat{\psi}(a^j \gamma + k/b)| &\le 2C^2 \frac{|a^j \gamma|}{(1 + |a^j \gamma|^2)^{3/2}} \left(\frac{1 + |a^j \gamma|^2}{1 + |k/b|^2} \right)^{2/3} \\
&\le 2C^2 \frac{|a^j \gamma|}{(1 + |a^j \gamma|^2)^{5/6}} \left(\frac{1}{1 + |k/b|^2} \right)^{2/3}
\end{aligned}
$$

In this last estimate, j and k appear in separate terms. Thus

$$\sum_{k \neq 0} \sum_{j \in \mathbb{Z}} |\hat{\psi}(a^j \gamma) \hat{\psi}(a^j \gamma + k/b)|$$

$$\leq 2C^2 \left(\sum_{j \in \mathbb{Z}} \frac{|a^j \gamma|}{(1 + |a^j \gamma|^2)^{3/2}} \right) \left(\sum_{k \neq 0} \left(\frac{1}{1 + |k/b|^2} \right)^{2/3} \right) \qquad (14.8)$$

For the sum over $k \neq 0$

$$\sum_{k \leq 0} \left(\frac{1}{1 + |k/b|^2} \right)^{2/3} = 2 \sum_{k=1}^{\infty} \frac{b^{4/3}}{(b^2 + k^2)^{2/3}}$$

$$\leq 2 b^{4/3} \sum_{k=1}^{\infty} \frac{1}{k^{4/3}}$$

$$\leq 2 b^{4/3} \left(\int_1^{\infty} t^{4/3} dt + 1 \right)$$

$$= 8 b^{4/3}$$

In order to estimate the sum over $j \in \mathbb{Z}$ in (14.8), we define the function

$$f(\gamma) = \sum_{j \in \mathbb{Z}} \frac{|a^j \gamma|}{(1 + |a^j \gamma|^2)^{5/6}}, \quad y \in \mathbb{R}.$$

We want to show that f is bounded. Note that $f(a\gamma) = f(\gamma)$ for all γ; it is therefore enough to consider $|\gamma| \in [1, a]$, so we can use that

$$|a^j \gamma| \leq a_{j+1}, \; 1 + |a^j \gamma|^2 \geq 1 + a^{2j}$$

Thus

$$\sum_{j \in \mathbb{Z}} \frac{|a^j \gamma|}{(1 + |a^j \gamma|^2)^{5/6}} \leq \sum_{j \in \mathbb{Z}} \frac{a^{j+1}}{(1 + a^{2j})^{5/6}}$$

$$= \sum_{j=-\infty}^{0} \frac{a^{j+1}}{(1 + a^{2j})^{5/6}} + \sum_{j=1}^{\infty} \sum_{a^{j+1}} (1 + a^{2j})^{5/6}$$

$$\leq \sum_{j=-\infty}^{0} a^{j+1} + \sum_{j=1}^{\infty} \frac{a^{j+1}}{a^{5/3 j}}$$

$$= a \sum_{j=0}^{\infty} a^{-j} + a \sum_{j=1}^{\infty} (a^{-2/3}) j$$

$$= a\frac{1}{1-a^{-1}} + a\frac{a^{-2/3}}{1-a^{-2/3}}$$

$$= \frac{a^2}{a-1} + \frac{a}{a^{2/3}-1}$$

That is, f is bounded as claimed. Putting all information together, and using (14.8)

$$\sum_{k\neq 0}\sum_{j\in\mathbb{Z}}|\hat{\psi}(a^j\gamma)\hat{\psi}(a^j\gamma+k/b)|$$

$$\leq 2C^2\left(\sum_{j\in\mathbb{Z}}\frac{|a^j\gamma|}{(1+|a^j\gamma|^2)^{5/6}}\right)\left(\sum_{k\neq 0}\left(\frac{1}{1+|k/b|^2}\right)^{2/3}\right)$$

$$\leq 16C^2b^{4/3}\frac{a^2}{a-1} + \frac{1}{a^{2/3}-1}$$

Proof (*Proof of Theorem* 14.2) We first prove that $\{a^{j/2}\psi(a^jx-k^b)\}_{j,k\in\mathbb{Z}}$ is a Bessel sequence for all $b > 0$. Arguments similar to the one used in the proof of Lemma 14.2 show that

$$\sum_{j\in\mathbb{Z}}|\hat{\psi}(a^j\gamma)|^2 \leq \frac{1}{a^4-1} + \frac{a^4}{a^2-1}$$

Via Lemma 14.2 it follows that

$$\leq \frac{1}{a^4-1} + \frac{a^4}{a^2-1}$$

by Theorem 14.1 we conclude that $\{a^{j/2}\psi(a^jx-k^b)\}_{j,k\in\mathbb{Z}}$ is a Bessel sequence. By choosing b sufficiently small, the assumption (*i*) implies that

$$\inf_{|\gamma|\in[1,\alpha]}\left(\sum_{j\in\mathbb{Z}}|\hat{\psi}(a^j\gamma)|^2 - 16C^2b^{4/3}\left(\frac{a^2}{a-1} + \frac{a}{a^{2/3}-1}\right)\right) > 0 \quad (14.9)$$

and in this case, by Lemma 14.2

$$\inf_{|\gamma|\in[1,\alpha]}\left(\sum_{j\in\mathbb{Z}}|\hat{\psi}(a^j\gamma)|^2 - \sum_{k\neq 0}\sum_{j\in\mathbb{Z}}|\hat{\psi}(a^j\gamma)\hat{\psi}(a^j\gamma-k/b)|\right) > 0$$

Theorem 14.1 now gives the desired conclusion.

Example 14.1 Let $a = 2$ and consider the function

$$\psi(x) = \frac{2}{\sqrt{3}}\pi^{-1/4}(1 - x^2)e^{(1/2)x^2}$$

Due to its shape, ψ is called the Mexican hat. It can be shown that

$$\hat{\psi}(\gamma) = 8\frac{2}{\sqrt{3}}\pi^{9/4}\gamma^2 e^{-2\pi^2\gamma^2}$$

A numerical calculation shows that

$$\inf_{|\gamma|\in[1,2]} \sum_{j\in\mathbb{Z}} |\hat{\psi}(2^j\gamma)|^2 > 3.27$$

Also, (14.6) is satisfied for $c = 4$, so a direct calculation using (14.10) shows that $\{2^{j/2}\psi(2^j x - k^b)\}_{j,k\in\mathbb{Z}}$ is a frame if $b < 0.0084$. This is far from being optimal: numerical calculations based on the expressions for A, B in Theorem 14.1 gives that $\{2^{j/2}\psi(2^j x - k^b)\}_{j,k\in\mathbb{Z}}$ is a frame if $b < 1.97$!

14.2 Dyadic Wavelet Frames

In this section, we consider the question of finding conditions on a system of the form

$$\psi_{j,k}(x) = 2^{j/2}\psi(2^j x - k), \quad j, k \in \mathbb{Z} \tag{14.10}$$

generated by translations and dilations of a single function $\psi \in L_2(R)$, so that it becomes a frame in $L_2(R)$. Obviously, every orthonormal wavelet is a frame of this type, but we shall show that the converse is not true. Nevertheless, they still have perfect reconstruction and they have been used in several applications.

If (14.11) is a frame, the general theory provides us with the dual frame $\tilde{\psi}_{j,k} = S^{-1}\psi_{j,k}$, where $S = \mathscr{F}^*\mathscr{F}$ and $\mathscr{F}$ is the frame operator. The operator $S = \mathscr{F}^*\mathscr{F}$ commutes with the dilations $(D^m f)(x) = 2^{\frac{m}{2}}f(2^m x)$, $m \in \mathbb{Z}$:

$$(\mathscr{F}^*\mathscr{F})((D^m f))(x) = \sum_{j\in\mathbb{Z}}\sum_{k\in\mathbb{Z}}\langle D^m f, \psi_{j,k}\rangle\psi_{j,k}(x)$$

$$= \sum_{j\in\mathbb{Z}}\sum_{k\in\mathbb{Z}}\langle f, \psi_{j-m,k}\rangle\psi_{j,k}(x)$$

$$= 2^{\frac{m}{2}}\sum_{j\in\mathbb{Z}}\sum_{k\in\mathbb{Z}}\langle f, \psi_{j-m,k}\rangle\psi_{j-m,k}(2^m x)$$

$$= 2^{\frac{m}{2}}(\mathscr{F}^*\mathscr{F} f)^*(2^m x) = D^m(\mathscr{F}^*\mathscr{F} f)(x)$$

Thus, S^{-1} also commutes with these dilations, and we have

$$\widetilde{\psi_{j,k}}(x) = (S^{-1}\psi_{j,k})(x) = (S^{-1}\delta_j\psi_{0,k})(x)$$
$$= \delta_j S^{-1}\psi_{0,k}(x) = \delta_j\widetilde{\psi_{0,k}}(X) = 2^{j/2}\widetilde{\psi_{0,k}}(2^j x)$$

Thus, for k fixed, the functions $\widetilde{\psi_{j,k}}$ are all dilations of a single function $\widetilde{\psi_{0,k}}$. Unfortunately, this is not the case for translations $(T_k)(x) = f(x-k)$.

$$((\mathscr{F}^*\mathscr{F})(T_l f))(x) = \sum_{k\in\mathbb{Z}}\sum_{j\in\mathbb{Z}}\langle T_l f, \psi_{j,k}\rangle \psi_{j,k}(x)$$
$$= \sum_{k\in\mathbb{Z}}\sum_{j\in\mathbb{Z}}\langle f, \psi_{j,k-2^j l}\rangle \psi_{j,k}(x)$$
$$= \sum_{n\in\mathbb{Z}}\sum_{j\in\mathbb{Z}}\langle f, \psi_{j,n}\rangle \psi_{j,n+2^j l}(x) = T_l(\mathscr{F}^*\mathscr{F}f)(x)$$

is valid if $2^j l$ is an integer, which is true for every l only when $j \geq 0$. Thus, in general, one cannot expect the dual frame to be generated by dilations and translations of a single function (for details see [59]).

We now address the problem of finding sufficient conditions on ψ for (14.11) to be a frame. It turns out that this problem shares some features with the one that led to the basic equations that characterize wavelets.

Define

$$t_m(\xi) = \sum_{j=0}^{\infty} \hat{\psi}(2^j\xi)\overline{(2^j(\xi+2m\pi))}, \ \xi \in \mathbb{R}, m \in \mathbb{Z}$$

and

$$S(\xi) = \sum_{j\in\mathbb{Z}} |\hat{\psi}(2^j\xi)|^2, \ \xi \in \mathbb{R}$$

Consider

$$\underline{S_\phi} = ess\inf_{\xi\in R} S(\xi), \ \bar{S}_\psi = ess\sup_{\xi\in R} S(\xi)$$

and

$$\beta_\psi(m) = ess\inf_{\xi\in R}\sup_{\xi\in R}\sum_{k\in\mathbb{Z}} |t_m(2^k\xi)|$$

Observe that all the expressions inside the infimum and suprema in the above definitions are invariant under the usual dilations (scaling) by 2, so that these infimum
and suprema need only be computed over $1 \leq |\xi| \leq 2$ (a dilation 'period').

Theorem 14.3 *Let $\psi \in L_2(R)$ be such that*

$$A_\psi = \underline{S}_\psi - \sum_{q \in 2\mathbb{Z}+1} [\beta_\psi(q)\beta_\psi(-q)]^{1/2} > 0$$

and

$$B_\psi = \overline{S}_\psi - \sum_{q \in 2\mathbb{Z}+1} [\beta_\psi(q)\beta_\psi(-q)]^{1/2} > \infty$$

Then, $\{\psi_{j,k} : j, k \in \mathbb{Z}\}$ is a frame with frame bounds A_ψ and B_ψ.

Remark 14.1 If $S(\xi) = 1$ for a.e. $\xi \in R$ and $t_m(\xi) = 0$ for a.e. $\xi \in R$ and all
$m \in 2\mathbb{Z}+1$, $A_\psi = B_\psi = 1$. If, in addition, $||\psi||_2 = 1$, then frame $\{\psi_{j,k} : j, k \in \mathbb{Z}\}$
is an orthonormal basis of $L_2(R)$, and thus a wavelet.

To prove Theorem 14.3, we need the following lemma.

Lemma 14.3 *Suppose that $\{e_j : j = 1, 2, \ldots\}$ is a family of elements in a*

$$A||f||^2 \leq \sum_{j=1}^{\infty} |\langle f, e_j \rangle|^2 \leq B||f||^2$$

*for all belonging to a dense subset $\mathscr{D}$ of H. Then, the same inequalities are true for
all $f \in H$; that is, $\{e_j : j = 1, 2, \ldots\}$ is a frame for H.*

Proof It can be verified that

$$\sum_{j=1}^{\infty} |\langle f, e_j \rangle|^2 \leq B||f||^2$$

for all $f \in H$. To show the other inequality, we choose $\varepsilon > 0$ and $g \in D$ such that
$||g - f|| < \varepsilon$. Then, by Minkowski' s inequality in l2, the above inequality, and

$$||f|| \leq ||g|| + \varepsilon \leq ||g|| + \sqrt{\frac{B}{A}}\varepsilon, \text{ we obtain}$$

$$||f|| - 2\frac{\sqrt{B}}{\sqrt{A}}\varepsilon \leq ||g|| - \frac{\sqrt{B}}{\sqrt{A}}\varepsilon \leq ||g|| - \frac{\sqrt{B}}{\sqrt{A}}||g - f||$$

$$\leq \left(\frac{1}{A} \sum_{j=1}^{\infty} |\langle g, e_j \rangle|^2 \right)^{1/2} - \frac{\sqrt{B}}{\sqrt{A}} \left(\frac{1}{B} \sum_{j=1}^{\infty} |\langle g - f, e_j \rangle|^2 \right)^{1/2}$$

$$(14.11)$$

This finishes the proof since ε is arbitrary.

Proof (Proof of Theorem 14.3) Let D be the class of all $f \in L_2(R)$ such that $\hat{f} \in L_\infty(R)$ and $\hat{f}$ is compactly supported in $R/\{0\}$. By Proposition 1.19 [59], we have

$$\sum_{j \in \mathbb{Z}} \sum_{k \in \mathbb{Z}} |\langle f, \psi_{j,k} \rangle|^2 = \frac{1}{2\pi} \int_R |\hat{f}(\xi)|^2 S(\xi) d\xi$$

$$+ \frac{1}{2\pi} \int_R \overline{\hat{f}(\xi)} \sum_{p \in \mathbb{Z}} \sum_{q \in 2\mathbb{Z}+1} \hat{f}(\xi + 2^p 2q\pi) t_q(2^{-q}\xi) d\xi$$

$$\equiv \frac{1}{2\pi} \int_R |\hat{f}(\xi)|^2 S(\xi) d\xi + \frac{1}{2\pi} R_\psi(f) \qquad (14.12)$$

for all $f \in D$. The Schwarz inequality gives us

$$|R_\psi(f)| \le \sum_{q \in 2\mathbb{Z}+1} \sum_{p \in \mathbb{Z}} \left(\int_R |\hat{f}(\eta)|^2 |t_q(2^{-p\eta})| d\eta \right)^{1/2} \cdot \left(\int_R |\hat{f}(\eta + 2^p 2q\pi)|^2 |t_q(2^{-p\eta})| d\eta \right)^{1/2}$$

In the second integral, we change variables to obtain

$$\left(\int_R |\hat{f}(\eta)|^2 |t_q(2^{-p\eta} - 2q\pi)| d\eta \right)^{1/2}$$

Since $t_q(\xi - 2q\pi) = \overline{t_{-q}(\xi)}$, we deduce, after applying Schwarz's inequality for series

$$|R_\psi(f)| \le \sum_{q \in 2\mathbb{Z}+1} \left(\sum_{p \in \mathbb{Z}} \int_R |\hat{f}(\eta)|^2 |t_q(2^{-p\eta})| d\eta \right)^{1/2} \cdot \left(\sum_{p \in \mathbb{Z}} \int_R |\hat{f}(\eta)|^2 |t_{-q}(2^{-p\eta})| d\eta \right)^{1/2}$$

$$\le \sum_{q \in 2\mathbb{Z}+1} [\beta_\psi(q)\beta_\psi(-q)]^{1/2} \|\hat{f}\|_2^2$$

Hence,

$$\sum_{q \in 2\mathbb{Z}+1} [\beta_\psi(q)\beta_\psi(-q)]^{\frac{1}{2}} \|\hat{f}\|_2^2 \le R_\psi(f) \le \sum_{q \in 2\mathbb{Z}+1} [\beta_\psi(q)\beta_\psi(-q)]^{\frac{1}{2}} \|\hat{f}\|_2^2$$

$$(14.13)$$

These inequalities, together with (14.12), give us

$$A_\psi \|f\|_2^2 \le \sum_{j \in \mathbb{Z}} \sum_{k \in \mathbb{Z}} |\langle f, \psi_{j,k} \rangle|^2 \le B_\psi \|f\|_2^2$$

for all $f \in \mathscr{D}$. Since $\mathscr{D}$ is dense in $L_2(\mathscr{R})$, the same inequalities hold for all $f \in L_2(R)$ by Lemma 14.3. This finishes the proof of Theorem 14.4.

Example 14.2 Let $\psi \in S$ be such that $supp(\hat{\psi})$ is contained in the set $\{\xi \in R : \frac{1}{2} \leq |\xi| \leq 2\}$ and

$$\sum_{j \in \mathbb{Z}} |\hat{\psi}(2^j \xi)|^2 = 1 \ for \ all \ \xi \neq 0$$

Then $t_q(\xi) = 0$ for all $\xi \in R$ and, consequently, $A_\psi = B_\psi = 1$. Thus, $\{\psi_{j,k} : j, k \in Z\}$ is a tight frame.

Example 14.3 An example of a frame of the type discussed in this section is the one generated by the Mexican hat function. This is the function

$$\psi(x) = \frac{2}{\sqrt{3}} \pi^{-1/4} (1 - x^2) e^{-1/2x^2}$$

which coincides with $-\frac{d^2}{dx^2}\left(e^{-1/2x^2}\right)$ when normalized in $L_2(R)$.

Daubechies has reported frame bounds of 3.223 and 3.596 for the frame obtained by translations and dilations of the Mexican hat function (see $\hat{\psi}(\xi)$). An approximate quotient of these frame bounds is 1.116, which indicates that this frame is 'close' to a tight frame.

14.3   Frame Multiresolution Analysis

We have presented the concept of multiresolution analysis in Chap. 12 (Definition 12.1) which was introduced in 1989 by Mallat. Frame multiresolution was introduced by Benedetto and Li [15] and an updated account can be found in [40], [Siddiqi03]. We discuss here the definition, a sufficient condition for a function of $L_2(R)$ to generate a multiresolution analysis and an example of frame multiresolution analysis.

Definition 14.2 A frame multiresolution analysis for $L_2(R)$ and a function $\phi \in V_0$ such that

A. $\ldots V_{-1} \subset V_0 \subset V_1 \ldots$
B. $\overline{\cup_j V_j} = L_2(R)$ and $\cap_j V_j = \{0\}$
C. $V_j = D^j V_0$
D. $f \in V_0 \Rightarrow T_k f \in V_0, \forall k \in Z$
E. $\{T_k \phi\}_{k \in \mathbb{Z}}$ is a frame for V_0

Theorem 14.4 *Suppose that $\phi \in L_2(R)$, that $\{T_k\phi\}_{k\in Z}$ is a frame sequence, and that $|\hat{\phi}| > 0$ on a neighborhood of zero. If there exists a function $H_0 \in L_\infty(T)$ such that*

$$\hat{\phi}(\gamma) = H_0\left(\frac{\gamma}{2}\right)\hat{\phi}\left(\frac{\gamma}{2}\right) \tag{14.14}$$

then ϕ generates a frame multiresolution analysis.

For the proof, we refer to [40].

Example 14.4 Define the function ϕ via its Fourier transform

$$\hat{\phi}(\gamma) = \chi_{[-\alpha,\alpha)}, \ for \ some \ a \in \left(0, \frac{1}{2}\right)$$

It can be seen that $\{T_k\phi\}_{k\in Z}$ is a frame sequence. Note that

$$\hat{\phi}(2\gamma) = \chi_{[-\frac{\alpha}{2}, \frac{\alpha}{2})}(\gamma)\hat{\phi}(\gamma)$$

For $|\gamma| < \frac{1}{2}$, let

$$H_0(\gamma) = \chi_{[-\frac{\alpha}{2}, \frac{\alpha}{2})} \tag{14.15}$$

extending H_0 to a 1-periodic function we see that (14.13) is satisfied. By Theorem 14.5, we conclude that ϕ generates a frame multiresolution analysis.

Given a continuous non-vanishing function θ on $[-\alpha, \alpha]$, we can generalize the example by considering

$$\hat{\theta}(\gamma) = \theta(\gamma)\chi_{[-\alpha,\alpha)}(\gamma)$$

Defining

$$H_0(\gamma) = \frac{\theta(2\gamma)}{\theta(\gamma)} \ \text{if} \ \gamma \in \left[-\frac{\alpha}{2}, \frac{\alpha}{2}\right)$$

$$= 0 \ \text{if} \ \gamma \in \left[-\frac{1}{2}, -\frac{\alpha}{2}\right[\cup \left[\frac{\alpha}{2}, \frac{1}{2}\right)$$

extending H_0 periodically, it again follows that $\hat{\phi}$ generates a frame multiresolution analysis.

14.4 Problems

Problem 14.1 Prove Lemma 14.1.

Problem 14.2 Prove Lemma 14.2.

Problem 14.3 Let $a = 2$ and $\psi(x) = \frac{2}{\sqrt{3}}\pi^{1/4}(1 - x^2)e^{-1/2x^2}$ Show that $\{2^{j/2}\psi(2^j x - k^b)\}_{j,k \in \mathbb{Z}}$ is a frame if $b < 1.97$.

Problem 14.4 Prove that

$$\sum_{j \in \mathbb{Z}} |\hat{\psi}(a^j \gamma)|^2 \leq \frac{1}{a^4 - 1} + \frac{a^4}{a^2 - 1}$$

Chapter 15
Gabor Analysis

Abstract A system based on translation and rotation called Gabor system was introduced by Gabor in 1946. In this chapter, we discuss basic properties of Gabor system.

Keywords Orthonormal Gabor system · Heil–Ramanathan–Topiwala conjecture (HRT conjecture) · HRT conjecture for wave packets Digital communication · Image representation · Biological vision

15.1 Orthonormal Gabor System

Dennis Gabor recepient of 1971 Physics Nobel prize introduced a system of functions, now known as the Gabor system, while studying shortcomings of Fourier analysis [79]. Gabor proposed to expand a function f into a series of elementary functions which are constructed from a single building block by translation and modulation. More precisely, he suggested to represent f by the series

$$f(t) = \sum_{n,m \in Z} c_{m,n} g_{m,n}(t) \tag{15.1}$$

where the elementary functions $g_{m,n}$ are given by

$$g_{m,n}(t) = g(t - na)e^{2\pi imbt}, \ m, n \in Z \tag{15.2}$$

for a fixed function g and time–frequency shift parameters $a, b > 0$. A typical g could be chosen as

$$g(x) = e^{-x^2} \tag{15.3}$$

Here, a denotes the time shift and b denotes frequency shift.

Detailed account of the Gabor system can be found in references [39, 40, 42, 74, 89] [Fe 98].

© Springer Nature Singapore Pte Ltd. 2018
A. H. Siddiqi, *Functional Analysis and Applications*, Industrial and Applied Mathematics,
https://doi.org/10.1007/978-981-10-3725-2_15

509

In this section, we discuss orthonormal Gabor system while Sects. 15.1–15.3 are respectively devoted to the introduction of Gabor frames, Heil–Ramanitation–Topiwala conjecture and applications of Gabor system.

We know that the functions $\left(\frac{1}{\sqrt{2\pi}e^{imx}}\right)_{m\in Z}$ form an orthonormal basis for $L_2(-\pi, \pi)$; since they are periodic with period 2π, they actually form an orthonormal basis for $L_2(-\pi + 2\pi n, \pi + 2\pi n)$ for any $n \in Z$. If we want to put emphasis on the fact that we look at the exponential functions on the interval $[-\pi + 2\pi n, \pi, \pi + 2\pi n[$, we can also write that

$$\left\{\frac{1}{\sqrt{2\pi}}e^{imx} \chi_{[-\pi+2\pi n,\pi+2\pi n[}(x)\right\}_{m,n\in\mathbb{Z}}$$

is an orthonormal basis for $L_2(-\pi + 2\pi n, \pi + 2\pi n)$. Now observe that the intervals $[-\pi + 2\pi n, \pi + 2\pi n[$, $n \in Z$, form a partition of R: They are disjoint and cover the entire axis $\mathbb{R}$. This implies that the union of these bases, i.e., the family

$$\left\{\frac{1}{\sqrt{2\pi}}e^{imx} \chi_{[-\pi+2\pi n,\pi+2\pi n[}(x)\right\}_{m,n\in\mathbb{Z}}$$

is an orthonormal basis for $L_2(R)$. We can write this orthonormal basis on a slightly more convenient form as

$$\left\{\frac{1}{\sqrt{2\pi}}\chi_{[-\pi+2\pi n,\pi+2\pi n[}(x - 2\pi n)\right\}_{m,n\in\mathbb{Z}}$$

The system in (15.4) is the simplest case of a Gabor system. Basically, (15.4) consists of the function $\frac{1}{\sqrt{2\pi}}\chi_{[-\pi+2\pi n,\pi+2\pi n[}(x)$ and translated versions and modulated versions, i.e., functions which are multiplied by complex exponential functions. It is called *orthonormal Gabor system*. A general Gabor system is obtained by replacing the function $\frac{1}{\sqrt{2\pi}}\chi_{[-\pi,\pi[}$ with an arbitrary function in $L_2(R)$ and allowing certain parameters to appear in the translation and in the complex exponential function:

Definition 15.1 Let $a, b > 0$ and $g \in L_2(R)$. Then, the family of functions

$$\{e^{2\pi imbx}g(x - na)\}_{m,n\in R} \tag{15.4}$$

is called a *Gabor system*.

The Gabor system in (15.4) corresponds to the choice

$$a = 2\pi, \; b = \frac{1}{2\pi}, \; g = \frac{1}{\sqrt{2\pi}}\chi_{[-\pi,\pi[}$$

Gabor systems are often used in time–frequency analysis, i.e., in situations where we want to know how a certain signal f changes with time, and also which frequencies appear at which times. In order to extract this information, one might try to expand the signal via a Gabor basis; in other words, to pick $a, b > 0$ and $g \in L_2(R)$ such that the system in (15.5) forms an orthonormal basis for $L_2(R)$ and then expand f via equation

$$f(x) = \sum_{m,n \in \mathbb{Z}} c_{m,n} e^{2\pi imbx} g(x - na) \qquad (15.5)$$

with

$$c_{m,n} = \int\limits_{-\infty}^{\infty} f(x) \overline{g(x - na)} e^{2\pi imbx} dx$$

If g has compact support, this expansion contains useful information about the function values $f(x)$. In fact, for a given $x \in R$, $g(x - na)$ is only nonzero for a finite number of values of $n \in Z$, and the size of the corresponding coefficients $c_{m,n}$ will give an idea about $f(x)$. This feature gets lost if g does not have compact support; but if g decays quickly, for example exponentially, the argument still gives a good approximation if we replace g by a function which is zero outside a sufficiently large interval.

Doing time–frequency analysis, we would also like to have information about f; using the rules for calculation with the Fourier transform indeed shows that

$$\hat{f}(\gamma) = \sum_{m,n \in \mathbb{Z}} e^{2\pi iabmnx} c_{m,n} e^{2\pi ibax} g(\gamma - mb)$$

Now, in order for this to be useful, we would like that $\hat{g}$ has compact support, or at least decays fast.

15.2 Gabor Frames

Definition 15.2 A Gabor frame is a frame for $L_2(\mathbb{R})$ of the form $\{E_{mb}T_{na}g\}_{m,n \in Z}$, where $a, b > 0$ and $g \in L_2(\mathbb{R})$ is a fixed function.

Frames of this type are also called *Weyl-Heisenberg frames*. The function g is called the *window function* or the *generator*. Explicitly

$$E_{mb}T_{na}g(x) = e^{2\pi imbx}g(x - na)$$

Note the convention, which is implicit in our definition: When speaking about a Gabor frame, it is understood that it is a frame for all of $L_2(R)$; i.e., we will not deal with frames for subspaces.

The Gabor system $\{E_{mb}T_{na}g\}_{m,n \in Z}$ only involves translates with parameters $na, n \in Z$ and modulation with parameters $mb, m \in Z$. The points $\{(na, mb)\}_{m,n \in Z}$ form a lattice in R^2, and for this reason, one frequently calls $\{E_{mb}T_{na}g\}_{m,n \in Z}$ a regular Gabor frame.

We now move to the question about how to obtain Gabor frames $\{E_{mb}T_{na}g\}_{m,n\in Z}$ for $L_2(R)$. One of the most fundamental results says that the product ab decides whether it is possible for $\{E_{mb}T_{na}g\}_{m,n\in R}$ to be a frame for $L_2(R)$:

Theorem 15.1 (Necessary Conditions) *Let $g \in L_2(R)$ and $a, b > 0$ be given. Then the following holds:*

(i) *If $ab > 1$, then $\{E_{mb}T_{na}g\}_{m,n\in Z}$ is not a frame for $L_2(R)$.*
(ii) *If $\{E_{mb}T_{na}g\}_{m,n\in Z}$ is a frame, then $ab = 1 \Leftrightarrow \{EmbTnag\}_{m,n\in Z}$ is a Riesz basis.*

Theorem 15.2 (Necessary and Sufficient Condition) *Let $A, B > 0$ and the Gabor system $\{E_{mb}T_{na}g\}_{m,n\in Z}$ be given. Then $\{E_{mb}T_{na}g\}_{m,n\in Z}$ is a frame for $L_2(R)$ with bounds A, B if and only if*

$$bAI \leq M(x)M(x)^\star \leq bBI \quad a.e. \ x$$

where I is the identity operator on l_2, $M(x)$ is given by

$$M(x) = \{g(x - na - m/b)\}_{m,n\in Z}, \ x \in \mathbb{R}$$

and $M(x)^\star$ is the conjugate transpose of the matrix $M(x)$.

Sufficient conditions for $\{E_{mb}T_{na}g\}_{m,n\in Z}$ to be a frame for $L_2(\mathbb{R})$ have been known since 1988. The basic insight was provided by Daubechies. A slight improvement was proved by Heil and Walnut.

Theorem 15.3 (Sufficient Condition) *Let $g \in L_2(R)$ and $a, b > 0$ be given. Suppose that*

$$\exists A, B > 0 : A \leq \sum_{n\in\mathbb{Z}} |g(x - na)|^2 \leq B \ for \ a.e. \ x \in \mathbb{R}$$

and

$$\sum_{k\neq 0} \left\| \sum_{n\in\mathbb{Z}} T_{na}g T_{na+k/b}\bar{g} \right\|_\infty < A \tag{15.6}$$

Then, $\{E_{mb}T_{na}g\}_{m,n\in Z}$ is a Gabor frame for $L_2(R)$. For the proof, we cite [39].

We discuss now some well-known functions and the range of parameters a, b for which they generate frames. First we consider the Gaussian

Theorem 15.4 *Let $a, b > 0$ and consider $g(x) = e^{-x^2}$. Then, the Gabor system $\{E_{mb}T_{na}g\}_{m,n\in Z}$ is a frame if and only if $ab < 1$. The Gaussian generates another function for which the exact range of parameters generating a frame is known, is the hyperbolic secant, which is defined by*

$$g(x) = \frac{1}{\cosh(\pi x)}$$

This function was studied by Janssen and Strohmer who proved that $\{E_{mb}T_{na}g\}_{m,n\in Z}$ is a frame whenever $ab < 1$. The hyperbolic secant does not generate a frame when $ab = 1$.

Let us now consider characteristic functions

$$g := \chi[0, c[, \ c > 0$$

The question is which values of c and parameters a, $b > 0$ will imply that $\{E_{mb}T_{na}g\}_{m,n\in Z}$ is a frame. A scaling of a characteristic function is again (multiple of) a characteristic function, so we can assume that $b = 1$. A detailed analysis shows [39] that

(i) $\{E_{mb}T_{na}g\}_{m,n\in Z}$ *is not a frame if $c < a$ or $a > 1$.*

(ii) $\{E_{mb}T_{na}g\}_{m,n\in Z}$ *is a frame if $1 \geq c \geq a$.*

(iii) $\{E_{mb}T_{na}g\}_{m,n\in Z}$‘ *is not a frame if $a = 1$ and $c > 1$. Assuming now that $a < 1, c > 1$, we further have*

(vi) $\{E_{mb}T_{na}g\}_{m,n\in Z}$ *is a frame if $a \notin Q$ and $c \in]1, 2[$.*

(v) $\{E_{mb}T_{na}g\}_{m,n\in Z}$ *is not a frame if $a = p/q \in Q$, $gcd(p,q) = 1$, and $2 - \frac{1}{q} < c < 2$.*

(vi) $\{E_{mb}T_{na}g\}_{m,n\in Z}$ *is not a frame if $a > \frac{3}{4}$ and $c = L - 1 + L(1 - a)$ with $L \in \mathbb{N}, L \geq 3$.*

(vii) $\{E_{mb}T_{na}g\}_{m,n\in Z}$ *is a frame if $|c - \lfloor c \rfloor - \frac{1}{2}| < \frac{1}{2} - a$.*

Heil-Ramanath-Topiwal Cojecture

Let (α_k, β_k) be distinct points in R^2, then $\{e^{2\pi i\beta_k t}g(t - \alpha_k)\}_{k=1}^N$ is linearly independent set of functions in $L_2(R)$.

Despite the striking simplicity of the statement of this conjecture, it remains open today in the generality stated. Some partial results were obtained there, including the following:

(i) If a nonzero $g \in L_2(R)$ is compactly supported or just supported on a half line, then the independence conclusion holds for any value of N.

(ii) If $g(x) = p(x)e^{-x^2}$ is a nonzero polynomial, then the independence conclusion holds for any value of N.

(iii) The independence conclusion holds for any nonzero $g \in L_2(R)$ if $N \leq 3$.

(vi) If the independence conclusion holds for a particular $g \in L_2(R)$ and a particular choice of points $\{(\alpha_k, \beta_k)\}_{k=1}^N$, then there exists an $\varepsilon > 0$ such that it also holds for any h satisfying $\|g - h\|L_2 \leq \varepsilon$, using the same set of points.

(v) If the independence holds for one particular $g \in L_2(R)$ and particular choice of points $\{(\alpha_k, \beta_k)\}_{k=1}^N$, then there exists $\varepsilon > 0$ such that it holds for that g and any set of N points in R^2 within ε of original ones.

Given $g \in L_2(R)$ and a sequence $A \subset R \times R^+$, the wavelet system generated by g and A is the collection of time scale shifts

$$W(g, A) = \{T_a D_b g\}_{(a,b) \in A} \tag{15.7}$$

Analogue of the HRT conjecture fails, for details see [95].

While the analogue of the HRT conjecture fails for wavelet system in general but Christensen and Linder [42] have interesting partial results on when independence holds, including estimates of the frame bounds of finite sets of time–frequency or timescale shifts.

For a comprehensive and update account of the HRT conjecture, we refer to Heil [95].

Theorem 15.5 *Let* $g \in L_2(R)$ *be nonzero and compactly supported, then* $\{e^{2\pi i \beta_k t} g(t - \alpha_k)\}_{k=1}^N$ *is linearly independent for any value of* N.

Theorem 15.6 *Let* $g(x) = p(x)e^{-x^2}$, *where* p *is a nonzero polynomial, then* $\{e^{2\pi i \beta_k t} g(t + \alpha_k)\}_{k=1}^N$ *is linearly independent in* $L_2(R)$ *for any value of* N.

Theorem 15.7 *Let* $g \in L_2(R)$ *and* $\wedge = \{(\alpha_k, \beta_k)\}_{k=1}^N$ *be such that* $\mathscr{G}(g, \wedge) = \{e^{2\pi i \beta_k t} g(t - \alpha_k)\}_{k=1}^N$ *is linearly independent in* $L_2(R)$ *for any value of* N. *Then, the following statement hold:*

(a) *There exists* $\varepsilon > 0$, *such that* $\mathscr{G}(g, \wedge')$ *is linearly independent for any set* $\wedge' = \{(\alpha_k', \beta_k')\}_{k=1}^N$ *such that* $|\alpha_k - \alpha_k'|, |\beta_k - \beta_k'| < \varepsilon$ *for* $k = 1, \ldots, N$.
(b) *There exists* $\varepsilon > 0$, *such that* $\mathscr{G}(h, \wedge)$ *is linearly independent for any* $h \in L_2(R)$ *with* $\|g - h\|_{L_2} < \varepsilon$, *where* $g \in \mathscr{G}(g, \wedge)$.

Proof (Proof of Theorem 15.5) Let $M_{\beta_k} T_{\alpha_k} g(x) = e^{2\pi i \beta_k t} g(t - \alpha_k)$ and $m_k(x) = \sum_{j=1}^{M_k} c_{k,j} e^{2\pi i \beta_{k,j} x}$.

Choose any finite set $\wedge \subset R^2$ and let $g \in L_2(R)$ be compactly supported on half line. Given scalar, $c_{k,j}$, let

$$0 = \sum_{k=1}^N \sum_{j=1'}^{M_k} c_{k,j} M_{\beta_{k,j}} T_{\alpha_k} g(x) = \sum_{k=1}^N m_k(x) g(x - \alpha_k) \quad a.e.$$

Since g is compactly supported, it can be argued that

$$m_k g(x - \alpha_k) = 0 \text{ a.e. for some single k}$$

We can find a subset of the support $g(x - \alpha_k)$ of positive measure for which this is true, inturn we find that trigonometric polynomials $m_k(x)$ vanishes on a set of positive measure. But this can only happen in $c_{k,j} = 0$ for all j. We can repeat this argument and get $c_{k,j} = 0$ for all k and j. For all details, see Heil [95] and references theorem, specially HRT {65}.

Proof (*Proof of Theorem 15.6*) Let $g(x) = p(x)e^{-x^2}$, where p is a nonzero polynomial. Further, let

$$M_{\beta_k} T_{\alpha_k}(x) = e^{2\pi i \beta_k x} g(x - \alpha_k)$$

$$and \qquad m_k(x) = \sum_{j=1}^{M_k} c_{k,j} e^{2\pi i \beta_k x}$$

Given scalars, $c_{j,k}$, let

$$s(x) = \sum_{k=1}^{N} \sum_{j=1}^{M_k} c_{k,j} M_{\beta_k} T_{\alpha_k} p(x)$$

$$= e^{-x^2} \sum_{k=1}^{N} \left(\sum_{j=1}^{M_k} c_{k,j} e^{-\alpha_k^2} e^{2\pi i \beta_{k,j}} e^{-2x\alpha_k} p(x + \alpha_k) \right)$$

$$= e^{-x^2} \sum_{k=1}^{N} E_k(t) e^{-2x\alpha_k} p(x - \alpha_k)$$

Since $N > 1$, we must have either $\alpha_1 < 0$ or $\alpha_N > 0$. Suppose that $a_1 < 0$. Then since $a_1 < a_2 < \cdots < a_N$ and p is polynomials, $|e^{-2x\alpha_k} p(t - \alpha_1)|$ increases as $x \to \infty$ exponentially faster than $|e^{-2t\alpha_k} p(t + \alpha_k)|$ for $k = 2, \ldots, N$.

Moreover, each E_k is a trigonometric polynomial. In particular, E_1 is periodic almost everywhere and $E_2, \ldots, E_N$ are bounded. Hence, if E_1 is nontrivial, then we can find a sequence $\{x_n\}$ with $n \to \infty$ such that $|E_1(x_n) e^{-2x_n \alpha_1} p(x_n + a_n)|$ increases exponentially faster than $|E_k(x_n) e^{-2x_n \alpha_n} p(x_n + \alpha_n)|$ for any $k = 2, \ldots, N$. Therefore, $s(x) \neq 0$ for large enough N. Since $s(x)$ is continuous, we find that $\{e^{2\pi i \beta_k x} g(x + \alpha_k)\}_{k=1}^{N}$ is linearly independent in $L_2(R)$ for any value of N.

Similarly, it can be proved for $\alpha_N > 0$.

Proof (*Proof of Theorem 15.7*)

(a) Let $\{Tx\}_{x \in R}$ and $\{M_z\}_{z \in R}$ be translation and modulation groups. They satisfy the conditions for all $f \in L_2(R)$,

$$\lim_{x \to 0} ||Tx - f||_{L_2} = 0 = \lim_{z \to 0} ||M_z f - f||_{L_2}$$

This implies that we can choose ε small enough such that $||T_x g - g||_{L_2} \leq \delta$, $||M_x g - g||_{L_2}$ follows from $|x| \leq \varepsilon$, where A, B are frame bounds for $\mathscr{G}(g, \wedge)$ as a frame for its span and $0 < \delta < A^{1/2}/(2N^{1/2})$.

Let $|\alpha_k - \alpha_k'| < \varepsilon$ and $|\beta_k - \beta_k'| < \varepsilon$ for $k = 1, 2, \ldots, N$. Then for any scalars $c_1, c_2, \ldots, c_N$, we have

$$\left\| \sum_{k=1}^{N} c_k M_{\beta_k} T_{\alpha_k} g \right\|_{L_2} \leq \left\| \sum_{k=1}^{N} c_k M_{\beta_k} (T_{\alpha_k} - T_{\alpha'_k}) g \right\|_{L_2}$$

$$+ \left\| \sum_{k=1}^{N} c_k (M_{\beta_k} - M_{\beta'_k}) T_{\alpha'_k} g \right\|_{L_2} + \left\| \sum_{k=1}^{N} c_k M_{\beta'_k} T_{\alpha'_k} g \right\|_{L_2}$$

$$\leq \sum_{k=1}^{N} |c_k| \left\| T_{\alpha_k} g - T_{\alpha'_k} g \right\|_{L_2} + \sum_{k=1}^{N} |c_k| \left\| (M_{\beta_k} - M_{\beta'_k}) T_{\alpha'_k} g \right\|_{L_2}$$

$$+ \left\| \sum_{k=1}^{N} c_k M_{\beta'_k} T_{\alpha'_k} \right\|_{L_2}$$

$$= \sum_{k=1}^{N} |c_k| \left\| T_{\alpha_k - \alpha'_k} g - g \right\|_{L_2} + \sum_{k=1}^{N} |c_k| \, \| M_{\beta_k - \beta'_k} g - g |_{L_2}$$

$$+ \left\| \sum_{k=1}^{N} c_k M_{\beta'_k} T_{\alpha'_k} g \right\|_{L_2}$$

$$\leq 2\delta \sum_{k=1}^{N} |c_k| + \left\| \sum_{k=1}^{N} c_k M_{\beta'_k} T_{\alpha'_k} g \right\|_{L_2}$$

Following a result of Christensen 2003, Theorem 3.6, we get

$$A^{1/2} \left(\sum_{k=1}^{N} |c_k|^2 \right)^{1/2} \leq \left\| \sum_{k=1}^{N} c_k M_{\beta_k} T_{\alpha_k} g \right\|_{L_2}$$

Combining these inequalities, we get

$$(A^{1/2} - 2\delta N^{1/2}) \left(\sum_{k=1}^{N} |c_k|^2 \right)^{1/2} \leq \left\| \sum_{k=1}^{N} c_k M_{\beta'_k} T_{\alpha'_k} g \right\|_{L_2}$$

Since $A^{1/2} - 2\delta N^{1/2} > 0$, it follows that if $\sum_{k=1}^{N} c_k M_{\beta'_k} T_{\alpha'_k} g = 0$ a.e., then $c_1 = c_2 = \cdots = c_N = 0$.

This proves the desired result.

(b) Let $\wedge = \{(\alpha_k, \beta_k)\}^N$ and define the continuous, linear mapping $T : C^N \to L_2(R)$ by

$$T(c_1, c_2, \ldots, c_N) = \sum_{k=1}^{N} c_k \, \rho(\alpha_k, \beta_k),$$

$$where \quad \rho(\alpha_k, \beta_k) = e^{2\pi i \beta_k t} f(t + \alpha_k).$$

T is injective as $\mathscr{G}(g, \wedge)$ is linearly independent.

Therefore, T is continuously invertible on the range of T. In particular, there exists $A, B > 0$ such that

$$A \sum_{k=1}^{N} |c_k| \leq \left\| \sum_{k=1}^{N} c_k \rho(\alpha_k, \beta_k) \right\|$$

$$\leq B \sum_{k=1}^{N} |c_k| \text{ for each } (c_1, c_2, \ldots, c_N) \in C^N.$$

Hence, if $\|f - g\| < A$ and $(c_1, c_2, \ldots, c_N) \in C^N$, then

$$\left\| \sum_{k=1}^{N} c_k \rho(\alpha_k, \beta_k) g \right\| \geq \left\| \sum_{k=1}^{N} c_k \rho(\alpha_k, \beta_k) f \right\| - \left\| \sum_{k=1}^{N} c_k \rho(\alpha_k, \beta_k)(g - f) \right\|$$

$$\geq A \sum_{k=1}^{N} |c_k| - \sum_{k=1}^{N} |c_k| \, \|\rho(\alpha_k, \beta_k)(f - g)\|$$

$$= (A - \|f - g\|) \sum_{k=1}^{N} |c_k|$$

If $\sum_{k=1}^{N} c_k \rho(\alpha_k, \beta_k) g = 0$, then c_k's are zero which in turn implies $\mathscr{G}(h, \wedge)$ is linearly independent if $\|g - h\| < \varepsilon$.

15.3 HRT Conjecture for Wave Packets

Given $g \in L_2(R)$ and a subset $\wedge \subset R \times R^+$, the collection of the type

$$\{D_b(M_{\beta_k} T_{\alpha_k} g)(t)\} = \left\{ b^{1/2} e^{2\pi i \beta_k b t} g(bt - \alpha_k) \right\} \tag{15.8}$$

is called the *wave packet system*.

The wave packet systems have been studied by Cordoba and Feffermen, Hogan and Lakey, Kalisa and Torrésani, Siddiqi and Ahmad, Hernandez et al. and Labate, Weis, Wilson.

Recently Siddiqi et al. have examined the HRT conjecture for wave packet systems. Analogous results to the ones given in (b), (d) and (e) above hold for wave packet systems under appropriate conditions.

15.4   Applications

Gabor systems are applied in numerous engineering applications, many of them without obvious connection to the traditional field of time–frequency analysis for deterministic signals.

Any countable set of test functions $\{f_n\}$ in a Hilbert space conveys a linear mapping between function spaces and sequence spaces. In one direction, scalar products $\langle f, f_n \rangle$ are taken (analysis mapping), and in the other direction, the members $c = \{c_n\}$ from the sequence space are used as coefficient sequences in a series of the form $\sum_n c_n f_n$ (synthesis mapping). In our concrete context, the analysis mapping is given by the Gabor transform and the synthesis mapping is given by the Gabor expansion. In principle, there exists two basic setups for the use of Gabor systems which pervade most applications:

- The overall system acts on the sequence space $\ell_2(\mathscr{G})$ where $\mathscr{G}$ has appropriate structure, and in particular, we can choose $\mathscr{G} = [0, 1]$ or unit sphere by (i) Gabor synthesis, (ii) (desired or undesired and mostly linear but probably nonlinear) modification, (iii) Gabor analysis. This setup underlies, e.g., the so-called multi-carrier modulation schemes in digital communication, but also applies to system identification and radar tracking procedures.
- The overall system acts on the function space $L_2(\mathscr{G})$ by (i) Gabor analysis, (ii) (desired or undesired and mostly nonlinear but also linear) modification, (iii) Gabor synthesis. Typical tasks where one encounters this setup may include signal enhancement, denoising, or image compression.

Speech Signal Analysis

Speech signals are one of the classical applications of linear time–frequency representations. In fact, the analysis of speech signals was the driving force that led to the invention of (a nondigital filter bank realisation of) the spectrogram. The advent of the FFT and the STFT up to now to the standard tool of speech analysts.

Representation and Identification of Linear Systems

The theory of linear time-invariant (LTI) systems and in particular the symbolic calculus of transfer functions is a standard tool in all areas of mechanical and electrical engineering. Strict translation invariance is however almost always a pragmatical modeling assumption which establishes a more or less accurate approximation to the true physical system. Hence, it is a problem of longstanding interest to generalize the transfer function concept from LTI to linear time-varying (LTV) systems. Such a time-varying transfer function was suggested by Zadeh in 1950. It is formally equivalent to the Weyl-Heisenberg operator symbol of Kohn and Nirenberg. Pseudo-differential operators are the classical way to establish a symbol classification that keeps some of the conceptual power which the Fourier transform has for LTI systems. Recently, Gabor frames have turned out to be a useful tool for the analysis of pseudo-differential operators.

Digital Communication

Digital communication systems transmit sequences of binary data over a continuous time physical channel. An ideal physical channel is bandlimited without in-band distortions. Under this idealized assumption, digital communication systems can be implemented by selecting a Gabor-structured orthonormal system, transmitting a linear combination of the elementary signals, weighted by the binary coefficients (Gabor synthesis) and the receiver recovers the coefficients by computing inner products with the known basis functions (matched filter receiver = Gabor analysis). However, in wireless communication systems which is one of the challenging research areas, the physical channel is subject to serve linear distortions and hundreds of users communicate over the same frequency band at the same time. Traditional orthogonal frequency division multiplex (OFDM) systems can be interpreted as orthonormal Gabor systems with critical sampling $ab = 1$ and come therefore with the well-known bad time–frequency localization properties of the building blocks.

Since completeness is not a concern here, recent works suggest the use of a coarser grid $ab > 1$, together with good TF-localized atoms to obtain more robustness.

Image Representation and Biological Vision

Gabor functions were successfully applied to model the response of simple calls in the visual cortex. In our notation, each pair of adjacent cells in the visual cortex represents the real and imaginary part of the coefficient $c_{m,n}$ corresponding to $g_{m,n}$. Clearly, the Gabor model cannot capture the variety and complexity of the visual system, but it seems to be a key in further understanding of biological vision.

Among the people who paved the wave for the use of Gabor analysis in pattern recognition and computer vision one certainly has to mention Zeevi, M. Porat, and their coworkers and Daugmn. Motivated by biological findings, Daugman and Zeevi and Porat proposed the use of Gabor functions for image processing applications, such as image analysis and image compression. Since techniques from signal processing are of increasing importance in medical diagnostics, we mentioned a few applications of Gabor analysis in medical signal processing. The Gabor transform has been used for the analysis of brain function, such as for detection of epileptic seizures in EEG signals, study of sleep spindles. The role of the Heisenberg group in magnetic resonance imaging has been recently analyzed.

Appendix

Key words: Set theoretic concepts, Topological concepts, Elements of Metric spaces, Notation and definition of concrete spaces, Lebesgue integration, Integral equations, Surface integrals, Vector spaces, and Fourier analysis are introduced in appendices. Results related to these fields provide foundation for Chaps. 1–15.

A.1 Set Theoretic Concepts

Definition A.1 A. Let X and Y be two nonempty sets. A function or mapping or transformation or correspondence or operator, which we denote by f, is a rule assigning to each element x in X a single fully determined element y in Y. The y which corresponds in this way to a given x is usually written as $f(x)$ or fx, and is called the *image* of x under the rule f, or the value of f at the element x. The set X is called the *domain* of f, the set of all $f(x) \forall x \in X$ is called the range of f, and it is a subset of Y. The symbol $f : X \to Y$ will mean that f is a function whose domain is X, and whose range is contained in Y.

B. Let $f : X \to Y$ and $g : Y \to Z$. Then gf, which is called the product of these mappings, is defined as $(gf)(x) = g(f(x))$ and $gf : X \to Z$.

C. Let $f : X \to Y$ and $A \subseteq X$. Then $f(A) = \{f(x)/x \in A\}$ is called the image of A under f.

D. Let $f : X \to Y$. f is called into if the range of f is not equal to Y. f is called onto if the range of f equals Y. f is called one-one if two different elements always have different images.

E. Let $f : X \to Y$ be both one-one and onto. Then we define its inverse mapping $f^{-1} : Y \to X$ as follows: For each y in Y we find a unique element x in X such that $f(x) = y$; we then define x to be $f^{-1}(y)$.

F. Let $f : XY$ and B be a subset of Y. Then $f^{-1}(B) = \{x/f(x) \in B\}$. $f^{-1}(B)$ is called the inverse image of B under f.

G. A function f is called an extension of a function g if $f(x) = g(x) \forall x$ belonging to K, the domain of g and $K \subset dom f$.

© Springer Nature Singapore Pte Ltd. 2018

A. H. Siddiqi, *Functional Analysis and Applications*, Industrial and Applied Mathematics, https://doi.org/10.1007/978-981-10-3725-2

Note A.1 A. The term function in (1) is generally used in real and complex analysis. If the range of a function consists of real numbers, it is called a real function; and if the range consists of complex numbers, it is called a complex function.
B. The term operator of (1) is generally used if X and Y are normed spaces.
C. The term transformation is generally used in linear algebra.
D. The term mapping is preferred in cases when Y is not necessarily a set of numbers.
E. The term operator is replaced by functional if X is a normed space and Y is a set of real or complex numbers or some other field.

Theorem A.1 *If $f : X \to Y$, $A_1, A_2, \ldots, A_n$ are subsets of X and $B_1, B_2, \ldots, B_n$ are subsets of Y, then*

A. $f(\phi) = \phi$
B. $f(X) \subseteq Y$
C. $A_1 \subseteq A_2 \Rightarrow f(A_1) \subseteq f(A_2)$
D. $f\left(\bigcup_i A_i\right) = \bigcup_i f(A_i)$
E. $f\left(\bigcap_i A_i\right) = \bigcap_i f(A_i)$
F. $f^{-1}(\phi) = \phi$
G. $f^{-1}(Y) \subseteq X$
H. $B_1 \subseteq B_2 \Rightarrow f^{-1}(B_1) \subseteq f^{-1}(B_2)$
I. $f^{-1}\left(\bigcup_i B_i\right) = \bigcup_i f^{-1}(B_i)$
J. $f^{-1}\left(\bigcap_i B_i\right) = \bigcap_i f^{-1}(B_i)$
K. $f^{-1}(B') = (f^{-1}(B))'$, B' and $(f^{-1}(B))'$ denote the complements of B and $f^{-1}(B)$, respectively.

Definition A.2 Let X be a nonempty set, then

1. A partition of X is a disjoint class $\{X_i\}$ of nonempty subsets of X whose union is the full set X itself. The $X_i's$ are called *partition sets*.
2. A relation $\sim$ (other symbols used in the literature for relations are $\leq, \subset, \subseteq, \equiv$ and γ) on (or in) the set X is a nonempty A subset of $X \times X$. We say that x is related by $\sim$ to y and we write $x \sim y$ if $(x, y) \in A$.
3. A relation $\sim$ on X is called an *equivalence relation* if it has the following properties:

    (a) $x \simeq x$ for every $x \in X$ (reflexive property),
    (b) if $x \simeq y$, then $y \simeq x$ (symmetric property), and
    (c) if $x \simeq y$ and $y \simeq z$, then $x \simeq z$ (transitive property).

4. A partial order relation on X is a relation which is symbolized by $\leq$ and satisfies the following properties:

    (a) $x \leq x$ for every x (reflexivity),

(b) if $x \le y$ and $y \le x$ then $x = y$ (anti-symmetry), and

(c) if $x \le y$ and $y \le z$ then $x \le z$ (transitivity). X is called a *partially ordered set*. Two elements of the partially ordered set X are called *comparable* if either $x \le y$ or $y \le x$.

Note A.2 The word "partial" in (4) emphasizes that there may be pairs of elements in X which are not comparable.

Definition A.3 A. A partially ordered set in which any two elements are comparable is called a *totally ordered set* or a *linearly ordered set* or a *chain* or a *completely ordered set*.

B. An element x in a partially ordered set X is called *maximal* if $y \ge x$ implies that $y = x$; i.e., if no element other than x itself is greater than or equal to x.

Definition A.4 Let A be a nonempty subset of a partially ordered set X. An element x in X is called a *lower bound* of A if $x \le a$ for each $a \in A$. A lower bound of A is called the *greatest lower bound (glb)* of A if it is greater than or equal to every lower bound of A. An element y in X is called an upper bound of A if $a \le y$ for every $a \in A$. An upper bound of A is called the *least upper bound (lub)* of A if it is less than or equal to every upper bound of A. [This means that if M is the set of all upper bounds of A, then an element m of M is lub of A provided $m \le b$ for all $b \in M$].

Remark A.1 A. The set R of real numbers is a partially ordered set, where $x \le y$ means that $y - x$ is nonnegative. It is well known that if a nonempty subset A of R has a lower bound, then it has a greatest lower bound which is usually called its *infimum* and written as inf A. Similarly, if a nonempty subset A of R has an upper bound, then it has a least upper bound and is usually called the *supremum* and written sup A.

B. If A is a finite subset of R, then the supremum is often called *maximum* and infimum is often called *minimum*.

C. If A is a nonempty subset of R which has no upper bound and hence no least upper bound in R, then we express this by writing $sup A = +\infty$ and if A is an empty subset of R, we write sup $A = -\infty$. Similarly, if A has no lower bound and hence no greatest lower bound, we write inf $A = -\infty$, and A is empty, then we say that inf $A = +\infty$. Let A be any subset of the extended real number system (the real number system with the symbols $+\infty$ and $-\infty$ adjoined). Then sup A and inf A exist. Thus, if we consider the extended real number system, no restriction on A is required for the existence of its supremum and infimum.

D. If A and B are two subsets of real numbers, then

$$\sup(A + B) \le \sup A + \sup B$$

E. Let C be the set of all real functions defined on a nonempty set X, and let $f \le g$ mean that $f(x) \le g(x)$ for every $x \in X$. Then C is a partially ordered set and $\sup_x |f(x) + g(x)| \le \sup_x |f(x)| + \sup_x |g(x)|$.

Theorem A.2 (Zorn's Lemma) *If X is a partially ordered set in which every chain or totally ordered set has an upper bound, then X contains at least one maximal element.*

Note A.3 Zorn's lemma is equivalent to the axiom of choice and therefore it can itself be regarded as an axiom. It is an exceedingly powerful tool for the proofs of several important results of functional analysis.

Definition A.5 A partially ordered set L in which each pair of elements has a greatest lower bound and a least upper bound is called a lattice. If $x, y \in L$ and L is a lattice, then $x \bigvee y$ and $x \bigwedge y$ represent the least upper bound and the greatest lower bound of x and y, respectively.

The set of all subsets of a set X is often denoted by 2^X.

A.2 Topological Concepts

Definition A.6 A. A family F of subsets of a set X is called a topology in X if the following conditions are satisfied:

A. Null set ϕ and X belong to $(\mathfrak{F})$.

B. The union of any collection of sets in $(\mathfrak{F})$ is again a set in $(\mathfrak{F})$.

C. The intersection of a finite collection of sets of $(\mathfrak{F})$ is a set in $(\mathfrak{F})$. $(X, (\mathfrak{F}))$ is called a *topological space*. However, for the sake of simplicity we write only X for a topological space, bearing in mind that a topology $(\mathfrak{F})$ is defined on X. The sets in $(\mathfrak{F})$ are called the open sets of $(X, (\mathfrak{F}))$.

D. A point p is called a *limit point* or a point of accumulation of a subset B of X if every neighborhood of p contains at least one point of B other than p.

E. The *interior* of a subset of X is the union of its all open subsets. A point of interior of a set is called its *interior point*.

F. A subset of a topological space $(X, (\mathfrak{F}))$ is closed if its complement is open.

G. The intersection of all closed sets containing a subset A of a topological space $(X, (\mathfrak{F}))$ is called the *closure* of A and is usually denoted by $\bar{A}$. The set of points in $\bar{A}$, which are not interior points of A, is called the boundary of A.

H. A collection $\mathfrak{P}$ of open subsets of a topological space $(X, (\mathfrak{F}))$ is called a base for the topology $\mathfrak{F}$ if every element of $(\mathfrak{F})$ can be written as a union of elements of $\mathfrak{P}$.

I. A collection $\mathfrak{P}x$ of open subsets of $(X, (\mathfrak{F}))$ is called a basis at the point $x \in X$ if, for any open set O containing x, there exists a set A in $\mathfrak{P}_x$ such that $x \in A \subset O$.

J. A nonempty collection $\mathfrak{P}$ of open subsets of a topological space $(X, (\mathfrak{F}))$ is called a *subbase* if the collection of all finite intersections of elements of $\mathfrak{P}$ is a base for $(\mathfrak{F})$.

K. If $(X, (F))$ is a topological space and $Y \subseteq X$, then the topology $(\mathfrak{F})_Y = \{A/A = B \cap Y, B \in (\mathfrak{F})\}$ is called the *natural relative topology* of Y generated by $(\mathfrak{F})$.

L. Let F_1, F_2 be two topologies on a set X. $(\mathfrak{F})_1$ is said to be weaker than $(\mathfrak{F})_2$ (or $(\mathfrak{F})_2$ is said to be stronger than F_1) if $F_1 \subset F_2$; i.e., if every open set of $(\mathfrak{F})_1$ is an open set in $(\mathfrak{F})_2$. Two topologies F_1 and $(\mathfrak{F})_2$ are said to be equivalent if $(\mathfrak{F})_1 = (\mathfrak{F})_2$; i.e., if they have the same open sets. The stronger topology contains more open sets.

M. A sequence $\{a_n\}$ in a topological space X is said to be convergent to $a \in X$ if every neighborhood of a contains all but a finite number of the points a_n.

Definition A.7 A. Suppose $(X, (\mathfrak{F})_1)$ and $(Y, (\mathfrak{F})_2)$ are two topological spaces, and $f : X \to Y$ is a function. f is called continuous if $f^{-1}(A) \in (\mathfrak{F})_1$, for every A in $(\mathfrak{F})_1$.

B. If f is a continuous one-to-one map of X onto Y such that the inverse function f^{-1} is also continuous, then f is called a *homeomorphism* or *topologically iso-morphic*. X and Y are called homeomorphic if there exists a homeomorphism between X and Y.

Theorem A.3 *Let $\mathfrak{P}$ be a collection of subsets of an arbitrary set X, and $(\mathfrak{F})$ the collection of arbitrary unions of elements of $\mathfrak{P}$. Then $(\mathfrak{F})$ is a topology for X if and only if*

A. *For every pair A, $B \in P$ and $x \in A \cap B$, there is a $C \in P$ such that $x \in C \subseteq A \cap B$.*

B. $X = \bigcup_{B \in \mathfrak{P}}$

Ways of Generating a Topology

Let X be any set. Then from the definition, it is clear that there are always two topologies on X: one consisting of X and the other consisting of all subsets of X; the first is known as an indiscrete topology and the second as a *discrete topology*. Methods of generating topologies between these two extremes are described as follows:

1. Let S be a collection of subsets of an arbitrary set X. The weakest topology containing S is the topology formed by taking all unions of finite intersections of elements of S, together with ϕ and X.

2. Let X be any set and let Y a topological space with the topology $F(Y)$ and $\{f_\alpha / \alpha \in A\}$ a collection of functions, each defined on X with range in Y. The weak topology generated by $\{f_\alpha / \alpha \in A\}$ is the weakest topology on X under which each of the function f_α is continuous. This requires that $f_\alpha^{-1}(O)$ be an open subset of X for each $\alpha \in A$, $O \in (\mathfrak{F})(Y)$. Let $S = \{f_\alpha^{-1}(O) / \alpha \in A, O \in (\mathfrak{F})(Y)\}$. Then as in (1), we can generate a topology. This topology with S as a subbase is then the weak topology generated by $\{f_\alpha\}$.

Definition A.8 (a) A topological space X is called a *Hausdorff space* if, for distinct points $x, y \in X$, there are neighborhoods N_x of x and N_y of y such that $N_x \cap N_y = \phi$.

(b) A topological space X is called *regular* if for each closed set A, and each $x \notin A$, there exist disjoint neighborhoods of x and A.

(c) A topological space is called *normal* if, for each pair of disjoint closed sets A, B, there exist disjoint neighborhoods U, V of A and B, respectively.

Note A.4 Here, the topological space satisfies the condition that sets consisting of single points are closed. Sets A and B are called disjoint if $A \cap B = \phi$.

Definition A.9 1. A covering of a set A in a topological space X is a collection of open sets whose union contains A. The space X is said to be compact if every covering of X contains a finite subset which is also a covering of X.
2. A topological space X is said to be *locally compact* if every point has a neighborhood whose closure is compact.
3. A topological space X is called *connected* if it cannot be represented as a union of two disjoint nonempty open sets. X is called disconnected if it is not connected.
4. A family of sets is said to have a *finite intersection property* if every finite subfamily has a nonvoid intersection.

Definition A.10 Let $X_1, X_2, \ldots, X_n$ be topological spaces, with $\mathfrak{P}_i$ a base for the topology of X_i. Their *topological* product $X_1 \times X_2 \times \cdots \times X_n$ is defined as the set of all n-tuples $(x_1, x_2, \ldots, x_n)$ with $x_i \in X_i$, taking as a base for the topology all products $U_1 \times U_2 \times \cdots \times U_n$ of U_i in $(\mathfrak{P})_i$.

Theorem A.4 1. *(Tychonoff theorem) The topological product of compact spaces is compact.*
2. *Every compact subspace of a Hausdorff topological space is closed.*
3. *A topological space is compact if and only if every family of closed sets with the finite intersection property has a nonvoid intersection.*
4. *If f is a continuous function defined on a topological space X into a topological space Y and A is a compact subset of X, then $f(A)$ is compact and if $Y = R$, then f attains its supremum and infimum on A.*
5. *Every closed subset of a compact topological space is compact.*
6. *A continuous one-to-one function from a compact topological space on to a Hausdorff topological space is a homeomorphism.*

A.3 Elements of Metric Spaces

Definition A.11 1. Let X be a set, and d a real function on $X \times X$, with the properties

1. $d(x, y) \geq 0 \; \forall x, y \in X$, $d(x, y) = 0$ if and only if $x = y$.
2. $d(x, y) = d(y, x) \; \forall \, x, y \in X$ (symmetry property).
3. $d(x, y) \leq d(x, z) + d(z, y)$ (the triangle inequality).

4. Then d is called a *metric*, or a *metric function* on or over or in X. (X, d) is called a *metric space*.

5. $S_r(x) = \{y/d(x, y) < r\}$ is called an open sphere with radius r and center x in X.

6. Let A and B be subsets of a metric space (X, d). Then

    1. $d(A, B) = \sup_{a \in A} \inf_{b \in B} d(a, b)$.
    2. The diameter of A, which we denote by $\delta(A)$, is the number, $\sup_{a, b \in A} d(a, b)$.
    3. $S(A, \varepsilon) = \{x/d(A, x) < \varepsilon\}$ is called the ε-neighborhood of A.
    4. A is bounded if $\delta(A) < \infty$.

5. A sequence in a metric space is a function whose domain is the set of positive integers and range is a subset of the metric space.

6. A sequence $\{A_n\}$ of subsets of a metric space is called a decreasing sequence if $A_1 \supseteq A_2 \supseteq A_3 \supseteq, \ldots$.

Theorem A.5 *1. The collection of open spheres in a metric space (X, d) forms a base for the metric topology.*

2. Every metric space is normal and hence a Hausdorff topological space.

Definition A.12 1. A sequence $\{a_n\}$ in a metric space (X, d) is said to be convergent to an element $a \in X$ if $lim_{n \to \infty} d(a_n, a) = 0$. $\{a_n\}$ is called a *Cauchy sequence* if $\lim_{m \to \infty, n \to \infty} d(x_m, x_n) = 0$.

2. A metric space is called complete if every Cauchy sequence in it is convergent to an element of this space.

3. Let (X, d_1) and (Y, d_2) be metric spaces, and f a mapping of X into Y. f is called *continuous* at a point x_0 in X if either of the following equivalent conditions is satisfied:

    1. For every $\varepsilon > 0$, there exists a $\delta > 0$ such that $d_1(x, x0) < \delta$ implies $d_2(f(x), f(x0)) < \varepsilon$.
    2. For each open sphere $S_\varepsilon(f(x_0))$ centered on f (x0), there exists an open sphere $S_\delta(x_0)$ centered on x_0 such that $f(S\delta(x_0)) \subseteq S_\varepsilon(f(x_0))$. f is called *continuous* if it is continuous at each point of its domain.

4. In (3), it can be seen that the choice of δ depends not only on ε but also on the point x_0. The concept of continuity, in which for each $\varepsilon > 0, a\delta > 0$ can be found which works uniformly over the entire metric space X, is called *uniform continuity*

Theorem A.6 *1. If a convergent sequence in a metric space has infinitely many distinct points, then its limit is a limit point of the set of points of the sequence. A convergent sequence in a metric space is bounded and its limit is unique.*

2. Let X and Y be metric spaces and f a mapping of X into Y. Then f is continuous if and only if $\{x_n\}$ in X converges to $x \in X$ implies that $f(x_n)$ in Y converges to $f(x) \in Y$.

3. A point p is in the closure of a set A in a metric space if and only if there is a sequence $\{p_n\}$ of points of A converging to p.

4. Let (X, d_1) and (Y, d_2) be metric spaces and f a mapping of X into Y. Then f is continuous if and only if $f^{-1}(O)$ is open in X whenever O is open in Y.

5. Principle of extension by continuity: Let X be a metric space and Y a complete metric space. If $f : A \to Y$ is uniformly continuous on a dense subset A of X, then f has unique extension g which is a uniformly continuous mapping of X into Y.

6. Every continuous mapping defined on a compact metric space X into a metric space Y is uniformly continuous.

7. Cantor's intersection theorem: Let X be a complete metric space, and $\{F_n\}$ a decreasing sequence of nonempty closed subsets of X such that $\delta(F_n) \to 0$. Then $F = \bigcap\limits_{n=1}^{\infty} F_n$ contains exactly one point.

8. Baire's category theorem: Every complete metric space is of the second category (if a complete metric space is the union of a sequence of its subsets, the closure of at least one set in the sequence must have a nonempty interior).

Definition A.13 1. Let B be a subset of (X, d). A subset A of X is called an ε-net for the set B if for each $x \in B$, there exists a $y \in A$ such that $d(x, y) < \varepsilon$.

2. A subset B of a metric space (X, d) is called totally bounded if for any $\varepsilon > 0$, there exists a finite ε-net for B.

3. A subset A of a metric space X is called compact if every sequence in A has a convergent subsequence having its limit in A.

Example A.1 1. In the metric space R^2, the subset $A = \{(m, n)/m, n = 0, \pm1, \pm2, \ldots\}$ is an ε-net for $\varepsilon > \sqrt{2}/2$.

2. In ℓ_2, $A = \{x \in \ell_2/d(x, 0) = 1\}$; i.e., the points of the surface of the unit sphere in ℓ_2 are bounded but not totally bounded.

Note A.5 Every totally bounded set is bounded.

Theorem A.7 *1. Every compact subset of a metric space is totally bounded and closed.*

2. *A metric space is compact if and only if it is totally bounded and complete.*

3. *Every compact metric space is separable.*

Definition A.14 1. A metric space in which the triangle inequality is replaced by a stronger inequality $d(x, y) \leq \max\{d(x, z), d(z, y)\}$ is called an *ultra-metric* space or a *nonarchimedean metric space*.

2. Let A be any nonempty set of real or complex functions defined on an arbitrary nonempty set X. Then the functions in A are said to be uniformly bounded if there exists a k such that $|f(x)| \leq k$ for every x in X and every f in A.

3. Let X be a compact metric space with metric d and A a set of continuous real or complex functions defined on X. Then A is said to be equicontinuous if, for each $\varepsilon > 0$, a $\delta > 0$ can be found such that $|f(x) - f(x')| < \varepsilon$ whenever $d(x, x') < \delta \, \forall f \in A$.

4. Let (X, d_1) and (Y, d_2) be metric spaces. Then a mapping f of X into Y is called an isometry if $d_1(x, y) = d_2(f(x), f(y))$ for all $x, y \in X$. If, in addition, f is onto, then X and Y are said to be isometric.

5. A subset A of a metric space X is called *everywhere dense* or *dense* in it if $\bar{A} = X$; i.e., every point of X is either a point or a limit point of A. This means that, given any point x of X, there exists a sequence of points in A that converges to x.

6. Let (X, d) be an arbitrary metric space. A complete metric space (Y, d_1) is called a completion of (X, d) if

   1. (X, d) is isometric to a subspace (W, d_1) of (Y, d_1), and
   2. the closure of W, $\bar{W}$, is all of Y; that is, $\bar{W} = Y$ (W is everywhere dense in Y).

3. Let X be a metric space. A real-valued function $f : X \to R$ is said to have compact support if $f(x) = 0$ outside a compact subset of X. The closure of the set $\{x / f(x) \neq 0\}$ is called the support of f. It is denoted by supp f.

Theorem A.8 (Ascoli's Theorem) *If X is a compact metric space and $C(X, R)$ denotes the set of all real-valued continuous functions defined on X, then a closed subspace of $C(X, R)$ is compact if and only if it is uniformly bounded and equicontinuous.*

Note A.6 The theorem is also valid if R is replaced by the set of complex numbers C.

Theorem A.9 (Completion Theorem) *Every metric space has a completion and all its completions are isometric.*

Example A.2 1. The sets R, C, R^n, c, m, ℓ_p, L_p, $BV[a, b]$, $AC[a, b]$, c_0, $Lip\alpha$, which are defined in Appendix D, are metric spaces with respect to the metric induced by the norm (see Sect. 1.2) defined on the respective spaces.

2. Let s be the set of all infinite sequences of real numbers. Then s is a metric space with respect to the metric d, defined in the following manner:

$$d(x, y) = \sum_{i=1}^{\infty} \frac{1}{2^i} \frac{|\alpha_i - \beta_i|}{1 + |\alpha_i - \beta_i|}$$

where $x = (\alpha_1, \alpha_2, \ldots, \alpha_n, \ldots)$ and $(\beta_1, \beta_2, \ldots, \beta_n, \ldots)$ belong to s.

3. Let K be the set of all nonempty compact subsets A, B of R^n and $d(x, A) = \inf\{d(x, a) / a \in A\}$. Then K is a metric space with metric d_1, where

$$d_1(A, B) = \frac{1}{2}[\sup_{a \in A} d(a, B) + \sup_{b \in B} d(b, A)]$$

A *metrizable space* is a topological space X with the property that there exists at least one metric on the set X whose class of generated open sets is precisely the given topology.

A.4   Notations and Definitions of Concrete Spaces

1. R, Z, N, Q, and C will denote, respectively, the set of real numbers, integers, positive integers, rational numbers, and complex numbers unless indicated otherwise.
2. The coordinate plane is defined to be the set of all ordered pairs (x, y) of real numbers. The notation for the coordinate plane is $R \times R$ or R^2 which reflects the idea that it is the result of multiplying together two replicas of the real line R.
3. If z is a complex number, and if it has the standard form $x + iy$ (x and y are real numbers), then we can identify z with the ordered pair (x, y), and thus with an element of R^2. When the coordinate plane R2 is thought to be consisting of complex numbers and is enriched by the algebraic structure indicated below, it is called the *complex plane* and denoted by C. Let $z_1 = a + ib$ and $z_2 = c + id$ belong to C. Then

   1. $z_1 = z_2$ if and only if $a = c$ (real parts of z_1 and z_2 are equal) and $b = d$ (imaginary parts of z_1 and z_2 are equal).
   2. $z_1 \pm z_2 = (a \pm c) + i(b \pm d)$.
   3. $z_1 z_2 = (ac - bd) + i(bc + ad)$.
   4. $\frac{z_1}{z_2} = \frac{ac + bd}{c^2 + d^2} + i \frac{bc - ad}{c^2 + d^2}$
   5. The conjugate of a complex number $z = a + ib$ is denoted by $\bar{z}$ or $z^\star$ and is equal to $a - ib$. The real part of z is given by $Rez = \frac{z + \bar{z}}{2}$. The imaginary part of $z = \frac{z - \bar{z}}{2}$

   $$Rez = \{z / Imaginary\ part\ of\ z = 0\} = \{z / \bar{z} = z\}$$

   The properties of the conjugate of a complex number are listed below:

   $$\overline{z_1 + z_2} = \overline{z_1} + \overline{z_2}$$
   $$\overline{z_1 z_2} = \overline{z_1} + \overline{z_2}$$
   $$|z| = |\bar{z}| \qquad\qquad (A.1)$$

   and

   $$\bar{z} = 0\ iff\ z = 0$$

   6. The absolute value of z or mod of z is denoted by $|z| = (a^2 + b^2)^{1/2}$, where $z = a + ib$. C is called the space of complex numbers.

4. Let n be a fixed natural number. Then R_n will denote the set of all ordered n-tuples $x = (x_1, x_2, \ldots, x_n)$ of real numbers.

Note A.7 Note For a detailed discussion of spaces R^2, C, and R_n, one may see Simmons [10, pp. 22, 23, 52–54 and 85–87].

5. A sequence of real numbers is a real-valued function whose domain is the set of natural numbers. A sequence $\{x_n\}$ of real numbers is called bounded if there exists a real number $k > 0$ such that $|x_n| \leq k \; \forall n$. The set of all bounded sequences of real numbers is usually denoted by m, and it is called the *space of bounded real sequences*. Every bounded sequence of real numbers always contains a convergent subsequence.

6. A sequence $\{x_n\}$ of real numbers is said to be convergent to $x \in R$ if for every $\varepsilon > 0$ there exists a positive integer N such that $|x_n - x| < \varepsilon \; \forall n > N$; i.e., it is convergent if $|x_n - x| \to 0$ as $n \to \infty$. The set of all convergent sequences in R is denoted by c and called the space of real convergent sequences. c_0 denotes the subset of c for which $x = 0$. c_∞ denotes the space of all real sequences $x = \{x_n\}$ for which the series $\sum_{n=1}^{\infty} x_n < \infty$ i.e., the series $\sum_{1}^{\infty} x_n$ is convergent.

Let x_n be a sequence. A real number M is called a superior bound of $\{x_n\}$ if $x_n \leq M \; \forall n$. A real number L is called an inferior bound of $\{x_n\}$ if $L \leq x_n \; \forall n$. The limit superior $\{x_n\}$ which we denote by $\overline{\lim}_{n \to \infty} x_n$ or $\limsup_{n \to \infty} x_n$ is defined as the greatest lower bound of the set of superior bounds of $\{x_n\}$. The limit inferior of $\{x_n\}$ which we denote by $\underline{\lim}_{n \to \infty} x_n$ or $\liminf_{n \to \infty} x_n$ is defined as the least upper bound of the set of inferior bounds of $\{x_n\}$. If x_n is convergent, then $\lim_{n \to \infty} x_n = \liminf_{n \to \infty} x_n = \limsup_{n \to \infty} x_n$.

The unit impulse signal δ_n is defined by

$$\delta_n = 1 \; if \; n = 0 = 0 \; if \; n \neq 0, \; \delta_{ij} = 1 \; if \; i \neq j \tag{A.2}$$

This is known as the Kronecker delta

Note A.8 1. For a bounded sequence, $\underline{\lim} x_n$ and $\overline{\lim} x_n$ always exist.
2. $\overline{\lim}_{n \to \infty} x_n = \sup\{\lim_{n_k \to \infty} x_{n_k} / \{x_{n_k}\}$ subsequence of $\{x_n\}$ and $\underline{\lim}_{n \to \infty} x_n = \inf\{\lim_{n_k \to \infty} x_{n_k} / \{x_{n_k}\}$ subsequence of $\{x_n\}$.

3. The set of all sequences $x = \{x_1, x_2, \ldots, x_n, \ldots\}$ of real numbers such that $\sum_{n=1}^{\infty} |x_n|^p < \infty$ for $1 \leq p < \infty$ is denoted by ℓ_p. Sometimes ℓ_2, in the case $p = 2$, is called the *infinite-dimensional Euclidean space*.

4. For $a, b \in R$ and $a < b$, $[a, b] = \{x \in R / a \leq x \leq b\}$ $(a, b) = \{x \in R / a < x < b\}$ and $(a, b] = \{x \in R / a < x \leq b\}$ are called, respectively, the closed, open, semiclosed, and semiopen-interval of R. R is a metric space with the metric $d(x, y) = |x - y|$, and therefore, a closed and bounded subset of R can be defined. As a special case, a real-valued function $f(x)$ on $[a, b]$ is called bounded if there exists a real number k such that $|f(x)| \leq k \; \forall x \in [a, b]$. It is said to be continuous at $x_0 \in [a, b]$ if, for $\varepsilon > 0$, there exists $\delta < 0$ such that $|f(x) - f(x_0)| < \varepsilon$ whenever $|x - x_0| < \delta$. It is called uniformly continuous on $[a, b]$ if the choice of does not depend on the point x0 in the definition of continuity. We can define

a continuous function in a similar fashion on an arbitrary subset of R. $C[a, b]$ denotes the set of all continuous real functions defined on $[a, b]$. It is called the space of continuous functions on $[a, b]$. Similarly, $C(T)$, where $T \subseteq R$, is called the space of continuous functions on T. If X is an arbitrary topological space, then $C(X)$ is called the space of continuous real functions on the topological space X.

5. $p(x) = a_0 + a_1 x + \cdots + a_n x_n$ for $x \in [a, b]$, where $a_0, a_1, \ldots, a_n$ are real numbers, is called a polynomial of degree n. It is a continuous real function. $P[a, b]$ denotes the set of all polynomials over $[a, b]$. $P_n[a, b]$ denotes the set of all polynomials of degree less than or equal to n over $[a, b]$. $C_n[0, \pi]$ denotes the set of all functions on $[0, \pi]$ of the form

$$f(x) = b_1 \cos x + b_2 \cos 2x + \cdots + b_n \cos nx$$

6. Let $f(x)$ be a real-valued function defined on $[a, b]$. The limits

$$f'_-(x) = \lim_{y \to x, y < x} \frac{f(y) - f(x)}{y - x}$$

and

$$f'_+(x) = \lim_{y \to x, y > x} \frac{f(y) - f(x)}{y - x}$$

are called left and right derivatives, respectively. The function f is said to be *differentiable* if the right and left derivatives at x exist and their values are equal. This value is called the first derivative of f at x and is denoted by f' or f^1 or $\dot{f}$ or $\frac{df}{dx}$. If the first derivative exists at every point of $[a, b]$, then it is said to be differentiable over $[a, b]$. If the first derivative of f is also differentiable, then f is called twice differentiable. In general, the nth derivative of the function f is the derivative of the $(n - 1)$th derivative. f is called n times differentiable over $[a, b]$ if nth derivative exists at every point of $[a, b]$. $C^{(n)}[a, b]$ denotes the set of all functions which have continuous derivatives up to and including the n-th order over $[a, b]$. [The derivatives of all orders on $[a, b]$, are called *infinitely differentiable*.] $C^\infty[a, b]$ denotes the class of all infinitely differentiable functions over [a, b].

7. Let $f(x)$ be a real-valued function defined on $[a, b]$. Let $P : a = x_0 \leq x_1 \leq x_2 \leq \cdots \leq x_n = b$ be a partition of $[a, b]$, $1()sup$ variation of $f(x)$ over $[a, b]$. $f(x)$ is called a function of *bounded variation* on $[a, b]$ if $V_a^b(x) < \infty$. $BV[a, b]$ denotes the space of all functions of bounded variation over $[a, b]$.

8. A real function $f(x)$ defined on $[a, b]$ is called absolutely continuous on [a, b] if, for $\varepsilon > 0$, there exists a $\delta > 0$ such that for any collection $\{(a_i, b_i)\}_1^n$ of disjoint open subintervals of [a, b], $\sum_{-\infty}$ holds whenever $\sum (b_i - a_i) < \delta$. $AC[a, b]$ denotes the class of all absolutely continuous functions on $[a, b]$.

9. A real function $f(x)$ defined on $[a, b]$ is said to satisfy a *Hölder condition of exponent* α over [a, b] or to be Hölder continuous or Lipschitz class if

$$\sup_{x,y \in [a,b]} \frac{|f(x) - f(y)|}{|x - y|^{\alpha}} < \infty$$

The class of Hölder continuous functions or Lipschitz class on [a, b] is denoted by $C_{\alpha}[a, b]$ or $Lip_{\alpha}[a, b]$.

10. We write $f(x) = O(g(x))$ if there exists a constant $K > 0$ such that $\frac{|f(x)|}{|g(x)|} \leq K f(x) = O(g(x))$ if $\frac{f(x)}{g(x)} \to 0$

These relationships are valid when $x \to \infty$, for $x \to -\infty$ or $x \to x_0$, where x_0 is some fixed number. If $b_n > 0$, $n = 0, 1, 2, \ldots$ and $\frac{a_n}{b_n} \to 0$ as $n \to \infty$, then we write $a_n = o(b_n)$. If $\frac{a_n}{b_n}$ is bounded, then we write $a_n = O(b_n)$.

Remark A.2 1. Every function of the Lipschitzian class over $[a, b]$ belongs to $AC[a, b]$.
2. $AC[a, b] \subset C[a, b]$.
3. $AC[a, b] \subset BV[a, b]$.
4. All continuous differentiable functions over $[a, b]$ are absolutely continuously over $[a, b]$.

Theorem A.10 *1. (Generalized Heine–Borel Theorem) Every closed and bounded subset of R^n is compact.*
2. *(Bolzano–Weierstrass Theorem) Every bounded sequence of real numbers has at least one limit point.*

Theorem A.11 *Let $f(x, t)$ be continuous and have a continuous derivative $\frac{\partial f}{\partial t}$ in a domain of the xt-plane which includes the rectangle $a \leq x \leq b$, $t_1 \leq t \leq t_2$. In addition, let $\alpha(t)$ and $\beta(t)$ be defined and have continuous derivatives for $t_1 < t < t_2$. Then for $t_1 < t < t_2$*

$$\frac{d}{dt} \int_{\alpha(t)}^{\beta(t)} f(x, t) dx = f[\beta(t), t]\beta'(t) - f[\alpha(t), t]\alpha'(t)$$

$$+ \int_{\alpha(t)}^{\beta(t)} \frac{\partial f}{\partial t}(x, t) dx$$

Inequalities

Theorem A.12 *Let a and b be any real or complex numbers. Then*

$$\frac{|a + b|}{1 = |a + b|} \leq \frac{|a|}{1 + |a|} + \frac{|b|}{1 + |b|}$$

Theorem A.13   *1. Hölder's inequality for sequences: If $p > 1$ and q is defined by $\frac{1}{p} + \frac{1}{q} = 1$, then*

$$\sum_{i=1}^{n} |x_i y_i| \leq \left[\sum_{i=1}^{n} |x_i|^p \right]^{1/p} \left[\sum_{i=1}^{n} |y_i|^q \right]^{1/q}$$

for any complex numbers $x_1, x_2, \ldots, x_n, y_1, y_2, \ldots, y_n$.

2. Hölder's inequality for integrals: Let $f(t) \in L_p$ and $g(t) \in L_q$. Then

$$\int_a^b |f(t)g(t)| dt \leq \left(\int_a^b |f(t)|^p dt \right)^{1/p} \left(\int_a^b |g(t)|^q dt \right)^{1/q}$$

where p and q are related by the relation $\frac{1}{p} + \frac{1}{q} = 1$

Theorem A.14   *1. Minkowski's inequality for sequences: If $p \geq 1$, then*

$$\left[\sum_{i=1}^{n} |x_i + y_i|^p \right]^{1/p} \leq \left[\sum_{i=1}^{n} |x_i|^p \right]^{1/p} + \left[\sum_{i=1}^{n} |y_i|^p \right]^{1/p}$$

for any complex numbers $x_1, x_2, \ldots, x_n, y_1, y_2, \ldots, y_n$.

2. Minkowski's inequality for integrals: Let $f(t) \in L_p$ and $g(t) \in L_q$. Then

$$\left(\int_a^b |f(t) + g(t)|^p dt \right)^{1/p} \leq \left(\int_a^b |f(t)| dt \right)^{1/p}$$

$$+ \left(\int_a^b |g(t)|^p dt \right)^{1/p} \quad \text{for } p \geq 1 \quad (A.3)$$

Lebesgue Integration

1. A σ-algebra $\mathfrak{P}$ on an abstract set Ω is a collection of subsets of Ω which contains the null set ϕ and is closed under countable set operations. A measurable space is a couple $(\Omega, \mathfrak{P})$ where Ω is an abstract set and $\mathfrak{P}$ a σ-algebra of subsets of Ω. A subset A of Ω is measurable if $A \in \mathfrak{P}$. A measure μ on a measurable space $(\Omega, \mathfrak{P})$ is a nonnegative set function defined on $\mathfrak{P}$ with the properties: $\mu(\phi) = 0$, $\mu = \left(\bigcup_{i=1}^{\infty} E_i \right) = \sum_{i=1}^{\infty} \mu(E_i)$ where E_i are disjoint sets in $\mathfrak{P}$. The triple $(\Omega, \mathfrak{P}, \mu)$ is called a *measure space*. If X is a normed space, $(X, \mathfrak{P})$ is a measurable space if $\mathfrak{P}$ is the smallest σ-algebra containing all open subsets of X. $\mathfrak{P}$ is called the *Borel algebra* and sets in $\mathfrak{P}$ are called Borel sets.

2. Let Ω be a set of infinitely many points, and $\mathfrak{P}$ the class of all subsets of Ω. Then $(\Omega, \mathfrak{P})$ is a measurable space and a measure on $(\Omega, \mathfrak{P})$ is defined by $\mu(E) =$ the number of points in E if E is finite and $\mu(E) = \infty$ otherwise. Consider the length function λ on the following class of subsets of R

$$\xi = \left\{ E/E = \bigcup_{i=1}^{\infty} C_i, C_i \text{ are disjoint half intervals} (a_i, b_i] \right\}$$

Then

$$\lambda(E) = \sum_{i=1}^{\infty} |b_i - a_i|$$

The smallest σ-algebra containing ξ is the algebra of Borel sets, $\mathfrak{P}$ of R. λ may be extended to the measurable space $(R, \mathfrak{P})$ as follows

$$\lambda(A) = \inf_{\bigcup_n I_n \supset A} \sum \lambda(I_n)$$

where I_n is a countable collection of intervals covering A. So $(R, \mathfrak{P}, \lambda)$ is a measure space. In fact, λ is a measure on a larger σ-algebra, called the *class of Lebesgue measurable sets* $\mathfrak{L}$. So $(R, \mathfrak{L}, \lambda)$ is a measure space, and λ is called the *Lebesgue measure*. Not all subsets of R are Lebesgue measurable but most reasonable sets are, e.g., all sets which are countable, are unions or intersections of open sets.

3. Let E_i be subsets of Ω, then the *characteristic function* of E_i is defined as

$$\chi_{E_i}(w) = 1 \text{ if } w \in E_i$$
$$= 0 \text{ if } w \in E_i$$

A function $f : \Omega \rightarrow R \bigcup \infty$ is called a simple function if there are nonzero constants c_i and disjoint measurable sets E_i with $\mu(E_i) < \infty$ such that

$$f(w) = \sum_{i=1}^{n} c_i \chi_{E_i}(w)$$

The integral of a simple function f is defined as

$$\int_E f d\mu = \sum_{i=1}^{n} c_i \tag{A.4}$$

for any $E \in \mathfrak{P}$. If u is a nonnegative measurable function on $(\Omega, \mathfrak{P}, \mu)$, then $u(w) = \lim\limits_{n \to \infty} u_n(w) \, \forall w \in \Omega$, for some sequence $\{u_n\}$ of monotonically increasing nonnegative simple functions. We define $\int_E u d\mu = \lim\limits_{n \to \infty} \int_E u_n d\mu \, \forall E \in \mathfrak{P}$. A measurable function f on $(\Omega, \mathfrak{P}, \mu)$ is integrable on $E \in \mathfrak{P}$ if $fd < \infty$. If in particular we consider the Lebesgue measure space $(R, \mathfrak{L}, \lambda)$, then we get a Lebesgue integral, and in such a case we write $\int_a^b f(t) dt$ or $\int_{[a,b]} f d\lambda$.

$L_2[a, b]$ is defined as

$$L_2(a, b) = \{ f : [a, b] \to R \; Lebesgue \; measurable / \int_a^b |f(t)|^2 dt < \infty \}$$

For $p \geq 1$

$$L_p(a, b) = \{ f \; Lebesgue \; measurable \; on \; [a, b] \; into \; R / \int_a^b |f(t)|^2 dt < \infty \}$$

Theorem A.15 (Lebesgue's Dominated Convergence Theorem)

1. *If the sequence $\{f_k\} \in L_1(a, b)$ has the property that $\lim\limits_{k \to \infty} f_k$ is finite a.e., on $(a, b]$ and if $|f_k| \leq h$ for some nonnegative function $h \in L_1[a, b]$ and for all $k \geq 1$, then $\lim\limits_{k \to \infty} f_k \in L_1(a, b)$ and*

$$\int_a^b \lim_{k \to \infty} f_k dx = \lim_{k \to \infty} \int_a^b f_k dx$$

2. *If the sequence $\{g_k\} \in L_1(a, b)$ has the property that $\sum\limits_{k=1}^{\infty} g_k$ converges a.e. on (a, b) and if $|\sum\limits_{k=1}^{n} g_k| \leq h$ for some nonnegative function $h \in L_1(a, b)$ and all $n \geq 1$, then $\sum\limits_{k=1}^{\infty} g_k \in L_1(a, b)$ and*

$$\int_a^b \left(\lim_{k \to \infty} g_k \right) dx = \lim_{k \to \infty} \int_a^b g_k dx$$

Theorem A.16   *1. Beppo Levi's Theorem: If the sequence $\{g_k\} \in L_1(a, b)$ has the property that*

$$\sum_{k=1}^{\infty} \int_a^b |g_k| dx < \infty$$

then $\sum_{k=1}^{\infty} g_k$ converges a.e. on (a, b) to an integrable function and

$$\int_a^b \left(\sum_{k=1}^{\infty} g_k \right) dx = \sum_{k=1}^{\infty} \int_a^b g_k dx$$

2. *Fatou's Lemma: If $\{f_n\}$ is a sequence of nonnegative measurable functions and $f_n(x) \rightarrow f(x)$ everywhere on a set E, then*

$$\int_E f dx \leq \lim \int_E f_n dx$$

3. *Monotone Convergence Theorem: If $f_1(x), f_2(x), \ldots, f_n(x) \ldots$ is a sequence of nonnegative functions such that*

$$f_1(x) \leq f_2(x) \leq f_3(x) \leq \cdots \leq f_n(x) \ldots$$

and $\lim_{n\to\infty} f_n(x) = f(x)$ on a set E, then

$$\lim_{n\to\infty} \int_E f_n(x) dx = \int_E f(x) dx$$

Theorem A.17 (Fubini's Theorem)

1. *If $f(x, y)$ is integrable in the rectangle $Q[a \leq x \leq b, c \leq y \leq d]$, then for all $x \in [a, b]$ except a set of measure zero, the function $f(x, y)$ is integrable with respect to y in $[c, d]$, and for all $y \in [c, d]$ except a set of measure space $f(x, y)$ is integrable with respect to x in $[a, b]$, the following equality holds*

$$\iint_E f(x, y) dx dy = \int_a^b dx \int_c^d f(x, y) dy = \int_c^d dy \int_a^b f(x, y) dx$$

2. *If f is Lebesgue measurable in R^2, say a continuous function except at finite number of points such that*

$$\int_{-\infty}^{\infty} \int_{-\infty}^{\infty} f(x, y) dx dy$$

and

$$\int\limits_{-\infty}^{\infty} \int\limits_{-\infty}^{\infty} f(x,y)dydx$$

exist and one of them is absolutely convergent, then the two are equal.

Integral Equations

Equations of the type:

1. $\varphi(x) = \int\limits_{a}^{b} K(x,t)f(t)dt$

2. $f(x) = \int\limits_{a}^{b} K(x,t)f(t)dt + \varphi(x)$

are called, respectively, Fredholm equations of the first kind and the second kind.

Surface Integral

Suppose $f(x,y,z)$ is a function of three variables, continuous at all points on a surface Σ. Suppose Γ is the graph of $z = S(x,y)$ for (x,y) in a set D in the xy plane. It is assumed that Γ is smooth and that D is bounded. The surface integral of f over a surface Γ denoted by $\int\limits_{\Gamma} f\, d\Gamma$ and is defined by

$$\int\limits_{\Gamma} f\, d\Gamma = \int\int\limits_{\Gamma} f(x,y,z)\, d\Gamma = \int\int\limits_{D} f(x,y,S(x,y))\, dS$$

where

$$dS = \sqrt{1 + \left(\frac{\partial S}{\partial x}\right)^2 + \left(\frac{\partial S}{\partial y}\right)^2}\, dxdy$$

$L_2(\ell)$ denotes the space of functions f for which $\int\limits_{\Gamma} |f|^2 d\Gamma$ exists.

A.5   Vector Spaces

A vector space or a linear space X over a field K consists of a set X, a mapping $(x,y) \to x+y$ of $X \times X$ into X, and a mapping $(\alpha,x) \to \alpha x$ of $K \times X$ into X, such that (1) $x+y = y+x$ (2) $x+(y+z) = (x+y)+z$ (3) $\exists 0 \in X$ with $x+0 = x$ for all $x \in X$ (4) for each $x \in X \exists x' \in X$ with $x + x' = 0$ (5) $(\alpha+\beta)x = \alpha x + \beta x$ (6) $\alpha(x+y) = \alpha x + \alpha y$ (7) $(\alpha\beta)x = \beta(\alpha x)$ (8) $1x = x$. A subset $Y \subset X$ is called a vector subspace or simply a *subspace* of X if it is itself a vector space with the same operations as for X. A subset $Y \subset X$ is subspace of X if and only if for every $x_1, x_2 \in Y$ and $\alpha, \beta \in K$, $\alpha x_1 + \beta x_2 \in K$. A subspace of a vector space X different from $\{0\}$ and X is called a *proper subspace*. If $K = R$, a vector space X over R is called the *real vector space*. If $K = C$, a vector space over C is called the *complex vector space*. Let S be any subset of X. Then the subspace generated by S or spanned by S, which we denote by $[S]$, is the intersection of all subspaces of X containing S.

It can be shown that $[S]$ is the set of all linear combinations $\sum_{i=1}^{n} a_i x_i$ of finite sets in S. Let $S = \{x_1, x_2, \ldots, x_n\}$ be a finite nonempty subset of a vector space X. Then S is called *linearly dependent* if there exist scalars α_i, $i = 1, 2, \ldots, n$ such that $\sum_{i=1}^{n} a_i x_i = 0$ does not imply that all $\alpha_i's$ are zero. If S is not linearly dependent, i.e., $\sum_{i=1}^{n} a_i x_i = 0$ implies that all α_i, $i = 1, 2, \ldots, n$, are zero, then it is said to be *linearly independent*. An arbitrary nonempty subset S is called linearly independent if its every finite nonempty subset is *linearly independent*. It can be seen that a subset S of a vector space X is linearly independent if and only if each vector in $[S]$ is uniquely expressible as a linear combination of the vectors in S. A subset S of a vector space X of linearly independent elements is called a *Hamel basis* or *algebraic basis* or *basis* in X if $[S] = X$. A vector space X may have many Hamel bases but all have the same cardinal number. This cardinal number is called the dimension of X. X is called *finite-dimensional* if its dimension is 0 or a positive integer and *infinite-dimensional* otherwise.

R^n is a vector space of dimension n while $C[a, b]$ is a vector space of infinite dimensions. Let Y be a subspace of a vector space X. We say that $x, y \in X$ are in relation denoted by $x \sim y$, if $x - y \in Y$. This is an equivalence relation. This equivalence relation induces equivalence classes of X which are called cosets. If the coset of an element x in X is defined by $x + Y = \{x + y/y \in Y\}$, then the distinct cosets form a partition of X. If addition and scalar multiplication are defined by

$$(x + Y) + (z + Y) = (x + z) + Y$$
$$\alpha(x + Y) = \alpha x + Y$$

then these cosets constitute a vector space denoted by X/Y and are called the quotient or factor space of X with respect to Y. The origin in X/Y is the coset $0 + Y = Y$, and the negative inverse of $x + Y$ is $(-x) + Y$.

A mapping T defined on a vector space X into a vector space Y is called linear if

$$T(x + y) = T(x) + T(y) T(\alpha x) = \alpha T(x) \text{ where } \alpha \text{ is a scalar} \qquad (A.5)$$

The set of all linear mappings defined on X into R is called the algebraic dual of X.

Let M and N be subspaces of a vector space X. We say that X is the direct sum of M and N and write $X = M \oplus N$ if every $z \in X$ can be written uniquely in the form $z = x + y$ with $x \in M$ and $y \in N$. The mapping P, defined on X into itself by the relation $P(z) = x$, is called an *algebraic projection* or *projection* on M along N. A linear mapping P of X into itself is a projection if and only if

$$P^2 = P[P(P(x)) = P(x) \ \forall x \in X]$$

A vector space X over the field $K (= R$ or $C)$ equipped with a topology is called a topological vector space if the mappings

1. $X \times X \to X : (x, y) \to x + y$, and
2. $K \times X \to X : (\alpha, x) \to \alpha x$ is continuous. If M and N are subspaces of a vector space X, then $M \oplus N$ is a subspace of X. $C, \ell_p, P_n[a, b], BV[a, b], L_2(a, b)$, and $AC[a, b]$ are examples of a vector space, where operations of addition and scalar multiplication are defined in a manner indicated below.

In the space of real-valued continuous functions defined on $[a, b]$, $C[a, b]$, the operations of addition and scalar multiplications are defined as follows:

1. $(f + g)(x) = f(x) + g(x) \ \forall \, x \in [a, b]$
2. $(\alpha f)(x) = \alpha f(x) \ \forall \, x \in [a, b], \alpha$ real scalar.

The operations of addition and scalar multiplication in $L_2(a, b)$, which is essentially the space of continuous functions on $[a, b]$ having possible finite number of discontinuities, are defined in a similar way.

These operations are defined similarly in $P_n[a, b]$, $BV[a, b]$, and $AC[a, b]$. In p and other spaces of sequences, the operations of addition and scalar multiplication are defined as follows: Let

$$x = (x_1, x_2, x_3, \ldots, x_n, \ldots), \ y = (y_1, y_2, y_3, \ldots, y_n, \ldots) \in \ell_2$$
$$x + y = (x_1 + y_1, x_2 + y_2, x_3 + y_3, \ldots, x_n + y_n, \ldots)$$
$$\alpha x = (\alpha x_1, \alpha x_2, \alpha x_3, \ldots, \alpha x_n, \ldots), \ \alpha \ is \ a \ scalar$$

A.6 Fourier Analysis

Let $f(t)$ be a periodic function with period T and Lebesgue integrable continuous functions having at most finite discontinuities over $(-T/2, T/2)$. Then the Fourier series of $f(t)$ is the trigonometric series

$$\frac{1}{2}A_0 + \sum_{k=1}^{\infty} \left(A_k \cos \frac{2\pi k}{T} x + B_k \sin \frac{2\pi k}{T} x \right)$$

where

$$A_k = \frac{2}{T} \int\limits_{-T/2}^{T/2} f(t) \cos \frac{2\pi kt}{T} dt, \ k = 1, 2, 3 \ldots$$

$$A_0 = \frac{1}{T} \int\limits_{-T/2}^{T/2} f(t) dt$$

$$B_k = \frac{2}{T} \int\limits_{-T/2}^{T/2} f(t) \sin \frac{2\pi kt}{T} dt, \ k = 1, 2, 3 \ldots$$

and we write it as

$$f \sim \frac{1}{2} A_0 + \sum_{k=1}^{\infty} \left(A_k \cos \frac{2\pi k}{T} x + B_k \sin \frac{2\pi k}{T} x \right) \tag{A.6}$$

Here we take

$$w_k = \frac{k}{T}, \ k = 0, 1, 2, 3 \tag{A.7}$$

Very often, we choose $T = 2\pi$. A_k and B_k are called cosine Fourier coefficient and sine Fourier coefficient, respectively. The set of triplex (A_k, B_k, w_k) where A_k, B_k, w_k are given by Equations $(F.2)$, $(F.3)$, and $(F.6)$, respectively, is called the *Fourier series frequency content*.

The complex form of the Fourier series of $f(x)$ is

$$\sum_{k=-\infty}^{\infty} C_n e^{2\pi int/T}$$

where

$$C_n = \frac{A_n + i B_n}{2} \ . \ n > 0$$
$$C_0 = A_0$$
$$\underline{C_n} = \frac{A_n - i B_n}{2} \ . \ n > 0$$
$$w_n = \frac{n}{T}, \ n = -2, -1, 0, 1, 2$$

Let

$$T = 2\pi \quad and \quad S_n(f)(x) = \sum_{k=-n}^{n} C_k e^{ikx}$$

$$= \frac{1}{2}A_0 + \sum_{k=1}^{n}(A_n \cos kx + B_k \sin kx)$$

be the nth partial sum of the Fourier series of f. Then

$$S_n(f)(x) = \frac{1}{\pi} \int_0^{2\pi} f(x-t)D_n(t)dt$$

$$= \frac{1}{\pi} \int_0^{2\pi} f(t)D_n(x-t)dt \tag{A.8}$$

where

$$D_n(x) = \frac{1}{2} + \sum_{k=1}^{n} \cos kx = \frac{\sin\left(n+\frac{1}{2}\right)x}{2\sin\frac{x}{2}} \tag{A.9}$$

is the "Dirichlet kernel," and

$$\sigma_n(x) = \frac{S_0(f) + S_1(f) + \cdots + S_n(f)}{n+1} = \frac{1}{\pi} \int_0^{2\pi} f(x-t)K_n(t)dt \tag{A.10}$$

where

$$K_n(x) = \frac{D_0(x) + D_1(x) + \cdots + D_n(x)}{n+1} = \frac{1}{n+1} \frac{\sin\left(\frac{n+1}{2}\right)(x)}{2\sin^2\left(\frac{x}{2}\right)} \tag{A.11}$$

is called the "Fejer kernel."

Theorem A.18 (Bessel's Inequality)

$$\sum_{k=-\infty}^{\infty} |C_k|^2 \le \|f\|_{L_2(0,2\pi)}^2$$

or

$$\frac{1}{4}A_0^2 + \sum_{n=1}^{\infty}(A_n^2 + B_n^2) \le \|f\|_{L_2(0,2\pi)}^2$$

This also means that $\{A_k\}$ and $\{B_k\}$ are elements of ℓ_2.

Theorem A.19 (Riesz-Fisher Theorem) *Let* $\{C_k\} \in \ell_2$. *Then there exists* $f \in L_2(-\pi, \pi)$ *such that* $\{C_k\}$ *is the kth Fourier coefficient of* f. *Furthermore*

$$\sum_{k=-\infty}^{\infty} |C_k|^2 \le \|f\|^2_{L_2(-\pi, \pi)}$$

For $\{A_k\}$, $\{Bk\}$ *belonging to* ℓ_2, *there exists* $f \in L_2(0, 2\pi)$ *such that* A_k, B_k *are, respectively, kth cosine and sine Fourier coefficients of* f. *Furthermore*

$$\frac{1}{2}A_0^2 + \sum_{n=1}^{\infty}(A_k^2 + B_k^2) = \left(\int_{-\pi}^{\pi} |f|^2 dt\right)$$

Theorem A.20 *Let* $f \in L_2(-\pi, \pi)$, *then*

$$\lim_{n \to \infty} \|f - S_n f\|_{L_2(-\pi, \pi)} = 0$$

Theorem A.21 *Let* $f \in C[0, 2\pi]$ *such that*

$$\int_{\delta}^{2\pi} \frac{w(f, t)}{t} < \infty, \ w(f, t) = \sup_t |f(x + t) - f(x)|$$

Then the Fourier series of f *converges uniformly to* f ; *that is*

$$\lim_{n \to \infty} \|f - S_n f\|_{L_\infty(-\pi, \pi)} = 0$$

If $w(f, \eta) = 0(\eta^\alpha)$, *then the condition of the theorem holds.*

Theorem A.22 *If* f *is a function of bounded variation, then*

$$S_n(f) \to \frac{f(x^+) - f(x^-)}{2} \ as \ n \to \infty$$

where

$$f(x^+) = \lim_{h \to 0} f(x + h)$$
$$f(x^-) = \lim_{h \to 0} f(x - h)$$

exists at every x, $a < x < b$.

Theorem A.23 *Let $f \in L_1(R)$, then the series $\sum\limits_{k=-\infty}^{\infty} (\infty) f(x + 2\pi k)$ converges almost everywhere to a function $\lambda(x) \in L_1(0, 2\pi)$. Moreover, the Fourier coefficients c_k of $\lambda(x)$ are given by $c_k = \frac{1}{2\pi} \hat{f}(k)$*

The Fourier Transform

Definition A.15 *(Fourier transform in $L_1(R)$)* Let $f \in L_1(R)$. Then the function $\hat{f}$ defined by

$$\hat{f}(w) = \frac{1}{\sqrt{2\pi}} \int\limits_{-\infty}^{\infty} e^{-iwx} f(x) dx$$

is called the Fourier transform of f. Very often, $\mathscr{F}\{f(x)\}$ is used as the notation for the Fourier transform instead of $\hat{f}$.

It can be verified that

$$\hat{f}(e^{-x^2}) = \mathscr{F}\{e^{-x^2}\} = \frac{1}{\sqrt{2}} e^{-w^2/4}$$

Theorem A.24 *Let $f, g \in L_1(R)$ and $\alpha, \beta \in C$. Then*

$$\mathscr{F}(\alpha f + \beta g) = \alpha \mathscr{F}(f) + \beta \mathscr{F}(g)$$

Theorem A.25 *The Fourier transform of an integrable function is a continuous function.*

Theorem A.26 *If $f_1, f_2, \ldots, f_n, \varepsilon L_1(R)$ and $\|f_n - f\|_{L_1} \to 0$ as $n \to \infty$, then $\hat{f}_n \to f$ uniformly on R.*

Theorem A.27 (Riemann-Lebesgue Theorem) *If $f \in L_1(R)$, then*

$$\lim_{|w| \to \infty} |\hat{f}(w)| = 0$$

Theorem A.28 *Let $f \in L_1(R)$. Then*

1. $\mathscr{F}\{e^{i\alpha x} f(x)\} = \hat{f}(w - \alpha)$ *(translation)*
2. $\mathscr{F}\{f(x - u)\} = \hat{f}(w) e^{-iwu}$ *(shifting)*
3. $\mathscr{F}\{f(\alpha x)\} = \frac{1}{\alpha} f\left(\frac{w}{\alpha}\right), \alpha > 0$ *(scaling)*
4. $\mathscr{F}(\bar{f}(x)) = \overline{\mathscr{F}(f(-x))}$ *(conjugate)*

Example A.3 If $f(x) = e^{iux - x^2/2}$, then $\hat{f}(w) = e^{-(w-u)^2/2}$.

Theorem A.29 *If f is a continuous piecewise differentiable function, $f, f' \in L_1(R)$, and $\lim \lim_{|x| \to infty} f(x) = 0$, then*

$$\mathscr{F}\{f'\} = iw \mathscr{F}(f)$$

Corollary .1 *If f is a continuous function, n-time piecewise differentiable, and* $f, f', \ldots, f^{(n)} \in L_1(R)$, *and*

$$\lim_{|x| \to \infty} f^{(k)}(x) = 0 \ for \ k = 0, \ldots, n - 1$$

then

$$\mathscr{F}\{f^{(n)}\} = (iw)^n \mathscr{F}(f)$$

Definition A.16 Let $f, g \in L_1(R)$ then the convolution of f and g is denoted by $f \star g$ and is defined by

$$\mathscr{F} = (f \star g)(x) = \frac{1}{\sqrt{2\pi}} \int\limits_{-\infty}^{\infty} f(x - u)g(u)du$$

Theorem A.30 *For $f, g \in L_1(R)$, $\mathscr{F}(f \star g) = \mathscr{F}(f)\mathscr{F}(g)$ holds.*

Theorem A.31 *Let f be a continuous function on R vanishing outside a bounded interval. Then $f \in L_2(R)$ and*

$$\|\hat{f}\|_{L_2(R)} = \|f\|_{L_2(R)}$$

Definition A.17 *(Fourier transform in $L_2(R)$)* Let $f \in L_2(R)$ and $\{\varphi_n\}$ be a sequence of continuous functions with compact support convergent to f in $L_2(R)$; that is,

$\|f - \varphi_n\|_{L_2(R)} \to 0$. The Fourier transform of f is defined by

$$\hat{f} = \lim_{n \to \infty} \hat{\varphi}_n$$

where the limit is with respect to the norm in $L_2(R)$.

Theorem A.32 *If $f \in L_2(R)$, then*

1. $\langle f, g \rangle_{L_2} = \langle \hat{f}, \hat{g} \rangle_{L_2}$ *(Parseval's formula)*
2. $\|\hat{f}\|_{L_2} = \|f\|_{L_2}$ *(Plancherel formula)*

In physical problems, the quantity $\|f\|_{L_2}$ is a measure of energy while $\|\hat{f}\|_{L_2}$ represents the power spectrum of f.

Theorem A.33 *1. Let $f \in L_2(R)$. Then*

$$\hat{f}(w) = \lim_{n \to \infty} \frac{1}{\sqrt{2\pi}} \int\limits_{-n}^{n} e^{iwx} f(x)dx$$

2. If $f, g \in L_2(R)$, then

$$\int\limits_{-\infty}^{\infty} f(x)\hat{g}(x)dx = \int\limits_{-\infty}^{\infty} \hat{f}(x)g(x)dx$$

Theorem A.34 (Inversion of the Fourier transform in $L_2(R)$) *Let $f \in L_2(R)$. Then*

$$f(x) = \lim_{n \to \infty} \frac{1}{\sqrt{2\pi}} \int\limits_{-n}^{n} e^{iwx} \hat{f}(w)dw$$

where the convergence is with respect to the norm in $L_2(R)$.

Corollary A.2 *1. If $f \in L_1(R) \cap L_2(R)$, then the equality*

$$f(x) = \frac{1}{\sqrt{2\pi}} \int\limits_{-\infty}^{\infty} e^{iwx} \hat{f}(w)dw$$

holds almost everywhere in R.
2. $\mathscr{F}(\mathscr{F}(f(x))) = f(-x)$ *almost everywhere in R.*

Theorem A.35 (Plancherel's Theorem) *For every $f \in L_2(R)$, there exists $\hat{f} \in L_2(R)$ such that:*

1. If $f \in L_1(R) \cap L_2(R)$, then $\hat{f}(w) = \frac{1}{\sqrt{2\pi}} \int\limits_{-\infty}^{\infty} e^{iwx} f(x)dx$

2. $\left\| \hat{f}(w) - \frac{1}{\sqrt{2\pi}} \int\limits_{-n}^{n} e^{iwx} \hat{f}(x)dx \right\|_{L_2} \to 0$ as $n \to \infty$

3. $\left\| f(x) - \frac{1}{\sqrt{2\pi}} \int\limits_{-n}^{n} e^{iwx} \hat{f}(w)dw \right\|_{L_2} \to 0$ as $n \to \infty$

4. $\|f\|_{L_2}^2 = \|\hat{f}\|_{L_2}^2$.
5. The map $f \to \hat{f}$ is an isometry of $L_2(R)$ onto $L_2(R)$.

Theorem A.36 *The Fourier transform is a unitary operator on $L_2(R)$.*

Example A.4 1. If $f(x) = (1 - x^2)e^{-x^2/2}$ = Second derivative of a Gaussian function, then

$$\hat{f}(w) = w^2 e^{-w^2/2}$$

2. If the Shannon function is defined by

$$f(x) = \frac{\sin 2\pi x - \sin \pi x}{\pi x} \tag{A.12}$$

then

$$\hat{f} = \frac{1}{\sqrt{2}} \, if \, \pi < |w| < 2\pi$$
$$= 0 \, otherwise$$

References

1. Adams R (1975) Sobolev spaces. Academic Press, New York
2. Aldroubi A, Unser M (1996) Wavelets in medicine and biology. CRC Press, Boca Raton
3. Antes H, Panagiotopoulos PP (1992) The boundary integral approach to static and dynamic contact problems. Birkhuser, Basel
4. Appell J, Deascale E, Vignoli A (2004) Non-linear spectral theory. De Gruyter series in nonlinear analysis and applications, Walter De Gruyter and Co
5. Argyris JH (1954) Energy theorems and structural analysis. Aircraft Eng 26:347–356, 383–387, 394
6. Attaouch H (1984) Variational convergence for functions and operators. Pitman, Advanced Publishing Program, New York, Applicable mathematics series
7. Averson W (2002) A short course on spectral theory: graduate texts in mathematics, vol 209. Springer, Berlin
8. Bachman G, Narici L (1966) Functional analysis. Academic Press, New York
9. Baiocchi C, Capelo A (1984) Variational and quasivariational inequalities application to free boundary problems. Wiley, New York
10. Balakrishnan AV (1976) Applied functional analysis. Springer, Berlin
11. Balakrishnan AV (1971) Introduction to optimisation theory in a Hilbert space. Lecture notes in operation research and mathematical system, Springer, Berlin
12. Banach S (1955) Theorie des operations lineaires. Chelsea, New York
13. Banerjee PK (1994) The boundary element methods in engineering. McGraw-Hill, New York
14. Benedetto J, Czaja W, Gadzinski P (2002) Powell: Balian-Low theorem and regularity of Gabor systems. Preprint
15. Benedetto J, Li S (1989) The theory of multiresolution analysis frames and applications to filter banks. Appl Comput Harmon Anal 5:389–427
16. Benedetto J, Heil C, Walnut D (1995) Differentiation and the Balian-Low theorem? J Fourier Anal Appl 1(4):355–402
17. Bensoussan A, Lions JL (1982) Applications of variational inequalities in stochastic control. North Holland, Amsterdam
18. Bensoussan A, Lions JL (1987) Impulse control and quasivariational inequalities. Gauthier-Villars, Paris
19. Berberian SK (1961) Introduction to Hilbert space. Oxford University Press, Oxford
20. Berg JC, Berg D (eds) (1999) Wavelets in physics. Cambridge University Press, Cambridge
21. Boder KC (1985) Fixed point theorems with applications to economic and game theory. Cambridge University Press, Cambridge

© Springer Nature Singapore Pte Ltd. 2018

A. H. Siddiqi, *Functional Analysis and Applications*, Industrial and Applied Mathematics, https://doi.org/10.1007/978-981-10-3725-2

22. Brebbia CA (1978) The boundary element methods for engineers. Pentech Press, London
23. Brebbia CA (1984) Topics in boundary element research, vol 1. Springer, Berlin
24. Brebbia CA (1990) In: Tanaka M, Honna T (eds) Boundary elements XII. Springer, Berlin
25. Brebbia CA (ed) (1988) Boundary element X. Springer, Berlin
26. Brebbia CA (ed) (1991) Boundary element technology, vol VI. Elsevier Application Science, London
27. Brebbia CA, Walker S (1980) Boundary element techniques in engineering. Newnes-Butterworths, London
28. Brenner SC, Scott LR (1994) The mathematical theory of finite element methods. Springer, Berlin
29. Brezzi F, Fortin M (1991) Mixed and hybrid finite element methods. Springer, Berlin
30. Brislaw CM (1995) fingerprints go digital. Notices of the AMS, vol 42, pp 1278–1283. http://www.c3.lanl.gov/brislawn
31. Brokate M, Siddiqi AH (1993) Sensitivity in the rigid punch problem. Advances in mathematical sciences and applications, vol 2. Gakkotosho, Tokyo, pp 445–456
32. Brokate M, Siddiqi AH (eds) (1998) Functional analysis with current applications to science, technology and industry. Pitman research notes in mathematics, vol 37. Longman, London
33. Byrnes JS, Byrnes JL, Hargreaves KA, Berry KD (eds) (1994) Wavelet and their applications, NATO ASI, series. Academic Publishers, Dordrecht
34. Cartan H (1971) Differential calculus. Herman/Kershaw, London
35. Chambolle A, DeVore RA, Lee NY, Lucier B (1998) Nonlinear wavelet image processing: variational problems, compression and noise removal through wavelet shrinkage. IEEE Trans Image Process 7:319–335
36. Chari MVK, Silvester PP (eds) (1980) Finite elements in electrical and magnetic field problems, vol 39. Wiley, New York. Chavent G, Jaffré, J (1986) Mathematical models and finite elements for reservoir simulation. North Holland, Amsterdam
37. Chen G, Zhou J (1992) Boundary element methods. Academic Press, New York
38. Chipot M (1984) Variational inequalities and flow in porous media. Springer, Berlin
39. Chipot M (2000) Elements of nonlinear analysis. Birkhuser Verlag, Basel
40. Christensen O (2003) An introduction to frames and Riesz basses. Birkhauser, Boston
41. Christensen O, Christensen KL (2004) Approximation theory from taylor polynomials to wavelets. Birkhauser, Boston
42. Christensen O, Linder A, 1–3, (2001) Frames of exponentials: lower frame bounds for finite subfamilies and approximation of the inverse frame operator. Linear Algebr Appl 323:117–130
43. Chui C, Shi X (2000) Orthonormal wavelets and tight frames with arbitrary dilations. Appl Comput Harmon Anal 9:243–264 (Pls. clarify 1–3, 2001)
44. Chui CK (ed) (1992) Wavelets: a tutorial in theory and applications. Academic Press, New York
45. Ciarlet PG (1978) The finite element methods for elliptic problems. North Holland, Amsterdam
46. Ciarlet PG (1989) Introduction to numerical linear algebra and optimization. Cambridge University Press, Cambridge
47. Ciarlet PG, Lions JL (1991) Hand book of numerical analysis finite element methods. Elsevier Science Publisher
48. Clarke FH (1983) Optimisation and non-smooth Analysis. Wiley, New York
49. Clarke FH, Ledyaev YS, Stern RJ, Wolenski PR (1998) Nonsmooth analysis and control theory. Springer, Berlin
50. Cohen A (2002) Wavelet methods in numerical analysis. In: Ciarlet PG, Lions JL (eds), Handbook of numerical analysis, vol VII. Elsevier science, pp 417–710
51. Coifman RR, Wickerhauser MV (1992) Entropy based algorithms for best basis selection. IEEE Trans Inf Theory 9:713–718
52. Conn AR, Gould NIM, Lt Toint Ph (2000) Trust-region methods. SIAM, Philadelphia

53. Cottle RW, Pang JS, Store RE (1992) The linear complimentarity problems. Academic Publishers, New York
54. Curtain RF, Pritchard AJ (1977) Functional analysis in modern applied mathematics. Academic Press, New York
55. Dahmen W (1997) Wavelets and multiscale methods for operator equations. Acta Numer 6:55–228
56. Dahmen W (2001) Wavelet methods for PDEs: some recent developments. J Comput Appl Math 128:133–185
57. Dal Maso G (1993) An introduction to Γ-convergence. Birkhauser, Boston
58. Daubechies I (1988) Orthonormal bases of compactly supported wavelets. Commun Pure Appl Math 4:909–996
59. Daubechies I (1992) Ten lectures on wavelets. SIAM, Philadelphia
60. Daubechies I, Jaffard S, Jaurne Wilson JL (1991) Orthonormal basis math exponential decay. SIAM J Math Anal 22:554–572
61. Dautray R, Lions JL (1995) Mathematical analysis and numerical methods for science and technology, vols 1–6. Springer, Berlin
62. Dautray R, Lions JL (1988) mathematical analysis and numerical methods for science and technology. Functional and variational methods, vol 2. Springer, Berlin
63. Debnath L, Mikusinski P (1999) Introduction to Hilbert spaces with applications, 2nd edn. Academic Press, New York
64. Dieudonné J (1960) Foundation of modern analysis. Academic Press, New York
65. Donoho DL (2000) Orthogonal ridgelets and linear singularities. SIAM J Math Anal 31:1062–1099
66. Duffin RJ, Schaeffer AC (1952) A class of nonharmonic Fourier series. Trans Am Math Soc 72:341–366
67. Dunford N, Schwartz JT (1958) Linear operators part I. Interscience, New York
68. Dupis P, Nagurney A (1993) Dynamical systems and variational inequalities. Ann Oper Res 44:9–42
69. Duvaut G, Lions JL (1976) Inequalities in mechanics and physics. Springer, Berlin
70. Efi-Foufoula G (1994) Wavelets in geophysics, vol 12. Academic Press, New York, p 520
71. Ekeland I, Tmam R (1999) Convex analysis and variational problems. Classics in applied mathematics, SIAM, Philadelphia
72. Falk RS (1974) Error estimates for the approximation of class of variation inequalities. Math Comput 28:963–971
73. Feichtinger HG, Strohmer T (eds) (1998) Gabor analysis and algorithms: theory and applications. Birkhauser, Boston
74. Feichtinger HG, Strohmer T (eds) (2002) Advances in gabor analysis. Birkhäuser, Boston
75. Finlayson BA (1972) The method of weighted residuals and variational principles. Academic Press, New York
76. Frazier M, Wang K, Garrigos G, Weiss G (1997) A characterization of functions that generate wavelet and related expansion. J Fourier Anal Appl 3:883–906
77. Freiling G, Yurko G (2001) Sturm-Liouville problems and their applications. Nova Science Publishers, New York
78. Fuciks S, Kufner A (1980) Nonlinear differential equations. Elsevier, New York
79. Gabor D (1946) Theory of communications. J IEE Lond 93:429–457
80. Gencay R, Seluk F (2001) An introduction to wavelets and other filtering methods in finance and economics. Academic Press, New York
81. Giannessi F (1994) Complementarity systems and some applications in the fields structured engineering and of equilibrium on a network. In: Siddiqi AH (ed) Recent developments in applicable mathematics. Macmillan India Limited, pp 46–74
82. Giannessi F (ed) (2000) Vector variational inequalities and vector equilibria. Kluwer Academic Publishers, Boston, Mathematical theories
83. Glowinski R (1984) Numerical methods for nonlinear variational problems. Springer, Berlin

84. Glowinski R, Lawton W, Ravachol M, Tenebaum E (1990) Wavelet solutions of linear and nonlinear elliptic, parabolic and hyperbolic problems in one space dimension. In Glowinski R, Lichnewski A (eds) Proceedings of the 9th international conference on computer methods in applied sciences and engineering (SIAM), pp 55–120

85. Glowinski R, Lions JL, Trmolieres R (1981) Numerical analysis of variational inequalities. North Holland Publishingl Co, Amsterdam

86. Goffman C, Pedrick G (1965) First course in functional analysis. Prentice-Hall, Englewood Cliffs

87. Gould NIM, Toint PL (2000) SQP methods for large-scale nonlinear programming. In: Powel MJP, Scholtes S (eds) System modeling and optimisation: methods, theory and applications. Kluwer Academic Publishers, Boston, pp 150–178

88. Griffel DH (1981) Applied functional analysis. Ellis Horwood Limited, Publishers, New York, Toronto

89. Groechening KH (2000) Foundations of time frequency analysis. Birkhuser, Basel

90. Groetsch CW (1980) Elements of applicable functional analysis. Marcel Dekker, New York

91. Groetsch CW (1993) Inverse problems in the mathematical sciences. Vieweg, Braunschweig

92. Hackbusch W (1995) Integral equations, theory and numerical treatment. Birkhäuser, Basel

93. Halmos P (1957) Introduction to hilbert space. Chelsea Publishing Company, New York

94. Hárdle W, Kerkyacharian G, Picard D, Tsybakov A (1998) Wavelets, approximations, and statistical applications. Springer, Berlin

95. Heil C (2006) Linear independence of finite gabor systems. In: Heil C (ed) Harmonic analysis and applications. Birkhauser, Basel

96. Heil C, Walnut D (1989) Continous and discrete wavelet transform. SIAM Rev 31:628–666

97. Helmberg G (1969) Introduction to spectral theory in hilbert space. North Holland Publishing Company, Amsterdam

98. Hernandez E, Weis G (1996) A first course on wavelets. CRC Press, Boca Raton

99. Hiriart-Urruty JB, Lemarchal C (1993) Convex analysis and minimization algorithms. Springer, Berlin

100. Hislop PD, Sigal IM (1996) Introduction to spectral theory: with applications to schr oedinger operators. Springer, Berlin

101. Hornung U (1997) Homogenisation and porous media. Interdisciplinary applied mathematics series, Springer, Berlin

102. Husain T (1964) Open mapping and closed graph theorems. Oxford Press, Oxford

103. Isozaki H (ed) (2004) Proceedings of the workshop on spectral, theory of differential operators and inverse problems contemporary mathematics. AMS, Providence, p 348

104. Istrătescu VI (1985) Fixed point theory. Reidel Publishing Company, Dordrecht

105. Jayme M (1985) Methods of functional analysis for application in solid mechanics. Elsevier, Amsterdam

106. Jin J (1993) The finite element method in electromagnetics. Wiley, New York

107. Kantorovich LV, Akilov GP (1964) Functional analysis in normed spaces. Pergamon Press, New York

108. Kardestuncer H, Norrie DH (1987) Finite element handbook, vol 16. McGraw-Hill Book Company, New York, p 404

109. Kelley CT (1995) Iterative methods for linear and nonlinear equations. SIAM, Philadelphia

110. Kelly S, Kon MA, Raphael LA (1994) Pointwise convergence of wavelet expansions. J Funct Anal 126:102–138

111. Kikuchi N, Oden JT (1988) Contact problems in elasticity: a study of variatinal inequalities and finite element methods. SIAM, Philadelphia

112. Kinderlehrer D, Stampacchia G (1980) An introduction to variational inequalities. Academic Press, New York

113. Kobyashi M (ed) (1998) Wavelets and their applications, case studies. SIAM, Philadelphia

114. Kocvara M, Outrata JV (1995) On a class of quasivariational inequalities. Optim Methods Softw 5:275–295

115. Kocvara M, Zowe J (1994) An iterative two step algorithm for linear complementarity problems. Numer Math 68:95–106

116. Kovacevic J, Daubechies I (eds) (1996) Special issue on wavelets. Proc IEEE 84:507–614

117. Kreyszig E (1978) Introductory functional analysis with applications. Wiley, New York

118. Kupradze VD (1968) Potential methods in the theory of elasticity. Israel Scientific Publisher

119. Lax PD, Milgram AN (1954) Parabolic equations, contributions to the theory of partial differential equations. Ann Math Stud 33:167–190

120. Lebedev LP, Vorovich II, Gladwell GMI Functional analysis applications in mechanics and inverse problems, 2nd edn. Kluwer Academic Publishers, Boston

121. Lions JL (1999) Parallel algorithms for the solution of variational inequalities, interfaces and free boundaries. Oxford University Press, Oxford

122. Liusternik KA, Sobolev VJ (1974) Elements of functional analysis, 3rd English edn. Hindustan Publishing Co

123. Louis Louis AK, Maass P, Reider A (1977) Wavelet theory and applications. Wiley, New York

124. Luenberger DG (1978) Optimisation by vector space methods. Wiley, New York

125. Mäkelä NM, Neittaan Mäki P (1992) Nonsmooth optimization. World Scientific, Singapore

126. Mallat S (1999) A wavelet tour of signal processing, 2nd edn. Academic Press, New York

127. Manchanda P, Siddiqi AH (2002) Role of functional analytic methods in imaging science during the 21st century. In: Manchanda P, Ahmad K, Siddiqi AH (eds) Current trends in applied mathematics. Anamaya Publisher, New Delhi, pp 1–28

128. Manchanda P, Mukheimer A, Siddiqi AH (2000) Pointwise convergence of wavelet expansion associated with dilation matrix. Appl Anal 76(3–4):301–308

129. Marti J (1969) Introduction to the theory of bases. Springer, Berlin

130. Mazhar SM, Siddiqi AH (1967) On FA and FB summability of trigonometric sequences. Indian J Math 5:461–466

131. Mazhar SM, Siddiqi AH (1969) A note on almost a-summability of trigonometric sequences. Acta Math 20:21–24

132. Meyer Y (1992) Wavelets and operators. Cambridge University Press, Cambridge

133. Meyer Y (1993) Wavelets algorithms and applications. SIAM, Philadelphia

134. Meyer Y (1998) Wavelets vibrations and scalings. CRM, monograph series, vol 9. American Mathematical Society, Providence

135. Mikhlin SG (1957) Integral equations. Pergamon Press, London

136. Mikhlin SG (1965) Approximate solutions of differential and integral equations. Pergamon Press, London

137. Moré JJ, Wright SJ, (1993) Optimisation software guide. SIAM, vol 143. Morozov, (1984) Methods of solving incorrectly posed problems. Springer, New York

138. Mosco U (1969) Convergence of convex sets. Adv Math 3:510–585

139. Mosco U (1994) Some introductory remarks on implicit variational problems. Siddiqi AH (ed) Recent developments in applicable mathematics. Macmillan India Limited, pp 1–46

140. Nachbin N (1981) Introduction to functional analysis: banach spaces and differential calculus. Marcel Dekker, New York

141. Nagurney A (1993) Network economics, a variational approach. Kluwer Academic Publishers, Boston

142. Nagurney A, Zhang D (1995) Projected dynamical systems and variational inequalities with applications. Kluwer Academic Press, Boston

143. Nashed MZ (1971) Differentiability and related properties of nonlinear operators: some aspects of the role of differentials in nonlinear functional analysis. In: Rall LB (ed) Nonlinear functional analysis and applications. Academic Press, London, pp 103–309

144. Naylor AW, Sell GR (1982) Linear operator theory in engineering and science. Springer, Berlin

145. Neunzert H, Siddiqi AH (2000) Topics in industrial mathematics: case studies and related mathematical methods. Kluwer Academic Publishers, Boston

146. Oden JT (1979) Applied functional analysis, a first course for students of mechanics and engineering science. Prentice-Hall Inc., Englewood Cliffs

147. Ogden RT (1997) Essential wavelets for statistical applications and data analysis. Birkhauser, Boston
148. Outrata J, Kocvara M, Zowe J (1998) Nonsmooth approach to optimisation problems with equilibrium constraints. Kluwer Academic Publishers, Boston
149. Percival DB, Walden AT (2000) Wavelet methods for time series analysis. Cambridge University Press, Cambridge
150. Polak E (1997) Optimization, algorithms and consistent approximations. Springer, Berlin
151. Polyak BT (1987) Introduction to optimization. Optimization Software Inc., Publications Division, New York
152. Pouschel J, Trubowitz E (1987) Inverse spectral theory. Academic Press, New York
153. Powell MJD (1986) Convergence properties of algorithms for nonlinear optimisation. SIAM Rev 28:487–500
154. Prigozhin L (1996) Variational model of sandpile growth. Eur J Appl Math 7:225–235
155. Quarteroni A, Valli A (1994) Numerical approximation of partial differential equations. Springer, Berlin
156. Reddy BD (1999) Introductory functional analysis with applications to boundary value problems and finite elements. Springer, Berlin
157. Reddy JN (1985) An introduction to the finite element method. McGraw-Hill, New York
158. Reddy JN (1986) Applied functional analysis and variation methods. McGraw-Hill, New York
159. Rektorys K (1980) Variational methods in mathematics, science and engineering. Reidel Publishing Co, London
160. Resnikoff HL, Walls RO Jr (1998) Wavelet analysis, the scalable structure of information. Springer, Berlin
161. Rockafellar RT (1970) Convex analysis. Princeton University Press, Princeton
162. Rockafellar RT (1981) The theory of subgradients and its applications to problems of optimization: convex and non-convex functions. Helderman Verlag, Berlin
163. Rockafeller RT, Wets RJ-B (1998) Variational analysis. Springer, Berlin
164. Rodrigue B (1987) Obstacle problems in mathematical physics. North Holland Publishing Co., Amsterdam
165. Schaltz AH, Thome V, Wendland WL (1990) Mathematical theory of finite and boundary finite element methods. Birkhäuser, Boston
166. Schechter M (1981) Operator methods in quantum mechanics. Elsevier, New York
167. Siddiqi AH (1994) Introduction to variational inequalities. mathematical models in terms of operators. In: Siddiqi AH (ed) Recent developments in applicable mathematics. Macmillan India Limited, pp 125–158
168. Siddiqi AH (1969) On the summability of sequence of walsh functions. J Austral Math Soc 10:385–394
169. Siddiqi AH (1993) Functional analysis with applications, 4th Print. Tata McGraw-Hill, New York
170. Siddiqi AH (1994) Certain current developments in variational inequalities. In: Lau T (ed) Topological vector algebras, algebras and related areas, pitman research notes in mathematics series. Longman, Harlow, pp 219–238
171. Siddiqi AH, Koçvara M (eds) (2001) Emerging areas of industrial and applied mathematics. Kluwer Academic Publishers, Boston
172. Siddiqi JA (1961) The fourier coeffcients of continuous functions of bounded variation. Math Ann 143:103–108
173. Silvester RP, Ferrari RN (1990) Finite elements for electrical engineers, 2nd edn. Cambridge University Press, Cambridge
174. Simmons CF (1963) Introduction to topology and modern analysis. McGraw-Hill, New York
175. Singer I (1970) Bases in banach spaces. I. Springer, Berlin
176. Smart DR (1974) Fixed point theorems. Cambridge University Press, Cambridge
177. Sokolowski J, Zolesio JP (1992) Introduction to shape optimisation, shape sensitivity analysis. Springer, Berlin

178. Stein E, Wendland WL (eds) (1988) Finite element and boundary element techniques from mathematical and engineering point of view. Springer, Berlin
179. Strang G, Nguyen T (1996) Wavelets and filter banks. Wellesley-Cambridge Press, Cambridge
180. Strang G (1972) Variational crimes in the finite element method. In: Aziz AK (ed) The mathematical foundations of the finite element method with applications to partial differential equations. Academic Press, New York, pp 689–710
181. Tapia RA (1971) The differentiation and integration of nonlinear operators. In: Rall LB (ed) Nonlinear functional analysis and applications. Academic Press, New York, pp 45–108
182. Taylor AE (1958) Introduction to functional analysis. Wiley, New York
183. Temam R (1977) Theory and numerical analysis of the Navier-Stokes equations. North Holland, Amsterdam
184. Teolis A (1998) Computational signal processing with wavelets. Birkhuser, Basel
185. Tikhonov AN, Senin VY (1977) Solution ill-posed problem. Wiley, New York
186. Tricomi F (1985) Integral equations. Dover Publications, New York
187. Turner MJ, Clough RW, Martin HC, Topp LJ (1956) Stiffness and deflection analysis of complex structures. J Aerosp Sci 23:805–823
188. Vetterli M, Kovacevic J (1995) Wavelets and subband coding. Prentice Hall, Englewood Cliffs
189. Wahlbin WB (1995) Superconvergence in Galerkin finite element methods. Springer, Berlin
190. Wait R, Mitchell AR (1985) Finite element analysis and applications. Wiley, New York
191. Walker JS (1999) A primer on wavelets and their scientific applications. Chapman and Hall/CRC, Boca Raton
192. Walnut DF (2002) An introduction to wavelet analysis. Birkhuser, Basel
193. Wehausen JV (1938) Transformations in linear topological spaces. Duke Math J 4:157–169
194. Weidmann J (1980) Linear operator in Hilbert spaces. Springer, Berlin
195. Weyl H (1940) The method of orthogonal projection in potential theory. Duke Math J 7:411–444
196. Whiteman J (1990) The mathematics of finite elements and applications, I, II, III. Proceedings of the conference on Brunel University, Academic Press, 1973 1976, 1979. Academic Press, Harcourt Brace Jovanovich Publishers, New York
197. Wickerhauser MV (1994) Adapted wavelet analysis from theory to software. M.A, Peters, Wellesley
198. Wilmott P, Dewynne J, Howison S (1993) Option pricing. Oxford Financial Press, Oxford
199. Wojtaszczyk P (1997) A mathematical introduction to wavelets. Cambridge University Press, Cambridge
200. Wouk A (1979) A course of applied functional analysis. Wiley Interscience Publication, Wiley, New York
201. Zeidler E (1990) Nonlinear functional analysis and its applications. Springer, Berlin
202. Zienkiewicz OC, Cheung YK (1967) The finite element method in structural and continuum mechanics. McGraw-Hill, New York
203. Zlamal M (1968) On the finite element method. Numer Math 12:394–409

Index

A
Abstract variational problem, 280
Adjoint operator, 106
Affine, 48, 50
Affine functional, 50
Algebra, 43
Approximate problem, 280

B
Banach space, 16
Banach–Alaoglu theorem, 165
Bessel sequence, 387
Bessel's inequality, 95
Bilinear form, 123, 124
Bilinear functional, 124
Biorthogonal systems, 390
Bochner integral, 215
Boundary element method, 279, 301
Bounded operator, 26
Burger's equation, 253

C
Cauchy sequence, 39
Cauchy–Schwartz–Bunyakowski inequality, 28, 74
Céa's Lemma, 281
Characteristic vector, 120
Closed, 4
Closed sphere, 23
Coercive, 124, 229
Collocation method, 298
Commutative, 43
Compact, 4, 212
Complete, 94
Complete metric space, 3

Continuous, 26
Contraction m apping, 5
Contraction mapping, 5
Convergence problem, 281
Convex functional, 50
Convex programming, 231
Convex Sets, 48

D
Dense, 21
Dirichlet boundary value problem, 250
Dual space, 33
Dyadic wavelet frames, 502

E
Eigenvalue, 68, 120
Eigenvalue problem, 267
Eigenvector, 68, 119
Energy functional, 281
Euclidean space, 74

F
Fréchet differentiable, 182, 228
Finite element, 290
Finite Element Method, 280
Finite element method, 280
Finite element of degree 1, 290
Finite element of degree 2, 291
Finite element of degree 3, 291
Fixed point, 5
Fourier series, 95
Frame multiresolution analysis, 506
Fréchet derivative, 178, 182
Friedrichs inequality, 209

© Springer Nature Singapore Pte Ltd. 2018
A. H. Siddiqi, *Functional Analysis and Applications*, Industrial and Applied Mathematics,
https://doi.org/10.1007/978-981-10-3725-2

Notational Index

© Springer Nature Singapore Pte Ltd. 2018
A. H. Siddiqi, *Functional Analysis and Applications*, Industrial and Applied Mathematics,
https://doi.org/10.1007/978-981-10-3725-2

Printed in the United States
By Bookmasters